PENGUIN REFERENCE BOOKS

THE PENGUIN DICTIONARY OF GEOGRAPHY

Audrey N. Clark is an author, editor and publisher. She was personal assistant to Sir Dudley Stamp from 1940 until his death in 1966; Secretary of the First Land Utilization Survey of Britain, and served in the Planning Branch, Ministry of Agriculture, Fisheries and Food. Audrey Clark is Director of Geographical Publications Ltd; editor of the World Land Use Survey Series, International Geographical Union; publisher to the Department of Land Economy, Cambridge, and a Fellow of the Royal Geographical Society.

Her books, which are all published by Longman, include the *Longman Dictionary of Geography: Human and Physical*; *A Glossary of Geographical Terms* (3rd edn); *Chisholm's Handbook of Commercial Geography* (19th and 20th edns); *The World* (19th edn); *A Commercial Geography* (9th edn); and she edited the *New Geography for Today* series.

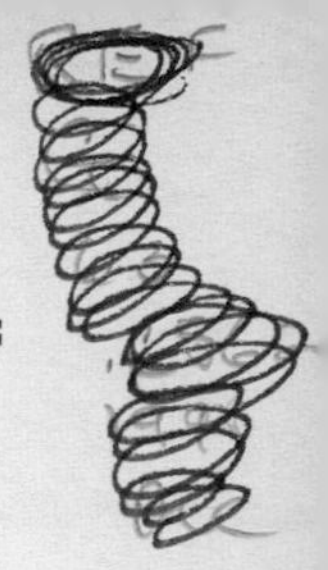

THE PENGUIN DICTIONARY OF GEOGRAPHY

Audrey N. Clark

SECOND EDITION

PENGUIN BOOKS

PENGUIN BOOKS

Published by the Penguin Group
Penguin Books Ltd, 27 Wrights Lane, London W8 5TZ, England
Penguin Putnam Inc., 375 Hudson Street, New York, New York 10014, USA
Penguin Books Australia Ltd, Ringwood, Victoria, Australia
Penguin Books Canada Ltd, 10 Alcorn Avenue, Toronto, Ontario, Canada M4V 3B2
Penguin Books (NZ) Ltd, 182–190 Wairau Road, Auckland 10, New Zealand

Penguin Books Ltd, Registered Offices: Harmondsworth, Middlesex, England

The *Longman Dictionary of Geography: Human and Physical*
published by Longman 1985
An abridged and revised edition published by arrangement with
Longman Group UK Ltd as *The New Penguin Dictionary of Geography*
in Penguin Books 1990
Reprinted as *The Penguin Dictionary of Geography* 1993
Second edition 1998
1 3 5 7 9 10 8 6 4 2

Typeset in 8.5/10pt Monotype Bembo
Typeset by Rowland Phototypesetting Ltd, Bury St Edmunds, Suffolk
Printed in England by Clays Ltd, St Ives plc

PREFACE

In the Preface to the first edition of this dictionary (1990) it was pointed out that geography as a discipline spans the humanities, the natural and the social sciences; has many specialist branches; is concerned not only with the physical phenomena of the planet (as in physical geography) but with the way in which people interact with them, and with each other (as in human geography). This breadth of view has led geographers to draw on some of the terminology of disciplines that impinge on their work (including anthropology, archaeology, economics, law, literature, medicine, philosophy, psychology, sociology etc., in the human sphere; and astronomy, biology, botany, geology, geomorphology, hydrology, soil science etc., in the physical), as well as on that of technical subjects such as cartography, remote sensing, surveying, mathematics and statistics.

Since 1990 swift progress in geography and these related disciplines has greatly extended the vocabulary in use in geographical literature and texts. To accommodate the advance this second edition of the dictionary has been updated and expanded, incorporating over 750 terms additional to those appearing in the first edition.

Geographers are, of course, supremely well placed to study environmental problems as well as present and potential effects of globalization. These aspects are well covered in this edition, which is designed to meet the needs of students from GCSE to first year college and university. The level of language used in the definitions is commensurate with the level of difficulty of the term involved.

It is obviously impossible to cover in detail in a concise dictionary such as this all the terms used by geographers (this has been done in the *Longman Dictionary of Geography: Human and Physical*, 724 pages, over 10 500 entries). But the most frequently used terms, with succinct definitions, appear here.

Constructive criticism and suggestions for additions will be warmly welcomed.

Audrey N. Clark
1998

LIST OF FIGURES

Cross references, amplifying the definitions, are indicated by SMALL CAPITAL LETTERS

A

aa (Hawaiian) a rough scoriaceous lava flow (SCORIA). PAHOEHOE. PILLOW LAVA, ROPY LAVA.

abandoned doorstep the original land-sea zone that had supported the PORT facilities of an urban area, left derelict as a result of the decline of the port and DE-INDUSTRIALIZATION.

abiocoen in ecology, all the non-living parts of the environment. BIOCOEN, HOLOCOEN.

abiotic without life, INORGANIC-1.

ablation the action or process of carrying away or removal, applied to **1.** the loss which a GLACIER undergoes through melting (the result of SOLAR RADIATION conducted by solid debris on the surface or flanking walls, or of rain falling on the surface, or of relatively warm water from melting streams), or through SUBLIMATION under favourable temperature, humidity and wind conditions, or through ABRASION (by sharp ice particles blown along the surface), or through the breaking away of ICEBERGS on the edge of tidal water **2.** the work of wind in removing fine rock debris in sandy areas (but for this process the term DEFLATION is usually preferred).

aborigine, aboriginal 1. a person who is believed, on contemporary evidence, to have been one of the original inhabitants of an area, or a descendant of such an individual, or one of the inhabitants occupying an area at the time of the arrival of European colonizers **2.** occasionally applied to a plant or animal believed on available evidence to have been one of the first or earliest in an area. The *adj.* is aboriginal.

abrasion the act or process of wearing down or wearing away by friction, e.g. by a river dragging stones along its bed, by the sea carrying sand and shingle along the coast, by glaciers with rocks frozen into their mass, by hard ice particles blown by wind over an ice surface. The process involved in the gradual reduction in the size of pebbles as one rubs against another is better termed ATTRITION. Abrasion results from the CORRASION of rock. TRANSPORTATION-2.

abrasion platform, abrasion plane a wave-cut platform, a flat, nearly smooth, rock platform extending from the foot of a sea-cliff and formed by the erosive action of BREAKERS in carrying shingle along the coast. The surface is usually bare, but may be covered with a thin layer of fine rock particles.

abscissa in mathematics, the horizontal or *x* coordinate in a plane COORDINATE system. ORDINATE.

absolute drought in UK, a period of at least fifteen consecutive days on none of which more than 0.25 mm (0.01 in) of rain falls.

absolute humidity the amount of WATER VAPOUR in a unit mass of air, usually measured by the grams of water vapour present in one cubic metre of the air (or in grains per cubic foot: 1 grain = 0.0648 gram). The amount of water vapour held by a body of air is governed by temperature and pressure: it becomes saturated at DEW-POINT. HUMIDITY, RELATIVE HUMIDITY, SPECIFIC HUMIDITY.

absolute instability the state of an air mass, holding any amount of WATER VAPOUR (CONDITIONAL INSTABILITY), in which the ENVIRONMENTAL LAPSE RATE is higher than the DRY ADIABATIC LAPSE RATE of the atmosphere surrounding it, with the result that it is unstable.

absolute range the difference between the extremes of temperature, rainfall etc., the highest and lowest values ever experienced at a place. RANGE-4.

absolute stability the state of an air mass in which the ENVIRONMENTAL LAPSE RATE is lower than the SATURATED ADIABATIC LAPSE RATE of the atmosphere surrounding it.

absolute zero the lowest temperature possible, the point at which all thermal molecular movement ceases, the zero point being 273.15 Celsius degrees below the ice point. This puts the ICE POINT (the melting point of pure ice at standard pressure) at 273.15 K. KELVIN SCALE.

absorption the physical process in which a material or system (the absorbent) takes into itself and holds another (the absorbate). ADSORPTION.

abstraction 1. the act of taking away, of reducing to a minimum, of extracting the essence of an idea. MODEL-2 **2.** the product of such acts **3.** the formation of an idea, separate from material things etc. **4.** a term regarded by some authors as synonymous with RIVER CAPTURE, which involves HEADWARD EROSION; but others apply it to the simplest type of river piracy, in which one stream, by its more vigorous action, widens its valley and captures another. MODEL-2.

abyssal *adj.* applied to **1.** the deepest regions (and associated phenomena) of the ocean, or, very loosely indeed, the ocean floor in general. DEEP **2.** less frequently, the deepest region of a lake. ABYSSAL ZONE, BENTHOS.

abyssal deposits deposits on the very deep parts of the ocean floor. ABYSSAL PLAIN, OOZE.

abyssal plain a deep-sea plain, a very large, relatively level area of the deep ocean floor, covered with a thin layer of sediment. OOZE. Fig 45.

abyssal zone 1. the deepest regions of the ocean, depth not precisely defined **2.** the zone in a lake not effectively penetrated by light. BENTHIC DIVISION.

accelerated erosion SOIL EROSION occurring at a rate exceeding the rate at which a SOIL HORIZON can be formed from the parent material. This is commonly due to human activity.

acceptable dose limit the greatest amount of an alien substance that can be released into an environment without causing serious harm, e.g. discharge of chemicals from a factory into a river. POLLUTION.

access means of approach, sometimes restricted to the physical means of approach. ACCESSIBILITY.

accessibility the state or quality of ease of approach, used in a physical sense (the ease of getting to a place, a facility or a service, measurable in terms of the distance to be travelled, the cost involved, the time taken) or a social sense (the extent to which different social groups are able to obtain goods, facilities, services, regardless of geographical location). INTERVENING LOCATION EFFECT, SHADOW EFFECT.

accessory mineral any of the many varied minerals occurring widespread and in very small quantity in a rock, which may reveal the origin of the rock but which do not affect its essential character. They are disregarded in the classification of the rock. ESSENTIAL MINERALS.

accidented relief rugged and irregular relief.

acclimatization the process by which plants and animals (including human beings) become, or are made to become, accustomed to climatic conditions unnatural to them (i.e. which are different from those of their native environment) so that they are able to live and reproduce.

accommodation a style of, or the stage reached in, INTEGRATION in society, in which conflict is reduced or avoided, the dominant group tolerating the differences of sub-groups such as ETHNIC MINORITIES. ASSIMILATION.

accordance of summit levels the general conformity of mountain peaks or hill tops to approximately the same level.

accordant drainage surface drainage directly related to the DIP of the underlying STRATA.

accordant junction a river junction in which a tributary joins the main stream at exactly the appropriate level, i.e. the tributary grades its course to the level of its outfall. DISCORDANT JUNCTION.

accretion the accumulation of particles, as in SEDIMENTATION-1, or in the process of growth by external addition, e.g. in the formation of crystals, or in the growth of ice particles by the addition of very small water particles.

acculturation **1.** a process of culture change in which more or less continuous contact between two or more culturally distinct groups or societies produces the results outlined under CULTURE CONTACT **2.** the state which results from such a process.

accumulated temperature the sum of the temperatures above or below a selected temperature measured over a period of time (DAY DEGREE). For example, 6°C (43°F) is the CRITICAL TEMPERATURE at which the growth of vegetation is commonly stimulated in midlatitude conditions. A temperature of 6°C (43°F) is therefore the basal crucial norm. If the mean daily temperature rises above this basal temperature each degree of the excess is known as the DAY DEGREE. The number of day degrees can be added together for periods of a week, a month or a growing season and the figures so obtained are referred to as the accumulated temperature.

accumulation in CAPITALISM, the building-up and reproduction of capital by continuing investment of profits. CAPITAL-2.

acid a compound containing HYDROGEN which, dissolved in water, provides hydrogen IONS (PROTONS); or a MOLECULE or ion which can give up protons to a BASE-2.

acid *adj.* **1.** relating to, or having the characteristics of, an ACID **2.** acid producing **3.** having an astringent, sharp, sour taste.

acid brown forest soil BROWN FOREST SOIL.

acidic *adj.* **1.** acid-forming **2.** ACID, *adj.*

acidify to increase the concentration of hydrogen IONS in a solution.

acidity the state or quality of being acid. ACID, pH.

acid lava a viscous, molten IGNEOUS material, with a high content of SILICA and a high melting point, which emerges slowly from a volcanic vent, cools and solidifies quickly, tending to form a plug or solid cone with steep sides (ACID ROCK, LAVA). Many acid lavas solidify into a glassy rather than a crystalline form.

acid peat BOG PEAT, PEAT.

acid rain precipitation, with a pH value of 5.6 or lower, charged with an excessive amount of acid droplets, formed particularly when oxides of SULPHUR and of NITROGEN released by combustion, especially by the burning of HYDROCARBONS, are converted to ACIDS in the atmosphere. Such precipitation may make over-acidic soils that are already acid (ACID SOIL), wash ALUMINIUM and other metals out of the ground, thereby polluting rivers and lakes, and greatly damage by chemical process the exterior of buildings. POLLUTION, SULPHUR DIOXIDE.

acid rock, acidic rock an IGNEOUS ROCK now defined as containing more than 10 per cent free QUARTZ. The typical acid PLUTONIC ROCK is GRANITE, the HYPABYSSAL is QUARTZ PORPHYRY, the VOLCANIC is RHYOLITE. ACID LAVA, BASIC ROCK, INTERMEDIATE ROCK, MAGMA, ULTRABASIC ROCK.

acid soil a soil in which the pH or HYDROGEN ION concentration is below 7.0 (neutrality). ACID RAIN, ALKALINE SOIL, ALKALI SOIL, pH.

acre a unit of area in British measures, formerly the area of land (customary acre) which could be ploughed in a day by a team of oxen, later defined by statute as 4840 sq yds (0.4 hectare). 640 acres equal one sq mi.

acre feet (American) in irrigation, measure of water needed to flood one acre of land to a depth of one foot.

acronym a name made up of the initial letters of an official title, e.g. United Nations Educational, Scientific and Cultural Organization, acronym Unesco.

action space the AREA-1 known to an individual, within which an individual acts, deciding where to live, to play, to shop, the LOCATION of each activity being perceived in relation to that of the others. ACTIVITY SPACE, SEARCH SPACE.

active *adj.* working, busy, energetic or effective, as opposed to passive, quiet, dormant or dead.

active glacier a GLACIER subject to a high rate of ALIMENTATION due to heavy snowfall in winter and a warm summer, resulting in rapid ABLATION and the transport of large amounts of debris. PASSIVE GLACIER.

active layer in soil, the annually thawed layer in PERMAFROST regions. Below it lies the permanently frozen layer.

active permafrost the layer of PERMAFROST which, having been thawed by artificial or natural causes, is able to return to permafrost under existing climatic conditions.

active volcano a VOLCANO which is liable to erupt, as opposed to a DORMANT VOLCANO.

activity space AWARENESS SPACE.

actual *adj.* existing in act or as a fact, as opposed to potential or theoretical.

actual isotherm an isotherm (ISO-) based on actual temperatures, not on temperatures adjusted to sea-level by taking into account the altitude of the recording station.

adaptation 1. in biology, the act or process by which an organism becomes fitted to its environment, thereby improving its chances of survival and of leaving descendants in the environment it is inhabiting **2.** a characteristic (an inherited or acquired structure or function) which makes an organism so fitted **3.** the condition resulting from the modifying act, process or characteristic.

adaptive behaviour BEHAVIOUR-1 that involves deliberate, rational choice to bring about satisfactory or OPTIMUM conditions, e.g. the behaviour of a firm in deliberately and rationally choosing an economically satisfactory or OPTIMUM LOCATION for an operation. ADOPTIVE BEHAVIOUR.

adiabatic *adj.* of the physical change during which no heat leaves or enters the SYSTEM-1. This commonly occurs in an ascending or a descending parcel of air in the ATMOSPHERE-1. An ascending body of air expands and cools, the constituent MOLECULES being dispersed; a descending body of air contracts and becomes warmer, the molecules being compressed; but there is no net gain or loss of heat. DIABATIC.

adiabatic gradient the rate at which the temperature of an AIR MASS, ascending or descending, changes in response to expansion or compression of that air mass.

adit a horizontal or nearly horizontal passage or working dug into a hillside from the surface for the purpose of extracting minerals; hence adit mine, adit mining.

administrative principle one of the principles used to account for the varying levels and distribution of CENTRAL PLACES in a CENTRAL PLACE SYSTEM. Assuming the size of the COMPLEMENTARY REGION of a central place to be governed by consideration of the most effective political or administrative control, there will be a clearcut separation of the higher order central place from the neighbouring lower order central places, i.e. each lower order central place will come within the complementary region of the single high order central place, and will thus be served entirely by this one, dominant higher order centre, which is the seat of administration. This results in a K-7 hierarchy. K-VALUE, MARKETING PRINCIPLE, TRAFFIC PRINCIPLE. Fig 9(a).

adobe (Spanish) **1.** a LOESS-like, in some cases CALCAREOUS, clay, brought by the wind from the rock waste of deserts and glaciated areas, deposited on the plains and in the basins of western USA and in arid parts of Mexico, and used for making unburnt, sun-dried bricks **2.** the sun-dried bricks made from such deposits **3.** a structure, e.g. a dwelling, made from such bricks or deposits **4.** soils derived from such deposits.

adoptive behaviour BEHAVIOUR-1 that displays a lack of conscious choice, e.g. the behaviour of a firm in not making an economically sound choice for the location of an activity (ADAPTIVE BEHAVIOUR) but in allowing its location to happen by chance or subjective choice, e.g. because the founder happened to live in the area.

adret (French; Italian adretto; German Sonnenseite) the sunny, warm slope of a hill or valleyside, as opposed to the UBAC or shady side. ASPECT, EXPOSURE.

adsorption a physical or chemical process by which MOLECULES of a gas, liquid or dissolved substance adhere to the surface of another. ABSORPTION.

advanced dune a small SAND DUNE formed round an obstacle and lying in front, i.e. to windward, of a larger dune from which it is separated by an eddying wind.

advanced industrial country AIC.

advection the horizontal movement of air, water or other fluid, applied especially to masses of air and water, in contrast to CONVECTION, vertical movement.

advection fog fog formed when a warm, moisture-laden air stream moves horizontally over a cooler surface, so that its lower layers are chilled below DEW-POINT.

adventitious *adj.* appearing casually or in unusual places, something added from without, applied since 1949 in the study of rural population, i.e. primary population (farmers and farm workers), secondary (people serving the primary group) and adventitious (the population living in rural areas solely from choice, e.g. retired people).

adventive crater lateral crater or parasitic crater, a CRATER which opens on the flanks of a great CONE-1.

adventure tourism TOURISM that involves elements of pioneering and physical effort (mountaineering, rock climbing, trekking, canoeing, rafting etc.) in order to experience unfamiliar environments. Accommodation is usually in tents or other basic shelter. ECOTOURISM.

aeolian, eolian *adj.* associated with Aeolus, the Greek god of the winds, hence related to wind action, i.e. borne, deposited, produced or eroded by wind.

aeration any process by which a substance becomes impregnated with air or another gas.

aeration zone the zone with PERMEABLE rocks, including the soil, which overlies the permanent WATER TABLE or zone of saturation (SATURATED-2) and in which pore spaces are partly filled with ground air.

aerial photograph a photograph taken from the air at an oblique or vertical angle, particularly useful in map making. PHOTOGRAMMETRY, REMOTE SENSING, STEREOSCOPE.

aerobe an organism able to live only in the presence of FREE OXYGEN. AEROBIC, ANAEROBE, ANAEROBIC.

aerobic *adj.* active, living or respiring only in the presence of FREE OXYGEN. AEROBE, ANAEROBE, ANAEROBIC.

aerobic respiration the process in which organisms use gaseous or dissolved oxygen (FREE OXYGEN) to release energy by the chemical breakdown of food substances. ANAEROBIC RESPIRATION, RESPIRATION.

aeroplankton minute organisms (spores, bacteria and other micro-organisms) floating freely in the atmosphere. PLANKTON.

aerosols **1.** ultra-microscopic solid or liquid colloidal particles suspended in a gas or in the ATMOSPHERE-1, appearing as mist, fog etc. In the atmosphere aerosols reduce the amount of SOLAR RADIATION reaching the earth's surface, and may thus have a cooling effect. POLLUTION **2.** a substance held under pressure in a container and released by a spraying device; or such a container itself.

afforestation **1.** historically, the result of the declaration of English medieval kings that a tract of land should be subject to forest laws (FOREST-2), or their action in making this declaration **2.** the clearing of land of sheep and cattle in Scotland in the mid-nineteenth century so that deer forests (FOREST-3) could be established **3.** the planting of land, not formerly so covered, with trees to make a forest for commercial or other purposes. FOREST, REFORESTATION.

aftershock one of a series of small shocks following a main EARTHQUAKE and originating at or close to its SEISMIC FOCUS.

agar-agar a gelatinous substance obtained from SEAWEEDS native to the waters around Sri Lanka, Malaysia, Java and Japan (especially from the seaweed *Gracilaria lichenoides*). Having been mixed with hot water and cooled, it sets to a firm jelly, suitable for use in cooking, the manufacture of fine silk and paper, in pharmaceutical products and, in bacteriology, as a culture medium.

age 1. the length of time for which an organism has lived or an object existed, exceptions being a life of markedly short duration (e.g. a butterfly living for a day or two is said to have a lifespan, not an age) and the life of subatomic particles and radioactive species, to which the term half-life is applied **2.** a very long time in the earth's history, a subdivision of an epoch (GEOLOGICAL TIME) **3.** a period of time based on cultural criteria, e.g. NEOLITHIC AGE.

agglomerate a rock consisting of angular fragments mainly exceeding 2 cm (0.75 in) in diameter thrown out of an erupting VOLCANO and more or less solidified in a matrix of ASH or TUFF. BRECCIA, CONGLOMERATE, PYROCLAST.

agglomeration 1. in general, a gathering together into a mass **2.** in soil science, a loose gathering together of soil particles which, when more closely united, form an AGGREGATE-3 **3.** in urban studies, an urban area, usually unplanned and formless, composed of formerly separate suburbs, villages or small towns which have expanded and coalesced **4.** a concentration of productive enterprises at a certain location, e.g. in a large town or industrial region, where each enterprise may enjoy the benefits of the readily available labour, good transport and service facilities, the large market, the proximity of allied and other enterprises, the concentration itself producing ECONOMIES OF SCALE. AGGLOMERATION ECONOMIES, COMPLEMENTARITY, DEGLOMERATION, DISECONOMIES OF SCALE, EXTERNAL ECONOMIES.

agglomeration economies the savings to the individual productive enterprise that come from operating in the same location as others (thereby sharing specialist servicing industries, specialist financial services and public utilities) or from serving a growing, large market occupying a small, compact geographical area. AGGLOMERATION-4, CENTRIPETAL AND CENTRIFUGAL FORCES, DEGLOMERATION.

aggradation 1. a process in which a land surface is built up by a deposition and accumulation of detritus, rock waste, sand, alluvium etc. derived from DENUDATION, thus the opposite of DEGRADATION, e.g. an alluvial river plain is an aggraded surface (PROGRADATION) **2.** a process in which PERMAFROST grows, the upper surface of the permafrost rising and accumulating ice on the under surface of the ACTIVE LAYER; termed aggradational ice, it appears as white horizontal bands interspersed with horizontal dirt bands.

aggregate **1.** a total derived by addition, e.g. in statistics, POPULATION-4 **2.** in geology, a cluster of mineral particles, i.e. a rock **3.** in soil science, a single mass or cluster of soil particles which acts as a unit. CRUMB STRUCTURE **4.** a construction material made from crushed stone or gravel and sand, used in making concrete, in road surfacing etc.

aggregate demand in Keynesian economic analysis, the value of the total planned expenditure in an economy. KEYNESIANISM.

aggressive water water capable of dissolving rock, e.g. as in CARBONATION.

agrarian *adj.* **1.** of or relating to land, especially to its cultivation, management and distribution in a system of LAND TENURE **2.** of, relating to, or connected with, landed property.

agrarian reform LAND REFORM combined with changes designed to improve rural life, e.g. by the provision of facilities for better education and social life as well as for more productive cultivation.

agribusiness **1.** all the operations and processes involved in running a FARM as a commercial enterprise **2.** specifically, a highly commercial, efficiently organized, business-like CAPITAL-INTENSIVE farming enterprise using up-to-date technology in equipment and production methods to achieve the highest possible output of produce of a consistently high standard which, in some cases, is sold under contract to large-scale customers.

agricultural *adj.* related to, characteristic of, AGRICULTURE.

agricultural area **1.** an area of land used for farming, including arable land, improved or unimproved grassland and other pasture **2.** in FAO statistics, arable land and land under tree crops, plus permanent meadows and pastures. Unimproved grassland presents a difficulty: where it corresponds to 'range land' (RANGE-3), or the open natural grassland of the tropical regions, it is normally excluded.

agricultural revolution a period in agricultural development characterized by change and INNOVATION in techniques, plant breeding, animal husbandry, etc. GREEN REVOLUTION, HIGH FARMING.

agriculture the science and art of cultivating the soil and the rearing of livestock, equivalent to FARMING. Some authors include FORESTRY, others exclude RANCHING; others restrict the term to the cultivation of the land. AGRIBUSINESS, AGRICULTURAL AREA, CROP FARMING, EXTENSIVE AGRICULTURE, EXTRACTIVE AGRICULTURE, INTENSIVE AGRICULTURE, LIVESTOCK FARMING, MIXED FARMING.

agro-forestal, agroforestal *adj.* pertaining to the use of land for a combination of AGRICULTURE and FORESTRY.

A horizon the surface layer in the SOIL containing HUMUS, an ELUVIAL LAYER from which minerals etc. are leached. The A horizon is subdivided by some soil scientists into A_{00}, a layer of litter; A_0, a layer with new organic matter; A_1, an organically rich layer; A_2, a leached layer; and A_3, the layer grading into the B HORIZON. The entry on SOIL HORIZON shows further refinements in the classification of the A horizon. L LAYER, MINERAL HORIZON, O HORIZONS, SOIL PROFILE.

AIC advanced industrial country.

air **1.** the mixture of gases enveloping the earth and forming the ATMOSPHERE **2.** that air considered as a medium for the transmission of radio waves or for the operation of aircraft.

air drainage the downward movement of cold AIR-I from higher to lower areas. FROST POCKET, KATABATIC WIND.

air frost air with a temperature at or below 0°C (32°F) as recorded at the level of a METEOROLOGICAL SCREEN. GROUND FROST.

air-gap WIND-GAP.

air mass, air-mass, airmass a mobile HOMOGENEOUS mass of air in the atmosphere, bounded by FRONTS. It may be very large, spreading over hundreds of sq km, or quite small and local; but it has distinct characteristics of LAPSE RATE and HUMIDITY (dry or moist) and temperature (hot or cold) derived from the region from which it originates, but which may be modified by long travel. There are many detailed classifications: a very broad general division identifies polar or tropical (based on temperature), maritime (on humidity, having taken up moisture from the ocean), continental (dry, of land, i.e. continental, origin). Such air masses, large and small, meet one another along fronts, the frontal surfaces which form the boundaries but which are often smoothed out into transition zones. Local weather is explained in terms of the movement of air masses.

air pollution POLLUTION.

air-stream a wind, a current of air flowing from an identifiable source.

alas, alas valley in THERMOKARST, a large flat-floored depression, with fairly steep sides, in some cases occupied by a shallow lake. As alases enlarge they may merge, to form alas valleys.

albedo the ratio of total solar electromagnetic radiation (SOLAR RADIATION) falling on a surface to the amount reflected from it, expressed as a percentage or a decimal. The average albedo of the earth is about 0.4 (40 per cent), i.e. four-tenths of solar radiation is reflected from the earth into space.

Aleutian low a sub-polar atmospheric low pressure area (LOW, ATMOSPHERIC) lying over the North Pacific, particularly in winter, characterized by a series of swiftly moving separate lows interspersed by occasional HIGH pressure systems. ICELANDIC LOW.

alfisols in SOIL CLASSIFICATION, USA, an order of soils which are relatively young and acid with a CLAY B HORIZON. Alfisols commonly occur under DECIDUOUS forest and are associated with HUMID, subhumid TEMPERATE and SUBTROPICAL climates.

algae pl. (sing. alga) simple PHOTOSYNTHETIC, non-vascular plants with unicellular organs of reproduction. The members vary widely in size and form, from the unicellular microscopic to the multicellular filamentous or ribbon-like with a more complex internal structure, most containing pigments in addition to the CHLOROPHYLL, which is often masked. The aquatic, e.g. DIATOMS, SEAWEEDS, live in fresh or salt water, the planktonic algae (PLANKTON) forming the bases of the aquatic food chains, and the seaweed providing a source of fertilizer and of human and livestock food as well as a source of substances used industrially (AGAR-AGAR). The land algae live in damp places, e.g. on walls or tree trunks, and in soil. LICHEN.

algal bloom, phytoplankton bloom the sudden increase of microscopic ALGAE in bodies of water due to an increased supply of nutrients. EUTROPHICATION.

alidade **1.** a surveying instrument used (with a PLANE-TABLE) to determine direction by viewing distant objects and noting angular measurements. It consists of a rule with sights at each end or with a telescope mounted parallel to it **2.** the index of any graduated surveying or measuring instrument.

alien a foreigner who is not a citizen of the country in which he/she is living.

alien *adj.* **1.** foreign, belonging to another country **2.** alien from, differing in nature or character **3.** alien to, so different as to be contrary or opposed in nature or character.

alienation an estrangement or separation between parts or the whole of the personality and those aspects of experience which are significant, e.g. a sense of estrangement or separation from society or the material environment, or of a lack of power to bring about social change; or a depersonalization of the individual in a large bureaucratic society.

alimentation the building-up of snow on an area of FIRN by snowfall and the effects of avalanches and the refreezing of melt-water, which may feed an outward moving GLACIER. A glacier advances if alimentation near its source exceeds ABLATION at the other end; it remains stationary if the two processes are in balance; and it retreats if alimentation is less than ablation. ACTIVE GLACIER.

alkali **1.** in chemistry, a usually soluble, strongly basic (BASE-2) hydroxide or carbonate of the alkali metals (i.e. of caesium, francium, lithium, potassium, rubidium, sodium) or of the ALKALINE EARTH METALS **2.** loosely applied to a substance with properties similar to those of a true alkali.

alkali flat a level area in an arid region with an incrustation of ALKALI salts formed as a result of the evaporation of a former lake. PLAYA.

alkaline *adj.* having the properties of an ALKALI-1.

alkaline earth metals a group of metals consisting of barium, beryllium, calcium, magnesium, radium and strontium, the hydroxides of which are weaker BASES-2 than those of the alkali metals (ALKALI-1).

alkaline rock, alkali rock an IGNEOUS ROCK low in CALCIUM but rich in alkali metals (ALKALI-1), indicated by the presence of SODIUM and POTASSIUM.

alkaline soil any soil which is alkaline in reaction, precisely above pH 7.0, in practice above pH 7.3, neutrality being 7.0. ACID SOIL, HYDROGEN ION, pH.

alkali soil soil containing ALKALI-1 salts, with a pH value of 8.5 and higher, likely to show surface incrustations of ALKALI-1, pH.

allochthon something (especially a rock) transported, not in its original place, the opposite of an AUTOCHTHON.

allochthonous *adj.* transported, not in place of origin, the opposite of autochthonous, applied to **1.** transported fossil plants or to organic deposits (e.g. some COALS) formed from them **2.** rocks which have travelled far from their place of origin, especially if moved by a tectonic process, e.g. overthrusting, recumbent folding, or sliding by gravity.

allogenic *adj.* applied **1.** in biology, having different genes **2.** in ecology, produced by external factors, the opposite of AUTOGENIC.

allogenic succession a plant SUCCESSION produced by changes in the environment, by external factors, not by changes produced by the plants themselves, as in AUTOGENIC SUCCESSION.

allometric growth the systematic differential growth of parts within a complex growth structure, so that as the SYSTEM-1,2,3 grows as a whole, the ratios between each part and the whole stay constant. Thus the growth rate of a part is proportional to the growth rate of the system as a whole, e.g. in urban studies, population growth increases up to a certain distance from a CENTRAL PLACE; in STREAM ORDER, as the number of stream segments grows the proportion falling into each stream order stays constant. GENERAL SYSTEMS THEORY.

allothigenic, allothigenous *adj.* originating at a distance, not AUTHIGENIC, AUTHIGENOUS, applied to **1.** constituents in a SEDIMENTARY ROCK which were formed outside and before the rock of which they are now a part, e.g. the pebbles in a CONGLOMERATE **2.** rounded crystals in an IGNEOUS ROCK which have come from some previously consolidated rock **3.** streams deriving much of their water from afar.

alluvial *adj.* of, pertaining to, or consisting of, ALLUVIUM.

alluvial cone an ALLUVIAL FAN with a steep slope, particularly likely to be built up if most of the water of the stream sinks into a porous deposit, so that nearly all the load is dropped and the structure gains height, e.g. in arid or semi-arid conditions.

alluvial fan a fan-shaped deposit of coarse alluvium (sand and gravel), the apex pointing upstream, laid down by a stream where it issues from a constricted course, e.g. from a gorge, or to a more open valley or to a plain. ALLUVIAL CONE.

alluvial flat a level, nearly horizontal, tract of land, bordering a river, which receives the ALLUVIUM deposited by the river in flood.

alluvial soil an AZONAL SOIL on newly deposited alluvium.

alluvial terrace part of an ALLUVIAL FLAT, in some cases paired by another on the opposite side of the river, left standing as the river cuts down its bed following REJUVENATION.

alluvium **1.** the unconsolidated, loose material (not only silt but also the gravel and sand) brought down by a river and deposited in its bed, floodplain, delta or estuary, or in a lake, or laid down as a cone or fan **2.** more specifically restricted to the fine-grained deposits, in texture the silt or silty-clay (GRADED SEDIMENTS), so laid down. These are rich in mineral content and so form some of the most fertile soils, of the highest agricultural value, in the world. ALLUVIAL.

almwind a strong and sometimes blustery FOHN-type wind, blowing from the south in the Tatra mountains down into the foreland of southern Poland, causing avalanches towards the end of winter and in spring.

Alonso model a MODEL-2 suggested by William Alonso in the 1960s to explain the variations in land values, land use and land use density in different districts within an urban area. Accepting VON THUNEN'S MODEL in explanation of the pattern of agricultural land use, he applied the principles of that model to the urban land use pattern. He assumed industry to be concentrated at the city centre, and physical ACCESSIBILITY (and thus transport costs) to be the prime consideration of those setting up industry as well as those working and living in the urban area. From this he suggested that a simultaneous resolution of BID PRICE CURVES, each related to a different category of land user, explains the variations in land values and land use, etc., and thus the urban land use pattern.

alp **1.** high mountain pasture, snow-covered in winter, usually above the TREE LINE, on a gently sloping BENCH or SHOULDER-2, commonly on the side of a U-SHAPED VALLEY where there is an abrupt change of slope. In summer it provides rich pasture, and in some cases a site for a seasonal dwelling for herdsmen. It may also afford a site for permanent housing or a wintersports centre **2.** pl. alps, high, especially snow-capped, mountains broadly similar to the Alps of Switzerland and adjacent European countries.

alpha index a measure of connectivity in a NETWORK-2:

$$\frac{E - V + S}{2V - 5}$$

where E represents the number of EDGES, V the number of NODES (vertices), S the number of graphs or subgraphs. 0% = no circuits, 100% = a completely connected network. BETA INDEX.

Alpides, Alpids a general term applied to the Alpine fold mountains. They have the same general trend as the ALTAIDES.

alpine *adj.* **1.** of, pertaining to, or characteristic of, the Alps or any similar high mountains **2.** the parts of a mountain above the TREE LINE and below permanent snow, or the plants and animals living in that zone. ALPINE OROGENY.

alpine glacier a VALLEY GLACIER, formed in an amphitheatre among mountain summits, descending a mountain valley and ending by melting or spreading out into a PIEDMONT GLACIER.

Alpine orogeny, Alpine earth movements, Alpine revolution the great mountain-building movements which took place mainly in the Tertiary period, culminating in the Miocene (GEOLOGICAL TIMESCALE) and resulted in the creation of the Alpides or Alpine systems of the Alps, Carpathians, Balkan Mountains, Pyrenees, Atlas and other great chains. OROGENESIS.

Altaides, Altaids, Altaid orogeny the great mountain chains stretching through central Europe and Asia, the result of the Altaid OROGENY which took place in late Carboniferous into Permian times (GEOLOGICAL TIMESCALE). The name is derived from the Altai mountains of central Asia, which are typical of the system. ALPIDES, ARMORICAN OROGENY, HERCYNIAN, VARISCAN.

alternative hypothesis any admissible HYPOTHESIS alternative to the one being tested. NULL HYPOTHESIS, STATISTICAL TEST.

alternative technology the TECHNOLOGY-3 concerned with the technical processes that make use of renewable rather than non-renewable NATURAL RESOURCES and which are therefore regarded as ecologically sustainable. APPROPRIATE TECHNOLOGY.

altimeter an instrument for measuring altitude, used in aircraft and by surveyors on land, which employs the fall in atmospheric pressure with height above sea-level (averaging 34 mb: 1 in of mercury for each 300 m: 1000 ft) as an index. Electronic techniques are used in a radio altimeter.

altimetric frequency curve, altimetric frequency graph a curve constructed to show generalized altitudes in an area. It may be made by dividing a map into small squares and taking the highest point, or the average of the highest and lowest points, in each; or by using specific heights above sea-level. If the measurements are shown on a graph, heights above sea-level appear on the horizontal scale, the percentage frequency on the vertical.

altiplano a high INTERMONTANE plateau in the Andes, specifically in western Bolivia.

altitude 1. the angular height of a star or other heavenly body measured in degrees from the plane of the observer's horizon **2.** the height of a point above mean sea-level, measured vertically **3.** the height of a mountain or hill above its base or surrounding plain, measured vertically from base to summit.

altitudinal zone VERTICAL ZONE.

altocumulus, altocumulus castellatus a cloud formation of middle altitudes (2400 to 6000 m: 8000 to 20 000 ft), consisting of small, fleecy, globular, relatively thin patches of CLOUD, the edges sometimes merging, and sometimes appearing as a MACKEREL SKY, usually indicating fine weather. Sometimes the tops of the small clouds become turrets, a formation termed altocumulus castellatus, indicating instability in the upper air, and thunderstorms.

altostratus a cloud formation of middle altitudes (2400 to 6000 m: 8000 to 20 000 ft), a wide expanse of continuous, uniformly grey, thick or thin, flat CLOUD, through which the sun or moon may usually dimly be seen, associated with an oncoming WARM FRONT and indicating rain. CIRROSTRATUS.

alumina ALUMINIUM oxide. SIAL.

aluminium a very light, white, ductile, malleable, metallic element, durable and resistant to corrosion, occurring widely in nature but difficult to separate from its ores. It is now extracted mainly from the ore BAUXITE. The ore is subjected to very high temperatures in an electric furnace to produce alumina (aluminium trioxide, Al_2O_3), smelted to produce aluminium (some 4 tonnes of bauxite producing 2 tonnes of alumina and 1 tonne of aluminium). Aluminium is extensively used pure or in alloys in manufacturing motor vehicles, lightweight containers, aircraft, electrical apparatus, cooking utensils, thin foil for wrapping purposes, etc.

ambient *adj.* surrounding, lying around, encompassing, or on all sides of, a phenomenon, organism, or object.

ambient temperature the TEMPERATURE of the atmosphere immediately surrounding something specific, e.g. a CLOUD or a THERMAL.

amelioration an improvement, e.g. climatic amelioration, an improvement or a modification of climatic conditions over a period of time.

amenity something in, or some quality of, the environment which is perceived as pleasant and attractive, which makes life agreeable, satisfying, for people.

ammonia a gaseous compound of NITROGEN and HYDROGEN, soluble in water, producing an ALKALINE solution, widely used in industry, and as a fertilizer directly or as the basis of fertilizer compounds. AMMONIUM NITRATE.

ammonium nitrate a crystalline salt used

in herbicides, insecticides, explosives; and as a fertilizer (AMMONIA). Excessive use as a fertilizer may lead to so much nitrate in the RUNOFF from the land that POLLUTION and EUTROPHICATION occur.

amorphous *adj.* formless **1.** in geology, unstratified **2.** of a mineral, noncrystalline.

amphidromic point a NODE-1 at the centre of an AMPHIDROMIC SYSTEM. It is theoretically possible for an amphidromic point to be located on land, where it is termed a degenerate amphidromic point. BINODAL TIDAL UNIT.

amphidromic system a unit area in the ocean in the OSCILLATORY WAVE THEORY OF TIDES. AMPHIDROMIC POINT.

anabatic *adj.* applied to air moving upward, due to CONVECTION, as opposed to katabatic, moving downward (KATABATIC WIND), e.g. applied to a local breeze which by day blows up a warm valley slope in mountain regions, the air on the slope being heated more rapidly and to a greater extent than air lying above the valley floor, thus creating a CONVECTION CURRENT as the cooler valley air moves up to replace the rising warmer, lighter air.

anaerobe an organism able to live in an absence of air or FREE OXYGEN. AEROBE, ANAEROBIC.

anaerobic *adj.* living or active only in the absence of FREE OXYGEN. AEROBE, AEROBIC, ANAEROBE.

anaerobic decomposition organic disintegration in the absence of air.

anaerobic respiration the process by which organisms, without taking in gaseous or dissolved oxygen (FREE OXYGEN), release energy by the chemical breakdown of food substances, e.g. the breakdown of sugar to ethyl alcohol and carbon dioxide by yeast (FERMENTATION). Many organisms are able to switch to anaerobic respiration when lack of oxygen inhibits AEROBIC RESPIRATION; but others never take in FREE OXYGEN. RESPIRATION.

anafront a cold FRONT in which warm air is in the main rising over a wedge of cold air.

analogue a thing that is similar to, corresponds in some respect to, some other thing, the implication being that the likeness is systematic or structured; e.g. in long-range WEATHER FORECASTING predictions of future repeating patterns of weather can be made by analysing and comparing synoptic patterns and sequences of weather already experienced.

analogue model (American, analog model) a MODEL-2 in which selected phenomena are represented by similar but different phenomena, e.g. clusters of people are represented by clusters of points. ICONIC MODEL, SYMBOLIC MODEL.

anastomosis 1. in biology, the intercommunication between vessels or channels in plants and animals by means of connecting branches **2.** in geomorphology, applied to rivers which divide and reunite continuously, producing a net-like mass of branches (BRAIDED RIVER COURSE), usually caused by excessive deposition of alluvial material in the stream.

anchor tenant in retailing, a tenant which, on account of its popular appeal and success elsewhere, is encouraged to open premises in a new shopping centre in the hope that it will, by its number of customers, dr ıw other retailers to the area.

andesite a pale grey, fine-grained

IGNEOUS ROCK of INTERMEDIATE composition, 52 to 65 per cent SILICA, composed essentially of a PLAGIOCLASE FELDSPAR. It is usually EXTRUSIVE, but also occurs in small INTRUSIONS; and it forms new continental crust at SUBDUCTION ZONES. ANDESITE LINE, PLATE TECTONICS.

andesite line a boundary passing through the Pacific Ocean from Alaska, Japan, Marianas, Bismarck Archipelago to Fiji, Tonga and New Zealand, separating two PETROGRAPHIC PROVINCES, the one to the west of the line being predominantly andesitic (ANDESITE) and intermediate, with rocks 52 to 65 per cent SILICA; the one to the east being basaltic (BASALT) and basic, with rocks less than 52 per cent silica. ATLANTIC SUITE, PACIFIC SUITE.

anecumene, anekumene, anoecumene, anoekumene, anokumene the part of the earth's surface which is uninhabited, or only temporarily inhabited, by people. ECUMENE, NEGATIVE AREA.

anemometer an instrument for measuring and recording the strength and direction of wind. There are two types in general use. The simpler is known as a pressure plate anemometer, in which a thin, small sheet of metal is suspended from a metal arm in such a way that if it is placed face (at right angles) to the wind the bottom edge is blown upward. The angle of its position is noted, and the speed (velocity) of the wind, in kph or mph, read off from tables. In the cup anemometer three or four cups are set at the ends of horizontal arms that cross and pivot at the top of a vertical rotating shaft. The wind rotates the arms, and their movement turns the shaft, to which a meter is attached to record the wind velocity in metres per second, or mph, or kph. A more elaborate, self-recording anemometer, known as an anemograph, continuously traces and records. The most common type of this uses the Dines tube, the opening of which is made as a vane, always faced to the wind. The ensuing pressure passes down the tube to a float that carries a pen which traces wind velocity on paper surrounding a rotating drum.

aneroid barometer BAROMETER.

angiosperm a plant producing seeds enclosed in an ovary (unlike a GYMNOSPERM). The angiosperms succeeded the gymnosperms as dominant land plants, having gained ascendancy by the end of the CRETACEOUS period. Most seed-bearing plants today are angiosperms.

angle of repose, angle of rest the steepest slope at which a mass of downward-moving unconsolidated rock debris remains stationary. REPOSE SLOPE.

angulate drainage a modified TRELLIS DRAINAGE pattern, the tributaries meeting the main streams at acute or obtuse angles as a result of jointing, sometimes of faulting, of the rocks. FAULT, JOINT.

animal **1.** any member of the kingdom Animalia (CLASSIFICATION OF ORGANISMS), the typical animal being a living organism with cell walls not composed of cellulose, with means of independent locomotion, with a nervous system which may or may not be centralized, that is incapable of PHOTOSYNTHESIS and usually needs complex organic food substances; but some animals (e.g. sponges) do not conform to those criteria, and exhibit plant-like characteristics. CARNIVORE, FOOD CHAIN, HERBIVORE, OMNIVORE, PARASITE, PLANKTON, PLANT **2.** any member of Animalia other than a human being.

anion ION.

annual *adj*. as applied to plants, existing for only one year, i.e. a plant that completes its life-cycle, from seed germination to seed production, within one season. BIENNIAL, EPHEMERAL, PERENNIAL.

annual growth rings the annual increase of secondary wood seen as concentric rings in a cross-section of the stem or trunk of a woody plant. The small-celled rings formed towards the end of one growing season, e.g. late summer, contrast with the succeeding wide-celled rings formed early in the next, e.g. spring. By counting these growth rings the age of the plant can be estimated; and by studying their width it is possible to infer the environmental (particularly the climatic) conditions prevailing at the time when the rings were being formed. DENDROCHRONOLOGY.

annular *adj*. shaped like a ring.

annular drainage a drainage pattern in which the CONSEQUENT STREAMS radiate, as the spokes of a wheel, and the SUBSEQUENT STREAMS, eroding their valleys in the weaker strata, seem to make a series of disrupted, concentric circles as they flow into the consequent streams. Annular drainage is especially likely to occur around dissected DOMES. Fig 17.

anomalous watershed a WATERSHED in a mountain region which does not run along the crest of the highest range of a mountain chain. NORMAL WATERSHED.

anomaly a departure from the normal, predicted, or uniform state or value.

anomie, anomy (Greek, without law) **1.** a state or condition (considered by some authors in the 1950s and 1960s to bear some similarity to ALIENATION) in which an individual, having lost traditional 'moorings', is prone to disorientation or psychic disorder **2.** a state or condition of a society in which commonly accepted social norms are either absent or unclear, conflicting or not integrated, leading to a lack of order and control in social life. This condition is sometimes assumed to occur in societies which are rapidly becoming urbanized and industrialized. INDUSTRIALIZATION, URBANIZATION.

Antarctic, antarctic strictly an adj. applied to the south polar regions, but used as a noun (the Antarctic) to denote the region lying within the ANTARCTIC CIRCLE (66°32'S). ANTARCTICA.

Antarctic *adj*. of, pertaining to, or characteristic of, the landscape, climatic conditions, animal and plant life etc. occurring roughly within the ANTARCTIC.

Antarctica the continental land area of approximately 11.5 mn sq km (4.4 mn sq mi) surrounding the SOUTH POLE. The overlying ICE CAP is so thick in some places that it is believed its base may be below sea-level; and it is possible that the land itself may be in two parts (East and West Antarctica). But in the prevailing conditions the ice may be considered as a rock, part of the land area.

Antarctic air mass an exceedingly cold air mass, symbol AA, originating over the ANTARCTIC ocean. It should not be confused with the POLAR AIR MASS.

Antarctic circle latitude 66°32'S, in use commonly 66°30'S, where, due to the inclination of the axis of the earth, the sun does not set about 22 December (SUMMER SOLSTICE in the southern hemisphere) and does not rise above the horizon about 21 June (WINTER SOLSTICE in the southern hemisphere). The number of days such as this, without sun in winter, increases southwards until at the SOUTH POLE six months of darkness follow six months of daylight. At any given date the conditions are the reverse of those in the ARCTIC. ARCTIC CIRCLE.

Antarctic convergence a well marked natural boundary in the oceans round ANTARCTICA (the position shifting a little with the seasons, being roughly parallel to the isotherm of 10°C (50°F) in the warmest month) where cold heavy waters from the south sink below warmer waters, with a consequent change of sea and air temperatures, and of animal and plant life.

antecedent *adj.* existing, happening, or going before in time or other sequence.

antecedent drainage, antecedent river a drainage system that was in existence before the present form of the land surface was established, and which maintains its original direction despite slow, localized uplift in its path which might turn it. It succeeds in the contest because it is able to cut its way through the barrier as rapidly as the barrier is formed. INSEQUENT DRAINAGE.

anthracite a hard, lustrous variety of COAL with a high proportion (about 85 to 93 per cent) of CARBON and a low proportion of volatile matter. It does not easily ignite, but when alight it burns with an almost clear, smokeless flame, giving out great heat.

anthropocentric *adj.* centred on human beings, regarding human beings as the reason for the existence of the world and/or as a measure of all value. EGOCENTRISM, ETHNOCENTRISM.

anthropocentrism the state or quality of being focused on human beings. ECOCENTRISM.

anthropogenic *adj.* **1.** of or relating to anthropogeny, the study of the origin of human beings **2.** having an origin in human activity.

anthropogeomorphology the study of landforms and processes resulting from human activities.

anthropomorphic soil an INTRAZONAL SOIL created by human activities.

anticlinal *adj.* of, or pertaining to, an ANTICLINE.

anticlinal ridge a ridge or uplifted tract of country which corresponds with an ANTICLINE in the underlying strata.

anticlinal valley a valley which follows an anticlinal axis. ANTICLINE, BREACHED ANTICLINE.

anticline an arch-shaped upfold caused by COMPRESSION in the rocks of the earth's crust, the beds dipping down and away from the central line (termed the anticlinal axis) but not always symmetrically. The term OVERFOLD is applied to an anticline pushed over on its side. ANTICLINORIUM, AXIS OF FOLD, BREACHED ANTICLINE, PITCH-3, SYNCLINE. Fig 1.

anticlinorium a great arch of STRATA composed of numerous subordinate wrinkles or ANTICLINES and SYNCLINES, the opposite of SYNCLINORIUM. Fig 1.

anticyclone a high pressure system (high in relation to the pressure of the surrounding air), commonly termed a high, a slow-moving atmospheric condition in which the barometric pressure is high towards the centre of the system and progressively declines outwards. It generally appears on a weather chart as a series of concentric, closed isobars (ISO-), fairly widely spaced. Winds are light and variable, blowing round such a system in a clockwise direc-

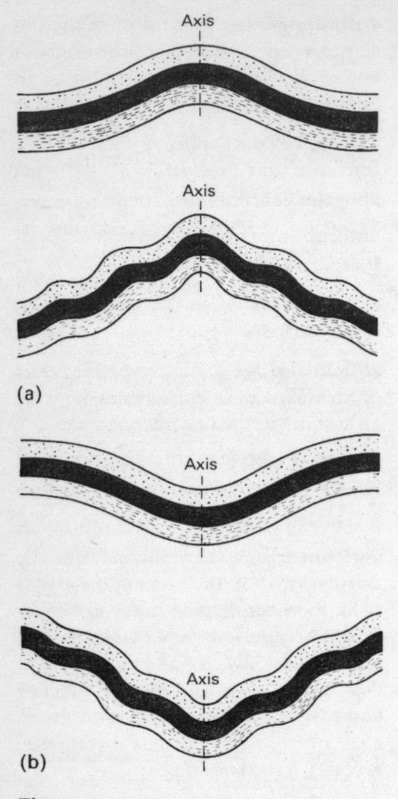

Fig 1
(a) Anticline and anticlinorium
(b) Syncline and synclinorium

tion in the northern hemisphere, counter-clockwise (anticlockwise) in the southern (BUYS BALLOT'S LAW). There may be a calm area at the centre, where pressure is highest. The temporary anticyclones of temperate latitudes (not to be confused with the permanent, subtropical belts of high atmospheric pressure of HORSE LATITUDES) bring generally settled weather with light winds, warm and sunny in summer, in winter cold and frosty or foggy (RADIATION FOG). DEPRESSION-3. Fig 2.

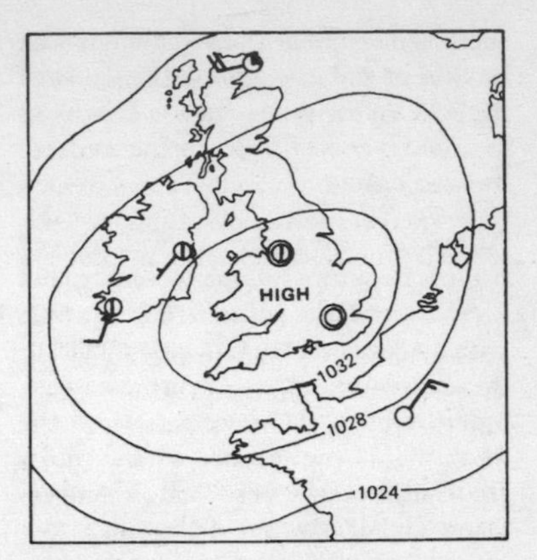

Fig 2 Anticyclone: the air moves clockwise in the northern hemisphere, anticlockwise in the southern

anti-dip stream a stream flowing in a direction approximately opposite to that of the DIP of the surface rocks of an area. It is frequently, but not necessarily, an obsequent stream (OBSEQUENT DRAINAGE).

antidune a sand wave, a transient form of ripple in the sand of the bed of a highly-loaded, swiftly-flowing stream (analogous to a SAND DUNE) which moves progressively upstream because the steeper slope, facing upstream, is constantly being built up by the addition of sediments carried by the flow of water which washes away sediments from the gentler, downstream slope.

antipodes places on the earth's surface diametrically opposite each other, or that part of the earth's surface which is diametrically opposite the observer.

anti-trade wind, anti-trade 1. a term now applied only to the westerly winds of the upper atmosphere (at a height exceeding 2000 m: 6500 ft), blowing above and in a direction (i.e. westerly) opposite to that of the surface winds of middle latitudes known as the TRADE WINDS **2.** formerly, but no longer, applied to the surface westerly winds (now termed WESTERLIES) of middle latitudes, also termed counter-trades (also an obsolete term).

anti-urbanism a very strong dislike of, an aversion to, large towns and cities. URBANISM.

anvil cloud, incus cloud a wedge-shaped cloud formed at the head of a large CUMULONIMBUS cloud, levelled by winds at a height of some 6000 m (20 000 ft), the lowest level of the STRATOSPHERE. ICE ANVIL.

AONB Area of Outstanding Natural Beauty, in British planning, an area designated under the National Parks and Access to the Countryside Act 1949 in order to 'preserve and enhance its natural beauty'. An AONB is usually smaller than a NATIONAL PARK, and is not managed in the same way; but the designation gives local planning authorities the power to operate some development control.

apartheid racial segregation, the policy of the White government of the former Republic of South Africa, under which White, African, Asian and Coloured (people of mixed ethnic groups) communities were segregated. Apartheid was a factor leading to the withdrawal of the Republic from the Commonwealth when other members expressed their disapproval of the policy. It also led to the Republic's setting up of nominally independent Bantustans, i.e. African homelands within the Republic, which were not in fact economically viable, and were not officially recognized internationally.

aphelion (Green apo, from; helios, the sun) that point farthest from the sun in the ORBIT of a comet or planet. The earth arrives at aphelion on 4 July, when it is some 152 mn km (94.5 mm mi) distant from the SUN. APSIS, PERIHELION.

apogean tide the tidal condition when the moon is at its APOGEE, so that its gravitational pull is reduced, resulting in lower high tides and higher low tides, with a tidal range smaller than is usual. PERIGEAN TIDE, TIDE.

apogee 1. the point farthest from the earth in the orbit of a planet (especially of the moon) or satellite **2.** the greatest distance of the sun from the earth when the earth is in APHELION. APSIS, PERIGEE.

Appalachian orogeny, revolution a period of intense earth movements which occurred between the Permian and Triassic periods (GEOLOGICAL TIMESCALE) in North America, equivalent to but slightly later than the ARMORICAN and HERCYNIAN movements in Europe.

apparent dip the DIP-2 as seen by the observer (which is not necessarily the TRUE DIP, the angle of maximum slope). Thus apparent dip may be defined as the amount and direction of DIP-2. Fig 16.

apparent time local solar time, or LOCAL TIME, as shown by the apparent diurnal movement of the real sun, e.g. as by a sundial. Apparent noon is the instant when the sun's centre reaches the highest point in its apparent daily course over any place on the earth's surface (i.e. when the sun crosses the MERIDIAN of that place) and the shadows of vertical objects are at their shortest. Because the sun's orbit is elliptical and inclined towards the equator, the period between two successive diurnal crossings is not constant; thus MEAN SOLAR TIME is a more useful measure.

Appleton layer F2 layer, the upper stratum of the IONOSPHERE (i.e. above the HEAVISIDE layer, the E layer) about 300 km (190 mi) above the surface of the earth, the layer in which ionization by solar radiation results in the refraction and reflection of short radiowaves back to earth. It was identified by Sir Edward Victor Appleton, 1892–1965, an English physicist and 1947 Nobel prizewinner.

applied geography the application of geographical knowledge and techniques to the solution of economic and social problems on a local to a world scale, in such fields as town and country planning, land use, location policy, underdevelopment, and population studies etc.

apposed glacier a GLACIER resulting from the merging of two separate glaciers.

appropriate technology **1.** the TECHNOLOGY-3 concerned with technical processes which are suited to the degree of INDUSTRIALIZATION in an area or country **2.** sometimes used as synonym for ALTERNATIVE TECHNOLOGY.

apron OUTWASH APRON, FAN, or PLAIN, a spread of ALLUVIUM deposited by streams, especially by those from a melting glacier.

apsis pl. apsides, a critical point on an ORBIT in relation to the centre of attraction, the higher apsis being the farthest point from that centre, the lower apsis the nearest to it, a straight line joining the two points being termed the line of apsides. Thus for a planet in the SOLAR SYSTEM the apsides are the APHELION and PERIHELION; and for the moon (in relation to the earth) the APOGEE and PERIGEE.

aquaculture the management of aquatic environments for the production of organic materials, mainly concentrated at present on PISCICULTURE, fish farming, the controlled breeding and rearing of fish (freshwater or marine) for commercial purposes. FISH FARM.

aquatic *adj.* living in, growing in, or frequenting, water. ESTUARINE, MARINE, RIPARIAN, TERRESTRIAL.

aqueduct an artificial channel built to carry water from one place to another, especially such a channel supported on a series of arches and spanning a valley or river. VIADUCT.

aquiclude a porous rock which, although usually PERMEABLE, becomes impermeable because of the saturation of its pores by water, e.g. a SATURATED CLAY or SHALE. AQUIFER, POROSITY.

aquifer, aquafer a water-bearing stratum of rock, sufficiently porous to carry the water and sufficiently coarse to release the water and permit its use. Hence an aquiferous rock, one conveying or yielding water. AQUICLUDE, ARTESIAN BASIN, ARTIFICIAL RECHARGE.

aquifuge an IMPERMEABLE rock stratum which not only obstructs the passage of water but cannot absorb it, e.g. GRANITE.

arable farming ARABLE LAND, CROP FARMING, FARMING.

arable land land capable of being ploughed or, more usually, agricultural land which is tilled for crops, though not necessarily each year, e.g. ploughland, market gardens, vineyards, temporary FALLOW land, and ROTATION GRASS; orchards are sometimes included, gardens attached to houses are usually excluded. In British agricultural statistics land is distinguished as arable, permanent grass, rough grazing and woodland. AGRICULTURE, CROP FARMING, FAO, FARMING.

arboretum a place set aside for the cultivation, display or study of trees, in some cases including rare SPECIES.

arboriculture the cultivation of TREES and SHRUBS.

arc furnace any furnace in which heat is supplied by an ELECTRIC ARC, as distinct from one heated externally or by a fuel, as in a BLAST FURNACE.

arch a curved opening formed in a rock mass by any of several processes, e.g. by the collapse of the roof in a limestone CAVERN, by marine erosion of projecting rocks (CAVE-1, STACK), or by weathering of a relatively soft layer in a rock mass.

archaeo-thanatology the study of deaths occurring in the distant past.

Archimedes' screw a simple device used in PERENNIAL IRRIGATION. A spiral screw revolves inside a close-fitting sleeve, the lower end of which is put in water at a low angle, the screw being turned at the higher end so that the water travels up the screw to flow out at that, higher, end.

archipelago applied initially to the Aegean Sea which is studded with islands, then to any sea studded with islands, now applied solely to a group of islands.

Arctic, arctic strictly an adj. applied to the north polar regions but used as a noun (the Arctic) to denote the region lying within the ARCTIC CIRCLE, or to the landscape, climatic conditions, animal and plant life found roughly within that area; and often used loosely just to mean very cold.

Arctic air mass an exceedingly cold air mass, symbol A, originating over the Arctic Ocean. It should not be confused with the POLAR AIR MASS. ARCTIC FRONT.

Arctic circle latitude 66°32'N, in use commonly 66°30'N, where, due to the inclination of the axis of the earth, the sun in the northern hemisphere does not set about 21 June (SUMMER SOLSTICE) and does not rise above the horizon about 22 December (WINTER SOLSTICE). The number of days such as this, without sun in winter, increases northwards until at the NORTH POLE six months of darkness follow six months of daylight. At any given date the conditions are the reverse of those in the ANTARCTIC.

Arctic front a not very active, almost permanent frontal zone (FRONT), lying to the north of the POLAR FRONT, in which cold air from the ARCTIC in the northern hemisphere meets less cold air in latitudes 50°N to 60°N. Its inactivity is due to the fact that the temperature difference between the ARCTIC AIR MASS and the POLAR AIR MASS is small.

Arctic prairies TUNDRA.

Arctic smoke fog occurring in high latitudes when icy-cold air from the land passes over warmer air lying over water and causes the water in the air to condense. FROST SMOKE.

arcuate delta a fan-shaped DELTA with the rounded outer margin, the arc of the fan, spreading into the sea, e.g. the Nile delta. Fig 13.

area 1. a part, space, tract or region of the earth's surface, of any size, a term used loosely or sometimes precisely (e.g. in British planning law, an Area of Outstanding Natural Beauty: AONB) **2.** a sunken, small piece of enclosed land, a yard or court adjoining and giving access to the basement of a dwelling **3.** in measures, the extent in two dimensional space, i.e. the extent of a surface contained within given limits, calculated by the use of any of various formulae, e.g. for a square, circle, ellipse, etc. **4.** a sphere of operation, e.g. sterling area **5.** a mental image of extent, possibilities, or range.

areal differentiation the varied nature of the earth's surface, apparent in the character, pattern, and interrelationship of relief, climate, soil, vegetation, land use, population distribution, and so on, which together produce a mosaic of dissimilar units. That concept formed the basis of regional geography (REGION-1), but in the 1980s the study of regional (areal) differentiation came to the fore in HUMAN GEOGRAPHY. GLOBALIZATION, HUMAN AGENCY, HUMAN ECOLOGY-1, HUMANISTIC GEOGRAPHY, INTERNATIONAL DIVISION OF LABOUR, UNEVEN DEVELOPMENT.

Area of Outstanding Natural Beauty AONB.

areic, aretic *adj.* without flow, applied to desert regions where the rainfall (if any) is so slight that it sinks into the ground, or evaporates, so that there are no flowing streams. ENDOREIC, EXOREIC.

arena a shallow roughly circular basin, the resistant rocks of the rim enclosing the central area where less resistant rocks have been eroded.

arenaceous *adj.* applied to **1.** a rock of sandy texture, consisting mainly of grains of SAND which may be loose or cemented; or a rock basin with a sandy floor. ARGILLACEOUS **2.** a plant growing in sand or on sandy soil.

arête (French) a sharp, narrow, steep-sided mountain ridge, especially one formed when two CIRQUES have been developed back to back. An alternative term is comb-ridge.

argillaceous *adj.* clayey, containing clay, or clay-like in composition or texture, applied especially to a SEDIMENTARY ROCK in which clay minerals (mainly of aluminium and iron silicates, CLAY-1) predominate. ARENACEOUS.

argillic horizon a SOIL HORIZON (commonly the B horizon) in which CLAY-1 minerals have accumulated by ILLUVIATION.

arid *adj.* dry, parched, lacking moisture, applied especially to climate or land (ARID LAND), the main factors being insufficient rainfall and a rate of EVAPORATION exceeding that of the PRECIPITATION.

aridisols, aridosols in SOIL CLASSIFICATION, USA, an order of soils with generally mineral profiles, with or without ARGILLIC HORIZONS, in which an accumulation of soluble salts or carbonates is common. Such soils are characteristic of DESERT and ARID regions.

aridity the state or quality of being ARID. In attempts to give arid and aridity a precise definition in relation to climate or land, several authors have suggested an index of aridity, based mainly on values of TEMPERATURE-2 and PRECIPITATION, including its REGIME.

arid land, arid zone land where, due to insufficient PRECIPITATION and a rate of EVAPORATION exceeding that of the precipitation, there is little or no natural vegetation, and agriculture is possible only with the aid of IRRIGATION.

arithmetic mean commonly termed the MEAN, an AVERAGE, calculated by adding together several quantities and dividing by the number of those quantities, e.g. the total of the values of a variable for all the observations in a data set divided by the total number of observations. CENTRAL TENDENCY, MEDIAN, MODE.

arithmetic progression a series of numbers each of which is greater or smaller than the one before it by the same amount, e.g. 2, 4, 6, 8 or 8, 6, 4, 2. GEOMETRIC PROGRESSION, LINEAR GROWTH.

arkose a coarse-grained SANDSTONE or GRIT derived from the swift breaking down of GRANITE or GNEISS and characterized by a considerable proportion of fragments of FELDSPAR, which have been altered little, if at all, by WEATHERING.

Armorican *adj*. applied to the great earth-building movements and mountain-building period (corresponding to the VARISCAN of central Europe), starting in Carboniferous times and continuing into Permian, which created mountain chains (sometimes considered as part of the ALTAIDES) in Brittany, southwest England and Ireland and throughout northern Europe. Through large areas the folds were east to west, which is accordingly termed the Armorican trend; where later folding took place along the same lines, the term Armoricanoid is used. HERCYNIAN, from the Harz Mountains of Germany, is used by some authors as synonymous with Armorican; but others consider that the Armorican mountains constitute the western sector of the Hercynian system, the Variscan the eastern.

array in statistics, an explicit display of a set of observations, all the values in a set of data displayed together. MATRIX-5.

arroyo (Spanish) a periodically dry water-course. NALA, WADI.

artesian basin a synclinal (SYNCLINE) basin in the earth's crust, in some places very large (e.g. in Australia), in which one or more PERMEABLE, POROUS water-bearing strata or AQUIFERS lie between IMPERMEABLE strata, the whole being folded to form the synclinal basin. ARTESIAN WELL, RECHARGE. Fig 3.

artesian well **1.** a perpendicular boring sunk through the upper IMPERMEABLE layer in an ARTESIAN BASIN to reach the AQUIFER. If the outlet of the well lies at a level lower than the WATER TABLE of the aquifer at the margins of the basin, water will rise in the well under HYDROSTATIC PRESSURE (PIEZOMETRIC LEVEL). The name is derived from Artois, France, where such wells were sunk in the twelfth century **2.** a term loosely applied to any

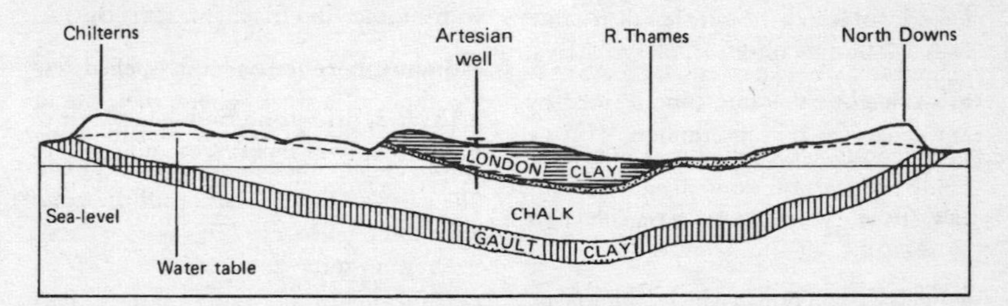

Fig 3 The London basin: a typical artesian basin and well. The London Clay and the Gault Clay are impermeable, the Chalk is an aquifer

deep well in which water rises under pressure, but not necessarily to the surface. Such wells are sometimes distinguished as sub-artesian, the water having to be pumped to the surface, as distinct from the true artesian, naturally flowing, wells.

artifact, artefact a simple object made by human workmanship, especially one related to prehistoric time, e.g. a stone tool.

artificial *adj*. made by a person, by human activity, as opposed to NATURAL.

artificial recharge the introduction of surface water into an AQUIFER by means of recharge wells. RECHARGE.

Asean, Asean+ the Association of South-East Asian Nations was established in 1967 by Indonesia, Malaysia, the Philippines, Singapore and Thailand to improve regional security, joined by Brunei in 1984. It developed into a trading bloc. Later the six Asean countries were joined by Hong Kong, Japan, South Korea and Taiwan to make an even larger trading bloc.

ash **1.** the powdery residue left after a substance has been burnt **2.** the material known as VOLCANIC ASH ejected from the crater of a VOLCANO in eruption. This is not true ash: it is not produced by burning, but consists of fine particles (the majority of which measure less than 4 mm in their long dimension) of pulverized lava.

ash cone the volcanic cone formed by ASH-2 ejected during eruption. CINDER CONE.

ash flow NUEE ARDENTE, PELEAN ERUPTION.

aspect the direction in which a thing faces, particularly applied to slopes in relation to the sun on account of its affect on settlement and plant growth. ADRET, UBAC.

asphalt pitch, a viscous, brownish-black bituminous substance, a mixture of HYDROCARBONS, occurring naturally (e.g. in Trinidad Pitch Lake, or the Athabaska tar sands, Canada) or as an industrial residue in the refining of some varieties of petroleum. It is used mainly as a surfacing material for roads, a waterproofing material for flat roofs, and in some fungicides and paints. BITUMEN.

assimilation in society, the process by which various groups in society merge, lose their distinguishing characteristics, their separate identities, are absorbed one with another in a 'melting pot', and become culturally HOMOGENEOUS-1. ACCOMMODATION, ACCULTURATION, CULTURE CONTACT, INTEGRATION, PLURAL SOCIETY.

Association of South-East Asian Nations ASEAN.

asteroid a planetoid, one of several thousand small planets in stable orbit round the sun in the SOLAR SYSTEM, occurring mainly in the asteroid belt between Mars and Jupiter. They are rocky, metallic or carbon-rich bodies, comprising the matter remaining after the formation of the PLANETS, and occasionally collide one with another and fragment. METEORITE.

asthenosphere (erroneously spelled aesthenosphere) a weak sphere, the zone of hot rock, believed to be in a plastic condition, underlying the solid LITHOSPHERE (the earth's CRUST), the top of the zone lying some 70 to 150 km (45 to 95 mi), the bottom some 200 to 360 km (125 to 225 mi), below the earth's surface. It is sometimes termed the soft layer of mantle, or the low velocity zone (LVZ), the latter because earthquake shock waves travel in it at reduced speed. Horizontal currents in this zone may be associated with plate movements. PLATE TECTONICS, GUTENBERG DISCONTINUITY.

astronaut one who travels outside the earth's atmosphere, a space traveller. The alternative term, cosmonaut, is preferred in Russia.

asymmetric, asymmetrical *adj.* showing lack of symmetry. ASYMMETRY.

asymmetrical fold a FOLD with one limb dipping away from the axis (AXIS OF FOLD) more steeply than the other. OVERFOLD. Fig 24.

asymmetrical valley a valley with one side sloping more steeply than the other.

asymmetry without SYMMETRY, the state of lacking the capability of being divided into two or more exactly similar and equal parts.

Atlantic polar front the POLAR FRONT between the polar maritime and tropical maritime (TROPICAL AIR MASS) air masses over the North Atlantic ocean.

Atlantic stage, of climate the sudden development of a mild moist phase following the dry cold Boreal phase during the retreat of the Great ICE AGE, lasting from about 5500 to 3000 BC, sometimes termed the Megathermal Period. In Britain a general rise of about 3 m (10 ft) in sea level resulting from the melting of the ice sheets, led to the formation of the Strait of Dover c.5000 BC, to the growth of mixed oak forest, and to the formation of peat. FLANDRIAN, PRE-BOREAL, SUB-BOREAL.

Atlantic suite a PETROGRAPHIC PROVINCE distinguished by the ATLANTIC TYPE OF COASTLINE, areas of BLOCK FAULTING and rocks rich in ALKALIS. ANDESITE LINE, PACIFIC SUITE, SPILITIC SUITE.

Atlantic type of coastline a type of coastline developed where the main folds and trend lines, often faulted, run at right angles or obliquely to the coastline, in contrast to the Pacific type (CONCORDANT COAST, LONGITUDINAL COAST) where they run parallel to the coastline.

atlas a uniform collection of maps bound in a volume or (rare use) a bound volume of pictures, engravings, etc., apparently so named because some early collections (notably one by Mercator in the sixteenth century) showed on the title-page a representation of a member of the older family of Greek gods, named Atlas in Latin, supporting the heavens on his shoulders. The term atlas first appeared on the general title-page of Mercator's *Atlas* 1595.

atmosphere 1. the air or mixture of gases, roughly (by volume) 20 per cent OXYGEN, 79 per cent NITROGEN, with 0.03 per cent CARBON DIOXIDE and traces of argon, krypton, xenon, neon, helium, as well as water vapour, ammonia, ozone, organic matter, some salts and solid particles in suspension, which envelops the earth (ASTHENOSPHERE, HYDROSPHERE, LITHOSPHERE). Various concentric layers are identified, based on such criteria as rate of temperature change, composition, electrical nature. The lower atmosphere is the TROPOSPHERE, then comes the TROPOPAUSE, above which is the upper atmosphere or STRATOSPHERE, passing into the STRATOPAUSE, MESOSPHERE, MESOPAUSE, IONOSPHERE (thermosphere), with the upper zone, the EXOSPHERE and the MAGNETOSPHERE **2.** a unit of pressure, a measure of air pressure, one atmosphere being equated with the pressure exerted by the weight of a column

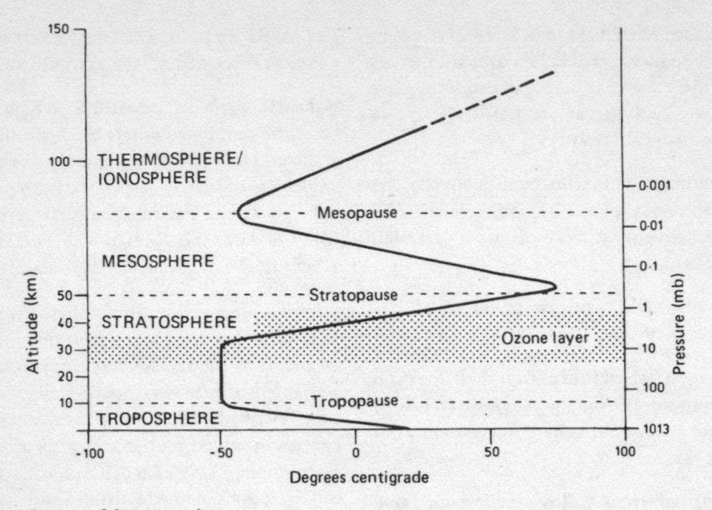

Fig 4 Layers of the atmosphere

of 760 mm (29.92 in) of MERCURY at 0°C (32°F) under standard gravity, at sea-level, or to a weight of air of 1033.3 gm per sq cm (14.66 lb per sq in), the average pressure over the earth's surface under those conditions. ATMOSPHERIC PRESSURE. Fig 4.

atmospheric cell 1. a large, three dimensional air mass of HIGH or LOW atmospheric pressure created by the disturbance of the planetary pressure system which is due mainly to the unequal solar heating of the irregularly distributed continents and oceans **2.** a vertical circulation cell in the TROPOPAUSE associated with meridional circulation (MERIDIONAL FLOW) of the atmosphere, e.g. the HADLEY CELL.

atmospheric circulation the general circulation of the ATMOSPHERE-1, covering all the movement of the air enveloping the earth in the lower and upper atmosphere resulting from the unequal heating of the earth and its atmosphere brought about mainly by the imbalance of SOLAR RADIATION between lower and higher latitudes, the differences in energy distribution in the atmosphere and the tendency for these differences to be smoothed out, and the angular momentum of the earth and its atmosphere. Fig 5.

atmospheric instability a state of the ATMOSPHERE-1 occurring when a body of air with a LAPSE RATE higher than that of the DRY ADIABATIC LAPSE RATE and warmer than overlying air, rises and expands. The adiabatic cooling of such moist air as it ascends disturbs the atmosphere, frequently gives rise to deep clouds, and leads to PRECIPITATION. ADIABATIC, SATURATED ADIABATIC LAPSE RATE.

atmospheric pressure the pressure exerted by the weight of the ATMOSPHERE-1 on the earth's surface, decreasing with height above sea-level and varying with weather conditions. The unit of measurement used in the ATMOSPHERE-2,

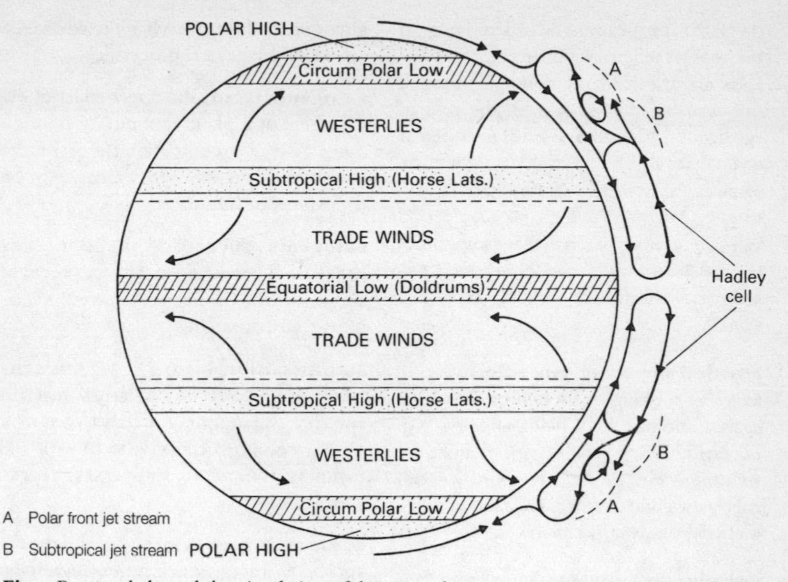

Fig 5 Pressure belts and the circulation of the atmosphere

expressed in millibars (mb): 1 atmosphere = 1013.25 mb (1000 mb = 1 BAR = 1 mn DYNES per sq cm), indicated by a BAROMETER.

atmospheric weathering WEATHERING.

atmospheric window WINDOW.

atoll a circular or almost circular CORAL REEF, the crest lying at a low height above sea-level, sometimes interrupted, enclosing a central LAGOON. Atolls are most common in the central and west Pacific ocean.

atom the smallest particle of an ELEMENT-6 which exhibits the properties of that element, consisting of a nucleus (the small, positively charged central core) with its orbiting electrons (negatively charged particles). MOLECULE, NUCLEAR ENERGY.

atomic energy NUCLEAR ENERGY.

attached dune a SAND DUNE formed around an obstacle lying in the path of blown sand in ARID LANDS, the obstacle forming the nucleus that 'fixes' the dune.

attitude 1. a state of mind, a relatively enduring tendency to perceive, feel or behave towards certain people or events in a particular manner. Attitudes play an important part in PERCEPTION, and they affect preferences and choice of goals **2.** in geology, the disposition of a rock STRATUM, horizontal or tilted, e.g. the relationship of a BEDDING PLANE to the horizontal plane in terms of its DIP and STRIKE.

attribute 1. a quality proper to, or characteristic of, a person or thing **2.** in statistics, a QUALITATIVE VARIABLE which can take only certain fixed values, sometimes termed a discrete variable. If the

CATEGORIES-3 cannot be ordered (e.g. in the case of religious affiliation), the variables are measured at nominal value; if they can be ordered (e.g. in the case of age groups), the variables are measured at ordinal level, and in this case they are sometimes termed ORDERED ATTRIBUTES. Attributes which take only two values (e.g. male or female) are sometimes termed dichotomous variables. MEASUREMENT-2, NOMINAL SCALE, ORDINAL SCALE.

attrition the act or process of wearing away by rubbing or friction of one thing against another, each being affected, e.g. occurring when pebbles carried along by running water, by sea waves, or by wind, rub against one another. ABRASION, CORRASION, TRANSPORTATION-2.

aureole METAMORPHIC AUREOLE.

aurora australis, aurora borealis (the former in the southern, the latter in the northern hemisphere) the spectacular coloured lights, probably electromagnetic in origin in the IONOSPHERE, seen near the horizon in the night sky in high latitudes. They are red, green and white, and they ascend in streaks and sheets, roughly in the shape of a fan. The phenomenon is known as Northern Lights in the northern hemisphere.

authigenic, authigenous *adj.* formed in place, applied especially to a mineral formed in situ during or after the formation of the rock of which it forms a part, e.g. the crystals in an IGNEOUS ROCK, in contrast to ALLOGENIC, ALLOTHIGENIC.

autogenic, autogenous from autogenesis, spontaneous generation) *adj.* produced by an organism itself, not by an external influence.

autogenic change a self-induced change produced by natural processes.

autogenic factor the direct effect of the members of a plant community on each other (e.g. by competition), or of the community on its own habitat (e.g. by deposition of humus).

autogenic succession a plant SUCCESSION-2 produced by changes brought about by the plants themselves. ALLOGENIC SUCCESSION.

autometamorphism the METAMORPHISM taking place in IGNEOUS material and the surrounding COUNTRY ROCK during cooling in PNEUMATOLYSIS and in HYDROTHERMAL PROCESSES. METASOMATISM.

autonomous *adj.* self-governing, usually applied to a state which is not a sovereign state or absolutely independent, but one which is self-governing in home affairs while being under the control of a larger (and sovereign) state. AUTONOMY.

autonomy self-government, the right of self-government. AUTONOMOUS.

autotrophic *adj.* applied to an organism capable of producing its own organic substances (i.e. without recourse to outside organic substances) from inorganic compounds, using energy from the sun (PHOTOTROPHIC) or from chemical reaction (CHEMOTROPHIC). Most CHLOROPHYLL-containing plants and some BACTERIA are autotrophic.

autumn (American, fall) the third SEASON of the year in MIDLATITUDES, variously defined. Properly from the autumnal EQUINOX to the winter SOLSTICE, i.e. from 21–22 September to 21–22 December in the northern hemisphere, from 21–22 March to 21–22 June in the southern. But in the northern hemisphere popularly regarded as September, October and November; in the southern hemisphere, February, March and April.

available relief local relief, the vertical distance from the original, fairly flat upland surface (i.e. from the height of a surface which is being dissected) to the local base level (the valley floors) of the dissecting streams. RELIEF.

avalanche a French dialect term originally applied to a large mass of snow mixed with earth, stones and ice loosened from a mountainside and falling swiftly by gravity to the valley below (SNOW AVALANCHE). The term avalanche is now usually restricted and applied to a fall of a mass of snow, ice and FIRN, being qualified if used to cover similar movements of other materials, e.g. rock avalanche (better termed a LANDSLIDE), sand avalanche (PLINTH).

avalanche cone a mass of material deposited by an AVALANCHE, including not only the SNOW, ICE, or FIRN but also everything torn away and carried along by the avalanche.

avalanche wind the rush of air produced by and preceding an AVALANCHE in its descent.

aven 1. a vertical or inclined shaft in limestone leading down, generally from the land surface, to a cave passage **2.** in England, an enlarged vertical joint in the roof of a cave passage, narrowing upwards.

average 1. ARITHMETIC MEAN **2.** an undefined measure of CENTRAL TENDENCY of the values in a DATA SET or FREQUENCY DISTRIBUTION.

awareness space all the locations about which a person has knowledge above a minimum level even without visiting some of them (INDIRECT CONTACT SPACE). Awareness space includes activity space (the area within which most of a person's activities are carried out, within which the individual comes most frequently into contact with others and with the features of the environment), and its area enlarges as new locations are discovered and/or new information is gathered. SEARCH SPACE.

axial plane the imaginary surface dividing the limbs of a FOLD as symmetrically as possible, and passing through the axis of the fold. Different types of fold are identified by the INCLINATION-2 from the vertical of the axial plane. Fig 24.

axis 1. a real or imaginary line around which a thing rotates **2.** one of the reference lines in a COORDINATE system **3.** an alliance between countries which aims to ensure a common policy.

axis of a fold the central line of a FOLD, the crest from which STRATA dip downwards and away in an ANTICLINE, or the central line of the lowest depth of the trough from which strata rise in opposite directions in a SYNCLINE. AXIAL PLANE. Fig 24.

axis of the earth the diameter between the NORTH POLE and the SOUTH POLE, tilted at an angle of about 66°30' to the plane of the earth's ORBIT, around which the earth rotates (anti-clockwise) once in every 24 hours. ORBIT OF THE EARTH, PRECESSION OF THE EQUINOXES.

azimuthal map projection a MAP PROJECTION in which all bearings are laid off correctly from the central point of the map, so that all points on the map are true in distance and direction from the centre.

azonal soil a young soil that lacks marked

horizons, commonly because insufficient time has elapsed for climate and vegetation to create them, e.g. a recent alluvial soil. In contrast, mature soils (ZONAL SOIL) group themselves according to the great climatic zones of the earth's surface. IMMATURE SOIL, INTRAZONAL SOIL, SOIL CLASSIFICATION, SOIL PROFILE.

Azores high a subtropical ANTICYCLONE stationed generally over the eastern sector of the North Atlantic ocean, more persistent in the summer than in the winter of the northern hemisphere, and in summer usually extending so far northeast as to affect western Europe, including the British Isles.

B

backing of wind a change of direction of the wind in a cyclonic (anti-clockwise) direction, i.e. from north through west to south in the northern hemisphere (the opposite of VEERING), where the wind is said to 'back' at a place north of the centre of a DEPRESSION-3 travelling eastward. In the southern hemisphere the direction of the wind in relation to pressure systems of closed isobars is reversed, so that a backing wind in the southern hemisphere is equivalent to a veering wind in the northern hemisphere.

backset bed a term applied mainly in the USA to a deposit of sand on the windward slope of a SAND DUNE, commonly fixed by vegetation.

backshore the land lying inland from the average HIGH-WATER line to the COASTLINE, bordered seawards by the FORESHORE (the land between the lowest LOW-WATER mark and the average high-water mark), and to seaward of which lies the offshore zone (below low-water mark to the depth at which substantial movement of beach material ceases). OFFSHORE-3.

back slope, back-slope the gentler slope of a CUESTA. Fig 42.

backswamp, backswamp deposits the tract of low, swampy land lying on the FLOODPLAIN of an alluvial river between the natural LEVEES or banks of the river and the BLUFFS; and the layers of silt and clay deposited there. In the USA the term is used particularly in relation to the floodplain of the Mississippi.

back wall, back-wall the steep wall at the back of a CIRQUE.

backwash, back-wash **1.** the seaward flow of a body of water down the slope after a WAVE-3 has broken on the beach, in contrast to SWASH **2.** the drag of a receding wave. LONGSHORE DRIFT. Fig 28.

backwash and spread effects CIRCULAR AND CUMULATIVE GROWTH.

bacterium pl. bacteria, a member of a very large group of microscopic unicellular or multicellular organisms which live in very large numbers in favourable habitats. Their activities are of major importance to human beings, positively in soil in the breakdown of organic matter (NITROGEN CYCLE, NITROGEN FIXATION), in sewage disposal, and as source of antibiotics; and negatively, e.g. as agents of plant disease and as the cause of serious diseases in animals, including human beings, e.g. tuberculosis, typhoid fever. PHOTOSYNTHESIS.

badlands originally and specifically applied to a large, dry region in South Dakota, USA, where erosion of the nearly horizontal, unconsolidated sedimentary beds resulted in a land of narrow ravines, sharp crests and pinnacles, devoid or almost devoid of vegetation. This highly dissected landscape was well described by early French travellers as 'mauvaises terres à traverser' (bad lands to cross). The term is now applied to similar lands elsewhere, e.g. in Algeria and Morocco.

bajada, bahada (Spanish) a continuous apron of gently sloping sediments, e.g. gravel and coarse sand, formed by the merging of ALLUVIAL FANS laid down by swollen streams from a series of mountain streams where they debouch on a plain at the base of a mountain range in an arid or semi-arid area.

balance of nature the relationship between all the component parts of the BIOSPHERE which, by interaction one with another, ensure that it is in a state of equilibrium. The balance is delicate and can be upset by human activity.

balance of payments 1. the relationship between the credits of one nation or group of nations against all other trading partners and the debits of that nation or group of nations to all other trading partners over a specified period of time **2.** a systematic record of all economic transactions between one nation or group of nations and all other nations with which it has contacts. Theoretically total debits and total credits should balance. Included are all goods (visible, such as manufactured goods, raw materials, bullion, etc.) and services (invisible, covering transport, banking, insurance, interest payments, tourism etc. as well as the flow of capital). BALANCE OF TRADE.

balance of trade the relationship between a nation's total visible imports and exports, i.e. of goods. BALANCE OF PAYMENTS.

balk, baulk a piece of unploughed land used for grazing and giving access to the ploughed parts of an open field under the medieval system, the minor ones being used as boundaries to separate one man's strip from that of another, the major to separate groups of strips. The larger balks, primarily grass-covered 'occupation roads', were usually termed town balks or common balks to distinguish them from the minor balks between strips.

balkanization the division of an area into small units, sometimes implying mutual hostility among such units, or between those units and others outside the area.

Baltic shield the PRECAMBRIAN platform of Finland and eastern Scandinavia. SHIELD.

bamboo a grass native to tropical and subtropical regions. The economically important varieties have edible shoots and grain and/or hollow, hard, durable stems used, especially in Asia, for tools, furniture, mats and papermaking, etc., and in buildings and their construction.

banana a giant herb native to tropical areas, now widely cultivated there for its nutritious, sweet fruit (a staple food in some tropical countries) which grows in bunches or stems about 10 to 12 half-spirals, known as 'hands' and markedly separated from each other, each hand consisting of 12 to 16 'fingers'. The export trade is important internationally and vital to some banana-growing countries. It is large-scale and highly organized. The bunches are cut when the fruit is green, before ripening, transported in chambers cooled at a constant temperature, as near as possible to 10° to 11°C (51°F) and carefully ripened on arrival at the importing country.

bank 1. sloping ground bordering a river, stream or lake **2.** in the north of England, a hill or hillside **3.** an elevation in the floor of a river or a shallow sea, usually of sand, mud, gravel (not of solid rock or CORAL), in some cases connected with the shore, but not sufficiently near the surface as to be dangerous to shipping.

bank caving the undercutting and erosion of the BANK-1 by water flowing on the outside curve of a river, resulting in the washing downstream of the material so dislodged. LATERAL EROSION.

bankfull stage in river flow, the stage when the channel is completely filled with water, from bank to bank, the stage before the river overflows. FLOOD STAGE, OVERBANK STAGE, STREAM STAGE.

banner cloud a CLOUD touching and flowing out like a banner on the LEE side of a mountain peak in clear sky. It occurs when water vapour in a forced up-draught of warm air from the mountainside condenses and the cooled air descends downwind, to be warmed, so that it rises again. Descent and ascent continue downwind until the ultimate complete evaporation of the water droplets. The best known example is the cloud of the Matterhorn. LEE WAVE.

bar 1. loosely applied to a marine deposit of mud, sand, shingle, covered by water at least at high tide (BARRIER), e.g. across an estuary, parallel to the shore (longshore or offshore bar), across a bay (BAY BAR), between an island and the mainland (TOMBOLO), across the access to a harbour (harbour bar) **2.** in USA, the deposits of ALLUVIUM etc. in streams, river mouths and some lakes **3.** in meteorology, the unit of ATMOSPHERIC PRESSURE equivalent to 1 mn DYNES per sq cm: 29.5306 in or 750.076 mm of mercury at 0°C in latitude 45°N. The unit commonly used is the millibar (mb), a thousandth part of a bar. In SI 1 bar = 10 NEWTONS. ATMOSPHERE, BAROMETER.

barbed drainage a pattern of drainage in which tributaries meet the main stream at obtuse angles, i.e. at such angles that their flow appears to be directed to the source of the main stream. It is caused by RIVER CAPTURE which has reversed the direction of flow of the main stream. Fig 17.

barchan a crescent-shaped dune of shifting sand formed when the direction of the wind varies only very slightly or not at all. The windward side is convex, with a gentle slope, the steeper leeside is concave, and the 'horns' point downwind. It travels if an adequate supply of sand is maintained; it occurs singly, or in groups; and the height ranges from quite low to over 30 m (100 ft). PARABOLIC DUNE. Fig 6.

bare fallow fallow land left without a crop for a whole season. FALLOW, GREEN FALLOW.

bar graph, bar chart a diagram drawn to display data. It consists of a series of bars or columns, representing categories, the length of each bar or column being proportional to the quantity represented. The bars are sometimes set horizontally (e.g. POPULATION PYRAMID, taking the form of a pyramid) or vertically (as columns), the relative importance of each category becoming immediately clear. All the bars have the same width, and may be separated from each other by small spaces, to emphasize that each category is distinct. The bars may show total values, or they may be divided to show the constituent parts of the total values. Fig 38.

barley a cultivated grass of the genus *Hordeum*. It is one of the food grains, but it does not make good bread. Flat barley cakes (the barley loaves of the Bible) are made and eaten in north Africa and some

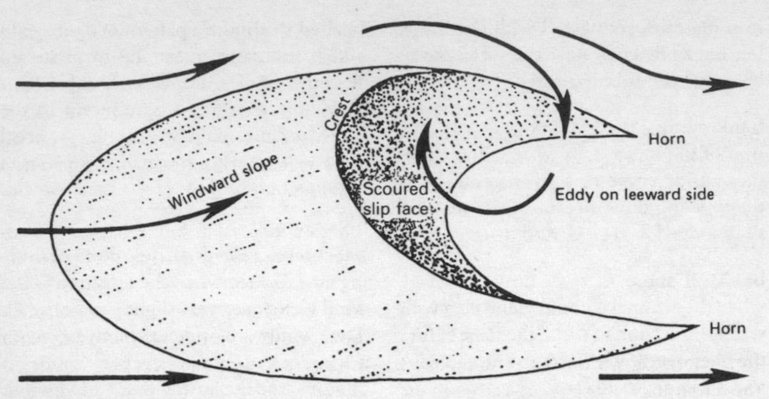

Fig 6 Barchan

eastern countries, but a major use of barley is, as formerly, in the brewing of beer and the making of whisky (a fact recognized by the old English terms 'bread corn', applied to wheat and rye; and 'drink corn', applied to barley). Today barley is grown principally for animal feed. Barley flourishes generally under conditions similar to those which suit WHEAT, but it tolerates poorer and lighter soils. It can take advantage of the long hours of daylight in northern latitudes, so 'arctic' barley grows farther north than any other grain. It will grow higher up mountain slopes than wheat, but it is even less tolerant than wheat of damp conditions.

barogram a continuous record of changes of ATMOSPHERIC PRESSURE as measured by an aneroid BAROMETER, indicated by a curve drawn on scaled paper.

barograph a self-recording BAROMETER, in which a pen linked to an aneroid barometer records changes in ATMOSPHERIC PRESSURE on a revolving cylinder which makes one complete revolution each week.

barometer an instrument, of which there are several forms, for measuring ATMOSPHERIC PRESSURE, used in estimating height above sea-level and in weather forecasting. In the mercury barometer the weight of a column of mercury is balanced against that of a column of the atmosphere, with adjustments made for latitude (standardized to 46°N), temperature (to 12°C: 53°F), altitude (average decrease of 33.9 mb: 1 in mercury for each 275 m: 900 ft for the first 1000 m: 3300 ft, progressively decreasing thereafter with height) and for any peculiarity of the (usually sensitive) individual instrument. A vertical tube is sealed at the top end, with the bottom, open end, standing in a container holding MERCURY; atmospheric pressure is measured by the vertical height of the column of mercury which the atmosphere will support. In the aneroid (without liquid) barometer, used in ALTIMETERS and BAROGRAPHS, a shallow metal box or cylinder is nearly exhausted of air. In one type, this metal box has flexible sides which expand and contract with changing air pressure; in another the thin metal upper face (the top), which is corrugated, is held up by an external clip spring; as the air pressure changes it upsets the balance between the spring and the pressure in the box/cylinder, so that the top moves. In either type, the movements are magnified and transmitted by a system of levers to a pointer moving over a calibrated scale. The aneroid barometer generally available is not capable of absolute accuracy, and is not to be compared with the precision models used in aviation and meteorology.

barometric gradient pressure gradient, the amount of change in ATMOSPHERIC PRESSURE between two points, indicated by the distance apart on a level surface of the isobars (ISO-) on a SYNOPTIC CHART. Closely spaced isobars indicate great differences in pressure and therefore a steep gradient, associated with strong winds, in some cases with TROPICAL REVOLVING STORMS; isobars with wide intervening spaces indicate only small differences, and thus a gentle gradient.

barometric tendency the character (increasing or decreasing) and amount of change in ATMOSPHERIC PRESSURE during a specified period, usually of three hours.

barrage 1. a large structure, usually of masonry or concrete, occasionally of earth, built to hold up a large quantity of water, especially for irrigation. A DAM similarly impounds water, but in some cases the term dam is used if the generation of power is involved, barrage if it is not. Sometimes the two are distinguished by the duration of water storage, i.e. a barrage serves annual storage of floodwater only, a dam has perennial use **2.** part of a tidal power station (TIDAL BARRAGE).

barrier something that hinders or prevents access or advance. Referring to coastal features, a bank of mud, sand, shingle, etc. is usually termed a BAR-1 if it is submerged at least at high tide, a barrier if it lies above high tide level. BARRIER BEACH, BARRIER CHAIN, BARRIER ISLAND, BARRIERS AND DIFFUSION WAVES.

barrier beach a long, narrow, sandy ridge, lying above high tide level and parallel to the coast from which it is separated by a LAGOON. The term is used sometimes as synonymous with SPIT. BARRIER CHAIN, BARRIER ISLAND.

barrier chain a series of BARRIER ISLANDS and BARRIER BEACHES extending a considerable distance along a coast.

barrier island a feature similar to a BARRIER BEACH but consisting of several ridges, commonly with DUNES, vegetation and swampy areas on the LAGOON side. The term barrier spit is sometimes used in the USA if the barrier island is at one end joined to the mainland. SPIT.

barrier lake a LAKE formed by a natural obstruction across a valley, e.g. by an AVALANCHE, ROCK FALL, ALLUVIAL deposits, TERMINAL MORAINE, or by a DAM formed by a build-up of VEGETATION, ICE, LAVA, etc.

barrier reef a CORAL REEF skirting a shore and some distance from it, so that it acts as a barrier between the open ocean and the sheltered LAGOON lying between the reef and the coast. On a large scale the Great Barrier Reef, stretching for over 1600 km (1000 mi) off the coast of Queensland, Australia, functions in that way.

barriers and diffusion waves barriers that act as a drag on the process of DIFFUSION-1 are commonly classified according to the decreasing amount of drag exhibited, i.e. the superabsorbing BARRIER (absorbing the message but destroying the transmitter), the absorbing (absorbing the message but not affecting the transmitter), the reflecting (not absorbing the message, but allowing the transmitter to transmit a new message in the same time period) or the direct reflecting (not absorbing the message but deflecting it to the available cell nearest to the transmitter).

barrio SHANTY TOWN.

barrow, tumulus a prehistoric mound of earth, piled over a burial ground, common in the British Isles and other parts of western Europe. NEOLITHIC.

barter 1. trade by exchange, without the use of money **2.** the thing so exchanged. BAZAAR ECONOMY.

barysphere loosely applied, sometimes to the dense mass (possibly of nickel-iron) believed to occupy the CORE of the earth below the MANTLE, sometimes to the mantle only, sometimes (and this is preferable) to the core and mantle together, i.e. all of the earth's interior beneath the LITHOSPHERE.

basal, basic *adj.* pertaining to, situated at, forming the BASE-1. BASIC.

basal complex BASEMENT COMPLEX.

basal conglomerate the CONGLOMERATE commonly formed at the beginning of a cycle of sedimentation, hence at the base of a series of strata which lie unconformably on older rocks.

basalt a fine-grained black or dark grey IGNEOUS ROCK belonging to the basic group (BASIC ROCK), with 45 to 52 per cent of SILICA, mainly sodium or potassium alumino-silicates with some iron (FELDSPAR). When erupted from volcanic vents or fissures it tends to be very fluid and to flood evenly over large areas before consolidating, hence many LAVA plains or plateaus are of basalt. It may solidify into perfect hexagonal columns (e.g. Giant's Causeway). SPHEROIDAL WEATHERING.

basaltic *adj.* of, pertaining to, or consisting of BASALT.

basaltic lava BASIC LAVA.

basal till TILL, in many cases with a high CLAY content, carried underneath or deposited by a moving GLACIER.

base 1. in general, the bottom, the lowest part or that on which something stands or rests **2.** in chemistry, a substance which reacts with an acid to form a salt and water only; or a substance which, dissolved in water, provides hydroxyl IONS from its own MOLECULES; or a molecule or ion which accepts PROTONS. **3.** in statistics, a number or magnitude used as a standard reference.

base flow a term applied particularly in North America to that part of a stream flow which comes from GROUND WATER, as distinct from the surface flow after rain has fallen.

base-level the lowest level to which a running stream can erode its bed under stable conditions of the earth's crust; or the level below which a land surface cannot be reduced by running water.

base line an accurately measured line on the earth's surface which is used as a base in trigonometric observations (TRIANGULATION), and so in the mapping of land.

base map a map of any kind on which additional information may be plotted for specific purposes.

basement, basement complex, basal complex a term loosely applied to the assemblage of ancient IGNEOUS and METAMORPHIC ROCKS which usually, but not always, underly the PRECAMBRIAN stratified rocks in any particular region.

basic *adj.* **1.** in chemistry, of the nature of a BASE-2, the opposite of ACIDIC, reacting chemically with acids (ACID) to form salts (pH) **2.** BASAL.

basic activity, basic function, basic industry in urban development, a manufacturing or service activity within a city or urban area which provides goods and services and thus earns revenue from outside the city or urban area, i.e. in this context a 'primary' or 'export' industry. BASIC INDUSTRY, ECONOMIC BASE THEORY, EXPORT BASE THEORY, LOWRY MODEL, NON-BASIC ACTIVITY, URBAN ECONOMIC BASE.

basic Bessemer process a method used to remove PHOSPHORUS from PIG IRON in the process of making cast STEEL, introduced in 1890 in the USA. A Bessemer converter (BESSEMER PROCESS) was lined with a material incorporating lime to serve as a 'base' with which the phosphorus, escaping from the iron, could combine (hence the name). If the proportion of phosphorus was too high to be removed in that way, more lime was added. The product was known as basic steel.

basic grassland grassland of chalk and limestone in which the grasses *Festuca ovina* and *Festuca rubra* are dominant.

basic industry **1.** a heavy industry of national economic importance, or an industry fundamental to other industries (e.g. iron and steel, or the manufacture of sulphuric acid) **2.** in the UK, sometimes officially applied to mining and quarrying; gas, electricity and water; transport and communication; agriculture and fishing.

basic lava, basaltic lava the most common type of LAVA, molten IGNEOUS ROCK, high in IRON, MAGNESIUM and other metallic element content, low in SILICA, and with a low melting point, which pours easily and quietly from a volcanic vent, spreading widely before hardening (commonly as BASALT) to form a SHIELD VOLCANO or broad plateau. HAWAIIAN VOLCANIC ERUPTION, MAGMA, PILLOW LAVA.

basic–non-basic ratio the proportion of BASIC ACTIVITY to NON-BASIC ACTIVITY in the economy of a city or region, measured either by numbers employed or by the value of production. The larger the city, the higher the proportion of non-basic workers.

basic rock a term loosely applied to an IGNEOUS ROCK which lacks QUARTZ and contains FELDSPAR with a content of CALCIUM higher than that of SODIUM. As the sodium content decreases the basic rocks become INTERMEDIATE ROCKS; as the feldspar decreases the basic rocks become ULTRABASIC. (A basic rock is not synonymous with an ALKALINE ROCK.)

basic slag waste material from blast furnaces, rich in minerals, especially phosphorus, used as a fertilizer.

basin a term applied loosely to some form of natural or artificial depression, varying in extent, in the earth's crust, e.g. **1.** the total area of land drained by a river and its tributaries (DRAINAGE BASIN, RIVER BASIN), termed WATERSHED in USA **2.** in geology, a circumscribed area where the strata dip inward towards the centre (SYNCLINE), or a stratified deposit (e.g. COAL) lying therein **3.** a hollow in the ground formed by surface settlement following the natural or artificial removal of underground deposits of salt or gypsum in solution **4.** a large or small depression occupied by a lake (lake basin) or pond **5.** a depression in the ocean floor (DEEP) **6.** an extensive depression occupied by an OCEANIC BASIN **7.** a hollow classified according to origin (TECTONIC basin, GLACIAL basin) **8.** a large area surrounded by high land, with or without access to the sea (Great Basin, USA) **9.** a dock or part of a canal or river widened for navigation and lined with wharfs, etc. **10.** a dock subject to tidal movement, or a depression (natural or artificial) filled with water at high tide (TIDAL BASIN).

basin and range a tract of country with a series of asymmetrical ridges separated by basins, the ridge consisting of tilted FAULT BLOCKS. The Basin and Range country of USA lies between the Sierra Nevada and Wasatch mountains.

basin cultivation the practice of dividing land by low earth ridges to form small or large basins where water can be retained and rapid runoff prevented. This technique is used especially near the equatorial margins of the tropics in Africa (e.g. in Nigeria and Ghana) to stem soil erosion by heavy rainfall.

basin irrigation a type of IRRIGATION in which floodwater from a river that annually and for a short time overflows on to its FLOODPLAIN is led off into prepared basins, varying greatly in size, and separated from one another by earth banks. PERENNIAL IRRIGATION.

basket of eggs relief rounded sandy mounds or DRUMLINS arranged in such a pattern that from a distance they resemble eggs in a basket, occurring in glaciated regions in valleys formerly occupied by ice.

batholith, bathylith a very large dome-shaped mass of IGNEOUS ROCK, usually of GRANITE, formed by a large-scale, deep-seated intrusion of MAGMA, the sides of which plunge down steeply to unknown depths. The domed upper surface may be exposed by DENUDATION over a long period, to form uplands, e.g. Dartmoor in southwest England. A batholith may be surrounded by a METAMORPHIC AUREOLE. Figs 7, 21.

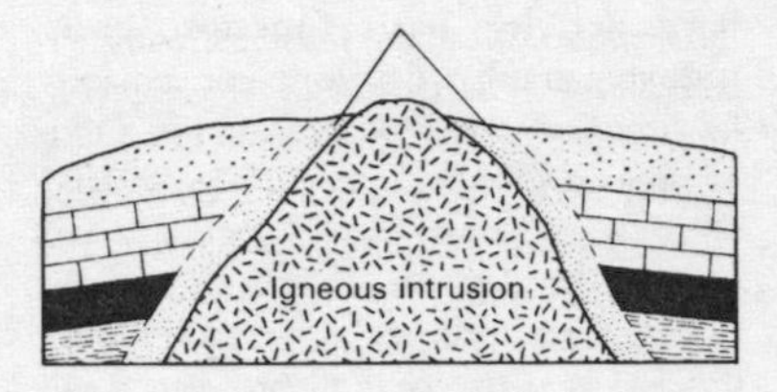

Fig 7 Batholith and metamorphic aureole

bathyal *adj.* applied loosely, of, or pertaining to, the deeper parts of the ocean, i.e. very broadly the CONTINENTAL SLOPE between the CONTINENTAL SHELF and the ABYSSAL ZONE; and the deposits and organic life present there.

bathyorographical *adj.* applied to maps showing both the relief of the land (OROGRAPHY) and the depths of the ocean.

bathysphere **1.** deep sphere, applied, incorrectly, by some writers to the interior mass of the earth as an alternative to BARYSPHERE **2.** a diving apparatus, a large spherical chamber, able to withstand great pressure, used for deep sea observations.

battery system in poultry farming, a CAPITAL-INTENSIVE system of husbandry common in industrialized countries, in which birds are reared and kept under cover in a carefully controlled environment (ENVIRONMENTAL CONTROL). The birds are confined to cages, fed and watered by automatic devices. They drop their eggs into channels running under the cages; the killing of the birds for meat is automated. FACTORY FARM.

bauxite (derived from deposits at Baux, near Arles, France) a naturally occurring ALUMINIUM oxide, an amorphous, earthy, granular mineral containing IRON oxides and PHOSPHATE, the colour ranging from grey-white through yellow, brown, red. Now the principal commercial ore of ALUMINIUM, it occurs widely in FELDSPARS and other SILICATES which readily break down in tropical conditions; hence bauxite is found as a surface crust in tropical lands.

bay **1.** a term applied loosely to a wide, curved indentation of the sea or of a lake into the land, especially one with a wide opening, or greater in width than in depth, usually considered to be larger than a COVE, smaller than a GULF **2.** in law, precisely defined in the 1958 convention on the delimitation of TERRITORIAL WATERS. Under the 1958 convention the existence of a bay is established by drawing a straight line to connect the seaward extremities of the natural PROMONTORIES on each side of the indentation. This line acts as the diameter of a semicircle which is drawn landwards: if the area of the water in the indentation limited by the straight line is as great as or greater than the area of the semicircle, it qualifies as a bay. If the length of the diameter is 38.6 km (24 mi) or less the waters 'enclosed' by the line can be treated as territorial. The waters in certain 'historic' bays (e.g. Hudson Bay, Canada) which do not meet these criteria are nevertheless considered to be territorial **3.** an elliptical, shallow depression in the coastal plain of the eastern USA (the Carolina Bays), origin uncertain **4.** in Germany (translation of Bucht), the spread of the lowland into an upland area along a river valley, e.g. Kölnische Bucht **5.** a recess in a range of hills or mountains.

bay bar, bay barrier (American, baymouth bar) a ridge of sand, mud or shingle extending across a BAY-1, linking the two headlands, caused by the lengthening of one SPIT or the convergence of two, or the moving of an OFFSHORE BAR towards the coast.

bay-head beach, pocket beach a crescent-shaped accumulation of sand and shingle piled up at the head of a small COVE on a sea coast.

bay-head delta a DELTA occurring at the head of a BAY-1.

bayou (southern USA term derived from French) a sluggish stream or a stagnant body of water such as an OXBOW LAKE or swampy backwater connected or associated with the lower Mississippi and its delta.

bazaar economy a commercial system in which a large number of buyers and sellers meet personally to transact their business without the aid of an intermediary such as a retail outlet. The transactions, usually unrelated one to another, are commonly centred on, but not necessarily restricted to, a MARKET-2.

beach the accumulation of loose material (mud, sand, shingle, pebbles) on the shore of a lake or of the sea at or near the limits of wave action, mainly between the low water SPRING TIDE line and the highest point reached by storm waves at HIGH TIDE. Beach material is classified by size, i.e. SAND, GRAVEL, PEBBLE and BOULDER. BEACH RIDGES, GROYNE, PROGRADATION, RAISED BEACH.

beach cusp a cone-shaped deposit of sand

and gravel with the apex pointing seawards, alternating with bay-like depressions, usually one of a series along a straight, open beach, resulting from the SWASH and BACKWASH of waves breaking at right angles to the coast. The distance between the points, increasing generally with wave height, varies from 9 to 60 m (30 to 200 ft).

beach ridges low sandy ridges on a coast representing a successive series of BERMS-1 produced in the PROGRADATION of a BEACH.

beaded drainage small pools joined by streams caused by the melting of the ground surface in PERMAFROST regions.

beaded esker an ESKER with a succession of mounds strung out along the ridge, like beads on a string, indicating pauses in the retreat of the GLACIER that fed the stream which formed the esker.

beaded lakes strings of long, narrow lakes between sand dunes.

beaded valley a VALLEY with alternating narrow and wide sections.

bearing the horizontal angle measured clockwise between a specific reference line and a point viewed by the observer **1.** for a true bearing the reference line is the MERIDIAN, so a true bearing is measured clockwise from TRUE NORTH **2.** a magnetic bearing is measured clockwise from MAGNETIC NORTH **3.** a compass bearing is measured clockwise from the north indicated by the COMPASS **4.** a grid bearing is measured from the north–south GRID-1 lines on a map **5.** a reverse (reciprocal) bearing is the reverse or reciprocal of a given bearing, i.e. a line drawn 180° from any bearing.

Beaufort scale a scale widely used for measuring and recording the strength of the wind, based on estimated velocity as 10 m (33 ft) above the ground, devised by Admiral Sir Francis Beaufort, RN, in 1805, used internationally since 1874, slightly modified 1926.

	Scale number	*Description*	*Force km/h*	*mi/h*
light winds –	0	calm	0	0
	1	light air	1.5–5	1–3
	2	light breeze	6–12	4–7
	3	gentle breeze	13–20	8–12
moderate winds –	4	moderate breeze	21–29	13–18
	5	fresh breeze	30–39	19–24
	6	strong breeze	40–50	25–31
	7	moderate gale	51–61	32–38
gales –	8	fresh gale	62–75	39–46
	9	strong gale	76–87	47–54
	10	whole gale	88–102	55–63
	11	storm	103–121	64–75
	12	hurricane	above 121	above 75

Notice that moderate gale, 7, is for statistical purposes classified under moderate winds.

bed **1.** the floor, the land at the bottom of a body of water (sea, lake, river, canal, pond), usually permanently covered by the water but possibly intermittently dried out **2.** in geology, a layer of STRATUM of rock, a feature of a SEDIMENTARY ROCK-1 distinguished from adjacent layers by its composition, structure or texture, and separated from the overlying and underlying layers by well marked BEDDING PLANES. INTERBEDDED.

bedding the arrangement of rock strata in bands of various thickness and character. FALSE-BEDDING, STRATIFICATION.

bedding plane the plane of stratification, the surface separating the successive distinctive layers of SEDIMENTARY ROCK, in many cases forming a line of weakness.

bedload traction load, the solid material, e.g. sand and gravel, and sometimes large boulders in time of flood, pushed or rolled by a STREAM-2 (TRACTION), or bouncing (SALTATION) along the BED-1 of a STREAM-1, as distinct from the material carried in SUSPENSION (SUSPENDED LOAD) or SOLUTION-1 (DISSOLVED LOAD).

bedrock **1.** the unweathered rock underlying the weathered superficial deposits (i.e. underlying the soil, subsoil and other loose unconsolidated rock, the REGOLITH) **2.** more specifically, the solid rock beneath PLACER deposits of GOLD or TIN.

beech-hanger HANGER.

beet sugar SUGAR.

behaviour **1.** the way in which an organism or a group of organisms (including a person or a group of people) reacts or responds to stimuli in the environment, e.g. to light, sound, touch, chemicals, the presence and activities of other organisms (including a person or people), or to a particular object, or to a particular event **2.** the way in which an organ, an organism, or a machine works, in terms of its efficiency. ADAPTIVE BEHAVIOUR, ADOPTIVE BEHAVIOUR.

behavioural *adj.* concerned with, or a part of, BEHAVIOUR-1,2.

behavioural approach in psychology **1.** the study of humans and ANIMALS-2 in terms of their BEHAVIOUR, the concepts of 'mental' or 'subjective' processes being considered of little importance, and usually excluded **2.** a synonym for BEHAVIOURISM.

behavioural environment the part of the ENVIRONMENT perceived (PERCEPTION) by the individual, to which the individual responds or to which BEHAVIOUR-1 is directed. It is the environment in which rational human behaviour begins, in which decisions are taken which may or may not result in conscious use or alteration of the PHENOMENAL ENVIRONMENT, or in a change in the individual's relationship with, or exposure to, that environment.

behavioural geography an approach in HUMAN GEOGRAPHY which uses the assumptions and methods of BEHAVIOURISM to determine the cognitive processes (COGNITION) involved in an individual's perception of, response and reaction to, his/her environment. The cognitive processes include the construction of mental maps (COGNITIVE MAPS) and the assessments of locations in the individual's ACTION SPACE.

behaviourism, behavioural approach a school of psychology based on the principle that psychological studies should, in order to be scientific, be confined to the observable, and preferably the measurable, reaction of human beings (and ANIMALS-2) to external stimuli, the study of 'mental' or 'subjective' processes (e.g. consciousness, introspection, freewill etc.) being excluded because these cannot be directly observed and measured. BEHAVIOUR.

beheading, of river RIVER CAPTURE.

bell pit an early type of mine in which a seam of chalk or coal or other deposit was worked from the base of a shallow shaft, the pit being abandoned when the roof became unsafe. Subsidence around the shafts later resulted in the formation of shallow depressions.

belt 1. a district with particular, distinctive characteristic(s) e.g. of climate (belt of calms), of vegetation (tundra), of prevalence of a mineral (coal), of a crop (cotton), of land use (green belt), etc. It is generally in the form of a broad, long strip which may or may not encircle something; but in some cases the term is used as a synonym for REGION-1, the shape being disregarded **2.** a long narrow stretch of water (Great Belt in the Baltic sea).

belt of calms ITCZ.

bench a natural step or terrace, usually narrow and backed by a steep slope, produced by structural change, natural (e.g. by EROSION, as is a WAVE-CUT BENCH) or artificial (e.g. by quarrying, mining). ALP-1, RIVER TERRACE.

bench mark BM, a surveyor's mark cut in some durable fixed material such as a rock, or the wall or face of a building, for which the height above the DATUM LEVEL (in Britain the Ordnance Datum (OD) at Newlyn, Cornwall) is accurately determined. Bench marks are recorded on British Ordnance Survey maps as BM, with height in metres/feet to one place of decimals.

beneficiation the first step in the removal of a commercially valuable mineral from the COUNTRY ROCK or GANGUE surrounding it after extraction from the ground. The process is usually a simple one (e.g. crushing, magnetic separation, flotation) carried out at or near the site of the mine or other working. The aim is to concentrate the ore to keep down the cost of its transport to the works where it is to be further processed.

Benelux countries Belgium, Netherlands and Luxembourg, the countries which formed a customs union on 1 January 1948. They joined the EEC in 1959, and the full economic union of the three came into force on 1 November 1960.

Benioff zone the seismically (SEISMIC) active zone at the bottom of an OCEAN TRENCH where an oceanic plate dives into the MANTLE. PLATE TECTONICS.

benthic *adj.* of, or relating to, the BENTHOS. DEMERSAL.

benthic division, benthic zone one of the two chief divisions of the aquatic environment based on depth of water (the other being the PELAGIC DIVISION), consisting of all the floor of the ocean or lake where BENTHOS live, irrespective of the depth of the floor. It is commonly divided into two systems, the NERITIC and the ABYSSAL, the division in the ocean being at the edge of the CONTINENTAL SHELF.

benthos the plants or animals living at or near the floor of the ocean or a lake, irrespective of the depth of the floor. The organisms are usually divided into LITTORAL benthos and deep water benthos. The abyssal benthos are plants and animals living on the very deep ocean or lake floor;

the phytobenthos are plants, the zoobenthos are animals, living on any other ocean or lake floor. NEKTON, PLANKTON, POTAMOBENTHOS.

bergschrund (German) a wide CREVASSE or series of crevasses occurring between the rocky mountain wall of a CIRQUE and the mass of ice which occupies it. As the ice, which will become a GLACIER, begins to move down its valley it pulls away from the wall and the ICE APRON attached to it, creating a crevasse with each wall of ice, i.e. the bergschrund; but if there is not an ice apron on the rock wall, the gap is termed a randkluft.

Bergwind, berg wind (Afrikaans and German) in general a mountain wind, but specifically in South Africa a hot, dry, FOHN-like wind blowing mainly in winter down from the plateau towards the coast, and thus warming adiabatically. ADIABATIC.

berm 1. a narrow ledge, shelf or terrace formed by material thrown up on the beach by storm waves to make a horizontal shelf above the FORESHORE **2.** a remnant flat surface, part of an earlier, broad valley floor, occurring above the present level of a river, originating from an interrupted CYCLE OF EROSION.

Bessemer process a method of producing cast steel (Bessemer steel) devised by Sir Henry Bessemer in 1860. In his original method molten PIG IRON was poured into a vessel known as a converter, lined with a highly refractory material (usually GANISTER), arranged so that cold air could be blown through the molten mass to burn away the carbon and silicon. The due proportion of CARBON was then added and mixed with the fused metal by a repetition of the blowing, resulting in a very brittle steel. But neither the original nor the later improved process (nor the OPEN HEARTH PROCESS) removed any PHOSPHORUS present in the pig iron (phosphorus makes steel brittle), so the Bessemer process was satisfactory only if the iron ore used (which came to be known as Bessemer ore) lacked phosphorus or contained it only in minute quantity. In the BASIC BESSEMER PROCESS, introduced in 1890, the phosphorus was extracted. The last Bessemer converter in the UK was closed in 1974. STEEL.

beta index a measure of the connectivity in a NETWORK 2. The number of EDGES are divided by the number of NODES. A value of less than 1 indicates no circuits; 1 indicates one circuit; greater than 1 more than one circuit. ALPHA INDEX.

betterment the fortuitous increase in the value of land which accrues to the owner as a result of the operation of a planning system or of public or private investment, sometimes termed unearned increment. BLIGHT.

B horizon the soil layer underlying the A HORIZON, an ILLUVIAL horizon into which minerals etc. from the A horizon are washed. It is sometimes divided into an upper layer, B_1, high organic content; B_2, the main depositional zone; and B_3, grading into the C HORIZON. The entry on SOIL HORIZON shows refinements in the classification of the B horizon. SOIL, SOIL PROFILE.

bias 1. in statistics, the distorting effect produced by a sample which does not accurately reflect the characteristics of the POPULATION-4 from which it is drawn owing to SYSTEMATIC ERROR rather than to RANDOM ERROR **2.** in sociological survey, the distorting effect produced either by questions which are framed in such a way that respondents are led to give particular answers, or by the researcher in interpreting and coding the answers given.

bid price curve, bid rent curve a curve on a GRAPH-1 relating price (rent) to distance, showing the price a land user would be willing to pay for a given area of land at various distances from a given point, especially from the city centre. Activities which depend on contacts and need to be located in the most accessible places (e.g. head offices of banks) must have a central location, and can afford the high prices and rents of the centre: they will have a steep curve. Those activities not much affected by their location, and which need to avoid high rents, will have curves with a gentle slope. Under those simple terms if the bid rent curve is superimposed on the actual rent curve for a given city, the best location (OPTIMUM LOCATION) for a particular activity will be where the actual rent curve just touches the lowest possible part of the bid rent curve, i.e. where the actual rent curve equals the bid rent curve. The bid price and the value and use of land are interrelated, mutually determining. ALONSO MODEL, TRADE-OFF THEORY. Fig 8.

biennial *adj.* as applied to plants, a plant which after seed germination vegetates for one year, storing food for the second year, when it flowers, fruits and dies, thus taking two years to complete its life-cycle. ANNUAL, EPHEMERAL, PERENNIAL.

bifurcation ratio the ratio between the number of streams of an order of magnitude and the number of streams in the next higher order of magnitude (STREAM ORDER). The term has also been applied to CENTRAL PLACES, but has not been generally adopted.

bight a crescent-shaped indentation of the coastline, usually of considerable extent, normally wider and with a shallower indentation than a BAY.

billabong (Australia: Aboriginal, dead river) an elongated waterhole in the bed of an intermittent stream, or a cut-off or OXBOW lake.

binodal tidal unit an AMPHIDROMIC (TIDAL) SYSTEM in which there are two NODES-1. AMPHIDROMIC POINT.

binomial distribution in statistics, a theoretical DISTRIBUTION-4 which predicts the PROBABILITY of a particular result occurring in a sample when the characteristics of the parent population are known and there are only two possible outcomes.

binomial nomenclature the universally accepted method of naming animals and plants which avoids the confusion arising from the use of local names. The generic name, designating the GENUS to which the animal or plant belongs, is written first, with a capital (upper case) initial letter; second is the specific (or trivial) name, that is the name peculiar to the SPECIES, printed with a small (lower case) initial letter. Those two names, the generic and the specific, are usually printed in italics. The author who named and described the species follows, not in italic; if a species originally allocated to one genus is later transferred to another, the name of the original author is put in brackets. In refinement, one of the species is commonly designated as the 'type specimen' (HOLOTYPE) and in later splitting of the species this type specimen is always included, followed by the name of the new specimen which resembles it. CLASSIFICATION OF ORGANISMS.

bio-catalyst a micro-organism or microbial enzyme used in a technological process

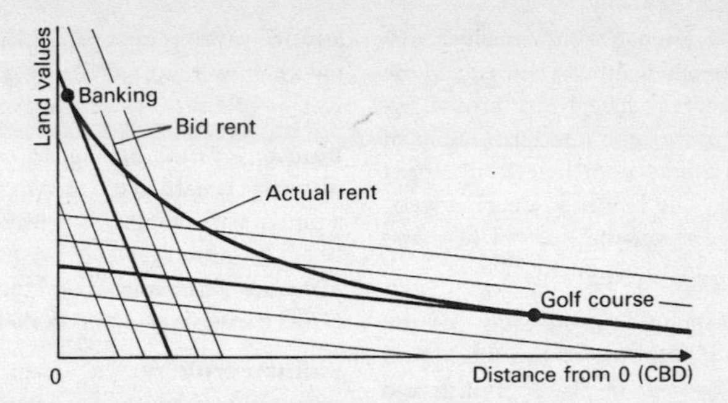

Fig 8 Bid price curve, bid rent curve, and land use. Banking, with its intensive use of land, can afford the high prices and rents of the city centre; golf courses, needing an expanse of land, seek the low rents of the city outskirts

(e.g. food processing, pharmaceuticals). BIOTECHNOLOGY, CATALYST.

biochore BIOSPHERE.

bioclastic *adj*. applied to a rock composed of fragmental organic remains. CLASTIC.

bioclimatology the study of climate as it affects the life and health of animals, plants and people.

biocoen in ecology, all the living parts of the environment. ABIOCOEN, HOLOCOEN.

biocoenosis, biocenosis 1. in biology, the participation of diverse organisms in all the resources of their environment; a BIOTIC community **2.** MUTUALISM between plants and animals **3.** in geology, a group of fossils consisting of the remains of organisms that once lived together, in contrast with THANATOCOENOSIS.

biocycle BIOSPHERE.

biodegradable *adj*. applied to a substance or material that can be decomposed by BIODEGRADATION.

biodegradation, biodeterioration the breaking down of a substance or material by the action of living organisms (mainly by AEROBIC BACTERIA).

biodiversity biological diversity, the great variety of all living organisms on earth, manifest at three different, interlinked levels: the GENES, the SPECIES-1, the ECOSYSTEM, the basic unit being the species. EARTH SUMMIT.

bioengineering the use of biochemical processes (the chemical reactions in living organisms) on a large, industrial scale, particularly in the recycling of waste materials to produce foodstuffs for people or livestock.

biofuels combustible material derived from CROPS-1. FOSSIL FUELS, NUCLEAR ENERGY.

biogenic sediment a sedimentary deposit, e.g. shelly LIMESTONE, formed by once-living organisms.

biogeochemical cycle the circulation of an ELEMENT-6 within ECOSYSTEMS. CARBON CYCLE, NITROGEN CYCLE, PHOSPHORUS CYCLE.

biogeography the geography of organic life, the study of the spatial distribution of plants and animals (but usually considered to exclude human beings) and the processes that produce the patterns of distribution, and of the interrelationship of plants and animals with their environment over time. The branches are phytogeography (concentrating on plants) and zoogeography (on animals).

biogeomorphology the study of the influence of landforms on the distribution and development of plants, animals and micro-organisms, and of their influence on the processes of the earth's surface and the development of landforms.

biogeosphere the outer part of the LITHOSPHERE down to the depth at which there is no organic life.

bioherm an ancient rock mass built up by sedentary organisms such as CORAL, molluscs, ALGAE and/or their remains, e.g. an ancient coral reef (BIOSTROME), especially one surrounded by rocks of different origin.

biological control the control of the population of a pest (including a PARASITE) by the use of its natural living enemies, e.g. by the introduction of a predator, or a virus, or of sterilized males of the pest or parasite.

biological indicator an organism, usually (but not always) a micro-organism, used to show the level of chemical activity, e.g. LICHEN, used to measure the level of sulphur dioxide in the atmosphere.

biological (biochemical) oxygen demand (BOD) in polluted water, the quantity of dissolved oxygen needed by aquatic organisms for the decomposition of organic matter.

biomass **1.** the total weight of organisms under consideration, e.g. in a specified area, or making up a particular TROPHIC LEVEL-1 OR POPULATION-3 **2.** the total weight of a SPECIES per unit area.

biome a major ecological community (BIOTIC COMMUNITY) of living plants and animals, occupying an extensive area, e.g. DESERT, GRASSLAND-1, RAIN FOREST, TUNDRA. The plants of land biomes comprise FORMATIONS. ECOSYSTEM.

biometeorology the science of the relationship between living organisms and the weather.

biosphere **1.** the parts of the earth's crust and atmosphere (LITHOSPHERE, HYDROSPHERE, ATMOSPHERE) occupied or penetrated by living organisms **2.** only the living organisms **3.** the living organisms together with the parts of the earth's crust and atmosphere that they occupy or penetrate **4.** the part of the earth's crust and atmosphere favourable to at least some form of life, divided into three biocycles (salt water, freshwater and land). Following this last definition, in biogeography the term biochore has been applied to the geographical environment with a distinctive plant and animal life adapted particularly to climatic factors, each biochore thus being characterized by a major type of vegetation. The biochore is subdivided, the smallest division being the NICHE-1. BIOTOPE.

biostrome a modern CORAL REEF, in the course of formation. BIOHERM.

biota a collective term for the animal and plant life of a specific area and/or period of time.

biotechnology the application of biological knowledge to industrial and other processes (e.g. GENETIC ENGINEERING), e.g. in agriculture, medicine etc. BIOCATALYST, GMO.

biotic *adj.* of, pertaining to, or relating to, living organisms.

biotic community a local association of interdependent plants and animals living in an area.

biotic factor an influence arising from the activities of living organisms, including people, which affects the environment, as distinct from such factors as the climatic (CLIMATIC ELEMENTS) and the EDAPHIC.

biotic pyramid a graph showing the number of individuals at each TROPHIC LEVEL-1 in a stable FOOD CHAIN, from the primary producers, to the primary, secondary and tertiary consumers. It is inevitably shaped like a pyramid because the number at each trophic level decreases, as explained under FOOD CHAIN.

biotite a common, rock-forming mineral, a silicate of iron, magnesium, potassium and aluminium, a form of MICA commonly occurring in IGNEOUS and METAMORPHIC rocks, especially as black crystals in GRANITE. It is glassy and transparent, the colour ranging from dark green through to brown-black. FERROMAGNESIAN MINERAL.

biotope 1. a term used by some ecologists to define a small community of plants and animals characteristic of a NICHE-1, the smallest division of an ECOSYSTEM. BIOSPHERE-4 **2.** a HABITAT in which there is uniformity in the main climatic, soil and BIOTIC conditions.

bipolar *adj.* pertaining to, occurring at, associated with, two poles or with the two polar regions.

bipolar distribution the distribution of certain species found in areas to the north and south of a median zone, but not in the intervening median zone itself.

bird's foot delta, birdfoot delta a DELTA with distributaries flanked by relatively narrow borders of sediments, projecting seawards in the pattern of a bird's foot, e.g. the Mississippi delta. Fig 13.

birth control the prevention of conception by various methods (protective devices, hormonal control, sterilization, induced abortion, avoidance of insemination at times of ovulation), important in attempts to limit family-size, especially in overpopulated countries (OVERPOPULATION). Official birth control programmes are difficult to implement and are not always successful. They may contravene religious beliefs; be unwelcome in cultures where large families are regarded as an economic resource in the family, a support for the aged and infirm; and in some societies may be viewed as an economic plot devised by industrialized countries to reduce the economic potential of a large work force in less developed countries.

birth-rate, birth rate in human population, the ratio of births to population within a given period, commonly measured by the average number of live births per 1000 of the population. This is the crude birth-rate, or natural increase, and currently ranges from 12 to 50, countries with a high standard of living recording between 15 and 20. The crude death-rate is similarly measured in deaths per 1000 of population, and varies from 6 to 25 or more. At present death-rates continue to fall in most countries, owing to improved medical skill and services; but there is only a slight tendency for birth-rates to fall, despite the spread of BIRTH CONTROL (the avoidance of unwanted pregnancies by prevention of fertilization). As a result the world population is still increasing. NATURAL CHANGE.

bitumen **1.** a general name for various viscous or solid mixtures of native HYDROCARBONS which have lost much of their gaseous material, e.g. ASPHALT **2.** tar, the residue from the distillation of COAL.

bituminous coal (American soft coal) humic coal, a COAL containing from 75 to 92 per cent CARBON, 4.5 to 5.6 per cent HYDROGEN and yielding from 15 to 45 per cent volatile matter when heated out of contact with air. The adjective bituminous, based on the long-standing incorrect assumption that these coals contain BITUMEN, is erroneous.

black box approach an approach in SYSTEMS ANALYSIS which ignores the internal structure and functioning present within the system under study (e.g. the human mind) and deals only with the nature of the output resulting from identified inputs. A grey box approach considers some of the sub-systems present, but is not concerned with their internal structure and functioning; a white box approach tries to identify in as much detail as possible the sub-systems, components, processes etc. present within the system in order to build up as complete an understanding as possible of the system's internal structure and functioning.

blackearth, black earth a general term covering CHERNOZEM and the dark plastic CLAYS of tropical regions.

black frost a hard FROST without RIME. GLAZE.

black ice a layer of GLAZE formed on roads, the ice being so clear that it is invisible and extremely dangerous to traffic.

black smoker a submarine jet of very hot water, gases (sulphur and methane) and particles bursting through an OCEANIC RIDGE as seawater drains into a crack in the spreading crust and is heated by the magma.

blanket bog, blanket peat a BOG occurring on a relatively horizontal land surface in regions of high rainfall and low evaporation, and covering the countryside like a blanket, except on steep slopes and rock outcrops. It is common in Ireland. RAISED BOG, VALLEY BOG.

blast furnace a furnace used to produce molten IRON. Poor quality iron ores are subjected to a preliminary roasting to remove volatile impurities. The furnace, a large, vertical steel shell lined with refractory bricks, is then charged with the ore, with COKE and LIMESTONE, through which hot air is blasted. Carbon monoxide from ignited coke reduces the oxides to iron, and the limestone acts as FLUX, so that molten iron flows to the furnace bottom, to be run off and cast into blocks known as pigs, hence the term PIG IRON. The process is now superseded by electric smelting. ARC FURNACE, STEEL.

blight in planning, the lowering of the value of land and buildings brought about by official planning proposals which indicate a change of land use or a shortening of the life of the existing buildings. The condition of property affected by planning blight may deteriorate as owners cease to care about its upkeep; and it may become unsaleable from the time when plans are first discussed or from the official designation, to the time of redevelopment, despite the fact that in Britain compensation becomes payable on official acquisition. BETTERMENT.

blind valley a valley in LIMESTONE

country, dry or with a stream, which ends in a steep wall, into the base of which the surface flow of water disappears underground.

blizzard any very cold, strong wind accompanied by falling or drifting snow, sometimes involving a WHITE-OUT.

block diagram **1.** a perspective drawing giving a three-dimensional impression, used particularly to show landforms **2.** a diagram showing the relationship between the surface form of the ground and the underlying geological structure by representing an imaginary block cut out of the earth's crust.

block disintegration the mechanical breaking-up of bedded, jointed rocks, usually by FROST action along lines of weakness.

block faulting, block-faulting faulting (FAULT) in which part of the earth's crust is divided into a number of small blocks by a series of faults, in many cases two sets roughly at right angles, some of the blocks being moved up, others moved down, others tilted. FAULT BLOCK, TILT BLOCK.

block mountain a mountain which is structurally an uplifted FAULT BLOCK, prominent because it has been thrown up by earth movements, or because the surrounding land has sunk. HORST, TILT BLOCK.

block slumping MASS MOVEMENT down the steep face of an escarpment or sea cliff of well-jointed rocks, e.g. of chalk or limestone slipping on an underlying clay stratum which has become wet through water seepage. The block usually breaks away sharply from the steep face and slumps downward with a rotational movement.

blowhole a nearly vertical, smallish hole on land near the seashore, the land opening of a funnel-shaped CAVE-1. Sea waves force air and water up from the cave through the small opening, so that a spout of spray is carried high in the air. A blowhole is formed when erosion occurs along a vertical or nearly vertical JOINT which passes from the land surface to the cave roof.

blowout, blow-out **1.** a hollow (deflation hollow) made by eddying wind in tracts of light or sandy soil, occurring especially in a coastal sand dune area or in an arid plain if vegetation cover is lacking (DEFLATION) **2.** a sudden, violent escape of gas or steam.

blue-collar worker a person who is engaged in and paid a wage for manual work. WHITE-COLLAR WORKER.

blue ground KIMBERLITE.

blue sky the apparent colour of the cloudless sky in daylight, due to the scattering of sunlight (with frequencies corresponding to the blue region of the visible solar SPECTRUM-2) by obstructing MOLECULES in the air.

bluff a steeply rising slope marking the outer margins of the FLOODPLAIN of a river, especially the almost perpendicular, steep slope cut by the stream as it erodes the concave side of a MEANDER. RIVER CLIFF. Fig 29.

BOD BIOLOGICAL (BIOCHEMICAL) OXYGEN DEMAND.

bog **1.** broadly, any soft, wet, spongy soil or ground into which the foot sinks **2.** precisely, an area of wet acid peat and the vegetation associated with its poorly drained or undrained surface, i.e. the natural wet peat-forming and peat-inhabiting plants (PEAT), SPHAGNUM being characteristic. Bogs occur in areas of poor drainage where lack of oxygen in the waterlogged soil inhibits the decomposition of dead plants, leading to the build-up of humic and other acids which modify plant structure and function. BLANKET BOG, FEN, MARSH, QUAGMIRE, RAISED BOG, SWAMP.

bogaz (Slavic) a long, narrow chasm in LIMESTONE (KARST), formed by CARBONATION-SOLUTION along a JOINT.

bog moss SPHAGNUM.

bog peat, moss peat acid, brown PEAT. The plant structure is visible, the cellulose content high, and it supports a vegetation of bog moss (SPHAGNUM) or SEDGES. FEN PEAT.

boiling point STEAM POINT.

bonitative map a map indicating land suitable or unsuitable for some specific economic development.

bora, borino (Italian) a very cold, often dry, violent north or northeasterly wind (but sometimes accompanied by rain or snow) blowing mainly in winter down from the mountains on to the eastern coast of the Adriatic (comparable with the MISTRAL). The borino is a weaker form blowing in summer. The term bora-type is applied to winds similarly blowing down moderately high mountains from a cold, continental high pressure area towards a low pressure area over warm sea shores or lowlands.

bore **1.** a tidal wave of some considerable height which regularly or occasionally rushes up certain rivers or narrowing gulfs. EAGRE **2.** a deep hole drilled in exploration for oil or water.

boreal *adj*. belonging to the north, applied especially to **1.** the northern CONIFEROUS FORESTS-1 **2.** the climatic zone with snowy winters and short summers **3.** the climatic period from 7500 to 5500 BC. PREBOREAL.

boss a small BATHOLITH with an upper surface, when exposed by denudation, roughly circular in cross-section.

boulder any large, detached, generally rounded mass of rock, larger than a COBBLE, especially one transported by ice, river, or sea, from its original home, but also in some cases one weathered by frost-shattering or EXFOLIATION in situ, specifically exceeding in diameter 200 mm (8 in) in UK, 265 mm (10.5 in) in USA.

boulder clay unstratified, unconsolidated GROUND MORAINE of mixed rock debris transported by ice and deposited when a former ice sheet or glacier has melted. Whilst usually defined as consisting of stiff clay enclosing boulders of various sizes, in some examples the matrix may be mainly sand instead of clay, and boulders may be few or even absent. For this reason the term has now been dropped in favour of TILL, because till does not specify the constituent materials.

boundary a line of demarcation, real or understood, visible or invisible, natural or artificial, of legal or of no legal significance, which may be perceived from either side (or both sides) of it, e.g. between countries (synonymous with FRONTIER-1) or administrative areas, between regions of various types, between market areas, between service areas.

bounded rationality the concept that a person cannot be completely rational, however hard that individual tries, because no-one can have perfect knowledge or a perfect ability to calculate. Thus the concept of ECONOMIC MAN becomes unreal.

bourgeois, bourgeoisie (French) **1.** broadly, the middle classes **2.** in Marxism, the capitalist, property-owning class. MODE OF PRODUCTION, PROLETARIAT, SOCIAL CLASS.

bourne a temporary or intermittent stream which may flow in a DRY VALLEY in chalklands, depending on the level of the WATER TABLE. In winter, when the water table rises above the height of the valley floor, there may be a surface stream, hence the term winterbourne. In summer, when the water table sinks below the level of the valley floor, the stream bed becomes dry. Bourne is often incorporated in place-names on the chalklands of southern England, e.g. Bournemouth, Eastbourne.

box-canyon a term applied in the western USA to a CANYON with more or less vertical walls, to distinguish it from canyon, a term commonly applied there to every young valley.

BP before the present day, used as a measure of time to avoid the necessity of using BC (before Christ, i.e. in the year before the reputed date of the birth of Christ) or AD (Latin Anno Domini, in the year of our Lord, i.e. in the year since the reputed date of the birth of Christ).

braided river course (to braid, to twist in and out, to interweave) an anastomosing river course (ANASTOMOSIS), a stream with a wide, shallow channel split into many small, shallow, interlaced channels separated by bars of alluvial material, visible when the water is low. It occurs particularly when a heavily laden shallow stream deposits so much sediment in its channel that the channel becomes too small, and part of the stream breaks out to follow a new course on the flat land of the valley.

brain drain the movement of the most capable, highly-skilled, technical and professional people from the country where they trained and gained their first work experience to another offering better career opportunities and/or higher rewards.

Brandt Report *North–South: a Programme for Survival*, the title of the report of the Independent Commission on International Development Issues, published 1980. The Commission was set up in December 1977 (at the invitation of Robert MacNamara, Chairman of the World Bank) under the chairmanship of Willy Brandt, German statesman and winner of the Nobel peace prize 1971. It consisted of people with varied political and professional experience outside the countries with a communist form of government. Eight members represented the North, ten the South. Under the terms of reference of the Commission 'global issues arising from economic and social disparities of the world community' were to be studied, and 'ways of promoting adequate solutions to the problems involved in development and in attacking absolute poverty' were to be suggested. The term North–South is accepted as a misnomer, a very broad generalization to stress the great social and economic imbalance between the rich, developed countries of the North, the northern hemisphere (i.e. North America, excluding Mexico, the countries of Europe, the then USSR, China, Japan, to which are added Australia and New Zealand from the southern hemisphere) and the poor, developing countries of the South (very broadly, the rest of the world). There are of course anomalies in each, e.g. the South under that definition includes the developing but rich oil-exporting countries of Arabia. Briefly to summarize the recommendations of the Commission, the members advised the setting-up of a five-year emergency programme to promote food production for the world's rapidly increasing population; to find new sources of energy; to deal with the transnational companies; to transfer financial resources from the rich to the poor countries; to start to reorganize the international institutions with the aim of establishing a reformed economic system.

Brave West Winds the westerlies, the planetary west or northwest winds blowing over the oceans of the southern hemisphere in midlatitudes (40°S to 65°S) where they blow with considerable force and regularity, swinging to north or south under the influence of seasonal change of world ATMOSPHERIC PRESSURE belts. ROARING FORTIES.

breached anticline an ANTICLINE in which the drainage, developed along the ridge (the axis) of the anticline, has eroded the overlying rocks along this line of weakness, revealing the underlying older rocks, and thus creating an ANTICLINAL VALLEY with escarpments facing inwards.

breaker a mass of turbulent water and foam, breaking violently against a rocky shore or passing over a reef or shallows, formed when a heavy ocean WAVE rushes from deep to shallow water, so that its CREST steepens, rolls over, and breaks. This occurs particularly when the ratio of wave height to wave length is greater than 1:7.

break-of-bulk point in a transport system, the point where CARGO (sometimes broken into smaller units) is transferred from one mode of transport to another, e.g. at a railway station, port or airport. PIGGYBACK TRANSPORT.

break of slope any more or less sudden change in a slope, e.g. of a hillside.

breccia a rock consisting of angular fragments of other rocks cemented together by some finer material. The term is not applied in English to a CONGLOMERATE in which the fragments are rounded. AGGLOMERATE.

breck, breckland 1. a tract of heathland (HEATH) with thickets **2.** a tract of land supporting such vegetation, cleared for cultivation from time to time, then allowed to revert.

brickearth 1. originally any earth, usually a loamy CLAY from which bricks could be made **2.** in current use, a fine-grained deposit overlying the gravels on river terraces, e.g. on certain of the Thames terraces. Originating from wind-blown material that has been re-worked, re-sorted and re-deposited by water, it has been likened to LOESS. It forms a fertile, friable soil.

brickfielder, brick fielder a hot, dry, dusty, squally wind, blowing in south-eastern Australia in summer southwards from the interior in front of a DEPRESSION-3.

bridge-point, bridging point a point at which a river is or could be bridged.

bridlepath, bridleway 1. a path fit for the passage of a horse or a pedestrian, but not for a vehicle **2.** a path which may in English law have a right of way for pedestrians and riders on horseback, but not for wheeled vehicles.

brigalow in Australia, scrub, mainly of *Acacia* species, bordering the MULGA in dry areas of Australia.

brine a very salt solution, commonly containing a higher proportion of a dissolved salt than that occurring in seawater.

brine pan a shallow pit or vessel used in the process of extracting salt from salt water by evaporation.

briquette, briquet a brick-shaped block of compressed coal fragments, usually of BROWN COAL or LIGNITE, the calorific value being high because water has been expelled in the compression.

British Summer Time BST.

Broad, Broadlands, the Broads a local term applied in East Anglia, England, to shallow freshwater lakes formed by the broadening out of a sluggish river, the sites where peat for fuel was dug out in the middle ages.

broadleaved trees any tree of Dicotyledonae, many with a leaf form generally wide in relation to length. Most are DECIDUOUS, but some are EVERGREEN.

bronze a metal, an alloy of COPPER and TIN, hard and resistant to moisture and weathering. It expands when solidifying, and thus makes good castings. It antedates iron smelting. BRONZE AGE.

Bronze Age an era in human development (succeeding the PALAEOLITHIC, MESOLITHIC and NEOLITHIC and preceding the IRON AGE) when BRONZE was used for tools and weapons. Writing and arithmetic developed in the Bronze Age, the plough, wheeled vehicles, and animals for riding and pulling, came to be used; towns were formed, work became specialized, there was trading and shipping.

brown coal a brown, fibrous deposit, intermediate between PEAT and BITUMINOUS COAL. The term is sometimes used as a synonym for LIGNITE, but brown coal is nearer bituminous coal: the plant fragments have been changed into an amorphous mass. Brown coal is usually worked OPENCAST. Heavy and soft, it is used mainly in thermal power stations near to the minehead. For domestic and industrial use it is made into BRIQUETTES.

brown earth, brown forest soil the rather unsatisfactory name for a range of ZONAL SOILS with merging horizons, generally associated with the lands in midlatitudes formerly covered with DECIDUOUS woodland, i.e. the region south of the BOREAL coniferous forest or TAIGA, in northeast USA, northern China, central Japan, northwestern and central Europe. There the humid climate and MULL-2 lead to the formation of a slightly leached, slightly acid A HORIZON, with a grey-brown lower layer (less leached than that of a PODZOL) and a B HORIZON that is granular, thick, dark brown, with BASES-2 and COLLOIDS from the A horizon. CALCAREOUS brown forest soils are included.

brown field site a built-up area with obsolete buildings and derelict land suitable for URBAN RENEWAL. GREEN FIELD SITE.

brown podzolic soils one of the subdivisions of PODZOLIC SOILS.

brown sands BROWN SOILS.

brown soils one of the seven groups in the 1973 SOIL CLASSIFICATION of England and Wales. It includes argillic brown earths (brown or reddish, with a loamy horizon overlying a clay layer), brown alluvial soils (non-calcareous, developed on new alluvium), brown calcareous soils (deep, organic, fertile soils of high agricultural quality, developed particularly on limestone, the A HORIZON being reddish-brown overlying a lighter B HORIZON) (MOLLISOLS), BROWN EARTHS, and brown sands (a group of brown earths developed on freely drained, non-alluvial deposits of sand and gravel). BROWN PODZOLIC SOILS.

brunizem PRAIRIE SOIL.

brush 1. SCRUB-1 or BUSH-3, or a thicket of small trees and shrubs **2.** vegetation of low, woody plants, especially SAGEBRUSH, in USA.

BST British Summer Time, usually one hour in advance of GMT (Greenwich Mean Time).

built-up area the part of a town where the land is so covered with buildings and roads, etc. that there is space for further similar development only if existing structures are demolished. Very small plots of land not built over (e.g. small gardens, school playgrounds, etc.) and derelict land awaiting redevelopment may be included. DEVELOPMENT-2.

bulrush millet a tall, drought-resistant MILLET with stems bearing long cylindrical seed-heads, generally resembling a bulrush, more widely grown than any other food crop in tropical areas with a low rainfall, an important food crop in Sudan, northern Nigeria and other countries on the southern Saharan border, as well as in the driest areas of the Indian subcontinent.

bunch grass any of the coarse grasses which grow in clumps or bunches (instead of forming a continuous cover of matted turf), in many cases separated by bare ground, e.g. in the semi-arid western plains of North America. It is also termed tussock grass, e.g. in New Zealand.

buran a strong northeast wind blowing in central Asia at all seasons but most frequently and fiercely in winter (then termed white buran or poorga) when it lifts and carries the snow, and ice particles. In the TUNDRA, especially in southern Russia and Siberia, it is termed purga. KARABURAN.

Burgess's concentric ring model CONCENTRIC ZONE GROWTH THEORY.

bush 1. a SHRUB or small TREE, especially one with branches arising near the ground **2.** uncleared or uncultivated country, especially that covered with trees of this type **3.** widely and variously used locally, e.g. natural vegetation of low woody plants, such as creosote bush (USA); wilder countryside as opposed to cultivated land (Africa); and further extended to the countryside as opposed to the town. BUSHVELD, VELD.

bushel a measure of capacity which varies for different commodities and in different countries. In British dry and liquid measures it is in general equal to 36.6 LITRES or 2219.36 cu in or 8 GALLONS; in American dry measures it is equal to 35.23 litres or 2150.42 cu in.

bush fallowing a farming practice common in equatorial forest areas in Africa, a modified form of SHIFTING CULTIVATION. A small part of the forest is cleared by cutting and burning, and crops are planted. When the fertility of the soil in that plot is exhausted, another clearing is made and the farmers cultivate it, but they continue to live in their village, they do not themselves move. The abandoned plot quickly becomes covered with such plants as bamboo and eventually, if left untouched, with trees. At this stage it may become managed fallow, the new trees supplying timber (and possibly fruit). But very often the plot is recultivated after a lapse of time. FALLOW.

bush veld, bushveld the SAVANNA of tropical and subtropical south Africa, sometimes open grassland with scattered trees (PARKLAND), grading to close woodland.

business park OFFICE PARK.

butte **1.** small, flat-topped, isolated hill with steep sides (its upper layers consisting of resistant rock overlying weaker layers, CAP-ROCK) which remains after partial denudation of the surrounding TABLE LAND. It may be a small, isolated part of a MESA **2.** in western USA, any isolated flat-topped hill with steep sides. BUTTE TEMOIN.

butte témoin the flat-topped outlier of an escarpment or plateau, of which it was once part, its height, being about the same as that of the escarpment or plateau, bearing witness (témoin) to its origin. Strictly every BUTTE is a butte témoin, but because the term butte has slipped into common usage in western USA the distinction of butte témoin becomes necessary.

Buys Ballot's Law a law postulated by C. H. D. Buys Ballot, Dutch climatologist, 1857, that if an observer in the northern hemisphere stands with back to the wind, the ATMOSPHERIC PRESSURE will be less to that individual's left than to the right, the reverse in the southern hemisphere. CORIOLIS FORCE, FERREL'S LAW.

bypass a road which skirts a place, especially a road designed to divert through-traffic from roads in a congested area, e.g. a town centre.

by-product **1.** a secondary product obtained during a specific process, of greater or less value than the product which is the primary objective of the operation **2.** an additional result, which may or may not have been intended or expected.

C

C 14 dating carbon 14 dating. RADIOCARBON DATING.

cacao, cocoa a small tree native to tropical America, now widely grown in warm tropical and equatorial regions with fairly high rainfall and a fertile soil. The flowers bloom directly on the main trunk and on the branches, the fruit (the pod) carrying seeds (beans) which after fermentation, drying and roasting are used in the manufacture of cocoa powder. Cocoa beans contain 50 per cent or more fat (cocoa butter) and extra cocoa butter is added in the making of chocolate. The term cocoa should be restricted to the beans and their products, but it is now also used for the tree (cocoa tree).

Cainozoic, Cenozoic, Kainozoic (Greek kainos, new; zoon, animal) *adj.* of, or pertaining to, the third of the main geological eras (GEOLOGICAL TIMESCALE), the era marked by the rapid evolution of mammals, subsequent to the Precambrian era. It is still termed the Tertiary era by some geologists.

cairn a pyramid of rough stones piled up as a monument or landmark of some kind.

calcareous, calcarious *adj.* **1.** of, pertaining to, consisting of, or containing CALCIUM CARBONATE, or limestone **2.** having the character of CHALK OR LIMESTONE.

calciferous *adj.* containing or producing CALCIUM, CALCIUM CARBONATE or other calcium compounds.

calcification 1. generally, the changing into CALCIUM CARBONATE or into a CALCIFEROUS state by the reaction of calcium salts **2.** of soil, the deposition of calcium carbonate near the surface of the soil (usually in the B horizon or C horizon) in ARID and semi-arid regions, the result of the rise, by CAPILLARY ACTIVITY, of CALCIUM salts in SOLUTION, followed by the evaporation of the water **3.** in geology, the replacement of organic or inorganic material in rocks by calcium minerals **4.** the hardening of plant or animal tissue by the deposition of calcium salts in it (FOSSIL).

calcination the process or action in which a physical change is brought about (usually in inorganic materials) by heating to a high temperature without fusing. It is used in oxidizing (especially metals), in converting a substance to powder form, or in releasing volatile constituents or products. SMELTING.

calcite a crystalline form of CALCIUM CARBONATE, colourless unless coloured by impurities, the main constituent of all LIMESTONES (including Iceland spar, the purest variety of calcite; and CHALK, MARBLE, STALACTITES, STALAGMITES), a GANGUE mineral in some HYDROTHERMAL deposits, a common cementing material in many coarse-grained SEDIMENTARY ROCKS. It forms when material from some weathered IGNEOUS ROCK is transported as a CALCIUM BICARBONATE solution, the bicarbonate decomposes, and the calcite remains as a deposit. CALCIFICATION.

calcium a soft, white element of the ALKALINE EARTH group, occurring mainly as CARBONATE (chalk, limestone, marble, coral). It is used in alloys, is widely used in industry, and is an essential nutrient for plants and animals. FELDSPAR, PEDALFER.

calcium bicarbonate a soluble salt formed when carbon dioxide from the air forms a solution of CARBONIC ACID with water, and this solution comes into contact with one of the forms of CALCIUM CARBONATE. Calcium bicarbonate causes the temporary hardness of water and acts as a bone-builder in vertebrates.

calcium carbonate an insoluble salt occurring, e.g. in CHALK, CORAL, LIMESTONE, MARBLE. It dissolves in water containing CARBON DIOXIDE to form soluble CALCIUM BICARBONATE. PEDALFER, PEDOCAL.

caldera **1.** a broad, shallow volcanic CRATER, formed by the blowing off of the top of a crater by PAROXYSMAL ERUPTION, or by subsidence, or by combined explosion and subsidence **2.** a large circular or amphitheatre-shaped DEPRESSION-2 of volcanic origin.

Caledonian folds, Caledonian orogeny the great mountain-building movements and associated geological phenomena of the late Silurian–early Devonian periods (GEOLOGICAL TIMESCALE), indicated by the northeast to southwest trend of folds, faults, hills, mountains and valleys, etc. in northwest Europe.

calendar a system of dividing time into fixed periods, the natural units being the day (the revolution of the earth on its axis) and the year (the revolution of the earth round the sun). The month (revolution of the moon round the earth) and the week are conventional divisions. There are difficulties if the month is regarded as a natural unit, a natural division of the year, because 12 lunar cycles represent 354 days, but the SOLAR YEAR consists of 365 days.

calf ice a piece of glacier ice, smaller than an iceberg, detached directly from a GLACIER or produced by the breakdown of an ICEBERG.

calm, calms a state of the atmosphere in which there is an absence of appreciable wind, such movement as there is registering Force 0 on the BEAUFORT SCALE. This may occur at any time, anywhere in anticyclonic (ANTICYCLONE) conditions; but periods of calm are common throughout the year in certain latitudes, i.e. the belt of calms, between 5°N and 5°S (DOLDRUMS) and in the HORSE LATITUDES.

calving of ice, the breaking away of a mass of ice from an ICE FRONT, ICEBERG or GLACIER. CALF ICE.

Cambrian *adj.* of, or pertaining to, the first geological period or system of rocks of the Palaeozoic era (GEOLOGICAL TIMESCALE) when such rocks as LIMESTONES, SANDSTONES, SHALES were formed under shallow seas, and the INVERTEBRATE animal was the characteristic form of life.

Campbell-Stokes recorder an instrument for measuring and recording the duration of bright sunshine by means of a graduated, sensitized card on to which the sun's rays are focused by a lens. As the sun and the position of the image move, a line burnt on the card records periods of continuous sunlight.

campo level, open grassland with scattered trees in Brazil, comparable with SAVANNA, probably not a natural CLIMAX vegetation but one arising from human activities, especially burning. Various types are distinguished: campo cerrado (closed grassland) with scrub woodland dominant; campo sujo (dirty grassland) with scattered trees or patches of forest; and campo limpo (clean grassland) open grassland without trees.

canal an artificial watercourse constructed **1.** to unite rivers, lakes, etc., for purposes of inland transport **2.** for water supply and irrigation **3.** to make a SHIP CANAL or seaway, available to ocean-going vessels.

canopy the high, leafy, continuous, uppermost layer formed by the crowns of trees of approximately the same height, e.g. in RAIN FOREST.

canyon 1. a deep valley with very steep sides, with a stream flowing at the bottom, common in arid and semi-arid lands where the downward cutting power of the stream exceeds the rate of WEATHERING of the rocks of the valley sides. The form becomes exaggerated if uplifting of the land occurs at the same rate as the down-cutting of the river. **2.** a submarine canyon, a deep, steep-sided trough in the ocean floor, in some cases very wide, in some winding.

CAP the Common Agricultural Policy of the European Economic Community (EEC). The basic features, adopted in January 1962, aimed to achieve more efficient agricultural production, a fair return for farmers, reasonable prices for consumers, and stable market conditions; common price levels were to be agreed and national protection systems were to be replaced by a Community system incorporating variable levies on imports of some farm products. Most arrangements were operating by July 1968, but eventually led to overproduction and excessive payments to farmers. Reforms introduced in 1992 aimed to reduce overproduction by limiting support payments to fixed quotas, and by bringing prices (particularly of cereals) down to world market prices. Cereal producers received area payments to compensate for lower prices, provided they 'set aside' 15 per cent of their arable acreage in rotation, or a higher proportion (to be specified) permanently. It was hoped that SET ASIDE would also enhance the conservation of land in the European Union.

capability constraints the limitation imposed on an individual's actions by biological needs (e.g. food, sleep) and/or inadequate access to desired facilities (e.g. lack of transport). TIME-SPACE CONSTRAINTS.

capacity 1. the ability to contain, accommodate **2.** the amount so contained or accommodated **3.** the ability of a factory, society, etc. to manufacture or process its product, especially this as a maximum **4.** a measure of the ability of energy to do work. CAPACITY OF A STREAM, CARRYING CAPACITY, CONGESTION.

capacity of a stream the maximum load of stones, pebbles, sand, etc. a stream can carry, measured in grams per second. COMPETENCE.

cape a piece of land jutting into the sea; a prominent headland or promontory.

capillarity a phenomenon occurring when the surface of a liquid touches a solid. The surface of the liquid is either raised or depressed, depending on the difference between intermolecular attraction in the liquid and between the liquid and the solid. CAPILLARY FLOW, CAPILLARY MOISTURE, MOLECULAR ATTRACTION.

capillary flow the rise of water through the soil spaces above the WATER TABLE by means of pore-surface attraction. CAPILLARITY.

capillary fringe the soil layer lying immediately over the WATER TABLE in which water drawn up from the ground water level is held by CAPILLARITY.

capillary moisture the water held by SURFACE TENSION in pores around soil particles and available to plant roots. CAPILLARY FRINGE, FIELD CAPACITY, HYGROSCOPIC MOISTURE.

capital 1. the head, the chief town of a country, state or province, and usually the seat of government. The term is often used loosely in the sense of the chief town or city, as in the title 'commercial capital' **2.** accumulated wealth used to finance production, or any form of wealth used to help in producing more wealth. ACCUMULATION **3.** the stock of goods and COMMODITIES in a country.

capital goods the machinery and equipment, and the primary and partly processed raw materials used in the manufacture of other goods, contrasting with CONSUMER GOODS. PRODUCER GOODS.

capital-intensive *adj.* needing a large investment of CAPITAL-2 for higher earnings or increased productivity, as opposed to LABOUR-INTENSIVE.

capitalism broadly, an economic system characterized by private ownership of, and private investment in, the production of goods; and by private enterprise, competition, profit-making and a MARKET ECONOMY, the allocation of resources and wealth being dependent on market forces. COMMUNISM, STATE CAPITALISM.

cap-rock 1. a layer of resistant rock covering another or others of less-resistant material. BUTTE **2.** an impermeable layer overlying an AQUIFER or SALT-DOME **3.** unproductive rock covering valuable ore.

carat 1. the international measure of weight used for precious stones and gemstones equivalent to 1/142 oz or 200 milligrams **2.** a measure of purity of GOLD, pure gold being 24 carat; 22 carat having 22 parts gold, 2 parts of alloy; 18 carat having 18 parts gold, 6 parts of alloy etc.

carbon an ELEMENT-6 which, combined with other elements, occurs in all living things and in CARBONATES in the earth's crust. Organic chemistry is the study of carbon compounds. The radioactive isotope, carbon 14, is used in RADIOCARBON DATING. CARBON CYCLE, HYDROCARBON, ORGANIC CARBON.

carbonaceous *adj.* containing CARBON, applied to rocks (e.g. COAL, SHALE) or other sedimentary material (e.g. PEAT) consisting largely of carbon usually derived from organic matter.

carbonate *adj.* applied to a rock consisting mainly of carbonate minerals, i.e. minerals containing the carbonate group CO_3, e.g. CALCIUM CARBONATE.

carbonation saturated with or reaction with CARBON DIOXIDE.

carbonation-solution the WEATHERING of rocks by a chemical process in which rainwater charged with CARBON DIOXIDE (forming CARBONIC ACID) reacts with and dissolves LIMESTONE and rocks with other basic (BASE-2) oxides. CALCIUM BICARBONATE, CORROSION, DECALCIFICATION, GRIKE.

carbon cycle the movement of CARBON in ECOSYSTEMS. The carbon occurring in the atmosphere as CARBON DIOXIDE is absorbed and stored by plants (PHOTOSYNTHESIS). The plants, some bacteria and animals then oxidize these photosynthetic products, having obtained nourishment from them directly or indirectly, thus giving back some carbon dioxide to the atmosphere. Decay and the burning of organic matter (especially FOSSIL FUELS) also contribute to carbon dioxide in the atmosphere.

carbon-dating RADIOCARBON DATING.

carbon dioxide a colourless, heavy gas present in the ATMOSPHERE and in solution in the HYDROSPHERE, formed by the OXIDATION of compounds containing carbon and by the action of acid on CARBONATES. It does not burn and dissolves in water to form CARBONIC ACID. PHOTOSYNTHESIS.

carbonic acid a weak acid formed by the solution of CARBON DIOXIDE from the atmosphere and water. CARBONATION, CARBON DIOXIDE, DECALCIFICATION.

Carboniferous *adj.* carbon-bearing, i.e. coal-bearing, applied in Britain to the period between the Devonian and the Permian (GEOLOGICAL TIMESCALE), the three main groups of rock being the Carboniferous Limestone (the lowest, not normally carrying coals), the Millstone Grit and the Coal Measures.

cardinal points the four main points of the compass: north, south, east, west.

cargo FREIGHT-3.

CARICOM Caribbean Community, an organization of Caribbean states established August 1973 by Barbados, Guyana, Jamaica and Trinidad and Tobago, joined by Belize, Dominica, Grenada, St Lucia, St Vincent and Montserrat in May 1974 and by Antigua (4 July 1974) and Associated State of St Kitts–Nevis–Anguilla (26 July 1974) with the aim of achieving economic integration through the Caribbean Common Market, cooperation in non-economic areas, the operation of certain common services and the coordination of the foreign policies of the independent member states.

carnivore a flesh-eating animal or plant, a secondary consumer in a FOOD CHAIN. HERBIVORE, OMNIVORE.

carrying capacity **1.** the maximum BIOMASS which an area can support for an indefinite period **2.** the maximum number of species that an area can provide food for during the annual period when conditions (e.g. of weather) are hostile **3.** the maximum POPULATION-1 of people or of a given species for which an area can provide food **4.** of agricultural land, the maximum number of grazing animals and/or the maximum amount of food crops that the land can support under a given level of management without suffering deterioration **5.** in planning, the maximum use or number of users that a natural or artificial RESOURCE can sustain under a given level of management without the character and quality of the resource suffering unacceptable deterioration, e.g. the maximum human population that a particular area can carry or support without suffering unacceptable deterioration. When, in such an area, the number of people exactly equals this carrying capacity, the area is said to have reached SATURATION LEVEL, i.e. to be completely filled.

cartel **1.** an arrangement made between firms (particularly those in international trade organizations) whereby each keeps control of its own organization but agrees to some form of joint action in restricting production and competition, e.g. in buying raw materials, distributing products, allocating markets and quotas, and price fixing. Cartels are especially effective in controlling production, distribution and pricing of goods that lack substitutes **2.** the firms so linked.

Cartesian coordinate system GRID-1.

cartogram a simplified map presenting statistical information in a diagrammatic form by the use of symbols such as dots, circles, shading, range of colours, etc.

cartography the science and art of drawing maps and charts.

cascade **1.** a rush of water falling from a height **2.** a waterfall or section of a large waterfall **3.** a waterfall in which the water tumbles naturally over rocks, or down a series of artificial shallow steps **4.** steep RAPIDS.

cash crop a crop grown primarily for sale, as contrasted with a SUBSISTENCE CROP, grown for the use of the grower and/or the grower's family.

cassava MANIOC.

castellanus cloud a CLOUD formation which presents a mass of turrets when viewed from the side.

cast iron iron-carbon ALLOY, 4 per cent carbon, produced in a BLAST FURNACE. It is brittle, but easily fused.

catalyst a substance capable of increasing the rate of chemical reaction without itself suffering permanent chemical change, e.g. an ENZYME, PLATINUM.

cataract formerly applied to a large waterfall, now mainly to a series of rapids of the type occurring in the river Nile.

catastrophe theory a theory concerned with the relationship between QUALITATIVE and QUANTITATIVE change within a SYSTEM-1,2,3, with the fact that a sudden qualitative change within the system can abruptly disrupt, and change the form of, a hitherto smooth continuous process produced by a quantitative change. SYSTEMS ANALYSIS.

catch crop a fast-maturing crop grown when the ground would otherwise be lying fallow or idle, i.e. between two main crops in a rotation (ROTATION OF CROPS), or as a substitute for a regular crop which has failed, or between the rows of a main crop.

catchment area strictly, the area over which rain falls and is caught to serve a natural drainage area, a river basin. WATERSHED-2 (American usage).

categorical data analysis the statistical methods used in the analysis of data measured on a NOMINAL SCALE, resembling those used in REGRESSION ANALYSIS, but unlike that technique in that either the DEPENDENT or the INDEPENDENT variables are measured at the categorical (nominal) level, or they both are, not (as in regression analysis) at the interval or ratio level. MEASUREMENT IN STATISTICS.

category **1.** any division which serves to classify **2.** any one of the divisions in a system of classification, e.g. genus in the CLASSIFICATION OF ORGANISMS. **3.** in philosophy, a division which serves to classify (as in **1.**), but only in certain general classes of things or ideas, these classes varying according to the personal theory of the philosopher **4.** in statistics, a homogeneous

CLASS-1 or group of a POPULATION-4 of objects or measurements. If the category is given an identifying number or letter it is usually termed a code.

cation ION.

cattle large, cud-chewing bovine mammals, the domesticated species being the European ox and the Indian ox (termed Zebu cattle, widespread in the Indian subcontinent and the Far East). They are kept in all temperate and tropical lands, except in areas which are too closely forested, too rugged or too dry for the adequate growth of fodder, or elsewhere where disease makes their keeping impossible (e.g. in the parts of Africa infested by TSETSE FLY). In many parts of the tropics, especially in humid areas, they are replaced by water buffalo. Cattle are used as working animals (drawing ploughs, and carts for local transport in tropical lands), for the production of meat and of milk (for liquid consumption or for the making of cream, butter, cheese). The distinction between beef and dairy breeds, traditional in midlatitude areas, is steadily disappearing. The tendency is to produce rapidly maturing dual-purpose animals, the males providing meat, the cows producing a high yield of medium quality MILK.

causality **1.** the operation or relation between two events, states of affairs, objects, in which one brings forth, produces, the other, i.e. one is the cause, the other the effect, an essential concept in DETERMINISM **2.** the state of being that which brings forth a result, i.e. of being a cause.

cave **1.** a natural cavity, recess or chamber under the earth's surface with an entrance at the surface, caused by water erosion or volcanic action. Caves occur particularly in weak areas of seashore cliffs, caused by the eroding action of waves or their load of pebbles, etc. (STACK, BLOWHOLE), or in LIMESTONE regions when water charged with CARBON DIOXIDE dissolves underground channels along a bedding joint to produce a bedding-cave, or a much larger chamber **2.** an artificial cavity, such as that caused by quarrying, or deliberately constructed for wine storage.

cavern **1.** most commonly, a large CAVE **2.** synonym for cave **3.** a large chamber within a cave **4.** (American) a cave formed in limestone country by solution by underground water and streams.

caving (American) the slumping of river banks.

cavitation a process in which bubbles in a liquid are formed and then collapse in the path of a fast-moving body. Cavitation caused by the sudden increase of velocity in a fast-running stream results in the erosion of rocks because the collapsing bubbles make little shock waves which strike the bed and banks of the stream.

cay KEY.

CBA COST BENEFIT ANALYSIS.

CBD CENTRAL BUSINESS DISTRICT.

celestial *adj.* pertaining to the sky, or the heavens.

celestial sphere the 'bowl' of the heavens, an imaginary sphere of infinite radius, with the earth (and the terrestrial observer) at its centre, on the interior surface of which heavenly bodies appear to be placed. The plane of the earth's equator, when produced, crosses the celestial sphere at the celestial equator. Similarly when the axis of the earth is extended it touches the celestial sphere at its North and South poles (celestial poles). NADIR, ZENITH.

cell **1.** in biology, the smallest individual structural unit of every living organism, consisting of translucent, jelly-like, granular material (protoplasm) surrounded by a thin membrane (plasma membrane), in plants surrounded by a cell wall (usually of CELLULOSE); and containing a NUCLEUS or nuclei. A cell may have all the characteristics of a living organism; or it may be highly specialized for a particular function, e.g. the cells of multicellular organisms, which are not only highly specialized but also vary greatly in structure. Many MICROORGANISMS are unicellular (consisting of one cell). ATMOSPHERIC CELL **2.** in statistics, a CATEGORY defined by specific values on several variables simultaneously.

cell frequency in statistics, the frequency with which observations fall into a particular CELL-2, i.e. the number in a particular cell.

cellulose the fibrous constituent of the CELL-1 wall in higher plants, many ALGAE and some FUNGI. For industrial purposes cellulose is obtained mainly from wood pulp, cotton and flax, to make paper, rayon, plastics, explosives, etc.

Celsius scale the internationally accepted name for the Centigrade TEMPERATURE SCALE with 99 divisions between the ice point (ABSOLUTE ZERO), the freezing point of pure water (0°C), and STEAM POINT, the boiling point of pure water at sea-level with a standard pressure of atmosphere of 760 mm (100°C). Thus one Celsius degree is 1/100 of the temperature interval between ice point and steam point. CENTIGRADE SCALE, FAHRENHEIT SCALE, KELVIN SCALE, REAUMUR SCALE.

Celtic Sea in British Isles, a sea extending southwest from St George's Channel, the Bristol Channel to the edge of the CONTINENTAL SHELF, so-named because it was navigated by ancient Celts from the bordering coasts of Ireland, Wales, Cornwall and Brittany. Name introduced in the 1970s with the advent of exploration for oil and gas fields, first appearing on Admiralty charts in 1974.

cement **1.** a manufactured substance widely used in building to bind together other building materials, such as bricks or stones, to cover floors, to make walls etc. It is produced by heating together and then grinding chalk or limestone with clay or shale, the resultant grey, powdery material consisting of silicates of calcium and aluminates which, when mixed with water, crystallize to a dry solid. Some types of cement (hydraulic cement) harden under water **2.** a natural SILICEOUS, CALCAREOUS or FERRUGINOUS material, deposited from circulating water and able to convert loose deposits (e.g. sand, gravel) into a hard compact rock.

census **1.** all the processes involved in an official complete counting of the total number of persons inhabiting a given area at a particular time on a given day, usually conducted at stated (commonly ten-yearly) intervals. Such a census usually incorporates social data relating to the persons counted **2.** a similar count conducted through SAMPLING procedure **3.** a similar complete (or by sampling) count of items in some other field, e.g. traffic in a particular area **4.** the data so collected **5.** the published results of the count.

centi- c, prefix, one hundredth, attached to SI units to denote the unit $\times 10^{-2}$, e.g. centigram (one hundredth part of a GRAM-2), centimetre (cm). HECTO-, KILO-, MILLI-.

Centigrade scale the name formerly

applied to the CELSIUS SCALE, still used sometimes but not in SI units.

central business district CBD (origin in USA, a term not always applicable elsewhere) the heart of a city (DOWNTOWN in the USA), the part in which there is the greatest concentration of financial and professional services and major retail outlets, the focus of transport lines, where land use is the most dense and land values are at their highest. It is characterized by tall buildings, a high daytime population and high traffic densities. The importance of many central business districts has declined with the spread of the city and the practice of DECENTRALIZATION. BID PRICE CURVE, INNER CITY. Fig 8.

central eruption a volcanic eruption from a single vent or from a tight group of vents, producing a CONE, in contrast to a FISSURE ERUPTION.

central good, central service, central function any good sold or service offered or function performed at any CENTRAL PLACE. ORDER OF GOODS.

centrality **1.** the quality or state of being central **2.** a CENTRAL TENDENCY or central position **3.** in CENTRAL PLACE THEORY, the relative importance of a place with regard to its surrounding area, or the degree to which a centre serves its surrounding area. Christaller applied the term to the 'surplus importance' of a place (i.e. of a town), expressing the centrality of a town as the ratio between all the services provided there (for its own residents and visitors from its COMPLEMENTARY REGION) and the services needed by its own residents only. Centres with a high degree of centrality provided many services per resident; those with low centrality only a few services per resident.

centralization **1.** a concentration at one central point **2.** the bringing or putting (e.g. of administration, of a country, of an institution, or of a firm, etc.) under central control. This often puts the minor units on the periphery at a disadvantage, and may hasten their economic decline. CENTRIPETAL AND CENTRIFUGAL FORCES, DECENTRALIZATION.

central place any LOCATION-1 which provides goods, services, administrative functions for the consuming population of its HINTERLAND, i.e. the surrounding area (termed the COMPLEMENTARY REGION, trade area, or tributary area). The CENTRALITY-3 of the central place is determined by its various localized, specialized functions. CENTRAL PLACE HIERARCHY, CENTRAL PLACE SYSTEM, CENTRAL PLACE THEORY, ORDER OF GOODS.

central place hierarchy in CENTRAL PLACE THEORY, the arrangement of CENTRAL PLACES in a series of discrete classes, the rank of each being determined by the level of specialization of functions. The central places in each class perform all the functions of centres in the classes below them in the hierarchy (i.e. the lower order centres) but in addition perform a group of functions that differentiate them from, and place them above, those lower order centres. Higher order centres stock a wide array of goods and services, and provide specialist goods and services to a wide area; lower order centres stock a limited part of the array of the higher order centres and provide day-to-day goods and services to a smaller area. CENTRAL PLACE THEORY, COMPLEMENTARY REGION, K-VALUE, ORDER OF GOODS, THRESHOLD POPULATION.

central place system the spatial distribution of any set of CENTRAL PLACES which are of different sizes and different spacing and which satisfy the daily, weekly, monthly or yearly needs of the general consuming population. The pattern of this distribution is usually termed a network of central places. ADMINISTRATIVE PRINCIPLE, CENTRAL PLACE THEORY, COMPLEMENTARY REGION, ISOTROPIC SURFACE, MARKETING PRINCIPLE, ORDER OF GOODS, TRAFFIC PRINCIPLE.

central place theory a theory expounded by W. Christaller, an economic geographer, in 1933, which asserts that the numbers, sizes, and patterns of spatial distribution of CENTRAL PLACES can be explained by the operation of the forces of supply and demand, by the way in which and the extent to which these centres provide goods and services to their surrounding areas. The theory, concentrating on the retailing of goods and services, assumes that both suppliers and consumers wish to derive the greatest economic benefit from their decisions. The suppliers (the profit maximizers) wish to earn the maximum profit from the sale of goods and services. The consumers (the distance minimizers) wish to satisfy their needs by obtaining goods and services with the minimum of effort and cost. The suppliers, having confirmed that the THRESHOLD POPULATION is sufficient to make their enterprise economically viable, therefore locate their establishments as close as possible to the consumers, taking into account the threshold of success as well as range. Threshold of success is the smallest volume of sales necessary for an establishment to be economically viable, range is the greatest distance consumers are willing to travel to obtain a good or service. Establishments can be classified according to this threshold and range: low order establishments, with a low threshold and range and a fairly compact sphere of influence, meet a daily need for which the consumer is not prepared to travel far; middle order, with a medium threshold and range and a more extensive sphere of influence, supply goods and services less frequently in demand, for which the consumer is willing to make greater effort; and high order establishments, with a high threshold and range and the most extensive sphere of influence, provide specialist goods and services even less frequently in demand, for which the consumer is prepared to travel a considerable distance. In any given area there will thus be many centres with low order establishments, fewer with middle order, fewer still with high order, giving rise to a CENTRAL PLACE HIERARCHY of higher and lower order centres (K-VALUE). In any network of central places (CENTRAL PLACE SYSTEM) there will thus be many smaller, lower order centres, forming a dense network close together, but fewer larger and more widely-spaced higher order centres. Christaller used three principles (ADMINISTRATIVE, MARKETING, TRAFFIC) to account for the varying levels and distribution of central places in a central place system. CENTRAL GOOD, CENTRALITY, CONVENIENCE GOOD, ISOTROPIC SURFACE, NESTING, ORDER OF GOODS, SPHERE OF INFLUENCE, ZONE OF INDIFFERENCE. Figs 9(a) and (b).

central tendency in statistics, the tendency of observations to cluster around a particular value, or to pile up in a particular category, the position of the central value being determined by one of the measures of location, i.e. the MEAN, MEDIAN or MODE, each of which makes different assumptions about the data, is calculated in different ways, and for different reasons. AVERAGE.

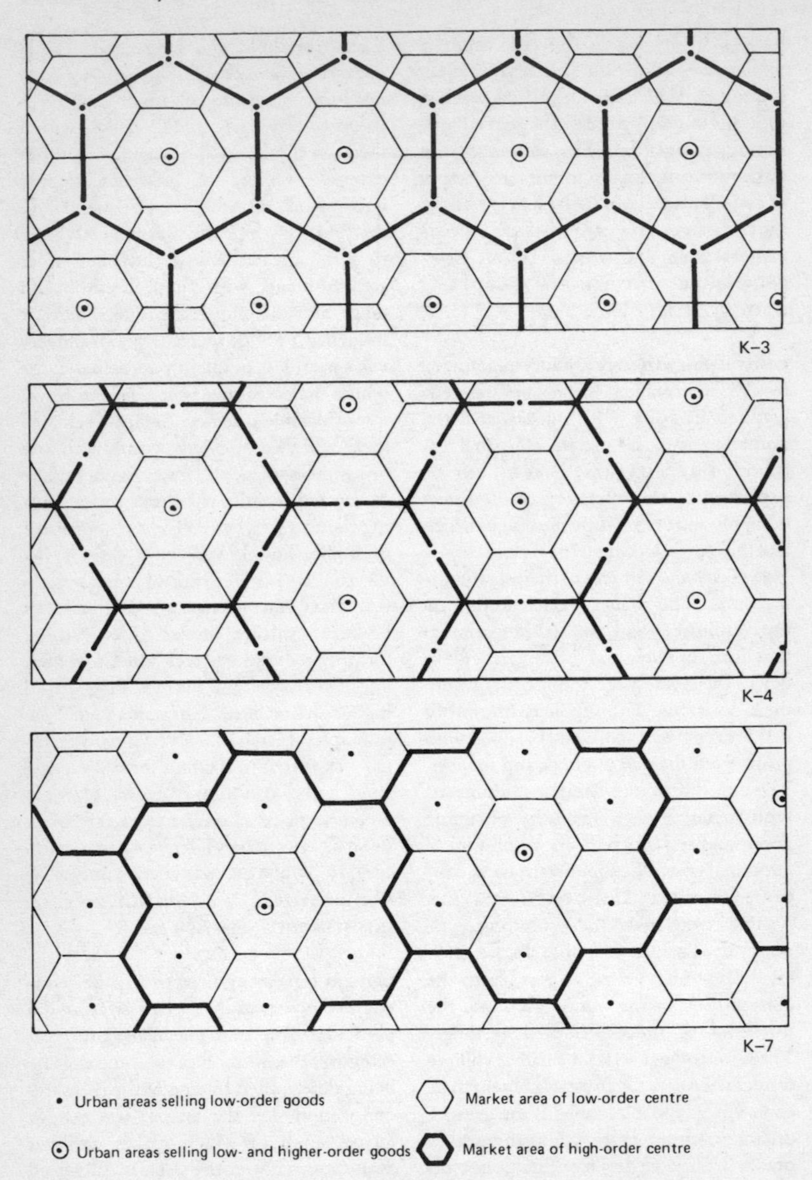

Fig 9 (a) Central place theory market areas: K-3 (marketing principle); K-4 (traffic or transportation principle); K-7 (administrative principle)

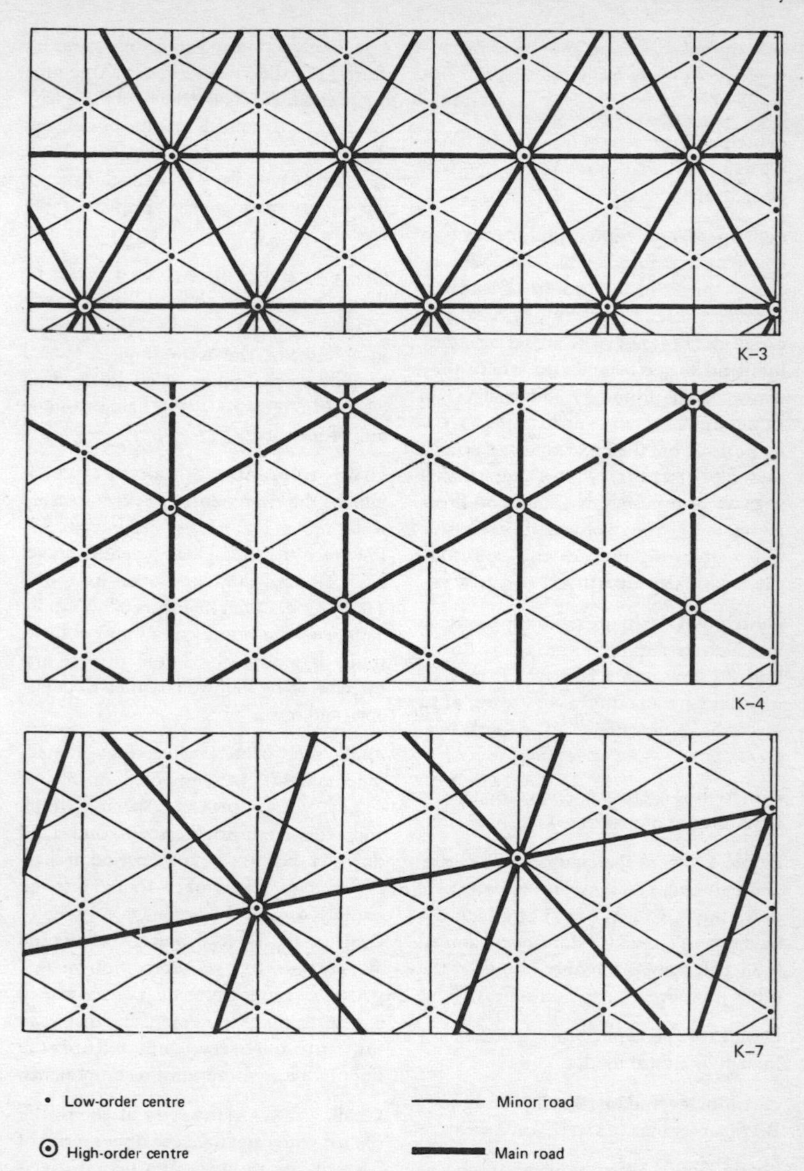

Fig 9 (b) Central place theory: transport network corresponding to hierarchies in Fig 9 (a)

centrifugal *adj.* acting, moving, or tending to move away from a centre, the opposite of CENTRIPETAL.

centripetal *adj.* acting, moving, or tending to move towards a centre, the opposite of CENTRIFUGAL.

centripetal and centrifugal forces two counteracting forces which are said to cause changes in the pattern of land use in urban areas. The CENTRIPETAL FORCE causes CENTRALIZATION, attracting establishments to the central area where they may benefit from the advantages of ACCESSIBILITY and AGGLOMERATION ECONOMIES; but the CENTRIFUGAL FORCE causes DECENTRALIZATION and URBAN SPRAWL as it pushes dwellings and businesses away from congested, expensive inner city areas towards the SUBURBS. CIRCULAR AND CUMULATIVE GROWTH.

centripetal drainage a drainage pattern in which streams flow from many directions to flow into a lake which may or may not have an outflowing stream, or to meet the major stream at a focal area. DRAINAGE.

centrocline a term sometimes used in the USA instead of PERICLINE.

cereal **1.** any of the cultivated flowering grasses (BARLEY, GRASS, MAIZE, MILLETS, OATS, RICE, RYE, WHEAT) of which the seed (grain) is used for human and animal food **2.** loosely, a member of one of the other plant orders, e.g. beans, peas.

cereal *adj.* pertaining to grain used for human or animal food.

cerrado, cerradão (Brazil: Portuguese) Brazilian savanna. CAMPO, SAVANNA.

chaco (South America: Spanish) the vast almost level alluvial plain with poor soil supporting grassy SAVANNA OR DECIDUOUS scrub woodland in South America north of the PAMPAS of Argentina between the Andes and the river Paraguay. There is heavy rainfall in some places, and the chaco is marshy nearly everywhere. The scrub woodland is characteristic of the eastern part, grassy savanna of the western.

chain a number of connected events or things, e.g. of like physical features such as islands, or lakes, etc. When applied to mountains the term implies several roughly parallel ranges, e.g. the Andean chain (the term RANGE-2 being a single line of mountains).

chain migration MIGRATION-1,2 in which the individual worker (usually male) moves and his/her dependants follow once any initial loan given for travel has been repaid and the worker has found a secure job. Other members of the family and/or of the home community follow, using as a base the home of the first migrant, to be followed by their dependants, and so on.

chalk soft, white, friable, fine-grained, pure LIMESTONE, composed mainly of CALCIUM CARBONATE, especially thick and extensive in southeast England. Laid down in the CRETACEOUS period in shallow water, it was once thought to be entirely organic in origin, consisting of shells or the skeletal remains of marine MICROORGANISMS. Current thought suggests that chemical precipitation has played a large part in its formation, and that TERRIGENOUS DEPOSITS such as sand from the floor of the seashore are also components.

Chalk (in STRATIGRAPHY it should be spelled with capital C) the upper series of rocks in the CRETACEOUS period (GEOLOGICAL TIMESCALE). In England the Chalk is divided into Upper Chalk (white chalk with abundant flints overlying Chalk rock, a bed of hard nodular chalk). Middle Chalk (soft, white chalk with fewer flints and some MARL) and Lower Chalk (grey chalk with more Chalk marl).

channel 1. a course for running water, either artificial, as in a canal or irrigation ditch; or natural, as in the deepest part of a river or stream **2.** a narrow stretch of water, wider than a STRAIT, connecting two larger stretches of water (e.g. two seas) or two land areas **3.** a deep, navigable waterway (natural or dredged) in a BAY, ESTUARY or SHALLOWS, which affords a safe passage for vessels.

chaos *adj*. chaotic, complete confusion, e.g. the state of a SYSTEM-1 when the interconnections and relationships between its elements are disturbed so that they lack stability and coherence, and the future of the system becomes unpredictable. A chance happening affecting one of the elements may trigger such a chaotic state.

characteristic sheet a reference sheet, a sheet providing the key to the system of CONVENTIONAL SIGNS used on a map. SYMBOL.

charcoal an amorphous form of CARBON, the solid residue obtained by the imperfect combustion (DISTILLATION) of animal or vegetable matter (usually wood) in a restricted supply of air. It is useful as a fuel, giving out intense heat; it is used in many scientific and industrial processes and, in stick form, in sketching and drawing.

chase 1. originally, in medieval England, an unenclosed private hunting ground held by a subject of the Crown and therefore subject to Common Law, as distinct from a Royal Forest, held by the King, and subject to Forest Law (FOREST-2, PARK-1) **2.** surviving in place-names, indicating a former use, e.g. Cannock Chase.

chatter mark a crescent-shaped mark, consisting of a series of minute cracks, so finely packed together that they resemble a bruise on the underside of firm but brittle rocks (e.g. granite) insecurely embedded in a GLACIER, the 'horns' of the crescent pointing to the direction of movement of the ice. It is caused by COMPRESSION and by the vibration arising from the looseness of the rocks.

chelation the process whereby organisms or organic substances bring about the decomposition and disintegration of soils and rocks.

chemical weathering CORROSION, HYDROLYSIS, MECHANICAL WEATHERING, ORGANIC WEATHERING, WEATHERING.

chemoautotrophic *adj*. applied to an organism which is able to produce organic material from inorganic compounds, the energy being supplied by simple inorganic reactions, e.g. by the oxidation of ferrous salts to FERRIC form, the energy source used by some BACTERIA. AUTOTROPHIC.

chemotrophic *adj*. applied to an organism that obtains energy from a source other than light (PHOTOTROPHIC), i.e. by chemical reactions associated with organic or inorganic substances. AUTOTROPHIC, CHEMOAUTOTROPHIC, HETEROTROPHIC.

chernozem (Russian, black earth) a group of fertile ZONAL SOILS suited to the cultivation of cereals. It is granular, well-drained, rich in humus and BASES-2, and develops under tall and mixed grasses in a temperate to cool, subhumid climate, i.e. in MIDLATITUDES where grassland is/was the natural vegetation, e.g. the

Ukraine or central Canada. Typically the dark coloured, nearly black A HORIZON is thick with MULL or mull-like humus, grading to a B HORIZON (if it is present) which is lighter brown, with or without a concentration of clay, beneath which lies the CALCAREOUS C HORIZON.

chestnut soils a group of ZONAL SOILS, usually dark brown over lighter coloured soil, overlying a calcareous horizon. Such soils are friable, formed under conditions drier than those resulting in CHERNOZEM, and are thus less leached than chernozem, but similarly cover wide areas of land where grassland (but of a drier type) was/ is the natural vegetation, e.g. the High Plains of USA, part of the PAMPAS of Argentina, the south African VELD, Hungary, the STEPPE to the south of the chernozem in Russia. The A HORIZON is dark brown and, becoming paler in colour, lies over a B HORIZON in which there is an accumulation of lime, overlying the calcareous C HORIZON. In some places there is little or no B horizon. SOIL, SOIL ASSOCIATION, SOIL HORIZON.

chimney 1. a narrow cleft in a vertical rock wall that can be used by rock-climbers **2.** a volcanic vent.

china clay KAOLIN.

chinook a warm dry, southwest wind, similar to the FOHN, which blows down the eastern side of the Rocky mountains, having been warmed adiabatically (ADIABATIC) as it blew from the west across the mountains. It usually occurs suddenly, accompanied by a rapid rise in temperature which melts the snow in winter.

chi-squared test a NONPARAMETRIC statistical test used to measure the extent of the agreement between observed and expected frequencies. Being nonparametric it can be used with nominal or ordinal data (NOMINAL SCALE, ORDINAL SCALE), and being 'distribution free' it does not assume that the data being analysed is normally distributed (NORMAL DISTRIBUTION). The formula for the measure, termed chi-squared, is:

$$\chi^2 = \sum_{i=1}^{k} \frac{(o_i - e_i)^2}{e_i}$$

k denotes the number of CELLS o_i and e_i the observed and expected frequencies respectively, for the ith cell. The value of χ will be 0 if there is perfect agreement between observation and expectation; and the value increases with increasing differences between the observed and expected frequencies. SIGNIFICANCE TEST.

chlorophyll a green pigment in the cells of all ALGAE and higher plants (apart from a few PARASITES and SAPROPHYTES), formed only in the presence of light. Several types occur, with small differences in chemical structure, but each contains MAGNESIUM and IRON, and each is essential in absorbing light energy in PHOTOSYNTHESIS. In some plants, e.g. some SEAWEEDS, the green of the chlorophyll is masked by other pigments. AUTOTROPHIC.

C horizon a distinct layer in the SOIL, underlying the A or B HORIZONS, or the organic or mineral horizons (O HORIZONS, MINERAL HORIZONS, SOIL HORIZON), consisting of the PARENT MATERIAL, i.e. the little altered but weathered bedrock, transported glacial or alluvial material, or an earlier soil, from which the soil is formed. Below the C horizon lies the D horizon, the unaltered bedrock. SOIL PROFILE.

chorography (Greek chōra, a place, a

district) a term much used in the seventeenth and eighteenth centuries to make a distinction between a (geographical) study of a special region or district (chorography) and geography, a study dealing with the earth in general. It is applied today by some authors to the identification of, or to a general account of, a large regional area (hence chorographic map) as distinct from TOPOGRAPHY, which deals with a detailed study of a small area. But some American authors use the term chorographic as relating to a very large area (say, a subcontinent) and a chorographic map as a map on a scale of between 1:500 000 and 1:5 mn. TOPOGRAPHIC MAP.

chorology the study of the causal relations of the phenomena present in a region, an explanatory study of a region.

chorometrics the statistical study of spatial distributions.

choropleth map a quantity in area map, presenting the subject under study in terms of average value per unit area within specific boundaries, e.g. density of population per sq km shown within local, regional or national administrative areas; or by dividing the unit area into squares or hexagons and calculating a mean value for each. Sometimes a range of stippling, shading or colouring is used to show orders of density.

chronology the science of measuring and adjusting time or periods of time, of recording and arranging events in order of time, and of assigning events to dates considered to be correct in the light of contemporary knowledge.

cinder the residue of incompletely burnt coal or similar combustible material. CINDER CONE.

cinder cone a cone formed round the vent of a VOLCANO by debris, usually of volcanic origin, cast up during eruption. As in the ASH CONE the fragmentary material has not been burnt. The majority of the particles exceed 4 mm in diameter or long dimension, the larger fragments being known as lapilli (5–10 mm) or SCORIAE. The slopes of a cinder cone are steeper than those of an ash cone because the angle of repose of its larger constituent fragments is greater.

circadian rhythm a term usually applied to the DIURNAL RHYTHM of a human being.

circular and cumulative growth, circular and cumulative feedback a process whereby growth feeds on and reinforces itself by the creation of new demands for goods and services, etc. It tends to reinforce major cities and favoured regions at the expense of less advantaged areas (CUMULATIVE UPWARD CAUSATION). As the central area prospers, the periphery (reinforcing the centre) suffers the backwash effect of the flow of skilled manpower, investment and locally generated capital to the centre; the periphery becomes poorer than the centre in social services and amenities, etc., and products from the centre flow to the periphery, flooding the market and inhibiting local enterprise. Thus unequal development is maintained. But ultimately the centre's successful development combined with the establishment of an efficient transport and communication network cause a spread effect, a centrifugal force (CENTRIPETAL AND CENTRIFUGAL FORCES), leading to decentralization accompanied by development in the periphery, thereby spreading development and reducing regional inequality. CORE-PERIPHERY MODEL.

circulation of people, short-term movements, many being repetitive, including daily commuting (COMMUTER), weekly movements, the following of seasonal work, travelling employment (e.g. of sales representatives) and short-term changes of residence (e.g. of employees in firms with works and offices in various locations).

circumdenudation, circumerosion 1. in general, DENUDATION all round **2.** the process by which hills or mountains, by DENUDATION all round them, were isolated from an original parent mass such as a plateau.

cirque (French; Gaelic coire; Scottish corrie; Welsh cwm) a steep-walled amphitheatre, or basin, of glacial origin at the head of a mountain valley (in some cases containing a small lake), resulting from frost and glacial action (NIVATION, ROTATIONAL SLIP). At the meeting of two cirques a knife edge or ARETE is formed. Fig 10.

cirque glacier a short-tongued small GLACIER which fills a separate, rounded basin glacially eroded on a mountainside.

cirrocumulus a type of high CLOUD (above 6000 m: 20 000 ft), usually formed by ice crystals, appearing as lines of small round puffs interspersed with blue sky, i.e. as a MACKEREL SKY.

cirrostratus a layer of milky, fibrous-looking high sheet CLOUD (above 6000 m: 20 000 ft), lightly veiling the sun, heralding the approach of a WARM FRONT. If it thickens it develops into ALTOSTRATUS.

cirrus high, wispy, fibrous CLOUD (6000 to 12 000 m: 20 000 to 40 000 ft), composed of tiny ice crystals, through which sunlight or moonlight may penetrate. Strong winds in the upper atmosphere may draw out the 'fibres' to form 'mare's tails' or 'stringers'. It is usually associated with fair weather, but if it thickens to CIRROSTRATUS it may signal the approach of a DEPRESSION-3. FALSE CIRRUS.

CIS Commonwealth of Independent States, the political system that came into

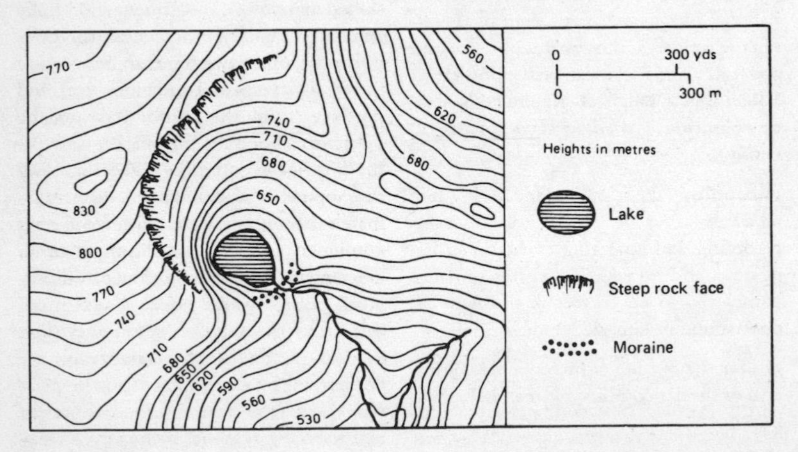

Fig 10 Cirque

being following the dissolution of the Union of Soviet Socialist Republics. SOVIET.

CITES Convention on International Trade in Endangered Species, a United Nations international convention set up to regulate trade in threatened or endangered species. 132 signatory countries agreed to establish a management and scientific authority to police such trade. TRAFFIC INTERNATIONAL, an NGO, acts as a watchdog to monitor the effectiveness of CITES.

city **1.** in general, in Britain, a large TOWN-1 **2.** more strictly, a town of any size which is or has been the seat of a bishop and has a cathedral **3.** in USA, city implies the incorporation of a settlement and the establishment of some form of local government, but a city may in this sense be only a few hundred people, and the word is used loosely as being synonymous with town. CENTRAL PLACE THEORY, CITY REGION, CITY STATE, CONCENTRIC ZONE GROWTH THEORY, ECUMENOPOLIS, GATEWAY CITY, INDUSTRIAL CITY, METROPOLIS, POST-INDUSTRIAL CITY, PRE-INDUSTRIAL CITY, PRIMATE CITY, RANK-SIZE RULE, WORLD CITY.

city region a city or large town and the area surrounding it (HINTERLAND-2), the surrounding area being so functionally linked to and dominated by that city or town that the two are interdependent and function together as one unit. CORE-PERIPHERY MODEL.

city state a state which has sufficient POWER-1,2 to constitute an independent, sovereign state, usually exerting its authority over the surrounding region, e.g. Athens, one of the city states of classical Greece, or Singapore.

cladistics a systematic method of classification used by some taxonomists (TAXONOMY), based on the theory of evolution by descent with modification, but excluding consideration of the speed or of the mechanism of evolution. PUNCTUATED EQUILIBRIUM.

class **1.** in general, a division based on quality or grade **2.** in biology, one of the groups used in the CLASSIFICATION OF ORGANISMS, consisting of a number of similar ORDERS-2, but sometimes of only one order **3.** a group of people of similar status, rank, or culture in a community. SOCIAL CLASS **4.** a concept or system of social division. SOCIAL CLASS **5.** in statistics, a group of occurrences with a common characteristic or set of characteristics formed when data is divided, each group being mutually exclusive. The variate values determining the upper and lower limits of a class are class boundaries; the interval between them is the CLASS INTERVAL; the frequency falling into the class is the class frequency.

classical economic theory the body of theory propounded by Adam Smith, David Ricardo, John Stuart Mill and others in the late eighteenth and the nineteenth centuries. To generalize broadly, the theory assumes that in a capitalist society a policy of individualism, of LAISSEZ-FAIRE, combined with the free operation of price in the market place, the investment of capital to promote economic growth, freedom from government intervention and freedom of international trade, would result in economic benefit to the whole community. The classical period of economics came to be dominated by Ricardo, who expounded his LABOUR THEORY OF VALUE. COMMODITY, NEOCLASSICAL ECONOMIC THEORY.

classification in general, arrangement in classes, putting into groups systematically, on the criteria of common characteristics or properties.

classification of lakes the systematic designation of lakes, commonly based on the origin of the depression in which the water accumulated, i.e. erosion (glaciation, solution, wind); enclosure by deposition or barrier (including deltaic and morainic deposits, sand bars, dams of ice or vegetation); structural (sagging of the earth's crust, rift-valley, down-faulted basins); volcanic (crater lakes, lava dam); artificial (especially by great dams built for hydro-electric power and irrigation schemes; and tanks in India). CRATER LAKE, ETANG, FINGER LAKE, LOCH, LOUGH, MERE, POND, SHOTTS, TANK, TARN.

classification of organisms the systematic designation of animals and plants based on groups which reflect evolutionary relationships, the rank of the group being measured by the number of organisms that belong to it. To qualify for group membership the organisms must resemble each other in some property or characteristic more than they resemble any other organism. The smallest group is the SPECIES-1, though in some cases subspecies and VARIETIES are identified; species that resemble each other more closely than they resemble other species are put together to form the genus (plural genera); genera that are similarly alike form the family; families form an ORDER-2, orders a CLASS-2; classes together form a phylum (botany or zoology) or a division (botany); and phyla or divisions form a kingdom, the highest rank.

class interval in statistics, the size of the interval used in the division of data into classes or categories, e.g. in the division of a group of people into age groups, the age groups selected being 0 to 10 years, 11 to 20 years, and so on, the size of the interval is 10 years.

clast a sedimentary particle, commonly a rock fragment or mineral grain produced by the disintegration of a larger mass or by volcanic explosion.

clastic *adj.* applied to a rock composed of broken fragments (CLAST) of pre-existing rocks or of shell fragments which have been converted into a consolidated mass, e.g. SEDIMENTARY ROCKS (CLAY, CONGLOMERATE, SANDSTONE, SHALE), PYROCLASTIC rocks (AGGLOMERATE, TUFF, VOLCANIC ASH).

clay **1.** a fine-textured SEDIMENTARY ROCK consisting mainly of hydrous silicates of ALUMINA, with other minerals and chemicals and organic material in varying amount, resulting from the weathering and decomposition of feldspathic rocks. Some of the constituents are usually in a colloidal state (COLLOID) and lubricate the grains and flakes of the non-colloidal constituents. When wet, clay is PLASTIC and IMPERMEABLE, because the water held by SURFACE TENSION round the particles fills the tiny interstices between them; when dry it loses its plasticity and develops cracks; when heated to a high temperature it becomes brittle and stone-like. **2.** the finest particles in a soil, with a diameter of less than 0.002 mm (0.005 mm in USA), and the resulting soils, a clay soil being one which has at least 30 per cent of such particles.

clay pan a layer of stiff CLAY-1 formed below the surface of the soil (HARD PAN), acting as a more or less IMPERMEABLE layer and leading to waterlogging.

clay-slate a SLATE formed from the

METAMORPHISM of CLAY-1, as distinct from SLATE derived from compacted VOLCANIC ASH.

clay soil CLAY-2, SOIL TEXTURE.

clay-with-flints a deposit consisting of a mixture of chalk-flints and reddish or brown CLAY-1, sometimes nearly black at the base, becoming lighter and sandier towards the surface, overlying the CHALK (e.g. in southern England) and also in PIPES-2 or POTHOLES. SUPERFICIAL DEPOSIT.

clear felling the felling of all the trees in an area, regardless of size. COPPICE.

cleavage usually applied to slaty cleavage, the fissile structure developed in certain fine-grained rocks (e.g. CLAY-SLATE or compacted VOLCANIC ASH) especially as a result of DYNAMIC METAMORPHISM. Minute flakes of micaceous minerals, formed as a result of the metamorphism, tend to arrange themselves at right angles to the direction of the pressure, and the rock thus splits in a direction quite different from that of the original bedding. As a result roofing slates can be prepared from rock with a well developed slaty cleavage.

cliff 1. a high, steep or perpendicular face of rock, the angle of slope dependent on the jointing, bedding and hardness of the rocks forming it, e.g. along a sea coast (sea-cliff, particularly subject to erosion by waves); or bordering a lake (lake-cliff). FREE FACE, RAISED BEACH **2.** a steep face of unconsolidated sediments carved by the river in a MEANDER (river-cliff). Fig 11.

climate the average WEATHER conditions throughout the seasons over a fairly wide or very extensive area of the earth's surface and considered over many years (usually 30 to 35 years) in terms of CLIMATIC ELEMENTS. CLIMATOLOGY, LOCAL CLIMATE, MACROCLIMATE, MESOCLIMATE, MICROCLIMATE, MICROCLIMATOLOGY.

climatic amelioration AMELIORATION.

climatic climax the CLIMAX developed in a LOCAL CLIMATE which differs from the climate normal to the area.

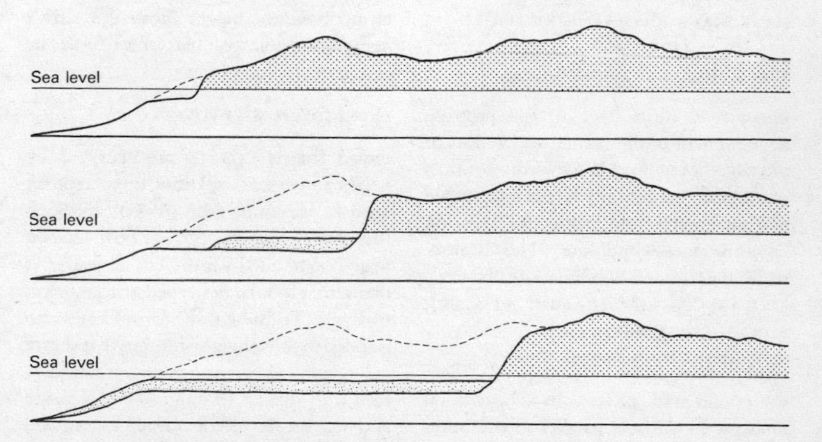

Fig 11 The formation of cliffs. The three sections show how cliffs are cut back and a submarine peneplane formed

climatic elements ATMOSPHERIC PRESSURE, HUMIDITY (covering CLOUDS, EVAPORATION, PRECIPITATION, WATER), TEMPERATURE (covering RADIATION), and WIND, resulting from the interrelationship of latitude, altitude, the spatial distribution of land and sea, ocean currents, relief, soil and vegetation. CLIMATE.

climatic formations vegetation groups classified according to the climatic factors (CLIMATE, CLIMATIC ELEMENTS) that determine them, as distinct from EDAPHIC FORMATIONS. The large vegetation formations tend to be determined by climatic factors, the smaller units by EDAPHIC FACTORS.

climatic geomorphology in geomorphology, the study of the association between CLIMATIC ELEMENTS and the development of landforms.

climatology the study of the CLIMATES of the earth, describing and where possible explaining them and the part they play in the natural environment. LOCAL CLIMATE, MACROCLIMATE, METEOROLOGY, MICROCLIMATE, MICROCLIMATOLOGY, SYNOPTIC CLIMATOLOGY.

climax the final stage in the possible development of the natural vegetation of a locality or region, when the composition of the plant community is relatively stable and in equilibrium with the existing environmental conditions. This is normally determined by climate (CLIMATIC CLIMAX) or soil (EDAPHIC CLIMAX). SERE, SUCCESSION-2.

clint, clent a low, flat-topped ridge, sometimes with LAPIES, in a horizontal LIMESTONE surface, parallel to the BEDDING PLANE and separated from another clint by furrows or fissures. GRIKE.

closed system an isolated SYSTEM-1,2,3 enclosed by a boundary through which neither energy nor material can pass (unlike an OPEN SYSTEM). GENERAL SYSTEMS THEORY.

cloud a visible mass of tiny particles floating in the atmosphere, consisting sometimes of ice crystals, more usually of water formed from CONDENSATION of WATER VAPOUR on nuclei of dust or smoke particles or ionized MOLECULES of the air itself. The formation of cloud thus depends on the cooling of moist air, which may result from the rising and expansion of such air, the mixing of warm air with cold air, or by a loss of heat by radiation. Very low cloud is termed FOG or MIST. CLOUD AMOUNT, CLOUD FORMS.

cloud amount cloud cover, estimated visually, expressed as the proportion of sky covered either in tenths or, more commonly, in eighths (oktas), 0 representing a cloudless sky, 8 total cloud cover. The lines drawn to show areas of equal cloudiness are termed isonephs.

cloud base the height above the earth's surface of the lowest part of a CLOUD or of a general cloud layer.

cloud forest MIST FOREST.

cloud forms CLOUDS are classified by height above sea-level (low, up to 2400 m: 8000 ft; medium, 2400 to 6000 m; 8000 to 20 000 ft; high, 6000 to 12 000 m: 20 000 to 40 000 ft) and by form (cirrus, feathery; cumulus, globular or heaped; stratus, sheet or layer). To these three form names alto is added to show height, nimbus to indicate falling rain. These main genera are subdivided into species (distinguished by shape and structure), varieties (arrangement and transparency, with additional features) and accessory cloud formations. The varieties of form include, among others, lenticular, lens-shaped; castellanus (formerly castellatus), turret-shaped; mammatus, breast-shaped; fracto-, ragged; banner, like a banner. Accessory cloud formations include arcus, arched; incus, anvil-shaped; tuba, column- or cone-like. ALTOCUMULUS, ALTOSTRATUS, ANVIL CLOUD, BANNER, CASTELLANUS, CIRROCUMULUS, CIRROSTRATUS, CIRRUS, CUMULONIMBUS, CUMULOSTRATUS, CUMULUS, FRACTUS, LENTICULAR, NIMBOSTRATUS, STRATOCUMULUS.

cloud seeding the introduction of dry ice, salt particles or silver iodide smoke into CLOUDS in order to promote rainfall. This technique is also being used experimentally to suppress lightning and to change the structure and movement of HURRICANES-2.

Club of Rome an unofficial association founded April 1968 by Aurelio Peccei, Italian manager and consultant, restricted to 100 members, comprising social and environmental scientists, educators, economists, civil servants, managers, philosophers, from different countries. The aims were to foster among policy-makers and the public alike a better understanding of the problems faced by the developing and industrialized world; and to promote new policy initiatives and action.

cluster a number of similar things growing or gathered together. A cluster may be shown on a map by means of symbols (DOT MAP) or on a graph by the flattening of a LORENZ CURVE.

cluster analysis in statistics, an analysis which aims to discover whether the individuals in a POPULATION-4 fall into groups or clusters. In the ideal CLUSTER the members should correlate highly with each other, but the cluster itself should have little correlation with items outside it.

CMEA Council for Mutual Economic Assistance, primarily an economic association, initiated 1949 by the USSR, working language Russian. The founder members were the USSR, Bulgaria, Czechoslovakia, Hungary, Poland, Romania. They were joined by Albania (1949–61), Cuba (1972), German Democratic Republic (1950), Mongolia (1962), Vietnam (1978). Yugoslavia agreed to participate partially (1964). Angola, Laos and North Korea participated as observers, and there were cooperative agreements with Finland, Iraq and Mexico. CMEA was the official acronym, but other popular abbreviations were COMECON and CEMA. The association ended in 1991 when the USSR became the Commonwealth of Independent States. CIS.

coal a carbonaceous FOSSIL FUEL, a brownish-black or black combustible mineral substance found in beds or seams in SEDIMENTARY ROCKS and derived from vegetable material growing in the CARBONIFEROUS era on level, swampy ground, compacted and hardened by pressure and heat arising from earth movements. A series based on percentage of fixed CARBON (from peats and brown coals with less than 55 per cent, through bituminous or humic coals, to anthracite with more than 93 per cent) may be distinguished, with a corresponding limitation in the amount of volatile material and moisture. ANTHRACITE, BITUMINOUS and BROWN COAL, COALFIELD, COAL GAS, COAL MEASURES, COKE, LIGNITE.

coalfield a tract of land underlain by COAL. If the workable coal is covered by younger deposits the coalfield is described as concealed.

coal gas a gas (in volume 50 per cent hydrogen, 30 per cent methane, 8 per cent carbon monoxide, 8 per cent carbon dioxide, oxygen and nitrogen, 4 per cent gaseous hydrocarbons other than methane) used as a fuel. It is obtained by heating suitable COAL in closed retorts (i.e. without air) whereby gas is driven off and COKE remains.

coal measures, Coal Measures the series of sediments, mainly SANDSTONES and SHALES, in which COAL is found. With initial capital letters, the term is applied specifically to the upper division of the CARBONIFEROUS. GEOLOGICAL TIMESCALE.

coal tar a viscous liquid obtained in the process of COAL distillation in retort or coke oven.

coast a term loosely applied to the zone of indeterminate width where land and sea (or other extensive tract of water) meet, considered as the boundary of the land (COASTLINE). More specifically, the meeting place of land and sea (width not specified) covering **1.** the narrow strip of land immediately landward of HIGH WATER, the line of the mean SPRING TIDES **2.** a more extensive zone stretching inland **3.** a zone which includes the SHORE **4.** a zone which excludes the shore. Various types of coast are identified, e.g. ATLANTIC, CONCORDANT, DISCORDANT, LONGITUDINAL, PACIFIC, TRANSVERSE.

coastal plain any comparatively level land of low elevation, sloping gently seaward and bordering the sea or ocean, resulting from the deposition of sediment washed down from the land, or from denudation by the sea, or by the emergence of part of the CONTINENTAL SHELF following a fall in sea-level.

coastline a term applied loosely to the continuous edge of the land, or the general appearance of the COAST, as seen from the sea; or to the zone between the BACKSHORE and the coast; or to the landward limit of the BEACH; or used as a synonym for COAST. More precisely it is applied to the line on the land indicated by **1.** the highest storm waves of the SPRING TIDES **2.** the high-water mark of medium tides **3.** the base of the sea-cliffs.

cobble, cobblestone a naturally rounded, water-worn stone, larger than a pebble, smaller than a boulder. The British Standards Institution ranks cobbles (60 to 200 mm: 2.4 to 8 in in diameter) between coarse gravel and a boulder. The USA WENTWORTH SCALE defines a cobble (64 to 256 mm: 2.5 to 10 in in diameter) as between a pebble and a boulder.

cockpit 1. a pit where gamecocks were set to fight for sport **2.** in KARST, any natural enclosed depression with steep sides **3.** a SINK-HOLE with steep sides, especially a star-shaped one with a conical or slightly concave floor, as in the Cockpit country of Jamaica.

cocoa CACAO.

coconut the coconut palm, native to tropical lands, tolerant of salty, sandy soils, and also bearing fruit in some warmer subtropical areas (e.g. the Bahamas). It is cultivated in the lowlands of both those areas for the sake of its edible fruit, which has a shiny waterproof skin, with a fibrous mass inside (the fibres yield coir, used in making coarse string, matting etc.). Inside the fibrous mass is the NUT which has a hard shell lined with a thin, white fleshy layer. There is a hollow within the nut which, before the nut is ripe, is partly filled with a nutritious sugary liquid, coconut milk. The milk is gradually absorbed into the flesh as ripening progresses. The flesh can be eaten directly, or dried and flaked to form desiccated coconut; when dried it forms copra, from which coconut oil is extracted. The oil is used in cooking and in making soap, margarine, and cosmetics. The residue left after the oil has been extracted from the copra is used as cattle cake. The shells of the nuts can be used as fuel. The tree trunks make good building timber, the leaves are used for thatching; the sap of the tree has a high sugar content, and can be evaporated to make crude sugar or fermented to make a drink (toddy), distilled to make the spirit arrack.

coefficient in mathematics, the non-varying factor of a variable product, i.e. the number or quantity usually placed before and multiplying another quantity (e.g. the 2 in $2x$).

coenosis a random assemblage of organisms held together by common ecological needs, as distinct from a COMMUNITY-2.

coffee a small tropical tree or shrub, grown for the sake of the 'berries' which contain aromatic beans (seeds), the beans being roasted, ground and brewed in hot water to produce a stimulating, non-alcoholic drink. It needs a rich, well-drained, slightly acid soil; moderate rainfall and equable heat; and protection from the direct rays of the sun; it can withstand slight frost.

cognition 1. the act or the faculty of knowing, a collective term covering all the psychological processes involved in the acquisition, organization and use of knowledge, including PERCEPTION, judgement, reasoning, remembering, thinking and imagining (COGNITIVE MAP) **2.** the product of the act of knowing.

cognitive *adj.* of, or pertaining to, COGNITION, to those aspects of mental life connected with the gaining of knowledge or the forming of beliefs.

cognitive consonance a condition of harmony in a COGNITIVE SYSTEM, when there is consistency and accord among the items of knowledge, ideas and beliefs in the system. COGNITIVE DISSONANCE.

cognitive dissonance 1. a condition of disharmony in a COGNITIVE SYSTEM, when there are contradictions and a lack of consistency among the items of knowledge, ideas and beliefs in the system **2.** perceived incongruity between the behaviour and attitudes of an individual.

cognitive map a mental map, an image of a place, of an environment, an organized representation of reality developed in the brain of an individual as a result of information's being received, mentally coded, stored, recalled, decoded and interpreted (COGNITION) and, in some cases, combined with sentiment, feelings, associated with the place or environment.

cognitive system the collection of interrelated items of knowledge, ideas and beliefs which an individual holds about other individuals, groups, events, objects, concrete or abstract subjects, etc. Each individual formulates a number of such systems, and these too are interrelated, the extent of the interrelationship varying widely. COGNITIVE CONSONANCE, COGNITIVE DISSONANCE.

cohort in demography, a group of individuals who experience a significant event during the same period of time, who thus have a common statistical characteristic, e.g. belonging to the same age group, entering hospital at the same time, etc. FERTILITY-3.

cohort analysis in demography, LONGITUDINAL ANALYSIS, the analysis concerned with the study of a COHORT over a long period of time, e.g. people born or married in a particular year who are studied at selected stages throughout their lives. It is used particularly in the study of FERTILITY-3.

coke the hard, porous, combustible residue, almost pure CARBON, produced when COAL is heated in a closed retort or oven so that COAL GAS and other volatile material is driven off.

col 1. a marked depression on a mountain ridge or range, commonly occurring where opposed CIRQUES meet, thus affording a PASS **2.** in meteorology, by analogy (higher pressure representing the ridge, lower pressure the valley), a region of relatively low pressure between two adjacent ANTICYCLONES or between two adjacent DEPRESSIONS-3.

colatitude the complement of the LATITUDE, i.e. the difference between 90° and the latitude.

cold desert a general term for areas with such low temperatures that plant and animal life are inhibited, e.g. POLAR-1 region, TUNDRA.

cold front the boundary zone between an advancing mass of cold air and a mass of warm air. The cold heavy air usually acts like a wedge, undercutting and forcing the lighter, warmer air upward, resulting in a drop in temperature, the formation of CLOUDS (especially CUMULONIMBUS and FRACTO-), rain (sometimes falling in heavy showers, sometimes with THUNDERSTORMS) and winds. ANAFRONT, FRONT, KATAFRONT, LINE SQUALL, OCCLUSION.

cold glacier polar glacier, a moving ice-mass, very rarely with surface melting, maintaining a constant temperature at −20°C (−4°F) or lower. WARM GLACIER.

cold occlusion an OCCLUSION in which the overtaking cold air is colder than the cold air ahead of it.

cold-water desert the continental west coast desert strip (e.g. of northern Chile, northwest and southwest Africa, or northwest Australia) where the climate is influenced by cold sea currents flowing towards the equator. The cool air flowing over the sea to the land reduces summer temperatures and produces fogs and heavy dew. FOG DRIP.

collective consumption 1. the services which can be consumed only collectively and are thus provided by the state, e.g. defence services **2.** the main services provided by the state, e.g. public transport, welfare **3.** in neo-Marxism, the collective ways by which the state works to create a labour force and sustain it, i.e. to provide goods and services for most of the population.

collective farming agricultural organization in which the farms are subjected to collectivization, i.e. they are acquired and amalgamated, e.g. by a village (as in the KIBBUTZ of Israel), or by the state (as in the KOLKHOZ of the former USSR), which assumes ownership of the land but leases it permanently to a large group of shareholder farm workers who cooperate in running the holding as a single unit. The farm workers usually have shares in the produce or in the revenue from sales, generally in proportion to the work done by the individual worker. In many cases the farm workers are allowed a small plot of land for their own, private use and personal benefit. LAND TENURE, SOVKHOZ, STATE FARMING.

collision zone in PLATE TECTONICS, a

zone where converging plates carrying continental crust meet, with the result that the edge of one plate dives under the other but the rocks of the continental crust pile up, crushed and buckled and mixed with material swept up from the floor of any ocean which may formerly have separated the plates, the ocean being squeezed out of existence. Such a collision produces chains of FOLD MOUNTAINS, e.g. the Alpine-Himalayan chain. OCEANIC TRENCH.

colloid a substance (gas, liquid or solid), finely divided and dispersed in a continuous gas, liquid or solid medium, the particles consisting of very large MOLECULES or aggregation of molecules which do not settle at all, or only very slowly. Thus the system is neither a SOLUTION (in which the dispersed particles are single molecules) nor a SUSPENSION (in which the particles are large enough to tend to fall by GRAVITATION and concentrate as a SEDIMENT-1). The electrical forces in the system are important in soil formation; colloids may loosen or dislodge rock particles from surfaces with which they are in contact, attracting IONS of dissolved substances, especially those that are basic (BASE-2); or some constituent particles of the soil may stick to each other, as in colloidal CLAY, COLLOIDAL PLUCKING, FLOCCULATION.

colloidal or colloid plucking a weathering process in which soil COLLOIDS loosen or pull off small fragments of rock from the surfaces with which they come into contact.

colluvial soil a soil formed from COLLUVIUM.

colluvium (American slope-wash) a collection of rock debris of varied origin which has accumulated at the base of a slope as a result of the movement of SCREES-2 and mud flows down the slope under gravity. MASS MOVEMENT.

colonial *adj.* **1.** of, or pertaining to, a COLONY **2.** in USA, of or belonging to the thirteen British colonies which became the United States, or to the period of time (seventeenth and eighteenth centuries) when they were still colonies, applied especially to the works of art, artifacts, furniture and architecture of that period.

colonial animal an animal which is a member of an association of incompletely separated individuals, e.g. CORAL.

colonialism **1.** the principle or practice of having or keeping colonies (COLONY-1,2) **2.** the economic, political and social policies by which colonies are governed by the sovereign METROPOLITAN country (the colonial power), usually based on the maintenance of a marked distinction between the governing country and the subordinate (colonial) population **3.** in a derogatory sense, an alleged policy of exploitation of weak peoples by a large, strong power, which has the effect of perpetuating the economic differences between the colonies and the governing power **4.** the belief that a colonial system (for policies see **2.**) benefits and promotes the welfare of the state colonized. NEO-COLONIALISM.

colonization **1.** the act or policy of bringing human settlers into a locality, or to what is to them a foreign country, of establishing a COLONY-1, of forming that territory into a COLONY-2. CONTACT-ZONE. DECOLONIZATION **2.** the spread of a group of animals, or of a plant species, into an area. CLOSED COMMUNITY.

colony **1.** a human settlement formed in a territory by people from other territory, usually from another country (to the government of which it, the colony, becomes in some degree subject) **2.** the territory so occupied **3.** in ecology, loosely applied to any collection of animals or of plants living together in one place (e.g. a SOCIETY-2), an isolated group, a group of individuals of a plant species migrant in a new habitat, a group of COLONIAL ANIMALS, or a culture of MICROORGANISMS.

columnar structure a geological structure comprising hexagonal columns formed in the cooling of IGNEOUS ROCKS, especially of BASALT. The contraction occurring in the cooling results in a series of regular JOINTS at right angles to the surfaces of cooling, thereby producing the columns, e.g. as in the Giant's Causeway, Northern Ireland. Similar hexagonal cracks develop when mud dries.

combe, coombe, coomb, coom 1. in southern England, a deep hollow or valley, especially if short and steep at the head, or closed in, common in CHALK country **2.** in southwestern England, a short steep valley opening to the sea **3.** a CIRQUE in the English Lake District.

COMECON CMEA.

comet a celestial body consisting of a gaseous cloud enveloping a bright nucleus, moving around the SUN in an elliptical or parabolic ORBIT so eccentrically that some comets escape from the SOLAR SYSTEM. On nearing the sun the pressure of the sun's RADIATION forces the gas of a comet into a tail, pointing away from the sun.

comfort zone the range of TEMPERATURE-2 and RELATIVE HUMIDITY in a climate within which human beings feel comfortable. Common standards are 20° to 21°C (68° to 72°F) and 55 to 60 per cent relative humidity, the latter preferably falling as temperature rises. SENSIBLE TEMPERATURE.

commensalism 1. in ecology, the close association between organisms of different species from which one benefits but the other is unharmed. COENOSIS, MUTUALISM, PARASITISM, SYMBIOSIS **2.** in urban geography, the association between an individual and a group of similar individuals operating in close proximity, the individual cooperating with the other members of the group and benefiting from the advantages derived from group activities and group membership while competing with the other members, e.g. specialized commercial or professional enterprises (e.g. the clothing industry, lawyers) in a particular district of a town or city.

commercial agriculture the growing of agricultural produce for sale. SUBSISTENCE AGRICULTURE.

commercial crops INDUSTRIAL CROPS.

commodity 1. in general, a GOOD which results from a production process (i.e. it is the product of labour), meets human needs, and has an exchange value, sometimes used as a synonym for ECONOMIC GOOD, i.e. a GOOD which has a price **2.** in the business community, raw materials, as in commodity exchange.

common, common land in England and Wales, land, usually unenclosed, over which certain persons or groups of people have various common rights, though they do not own the land. In general common land represents the poorer quality land which, when INCLOSURE of lands took place, was left unenclosed and provided grazing for the villagers or peasants who would otherwise have been left landless. Under later legislation, the general public has been given rights of access on certain commons, including those within boroughs or urban areas, but unless there is specific legislation the public has no rights on rural commons.

Common Agricultural Policy CAP.

common field, common arable in England and parts of western Europe until inclosure in the fifteenth to eighteenth centuries, one of the large, open arable fields worked by the village community. FIELD SYSTEM.

Common Market EEC.

common rights, rights of common the rights held by certain persons on COMMON LAND which is the property of another. The chief rights are to pasture certain animals, usually limited in number (common of pasture); of digging peat for fuel (common of turbary); of gathering, sometimes cutting, wood for fuel or house repairs (common of estovers); of fishing (common of piscary); of digging for sand and stone (common of soil and stone). Common rights may belong to an individual such as the tenant of a farm (right of common appendant), or may be attached to the dwelling itself (common appurtenant) or by a grant to an individual descending to that individual's heirs (common in gross).

commonwealth, the Commonwealth **1.** a free association of self-governing, individual territories organized in a federation, the federation government taking responsibility for certain common matters, such as defence, e.g. Australia **2.** the Commonwealth, a free association of Britain and certain independent SOVEREIGN STATES, each of which was formerly a DEPENDENCY or COLONY-2 within the British Empire.

Commonwealth of Independent States CIS.

communism **1.** historically, the common ownership of all property in a society, e.g. as in a non-literate society, extended today to some monastic establishments **2.** since 1848, a theory or practice linked especially to the ideas developed by Karl Marx from his interpretation of history. He advocated a classless society, organized on the basis of common ownership of property and the means of production, distribution and supply, the individual members contributing 'each according to his ability' and receiving 'each according to his need'.

community **1.** a group of people living in a particular area; or living near one another, with distinct social relationships; or a group of people sharing a common faith, culture, profession, life-style **2.** in ecology, a general term applied to any naturally occurring group of different organisms occupying a common environment, interacting with each other, particularly through food relationships, but relatively independent of other groups. The size may vary and the larger communities may contain smaller ones. SOCIETY.

commuter one who travels regularly, usually daily (but also at other regular intervals, e.g. weekly or monthly) from residence to place of work. Historically the greatest number of commuting trips were inward to central employment areas; but outward (or reverse) commuting is now common where the speed of DECENTRALIZATION has outpaced the shift of population to new locations.

commuter village dormitory village, a village in a rural area, formerly inhabited by people who worked in, or who had worked in, the village or close to it, now inhabited mainly by people who travel regularly to work in a nearby town. COMMUTER, DORMITORY TOWN.

commuter zone the area in which commuting takes place, from which COMMUTERS are drawn to work in a nearby town.

compaction **1.** in geology, the process in which fine rock particles, e.g. of silt or clay, are combined tightly together by pressure of earth movements or weight of later overlying deposits **2.** of soils, the pressing together of soil particles (e.g. by torrential rain, or by heavy mechanical equipment especially in wet conditions) so that the voids between them are reduced, with consequent loss of air, to the detriment of soil fertility.

comparative cost analysis an evaluation of the advantages or disadvantages of alternative locations, based on the cost of production at those locations. COST BENEFIT ANALYSIS, VARIABLE COST ANALYSIS, VARIABLE REVENUE ANALYSIS.

compass an instrument used to find direction. In a magnetic compass a free-swinging magnetized needle is fixed to, and swings freely over, a dial which is graduated in degrees and shows the cardinal points (north, east, south and west). Under the influence of the local line of magnetic force, the needle indicates the NORTH and SOUTH MAGNETIC POLES. A non-magnetic compass (GYROCOMPASS) points TRUE NORTH. BEARING-3, PRISMATIC COMPASS.

competence of a stream the ability of a stream to transport debris, measured in terms of the size (not the weight) of the largest pebble or boulder it can move. CAPACITY OF A STREAM.

competence of rocks, competent bed the relative strength of a bed of stratum when subjected to folding. If strong enough to bend without distortion when subjected to the stress of folding, it is said to be competent; if weak and thus liable to distortion, it is incompetent.

complementary region a trade area, a tributary area, the area served by a CENTRAL PLACE, that of a higher order centre being large and in many cases overlapping the smaller area served by a lower order centre. CENTRAL PLACE HIERARCHY, ISOTROPIC SURFACE.

components of change approach an approach to the study of the changing pattern of employment (usually in manufacturing) in a region or urban area. The changes that have occurred during a defined time period in the study area are broken down into four components, i.e. changes caused by (A) birth (the formation of new firms); by (B) death (the closing down of existing firms); by migration, i.e. movement of some firms into (C), and others out of (D), the area; or by (E) in situ change (the growth or decline of employment in firms existing in the area at the start of the period of study). The net change in employment in the area during the defined time period can then be calculated: A (birth) minus B (death) plus E (net change in employment in firms surviving through the period) plus C (immigrant firms) minus D (emigrant firms).

composite volcanic cone a VOLCANIC CONE composed of layers of ash, cinder and lava built up over a long period of time by a series of ERUPTIONS through the main PIPE which is topped by a CRATER-1, e.g. Vesuvius. VOLCANO.

compost a soil conditioner and FERTILIZER produced by the planned decomposition of organic material, such as vegetable remains.

compression forcing into smaller compass, reducing in volume, condensation by pressure, pressing together. The effect of compression on the rocks of the earth's surface contributes to FAULTING and FOLDING. CHATTER MARK, TENSION.

compressional wave PUSH WAVE.

concentric zone growth theory a theory introduced by E. W. Burgess in 1927 based on his studies of urban growth in the Chicago area. He saw Chicago as a city in an industrialized country, expanding radically from its centre in a series of concentric zones. He suggested that the expansion and the formation of these concentric zones were created by SUCCESSION AND INVASION, as the occupiers of each inner zone, seeking what they saw as more agreeable locations, moved outwards to colonize the next outer zone. From the centre outwards he identified the concentric zones as (1) the inner CENTRAL BUSINESS DISTRICT, (2) a transition zone (INNER CITY) with residential areas invaded by business and industry from the CORE-2, the run-down dwellings being subdivided and overcrowded and inhabited by poor immigrants, especially ETHNIC MINORITIES, (3) a low-income residential zone with second generation immigrant dwellings, (4) a middle-income residential zone with one-family dwellings, and (5) an outer commuting zone with higher income dwellings in suburban areas and satellite towns. He acknowledged that the general, simplified pattern would be modified if applied to other cities (e.g. by terrain, routes and other constraints); but he suggested that radial expansion along a broad front, stimulated by invasion and succession, was a dominant process in the shaping of the pattern of a city. INDUSTRIAL CITY. Fig 12.

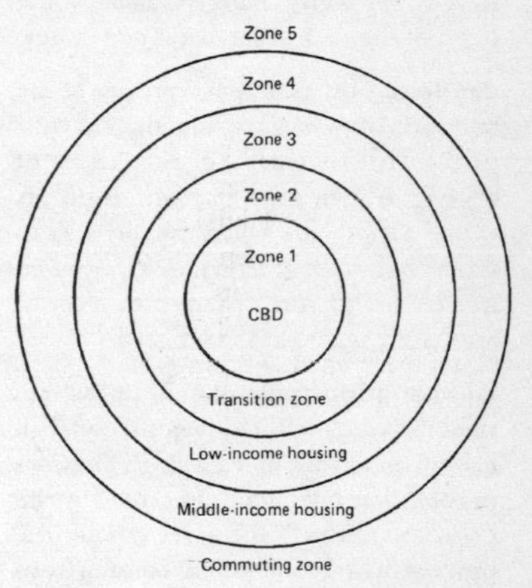

Fig 12 Burgess's concentric zone theory

concept a general notion, idea, or understanding, especially one constructed by generalization from particular examples.

concordance of summit-levels ACCORDANCE OF SUMMIT LEVELS.

concordant *adj.* agreeing or consistent with, thus lying or running parallel to the structural trend lines of the relief, of the general strata. DISCORDANT.

concordant coast a coast lying approximately parallel to the structural trend lines of the land, also termed a LONGITUDINAL OR PACIFIC COAST. If it is drowned by the sea a line of islands (the peaks of a former mountain range) may be formed, separated from the mainland by the drowned parallel valleys (SOUND), e.g. the eastern Adriatic coast.

concordant drainage the pattern of DRAINAGE-2 which arises from and closely follows the trends of the underlying STRATA.

concordant intrusion an INTRUSION of IGNEOUS material lying parallel to the trend of the rock strata that it penetrates.

condensation the physical process of the transition of a substance from the VAPOUR to the LIQUID state, e.g. as a result of cooling or increase of pressure. It occurs in the atmosphere when the air is SATURATED (by evaporation into it), or when it is cooled. CLOUD, DEW, DEW-POINT, FOG, MIST, RAINFALL, SNOW.

condensation trail contrail, the white, ribbon-like, cloud-like phenomenon seen behind an aircraft flying at high altitudes in cold, clear but humid air, caused by the CONDENSATION of the water vapour (the product of fuel combustion) coming from the engine's exhaust, and by the lowered pressure behind the wing-tips.

conditional instability the state of an air mass with an ENVIRONMENTAL LAPSE RATE greater than the SATURATED ADIABATIC LAPSE RATE but less than the DRY ADIABATIC LAPSE RATE, the instability depending on the amount of water vapour held in the air mass. A pocket of unsaturated air, if pushed upward from ground level, will gradually become cooler than the surrounding air and descend; but a pocket of saturated air will remain relatively warmer, and thus will continue to ascend. ABSOLUTE INSTABILITY, POTENTIAL INSTABILITY.

conductivity the ability of a substance to transmit or be capable of transmitting, heat or electricity. THERMAL CONDUCTION.

cone **1.** a volcanic peak, with a roughly circular base tapering to a point. ADVENTIVE CONE, ALLUVIAL CONE, ASH CONE, CINDER CONE, DOME VOLCANO, PARASITIC CONE, SPATTER CONE **2.** the fruit of some trees, e.g. pine trees. CONIFEROUS.

confluence **1.** the place where two streams, about equal in size, converge and unite **2.** the place at which a stream flows into another **3.** the body of water so produced **4.** by analogy, a junction of routeways.

conformable *adj.* applied in geology to STRATA deposited one on another in parallel planes in proper geological sequence, without breaks or interruption caused by denudation or earth movements, in contrast to UNCONFORMABLE. UNCONFORMITY.

congeliflucton, congelifluxion a flow of earth under PERIGLACIAL conditions. SOLIFLUCTION.

congelifraction the splitting of rocks etc. by frost action.

congeliturbation frost action, including FROST-HEAVING, churning of the ground and SOLIFLUCTION, affecting both the soil and subsoil, in some cases producing PATTERNED GROUND.

congestion the state of being packed closely together, clogged by overcrowding, e.g. the result of the use of some facility, such as a road network, in excess of its CAPACITY-2.

conglomerate a SEDIMENTARY ROCK, e.g. PUDDINGSTONE, consisting of round, waterworn pebbles in a matrix of natural cementing material such as CALCIUM CARBONATE, SILICA or IRON oxide. AGGLOMERATE, BRECCIA.

coniferous *adj.* applied to trees belonging to Coniferales, a large order of trees and shrubs, commonly (but not always) EVER-

GREEN, with slender leaves, and reproducing by means of seeds contained in a cone (a reproductive structure consisting of woody carpels closely grouped around a central axis, bearing pollen or ovules).

coniferous forest forest characterized by coniferous trees which occurs under many conditions of soil, climate, aspect, elevation (HYGROPHYTE, MESOPHYTE, XEROPHYTE). The largest continuous area is the northern coniferous forest (TAIGA), also known as the BOREAL forest. Away from very low temperatures, coniferous trees generally grow rapidly, so are much favoured there for afforestation and as a commercial crop, producing SOFTWOODS of varying quality for widely ranging purposes (wood-pulp for paper-making, timber for house construction, etc. as well as resin, turpentine, etc.). Most of the timber consumed commercially is that of coniferous trees. FOREST.

conjunction the position of two planets or other heavenly bodies which, when viewed from the earth, lie in line in the same direction (SYZYGY). The 'new moon' is the result of the earth, moon and sun being in conjunction, when forces bringing about tidal action are at maximum strength, giving rise to SPRING TIDES. MOON, OPPOSITION, QUADRATURE.

connate water fossil water, water trapped in SEDIMENTARY ROCK at the time of its deposition.

connectivity NETWORK CONNECTIVITY.

consequent drainage, consequent river or stream, consequent valley a natural water flow or valley directly related to the original slope of the land surface; also a river or stream flowing in the same direction as the DIP of the underlying rocks. A secondary consequent stream is a tributary of a SUBSEQUENT STREAM; it flows parallel to the main consequent stream. DRAINAGE, OBSEQUENT. Fig 42.

conservation protection from destructive influences. A term applied in general to the positive work of maintenance, enhancement and wise management, or reducing the rate of consumption to avoid irrevocable depletion, in order to benefit posterity, as in the conservation of nature, or of natural resources, or of buildings or works of art of special merit, etc. Specifically it is applied to the work of protecting and maintaining the soil (soil conservation) and wild life (nature conservation). PRESERVATION, RESOURCE CONSERVATION.

conservative plate margin PLATE TECTONICS, TRANSFORM FAULT.

constant capital in Marxism, labour from the past (dead or expended labour) incorporated in the MEANS OF PRODUCTION, qualified as constant because the value is fixed (only living labour being able to create new value). SOCIAL INVESTMENT, VARIABLE CAPITAL.

constant slope part of a SLOPE profile, the straight slope of the lower hillside, below the FREE FACE and above the WANING SLOPE.

constructive plate margin OCEANIC RIDGE, PLATE TECTONICS.

constructive wave one of a series of waves rolling regularly and gently on a coast, the SWASH of the wave being more powerful than the BACKWASH, with the result that shingle, etc. is pushed up the beach to form ridges.

consumer durables those CONSUMER GOODS which can be used many times over a period of time and are thus used up gradually (e.g. cars, clothes, furniture). Consumer goods which can be stored but used only once (e.g. canned or frozen food) are not classified as durable.

consumer goods, consumers' goods goods and services which directly satisfy the needs and desires of the individual person, e.g. food items. CAPITAL GOODS, CONSUMER DURABLES, ECONOMIC GOODS, FREE GOODS, PRODUCER GOODS.

contact field the distribution pattern of an individual's or a group's contacts with others, especially in activity space. AWARENESS SPACE.

contact metamorphism THERMAL METAMORPHISM; and IGNEOUS, MAGMA, METAMORPHIC AUREOLE, METAMORPHISM.

contact zone in COLONIZATION-1 times, the space within which communities previously geographically and historically separate meet and interrelate.

container, container transport a special large, strong, durable packing case of standard size, usually with internal volume exceeding 1 cu m (35.3 cu ft), made of steel, aluminium-alloy etc.; suitable for repeated use and for mechanical handling, into which goods are easily packed and locked for transport by road, sea, air, thus designed for easy switching from one form of transport to another. For example containers at a port may be moved from quay to ship by lift-on/lift-off (Lo-Lo) equipment; or they may be directly transferred by the practice of roll-on/roll-off (Ro-Ro), a wheeled vehicle, complete with containers, being driven along a ramp or link-span pontoon bridge joining the quay with the ship. PIGGYBACK TRANSPORT.

contiguity the state of being **1.** in contact, touching, neighbouring **2.** next in order or in time. TOPOLOGY.

contiguous zone TERRITORIAL WATERS. Fig 45.

continent one of the large continuous masses of land on the earth's surface. Seven are usually distinguished: North and South America, Europe, Asia, Africa, Australia, and Antarctica. Australia and New Zealand together are often referred to as 'Australasia'. 'Oceania' covers the Pacific islands. 'The Continent' in British writings denotes the mainland of Europe (CONTINENTAL). Some authors insist that Eurasia is one continent. The traditional boundary between Europe and Asia along the Ural Mountains and the Ural River does not coincide with any existing administrative boundary within the CIS.

continental *adj.* of, relating to, or characteristic of, a very large land mass, i.e. of a CONTINENT.

continental air mass an AIR MASS, usually of low humidity, the source of which is a HIGH pressure region over the interior of a continent. Distinguished by the symbol *c* it may be of high latitude, i.e. polar (*Pc*) or low latitude, i.e. tropical (*Tc*).

continental climate the climate associated with the interior of a CONTINENT or other places protected from or unaffected by the moderating influence of the sea, and so characterized by great extremes of temperature between summer and winter, low, variable, precipitation, occurring mainly in early summer, and low humidity. The effect of CONTINENTALITY is most marked in midlatitudes, but it is also significant in high and low latitudes. MARITIME CLIMATE, OCEANIC CLIMATE, TROPICAL AIR MASS, TROPICAL CLIMATE.

continental crust PLATE TECTONICS.

continental divide the main waterparting in a continent, e.g. in North America, where the streams flow on one side of the divide to the Pacific and on the other to the Atlantic.

continental drift the theory or hypothesis first postulated in 1858, re-stated by A. Wegener in 1911, that the present distribution of the continental masses is the result of fragmentation of one or more pre-existing masses which have drifted apart, the intervening hollows having become occupied by the oceans. PLATE TECTONICS.

continental island an island near to and closely associated structurally with the neighbouring land mass. OCEANIC ISLAND.

continentality the condition of being CONTINENTAL (*adj.*) as opposed to OCEANIC, applied especially to the measure of the extent to which the climate of a place is influenced by its distance from the sea. CONTINENTAL CLIMATE.

continental margin a zone comprising the CONTINENTAL SHELF and the CONTINENTAL SLOPE, extending from the coastline to depths of approximately 2000 m (1095 fathoms: 6560 ft), but distinct from the deep sea floor. TERRITORIAL WATERS.

continental platform a continent and its CONTINENTAL SHELF to the edge of the CONTINENTAL SLOPE, i.e. the part of the earth's crust carrying SIAL, as distinct from the oceanic parts (SIMA).

continental shelf a gently sloping submarine plain, usually of 1° slope or less, and of variable width (scarcely present along some CONCORDANT COASTS where FOLD MOUNTAINS lie close to the ocean), forming a border to nearly every CONTINENT, stretching from the coast to the CONTINENTAL SLOPE, i.e. to the point where the seaward slope inclines markedly to the ocean floor. The depth of this outer edge has been defined as lying approximately between 120 m (65 fathoms: 395 ft) and 370 m (200 fathoms: 1215 ft); but the continental shelf itself is, in general, considered to be covered by seawater usually less than 183 m (100 fathoms: 600 ft) deep. The precise definition and delimitation of the continental shelf is important in international law in connexion with the ownership of minerals and other resources lying on or under the shelf. CONTINENTAL MARGIN, TERRITORIAL WATERS.

continental slope the marked slope, commonly with an angle between 2° and 5°, lying between the edge of the CONTINENTAL SHELF and the deep ocean floor (ABYSSAL PLAIN). BATHYAL, CONTINENTAL MARGIN. Fig 45.

continental terrace the marine-built terrace consisting of material removed in the cutting of the marine-cut terrace which lies to landward. The ABRASION PLATFORM and the marine-built terrace together constitute the CONTINENTAL SHELF.

contour an imaginary line joining all the points on the ground that are at the same height above sea-level or, for submarine contours, below sea-level, usually based on an instrumental survey. FORM-LINE, RELIEF MAP.

contour interval the difference in height represented by adjacent CONTOURS shown on any given map. If the contours shown are at 50, 100, 150, 200 m or ft above sea-level, the contour interval would be 50 m or ft; if at 100, 200, 300 m or ft, the contour interval would be 100 m or ft. The contour interval is often varied on a map.

contour ploughing the farming practice of cutting furrows across a hillslope, following the CONTOURS rather than ploughing up and down the slope, the object being to reduce SOIL EROSION caused by the run-off of rainwater.

controlled environment the state within a building (or sometimes within a group of totally enclosed buildings) where the air temperature, the humidity, the rate of movement of and the particle content in the air are completely controlled, and lighting is artificial.

conurbation a continuously urban area formed by the expansion and consequent coalescence of previously separate urban areas. In some cases it may include enclaves of rural land in agricultural use. ECUMENOPOLIS, MEGALOPOLIS, RANDSTAD.

convection the process of heat transfer from place to place within a FLUID (i.e. a gas or liquid) caused by the circulatory movement of the fluid itself (due to differences in temperature and hence in density) and the pull of GRAVITY. Convection produces vertical movement (in contrast to ADVECTION, horizontal movement), as in the upward welling of cold water in the oceans or in CONVECTION RAIN. RADIATION, THERMAL CONDUCTION, UNSTABLE AIR MASS.

convection current a stream of FLUID, e.g. in the atmosphere or ocean, produced by CONVECTION.

convection rain precipitation caused by the warming of air (moist with water taken up from the ground and its vegetation) by THERMAL CONDUCTION from the heated land surface. The warmed, moist air expands, rises and cools (ADIABATIC) to DEW-POINT, to form CUMULUS and CUMULONIMBUS CLOUDS, which drop very heavy, torrential rain, often accompanied by thunder.

convenience good a good of low order (ORDER OF GOODS) which needs to be bought frequently but for which people are not willing to travel a great distance, and for which the THRESHOLD POPULATION is fairly low, e.g. bread. CENTRAL PLACE THEORY, NEIGHBOURHOOD UNIT.

conventional name an exonym, the name given by the people of one linguistic group or nation to important towns, localities, physical features etc. of another. Thus in English the exonyms Warsaw, Vienna, Rome are used for cities known to the inhabitants of the countries concerned as Warszawa, Wien, Roma. There is now international agreement that local names should be used.

conventional sign a standard SYMBOL used on a map, and explained in the LEGEND, to convey a definite meaning, e.g. dots of different size may represent towns of varying size.

convergence in general, the act of coming together to meet in a common result or at a point of operation (DIVERGENCE). There are many specialized applications: **1.** in climatology, with reference to air flow, ITCZ **2.** in geology, the thinning of rock formations so that the upper and lower horizons draw closer together **3.** in oceanography, the movement of waters of different salinity and temperature **4.** in PLATE TECTONICS, the movement of plates. COLLISION ZONE, OCEANIC TRENCH.

cooperative farm a farming enterprise in which individual farmers combine in a group for mutual benefit. Usually the group pays for the bulk purchase of machinery, seeds, fertilizers etc. and for services, and markets the produce, the costs and profits being shared by the members.

coordinate, co-ordinate in mathematics, each of a system of two or more magnitudes specifying the position of a point on a line or a surface or in space, e.g. LATITUDE and LONGITUDE are the coordinates of a point on the earth's surface.

copper a red, ductile, metallic ELEMENT-6, unaffected by water or steam, a TRACE ELEMENT, found native in nature, but mainly obtained from a variety of ores, copper pyrites and the carbonates (malachite and azurite) in particular. The ores occur especially in veins or scattered through METAMORPHIC ROCKS. It was the first metal to be used for ornaments and vessels, later alloyed with tin to form BRONZE. A good conductor of electricity and heat, it is now widely used in alloys, some of its salts being used as fungicides.

coppice a small WOOD-2 in which the trees are coppiced, i.e. cut off close to the ground and so encouraged to send up several shoots, each to serve in due course for fencing posts or poles. Trees used in this way include the oak, hazel, sweet chestnut.

copse a small wood consisting of trees and undergrowth. COPPICE.

coral a hard, calcareous, rock-like substance formed either by the continuous skeleton or fused skeletons of members of a group of sedentary marine animals that live in COLONIES-3 only in clear, warm, shallow seas, or by the skeleton or fused skeletons together with the animals (polyps) that secrete it/them. The colour and form of the coral depend on the habits of the species that builds it. COLONIAL ANIMAL, ZOOPHYTE.

coral reef an extensive REEF formed by CORAL, CORAL SANDS and other coral derivatives, commonly known as a BARRIER REEF if separated from the shore by a LAGOON, or a fringing reef if bordering the land. ATOLL.

coral sand sand composed of comminuted fragments of coral.

cordillera a mountain chain, a term applied particularly to the parallel ranges of the Andes in South America and their continuation through central America and Mexico, and to the ranges of the great western mountain system of North America, including the Rockies, Sierra Nevada, Cascades, Coast Ranges etc.

core the innermost part of anything: **1.** of the earth, the central part of the earth's interior (radius approximately 3476 km: 2160 mi) consisting of the outer layer of dense material (probably nickel-iron, with a density of about 12.0) which behaves as a liquid, and within which the EARTH'S MAGNETIC FIELD is generated by circulation, and containing the solid inner core (radius approximately 1380 to 1450 km: 860 to 900 mi, density about 17.0) at the centre. Between the outer core and the MANTLE that envelops it is the GUTENBERG DISCONTINUITY **2.** of a city or town, the functionally specialized centre, DOWNTOWN (CENTRAL BUSINESS DISTRICT), usually lacking residences. CORE AREA, CORE-FRAME CONCEPT, CORE-PERIPHERY MODEL.

core area of a CULTURE-1, REGION-1, state, or city, a term loosely applied to the central area, the heart, the nucleus or the place of birth and nurture of a culture, region, nation, or city, from which the culture/region/state/city expanded and spread. CULTURAL HEARTH, DOMAIN-2, OUTLIER-2, SPHERE-3,5.

core-frame concept in the structure of the central area of a city, the notion that within the central area there is an inner CORE-2 (with high land values, tall buildings, a concentrated daytime population, where strong functional links between various offices, and between offices and shops, create distinct clusters of functions which form particular small land use zones) and a 'frame' (the less intensively developed area surrounding the core, with relatively lower land values and widely distributed functions such as wholesaling, off-street parking, light manufacturing, etc. which have only their location in common).

core-periphery model a MODEL-2 or hypothesis concerned with the spatial structure of the relationship between a centre (a CORE-2, CORE AREA) and its peripheral areas in an economic system. The wealthy core, with its high potential for growth, productivity, technical progress and innovation, is seen as economically and socially dynamic. Its concentration of advantages encourages a steady influx of CAPITAL-2 and talent, some from the periphery, to the disadvantage of the peripheral areas, which are accordingly economically and socially weakened (CUMULATIVE UPWARD CAUSATION). The institutions in the core, becoming ever more powerful, come to dominate the peripheral areas, making decisions that not only determine their growth and development but favour the core itself, an imbalance that may be further reinforced by governmental, economic and political policies that support the core at the expense of the peripheral areas. CENTRIPETAL AND CENTRIFUGAL FORCES, CIRCULAR AND CUMULATIVE GROWTH.

Coriolis force (Corioli's is incorrect spelling) the effect of the force produced by the earth's rotation on a body moving on its surface. The body is deflected to the right of the path of movement in the northern hemisphere, to the left in the southern. BUYS BALLOT'S LAW, FERREL'S LAW.

cork the impermeable layer of dead cells forming the outermost protective layer of stems or roots, applied particularly to the thick bark of the cork oak, a small tree growing in Mediterranean lands. As soon as the tree is large enough the bark (cork) can be cut off the trunk and main branches. It soon begins to form again, so that after ten or twelve years it can again be cut. In nature the purpose of the bark is to prevent loss of moisture in the hot, dry Mediterranean summer.

corn in North America applied to Indian corn or MAIZE; but in Britain (and continental Europe) applied either to the CEREALS wheat, rye, oats, barley etc. collectively, or to the most important cereal of an area, e.g. barley in some districts, wheat in others.

Corn Belt the region in the USA, south and southwest of the Great Lakes, where corn (maize) is or was the dominant cultivated plant.

corrasion the process of mechanical EROSION of a rock surface by the friction of rock material with the surface, the rock material being moved under gravity, or carried by running water (streams, rivers and waves), by ice (glaciers), or by wind

(wind-blown sand). ABRASION, TRANSPORTATION-2.

correlation **1.** in general, the condition of two or more things mutually or reciprocally related, or the act of bringing two or more things into mutual or reciprocal relationship **2.** in statistics, the degree of relationship, the extent to which two measurable VARIABLES vary together, linear correlation being commonly measured by a correlation COEFFICIENT, of which there are several types, most varying between +1.0 and −1.0. The term positive correlation indicates that the variables have a tendency to increase or decrease together. Negative or inverse correlation indicates that there is a tendency for one variable to increase as the other decreases, and vice versa. If changes in one variable are negatively or positively proportional to changes in the other, the correlation is said to be linear, and non-linear if the changes are not proportional. A commonly used measurement of correlation which tests the degree to which any two sets of data are correlated is the SPEARMAN'S RANK CORRELATION COEFFICIENT.

corridor a strip of territory belonging to one country but running through the territory of another country in order to give the country owning the strip access to an international waterway, the sea, etc. The term has been extended to air corridor, an airway allowing a country or countries right of access by air over the territory of a foreign country.

corrie (Scottish) CIRQUE.

corrosion **1.** the wearing away of rock or soil by chemical and solvent action, i.e. by CARBONATION, HYDRATION, HYDROLYSIS, OXIDATION, SOLUTION **2.** the changes in the crystals of a rock produced by the solvent action of residual magma.

co-seismal line, co-seismic line homoseismal line, a line connecting points on the earth's surface simultaneously experiencing the arrival of an earthquake wave, not to be confused with an isoseismal line (ISO-) which is related to the amount of earthquake intensity.

cosmic *adj.* of, or pertaining to, the COSMOS (the universe seen as an orderly whole), applied to any phenomenon of space outside the atmosphere of the earth.

cosmic dust very small particles of solid matter present everywhere in the universe.

cosmic rays a continuous shower of very high-frequency subatomic particles that, originating in outer space (at that stage termed primary cosmic rays), collide with atoms and molecules of the earth's upper atmosphere and, as secondary cosmic rays, reach the earth's surface, sometimes penetrating it. The subatomic particles of the primary cosmic rays consist mainly of the nuclei of atoms, certainly those of hydrogen (PROTONS), and possibly of silicon and iron.

cosmography **1.** a description or representation of the general features of the earth and/or the universe (COSMOS) **2.** the science concerned with the structure of the earth and/or the universe, including astronomy, geography and geology **3.** historically, a synonym for physical geography. A term with a long, complex history, it is now rarely used.

cosmology the science or theory concerned with the physical universe viewed as an ordered entity and with the general laws that govern it.

cosmonaut an ASTRONAUT, especially one from the CIS.

cosmopolitan *adj.* belonging to all parts of the world and free from national limitations.

cosmos the world or the universe as an orderly whole, a harmonious system.

costa (Spanish) coast, applied particularly to the Spanish coasts bordering the Mediterranean which have been developed with holiday resorts, e.g. Costa Brava, Costa del Sol etc.

cost benefit analysis CBA, the evaluation in monetary terms of the costs and benefits accruing from a scheme or from alternative schemes, taking into account all the factors involved, i.e. not only the commercial but also the social and environmental factors for which value judgements must be made (e.g. social benefits, loss of AMENITY, traffic hazards, mental or physical strain, etc.).

cost curve a line on a two-dimensional graph which shows the relationship between cost of production and volume of output. SPACE COST CURVE.

cost space, cost distance, time space, time distance in the explanation of human spatial structure, relative space as opposed to absolute space (e.g. people travelling or moving goods between places are more concerned with the cost and the time involved than they are with the actual distance to be covered). Some authors prefer the term cost (or time) distance to cost (or time) space as being more precise, maintaining that the term 'space' encompasses too many dimensions. ACCESSIBILITY, TIME-SPACE CONVERGENCE.

cost surface a three dimensional representation of an area, resembling a CONTOUR map, but constructed to show spatial variations in production costs at varying distances from various locations, the contours joining, for example, points with equal total costs, or the cost of a single item (e.g. land, labour, transport, power, materials, or a particular individual material involved in an operation). A cost surface with its cost contours therefore shows not only the total costs an enterprise at a given location may incur in obtaining land, labour, power, materials etc. and in distributing its products to customers at any and all points throughout an area: it also identifies the LEAST-COST LOCATION. SPACE COST CURVE.

co-tidal line a line drawn on a tidal chart linking points with a common time of high water. Co-tidal lines radiate from the AMPHIDROMIC POINT (OSCILLATORY WAVE THEORY OF TIDES) where the water remains relatively level, the height of the tidal rise growing as the co-tidal lines spread to their extremities. STANDING WAVE.

cottage industry manufacturing wholly or partly carried on in the home of the worker.

cotton a small, shrubby plant native to tropical and subtropical regions, cultivated especially for the sake of its seed hairs, i.e. the fibres (used for textiles) attached to its seeds, the seeds and white fibres being contained by the 'boll' or seed case. According to the species and variety cultivated, the fibres vary in length (termed staple), and in texture (from harsh and coarse to fine and silky). Being a tropical or subtropical crop, cotton needs annually 200 frost-free days and a low rainfall, especially at picking time when rain would damage the ripe, open bolls. The hairs are separated from the seeds by mechanical 'ginning' (once done by hand). The cotton seed is crushed and the edible oil extracted for use in cooking and food preparations, such as margarine; the residue provides meal or oilcake, used as cattle feed or fertilizer. The cotton fibres (lint) are combed out into slivers, twisted to form yarn (spinning), then woven into fabric (weaving). In many manufacturing areas in Britain spinning and weaving were carried out in separate mills, even in separate towns. In clothing manufacture cotton gradually replaced wool in the eighteenth century, expanded greatly in the nineteenth century, declined with the advent of synthetic fibres; but cotton fibres, valued for their absorbent qualities, can satisfactorily be woven with synthetic fibres to make fabrics comfortable to wear, so cotton is still in demand.

Cotton Belt that part of the southeastern USA where cotton was the dominant crop.

Council for Mutual Economic Assistance CMEA, popularly abbreviated to COMECON.

counterfactual a construction of a hypothetical event, process, or state of affairs which does not accord with the facts known about an actual event, process, or state of affairs.

counter-urbanization the movement of people and industry away from major towns and cities. DE-INDUSTRIALIZATION, URBANIZATION.

country park in Britain, a tract of land, in some cases land and water, commonly of about 10 ha (25 acres), i.e. much smaller than a NATIONAL PARK, designated for the recreational use (without payment) by the public. Signposted nature trails are usually a feature, and in most country parks information about the plant and animal life etc. is supplied.

country rock a mass of rock traversed by later INTRUSIONS of IGNEOUS ROCK or penetrated by a mineral VEIN.

covariance in statistics, a measure of the extent to which values on two variables vary together, i.e. high values on one being associated with high values on the other, or low with low. It is not, as is a CORRELATION-2 coefficient, restricted to values between +1.0 and −1.0. The size of the covariance increases or decreases with the increasing or decreasing strength of the relationship. A zero value indicates no LINEAR-5 relationship (LINEAR MODEL), and positive or negative values indicate data showing evidence of corresponding to a positive or negative linear trend. VARIANCE.

cove **1.** a steep-sided, rounded hollow or recess in a rock **2.** a small inlet in a rocky sea coast with a narrow opening and a small curved bay **3.** a recess with precipitous sides on a steep mountainside **4.** a small rounded hollow at the head of a valley.

cover crop a fast-growing crop, planted on cleared land between main crops, to form a blanket of vegetation and protect the soil from erosion.

crag and tail a rock mass showing an abrupt and often precipitous face on one side (the crag, the stoss side or STOSSEND) which faced a pre-existing glacier and a long gentle slope or tail on the other where the hard rock of the 'crag' protected loose material so that it was not swept away by the ice.

crater **1.** a bowl-shaped depression or cavity in the earth's surface, especially that around the orifice of a VOLCANO **2.** a depression made by the impact of a meteorite or artificial explosive, e.g. a bomb **3.** a flaring or bowl-shaped opening of a GEYSER. ADVENTIVE CRATER.

crater lake a lake occupying a crater, usually the crater of an extinct VOLCANO.

craton, kratogen a resistant, very large, stable block of the earth's crust which has remained relatively unaffected by orogenic activity (OROGENESIS) for a very long period of time.

creek diverse uses, but most commonly applied to **1.** a comparatively narrow inlet of fresh or salt water which is tidal **2.** in USA and elsewhere, a branch of a main river bigger than a brook but smaller than a river **3.** in Australia, an intermittent stream.

creep the gradual movement downhill of rock debris and soil, primarily due to gravity helped by the presence of water, alternate freezing and thawing or wetting and drying, the growth of plant roots, the work of burrowing animals. MASS MOVEMENT.

crescentic dune BARCHAN.

crest of a wave, the position in a WAVE where the displacement or disturbance of particles is at its maximum. TROUGH.

Cretaceous *adj.* applied to the geological period when the beds of CHALK of southern England were deposited, reptiles were dominant on land and in the sea, and ANGIOSPERMS became the dominant plants. GEOLOGICAL TIMESCALE.

crevasse 1. a fissure or chasm in the ice of a GLACIER. LONGITUDINAL CREVASSE, TRANSVERSE CREVASSE **2.** in USA, a break in the natural LEVEE or bank of a river such as the Mississippi.

critical geography an approach to the study of the meeting-point of international relations and political geography. GEOPOLITICS.

critical isodapane the isodapane (ISO-) which indicates the point where additional transport costs balance savings in labour costs.

critical path analysis a technique used to discover the most efficient sequence of phases necessary to carry out and complete a complex task speedily. A schedule of phases is drawn up, and this reveals those places where a slight delay will have a knock-on effect, upsetting the whole schedule. These are the phases which mark the critical path, the identification of which is crucial to the smooth, speedy running of the operation.

critical temperature 1. the level of temperature above which a gas cannot be liquefied by pressure alone, the highest temperature at which a liquid and its vapour can co-exist **2.** a temperature of vital importance to a plant, e.g. the temperature below which growth cannot take place, in MIDLATITUDES 6°C (43°F) for most food crops. ACCUMULATED TEMPERATURE.

croft a small agricultural holding worked for subsistence by an hereditary tenant, especially in the Highlands of Scotland.

crofting a system of small hereditary tenant holdings, in which the tenant (the crofter) holds the arable land with a right of grazing, with others, on unenclosed hillsides.

crop 1. the annually or seasonally harvested produce resulting from the cultivation of grain, grass, fruit etc. **2.** cultivated produce while growing. CROP FARMING, ROTATION OF CROPS.

crop *verb.* **1.** to cultivate, sow, plant, reap or bear a crop **2.** by animals, to eat the tops of plants, grasses, to graze **3.** in geology, to crop out, to come to the surface.

crop farming arable farming, the growing of INDUSTRIAL CROPS, or of food crops such as cereals, vegetables etc. on ARABLE LAND (i.e. land which has been tilled), as distinct from LIVESTOCK FARMING or MIXED FARMING.

cross-cutting relationships, law of a law which states that an IGNEOUS ROCK is younger than any rock across which it cuts.

cross profile of a valley a transverse profile of a valley drawn approximately at right angles to the stream flowing through it. RIVER PROFILE.

cross section 1. a transverse SECTION-1, the surface resulting from the cutting of a solid at right angles to its length or to the axis of a cylinder. **2.** a sample that is typical or representative of the whole.

cross-sectional study LONGITUDINAL STUDY.

crude oil PETROLEUM.

crumb in soil science, a naturally formed rounded, porous AGGREGATE-3 of soil particles, up to 10 mm in maximum dimension. CRUMB STRUCTURE, PED, SOIL STRUCTURE.

crumb structure a soil in which the constituent particles are aggregated into CRUMBS which permit the percolation of air and water between them. SOIL STRUCTURE.

crust of the earth the LITHOSPHERE, the outermost shell of the earth, consisting of the surface granitic SIAL and the intermediate basic SIMA layers, separated from the underlying MANTLE by the MOHOROVICIC DISCONTINUITY. There are two kinds of crust: continental, which has an average density 2.7, average thickness 35 to 40 km (22 to 25 mi) but under high mountain chains ranging between 60 and 70 km (37 and 44 mi), with large areas older than 1500 mn years (some exceeding 3500 mn years), a complicated structure, and variable composition; and oceanic, which is heavier than continental, average density 3.0, average thickness only 6 km (3.7 mi), nowhere older than 200 mn years, with a simple layered structure of uniform composition. The junction of the two types of crust is usually obscured by recent SEDIMENTATION. PLATE TECTONICS.

cryopedology the scientific study of frost action and of permanently frozen ground, including studies of the processes and their occurrence and also of the engineering devices and techniques which may be employed to overcome the physical problems in such conditions.

cryophyte a plant growing on snow or ice, usually a micro-plant such as ALGAE, but it may also be a BACTERIUM, FUNGUS or MOSS. The algae produce RED SNOW.

cryosphere the permanently frozen region of the earth's surface.

cryostatic hypothesis the HYPOTHESIS-1 that debris, in freezing, is squeezed between downward freezing ground and the PERMAFROST table, upward injection taking place in areas where relief is easiest. MUD CIRCLE.

crystal the solid state of many simple substances and their natural compounds, created naturally in the aggregation of their constituent atoms in an ordered, regular pattern. Usually the external, symmetrical plane faces of the solid meet at angles peculiar to the substance.

crystalline *adj.* consisting of, or resembling, a CRYSTAL.

crystalline rocks rocks composed partly or wholly of CRYSTALS. The term applies both to IGNEOUS ROCKS (crystallized as a result of cooling of molten rock) and to METAMORPHIC ROCKS (in which the crystals have developed as a result of METAMORPHISM).

crystallization the solidification of MINERALS as CRYSTALS either through the cooling of molten MAGMA or as a result of METAMORPHISM. CRYSTALLINE ROCK.

cuesta (Spanish) a landform consisting of an asymmetrical ridge with an abrupt cliff or steep slope, the ESCARPMENT (the inface) and a gentle BACK SLOPE (dip slope) which lies almost parallel to the dip of the strata, the result of differential denudation of gently inclined strata where resistant beds overlie weaker layers. The term covers the whole landform and is preferable to the ambiguous terms formerly used, i.e. ESCARPMENT and SCARP, which are better applied not to the whole of the landform but only to the steep slope. Fig 42.

cultigen a plant SPECIES-1 or its equivalent produced by cultivation, a product of human action. CLASSIFICATION OF ORGANISMS, CULTIVAR.

cultivar cv, a plant VARIETY arising from cultivation, deliberately produced by human action, as distinct from a natural variety, occurring without human intervention. CLASSIFICATION OF ORGANISMS, CULTIGEN.

cultural geography a branch of HUMAN GEOGRAPHY concerned particularly with people – environment interrelationships and impact, the distribution and spatial DIFFUSION-1 of distinct cultures and the differences between them, and their effects on the landscape. CULTURAL LANDSCAPE, CULTURE AREA, ECOLOGY.

cultural hearth the centre, the cradle of a culture or of a cultural group from which, in favourable conditions, the successful culture or cultural group will spread, effecting the changes in the NATURAL LANDSCAPE which bring into being the particular CULTURAL LANDSCAPE of the culture or group. DOMAIN-2, OUTLIER-2, SPHERE-5.

cultural landscape the NATURAL LANDSCAPE as modified by human activities, i.e. most of the present landscape, there being very few parts of the world now unaffected by such activity. CULTURAL HEARTH, CULTURE, CULTURE AREA, LANDSCAPE.

cultural region CULTURE AREA.

culture 1. the collective mental and spiritual manifestations (aesthetic perception, beliefs, ideas, symbols, values etc.), the forms of behaviour and social structures (modes of organization, rituals, groupings, institutions etc.), together with material and artistic manifestations (tools, buildings, works of art etc.), formulated and created by people according to the conditions of their lives, characterizing a SOCIETY-2,3 and transmitted as a social heritage from one generation to the next, undergoing modification and change in the process (CULTURE AREA, CULTURE CONTACT, ENCULTURATION). Culture exerts a strong influence on the way in which the ENVIRONMENT is perceived. PERCEPTION **2.** the rearing of fish,. silkworms etc. **3.** the growing of microorganisms, tissues etc. in a prepared media, or the product of such cultivation.

culture area, cultural region a REGION-1 identified by the existence within it of a single, distinctive CULTURE-1 or of cultures similar to one another. CULTURAL HEARTH, CULTURAL LANDSCAPE, DOMAIN-2, SPHERE-5.

culture contact the meeting or mingling of groups with differing cultural traditions, brought about by migration, displacement, or enlargement of territory etc. Results vary: a dominant, technologically superior, stronger, more successful culture may swamp the weaker; the weaker may form itself into a self-sufficient, tightly-knit ETHNIC MINORITY within the larger group; cultural groups of equal strength may merge; or, as a result of the interaction of diverse cultural groups and traditions, a new society with its own particular cultural traditions may emerge. ACCULTURATION, ENCULTURATION, PLURALISTIC INTEGRATION, PLURALISM, PLURAL SOCIETY.

cumec CUSEC.

cumulative frequency distribution in statistics, a distribution produced by starting at one end of a range of scores and for each successive CLASS INTERVAL adding the frequency in that class interval to all the preceding class intervals. If percentages of the total cases falling in each class interval are used, it is termed a cumulative relative frequency distribution. FREQUENCY DISTRIBUTION.

cumulative upward causation the process by which economic activity leading to prosperity and increasing economic development tends to concentrate in an area with an initial advantage, draining investment and skilled labour from the peripheral area (part of the backwash effect). AGGLOMERATION ECONOMIES, CENTRIPETAL AND CENTRIFUGAL FORCES, CIRCULAR AND CUMULATIVE GROWTH, MULTIPLIER EFFECT.

cumulonimbus a low-based mass of CUMULUS cloud, dark grey when viewed from below, shining white from the side, developed to a great vertical height, the upper part spreading into the shape of an anvil (ANVIL CLOUD, ICE ANVIL, INCUS), usually associated with THUNDERSTORMS, heavy rain or HAIL. CLOUD.

cumulostratus STRATOCUMULUS.

cumulus a heaped mass of low-lying, rounded convection CLOUD with a large, white, domed crown, which develops vertically from a flat base, in some cases to a considerable height. It may disperse as CONVECTION currents die away, or develop into CUMULONIMBUS if convection currents grow in power.

current a body of water, air or other fluid moving vertically or horizontally in a definite direction, e.g. **1.** the vertical movement of fluid material within the earth, of water in the ocean, of air in an air mass. CONVECTION CURRENT **2.** the horizontal movement of water in certain channels of a river **3.** the permanent or semi-permanent, horizontal movement of the surface water of the OCEAN-1 (DRIFT) caused mainly by the dragging action of the PLANETARY WINDS. Fig 34. **4.** the horizontal movement of water through a STRAIT due to differences in temperature and salinity at each end. DENSITY CURRENT **5.** the horizontal movement of water through a restricted channel due to differing tidal regimes at each end, i.e. a TIDAL CURRENT (not to be confused with a TIDAL STREAM). LITTORAL CURRENT, LONGSHORE CURRENT.

current bedding, current-bedding, cross bedding, cross-bedding LAMINAE that lie parallel to each other for short distances but are inclined at angles oblique to the general stratification, occurring especially in SANDSTONE, due to a change in the direction of the currents in the water and wind depositing the sand grains. If such a change was very marked the beds may be truncated; and laminae deposited thereafter may lie at a different angle. FALSE BEDDING.

curvilinear relationship in statistics, the relationship between two VARIABLES which is shown by some kind of curved line rather than by a straight line when one is plotted against the other. LINEAR-4.

cusec the unit of measurement of flow of a fluid, an abbreviation of cubic feet per second (1 cusec = 102 cu m per hour), commonly used as a measure of river flow (the number of cubic feet per second passing a particular REACH). The abbreviation of cubic metres per second is cumec. DISCHARGE OF A STREAM.

cuspate delta a triangular-shaped DELTA with the apex projecting seawards, formed on a straight coast where strong wave action distributes the sediments evenly on each side of the river mouth, e.g. the delta of the River Tevere (Tiber), Italy.

cuspate foreland a large, approximately triangular-shaped FORELAND-1 with the apex pointing seawards, usually formed by the convergence of two curved SPITS of shingle and sand built up by the action of opposing, powerful constructive waves and currents.

cwm (Welsh) CIRQUE.

cyanobacteria blue-green ALGAE occurring naturally in EUTROPHIC conditions in persistent warm calm weather, representing the final stage in the biological succession: DIATOMS (spring), green algae and cyanobacteria (late summer/early autumn).

cycle 1. a series of events or phenomena recurring in the same order **2.** a period of time occupied by such events or phenomena **3.** an ordered series of phenomena in which a process is completed, e.g. a LIFE CYCLE.

cycle of erosion the geomorphic cycle defined by W. M. Davis, 1850–1934, American geologist and meteorologist, the hypothetical sequence of changes or stages through which an uplifted land surface would pass in its reduction to base level by the action of natural agencies in the processes of erosion. In YOUTH streams flow through steep-sided V-shaped valleys; in adolescence the characteristic features of maturity start to appear; in maturity (MATURE) or middle age valleys are broader with gentler slopes; in OLD AGE valleys are broad and flat, rivers sluggish, and the land surface becomes a PENEPLAIN, ready for REJUVENATION. This hypothesis is no longer uncritically accepted. RIVER.

cycle of poverty POVERTY CYCLE.

cyclogenesis an atmospheric process in which a local heat force causes horizontal convergence and violent vortical disturbance, the rotation of the air being intensified and upward motion speeded up, resulting in the formation of towering CUMULUS clouds and the development of intense tropical storms over the ocean.

cyclone a system of winds (circulating anticlockwise in the northern hemisphere, clockwise in the southern) round a centre of low barometric pressure, especially applied to a fast-moving tropical revolving storm where the LOW PRESSURE SYSTEM is small (diameter 80 to 400 km: 50 to 250 mi) and the barometric gradient steep, associated with strong winds, thunderstorms, heavy rainfall, known as a HURRICANE OR TYPHOON. The term used to be applied to any small, travelling low pressure system, but in midlatitudes such a system is now termed a DEPRESSION-3, low, or cyclonic disturbance. FILLING.

cyclonic (frontal) rain precipitation associated with the passage of a DEPRESSION-3 in middle and high latitudes, as the warm moist air mass of the depression meets and overrides colder, heavier air. Characteristically, widespread drizzle is followed by heavy rain and squalls as the COLD FRONT passes.

cyclothem in geology, a stratigraphical unit consisting of a series of beds deposited during a single cycle of sedimentation, beginning with shallow water deposits as the land sank, passing into deeper water deposits and again into shallow water or coastal deposits as the basins became infilled or the land rose.

D

dairy cattle cattle reared specifically for milk production.

dairy farm a farm specializing in DAIRY FARMING.

dairy farming farming devoted primarily to the keeping of cows for their yield of milk, whether for consumption as such, or conversion to butter, cheese and other milk products.

dale mainly in northern England, a broad, open valley, a term commonly used in place-names.

Dalmatian coast a CONCORDANT COASTLINE, a name derived from the eastern Adriatic coast where the coastline lies approximately parallel to the trend of the relief, the tops of former mountain ranges appearing as lines of islands, the parallel valleys having been drowned when the sea level rose. SUBMERGED COAST.

dam a barrier of earth, rock, masonry or concrete built across the course of a river to hold back or restrict the flow of the water for a specific purpose. BARRAGE.

dambo (Bantu) **1.** the floodplain of a river in central Africa, swampy in the wet season, but dry for most of the year, supporting long grass **2.** a shallow depression, lacking distinct drainage channels, at the head of a drainage system, a term originally applied to such a depression in tropical Africa, now extended to other areas.

Danelaw 1. the code of laws established in north and east England by Norse invaders in the ninth and tenth centuries **2.** the part of England governed by those laws.

dasymetric method a method used in drawing a density map (e.g. of population) which does not use average figures related to administrative units (which produces an unrealistic map with sharp contrasts between one area and another, CHOROPLETH MAP) but instead draws on all available geographical knowledge to draw up realistic categories for which densities can be estimated. Such dasymetric plotting gives a more realistic representation.

date palm a long-lived tall tree of dry subtropical areas, characteristic of the oases of hot deserts, usually grown without irrigation, and bearing a nutritious fruit with high sugar content. The fruit is sold loose as fresh dates, or semi-dried and packaged; if dried it keeps for a long time and can be ground into a flour, or soaked in water for eating.

dating the determining, so far as is possible, of dates for structures, events, ARTIFACTS. The techniques used include DENDROCHRONOLOGY, LICHENOMETRY, OSL, PALAEOMAGNETISM (archaeomagnetism) investigation, POTASSIUM argon dating, RADIOCARBON DATING, VARVE investigation.

datum (Latin, pl. data) a thing given, something known and made the basis of reasoning or calculation. Purists maintain that the plural, data, should always be used with a plural verb.

datum level, datum line, datum plane the zero altitude base from which the measurement of elevation starts. The Ordnance Datum (OD) from which heights for British official (Ordnance Survey) maps are calculated is the mean sea level at Newlyn, Cornwall, England.

day the time during which the SUN is above the horizon, the opposite of NIGHT.

day degree the measure of the duration of TEMPERATURES-2 above or below a selected basal temperature in a period of 24 hours. ACCUMULATED TEMPERATURE, CRITICAL TEMPERATURE.

daylength PHOTOPERIOD, one of the important environmental factors influencing the reproduction of plants, and the seasonal activities of animals (e.g. hibernation, MIGRATION-3, reproduction).

daylight saving a system in which time in an area is advanced, usually by one hour, in relation to the STANDARD TIME of that area in order to extend the period of daylight at the end of a normal working day.

dead *adj.* **1.** having no life **2.** inactive, in contrast to ACTIVE.

dead cave a cave in which excavation and deposition have finished. ACTIVE CAVE.

dead cliff a sea CLIFF no longer subject to EROSION by waves owing to the build-up of protecting beach material or to a fall in sea level. WEATHERING reduces the angle of slope, and ultimately the dead cliff is colonized by vegetation.

dead ice stagnant ice, usually covered with rock debris, at the edges of a motionless GLACIER or ice sheet.

death-rate BIRTH-RATE.

debris DETRITUS, an accumulation of rock waste consisting of disintegrated rocks, sand, clay, moved from their place of origin and redeposited. DENUDATION.

debris avalanche a rapid flow of rock debris, sliding in narrow tracks down a steep slope under the influence of gravity. MASS MOVEMENT.

debris fall the precipitate, nearly free, fall of earth debris from a vertical or overhanging face under the influence of gravity. MASS MOVEMENT.

debris slide the rapid downward rolling or sliding of unconsolidated earth debris, without backward rotation of the mass, under the influence of gravity. MASS MOVEMENT.

decalcification the removal of CALCIUM CARBONATE from a SOIL HORIZON or horizons as CARBONIC ACID reacts with the CARBONATE mineral material. CARBONATION.

decentralization **1.** the action or fact of moving away from a concentration at a central point **2.** the diminishing of central control or authority in administration in order to increase the authority of groups at places, branches etc. distant from the centre, or the pursuance of that policy. CENTRAL BUSINESS DISTRICT, CENTRIPETAL AND CENTRIFUGAL FORCES.

deciduous *adj.* applied to trees or shrubs which shed all their leaves at a certain season every year, as opposed to evergreen, applied to trees or shrubs which carry green leaves throughout the year. Although certain CONIFEROUS trees (notably the larch) are deciduous, deciduous is commonly used as synonymous with BROADLEAVED, applied to trees which shed their leaves in the AUTUMN or fall in MIDLATITUDES as the temperature falls. Deciduous trees of the MONSOON FORESTS drop their leaves in the hot dry season as a protection against excessive TRANSPIRATION.

decolonization the process whereby a METROPOLITAN-2 country gives up its authority over a dependent territory so that the dependent territory becomes a SOVEREIGN STATE. COLONIZATION-1.

decomposition 1. in biology, the breaking down, the separation, of organic matter into simpler compounds. FOOD CHAIN **2.** in geology, the breaking down of rocks, commencing with CORROSION of the component minerals.

deduction 1. reasoning from the general to the particular, i.e. in which the conclusion necessarily follows from the given premises **2.** the conclusion reached in this way. ECOLOGICAL FALLACY, INDUCTION.

deep in the ocean, a trough-like depression or trench in the sea floor, of limited extent and great depth (over 5500 m: 3000 fathoms: 18 050 ft). Deeps occur mostly at the convergence of plates in SUBDUCTION ZONES (PLATE TECTONICS). Thus there are many near the ISLAND ARCS in the Pacific, e.g. the Mariana Trench near Guam (11 033 m: 6000 fathoms: 36 198 ft), or the Emden Deep near the Philippines (10 794 m: 5900 fathoms: 35 413 ft).

deepening in meteorology, the decreasing atmospheric pressure at the centre of a low pressure system (DEPRESSION-3), as opposed to FILLING. Deepening usually gives rise to greater windspeed and PRECIPITATION.

deep focus of an earthquake, a SEISMIC FOCUS occurring at a depth below some 300 km (185 mi).

deer forest FOREST-4.

deer park PARK-1.

defended neighbourhood a NEIGHBOURHOOD with residents who, feeling threatened by outsiders who are seen as competitors for scarce resources (e.g. housing), try to exclude those outsiders from the neighbourhood, e.g. by raising house prices, racial attack, threatening graffiti (GRAFFITO).

deferred junction of a tributary with a river, a phenomenon occurring when a tributary is prevented by LEVEES from joining the river, and flows parallel with it for a considerable distance before the ultimate CONFLUENCE-2 which in many cases takes place on the convex side of a large MEANDER. A deferred junction usually occurs on a FLOODPLAIN.

deficiency disease a disease in humans caused by the lack of an essential food substance (e.g. rickets in children, due to a lack of vitamin D).

deflation 1. the removal of fine rock debris by wind, especially likely to occur in ARID or semi-arid areas lacking the protection of vegetation (DUST BOWL). **2.** in economic geography and economics, the reduction of the value of currency or of prices from an inflated condition. INFLATION.

deflation hollow BLOWOUT-1.

deforest, deforestation 1. the permanent removal of FOREST-1 and its undergrowth **2.** historically, the release of an area from the strict Forest Laws (FOREST-2).

deformation in geology **1.** a general term applied to the change in the shape, volume or structure of some region or feature of the earth's crust caused by STRESS-1 arising from tectonic activity (e.g. of rocks: COM-

PRESSION, FAULT, FOLD-2, SHEARING) **2.** the process producing those results.

deglaciation, deglacierization the withdrawal of an ice sheet from an area. Deglaciation sometimes refers to former times only, deglacierization to the present.

deglomeration the setting-up of productive enterprises at a distance from the centre of an established AGGLOMERATION-4, occurring when the advantages or economies of the agglomeration (AGGLOMERATION ECONOMIES) begin to diminish with its increasing size and land values, etc.

degradation **1.** the process of lowering a surface by EROSION and the removal of rock waste **2.** the general lowering of the surface of the land by erosive processes **3.** of soil, a change in the soil due to increased LEACHING.

degree (symbol °) **1.** the measure of temperature on any thermometric scale. TEMPERATURE **2.** 1/360th of the angle that the radius of a circle describes in a full revolution; thus a right angle is 90° **3.** the unit of angular measurement of latitude and longitude, 1/360th of the earth's circumference, measured for LATITUDE from the EQUATOR, for LONGITUDE from the PRIME MERIDIAN. In latitude and longitude each degree is divided into 60 minutes (symbol ') and each minute into 60 seconds (symbol "). A degree of latitude is roughly 111 km (69 statute mi).

degrees of freedom in statistics, the highest number of observations or categories it is possible to assign freely before all of those which remain are completely determined.

de-industrialization the process in which there is a marked movement of employment away from the production of goods (i.e. from SECONDARY INDUSTRY) usually to the provision of services (i.e. to TERTIARY INDUSTRY). COUNTERURBANIZATION, INDUSTRIALIZATION.

delta originally the tract of alluvial land traversed by the distributaries of the River Nile downstream from Cairo and so applied to the more or less triangular terminal floodplain of other rivers. A delta is formed when a river deposits solid material at its mouth at a rate faster than that by which it can be moved by tidal and other currents. As the deposits grow the river splits to make new channels and the distributaries thus formed divide and subdivide, each stream depositing its load. Small deltas may form where a river meets a lake (LACUSTRINE DELTA), or at the confluence of rivers where a swiftly flowing stream, carrying its load of deposits, meets a sluggish older, broader stream. Delta forms are classified as ARCUATE (fan-shaped, as the Nile delta), BIRD'S FOOT (lobate or DIGITATE, as the Mississippi) and CUSPATE (as the Tavere). Fig 13.

demersal *adj.* sometimes used as a synonym for BENTHIC, living in the lowest layer of a sea or lake, e.g. demersal fish, such as cod, haddock, halibut, sole, plaice, living near the bottom of a sea in comparatively shallow water and caught mainly by trawling. SEINE.

democracy **1.** a form of government in which political power lies in the hands of the people through elected representatives. The popular interpretation of democracy in western societies is that it is government for the people, of the people and by the people; and that it therefore gives rise to policies and decisions which accord with the general will of the people **2.** a state so governed.

demographic transition the gradual

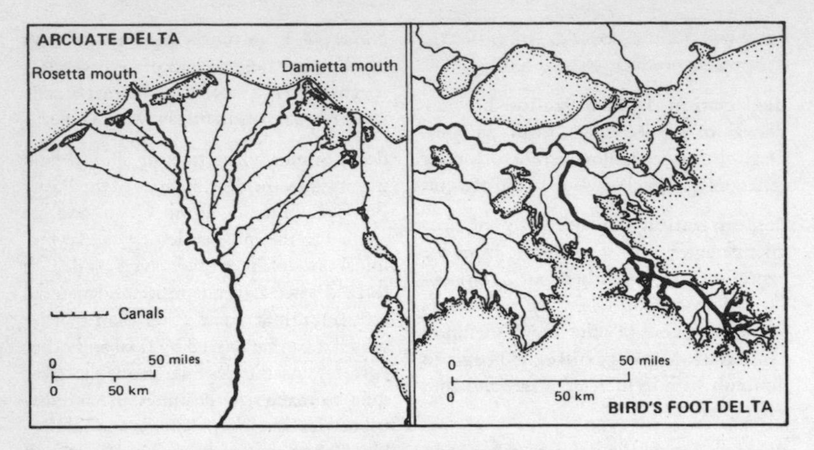

Fig 13 Delta forms: an arcuate and a bird's foot delta

change in the manner of population growth occurring over a period of time, particularly that associated with the effects of the spread of INDUSTRIALIZATION and URBANIZATION on FERTILITY-3 and MORTALITY. The classical stylized model shows four phases. The first is termed high stationary (high birth and death rates, a fluctuating but low population growth due to famine, disease, war); the second is early expanding (continuing high birth rate, but declining death rate, and increase in life expectancy resulting from better nutrition, sanitation and medical care, leading to an expansion of population); the third, late expanding, is characterized by the stabilization of the death rate at a low level, and a decline in the birth rate linked to the growth of an urban-industrial society with its high cost of child-rearing and the ready availability and use of BIRTH CONTROL techniques. The fourth phase, low stationary, is a period of very slow population growth, with birth and death rates stabilized at a low level, the former being more likely to fluctuate than the latter. In this it differs from the first phase, when the death rate is more likely to fluctuate than the birth rate. BIRTH RATE, POPULATION PYRAMID.

demography the scientific study of human populations primarily in respect of their size, structure and development. It is concerned not only with statistics of birth, disease, death, marriage, life expectancy, migration, the division of population into groups on the basis of sex, age, marital status and the changes in those structures, but also with all aspects of population studies, including relationships with social and economic factors. POPULATION GEOGRAPHY.

demoiselle (French) an EARTH PILLAR, weathered from VOLCANIC BRECCIA or similar material, but capped by a large boulder which has protected the material underlying it.

dendritic drainage a drainage pattern resembling the branching of a tree, developed especially on a gentle, nearly uniform slope where no control is exercised by the underlying geological structure, so that INSEQUENT STREAMS develop, and as each insequent cuts its own valley it receives its own insequent tributaries. DRAINAGE. Fig 17.

dendroarchaeology the investigation concerned with the analysis of wood remains found in archaeological excavations and the dating of the sites based on such evidence. DENDROCHRONOLOGY.

dendrochronology dating by the process of counting the ANNUAL GROWTH RINGS of a tree. DENDROARCHAEOLOGY.

denitrification the process in which nitrates are broken down by BACTERIA in the soil, resulting in the release of free NITROGEN (FREE-2) and a reduction in soil fertility. It usually occurs in ANAEROBIC conditions, e.g. in a WATERLOGGED soil. NITROGEN CYCLE.

density **1.** the quantity of anything per unit area, hence the density of persons, of houses, or of habitable rooms per sq km, per ha, per acre etc. **2.** the relation of mass (the amount of matter) to the space it occupies (its volume) expressed as gm per cu cm, the unit of measurement used being the density of water at 0°C (at that temperature 1 cc of water weighs nearly 1 gm). Thus density is an absolute quantity, unlike SPECIFIC GRAVITY which, while numerically the same, is a relative quantity. The density of water depends on temperature, salinity, particles in solution.

density current an ocean CURRENT resulting from the differences in DENSITY-2 of water caused by variations in salinity and temperature, cold or very saline water being more dense, and therefore sinking and flowing under less dense, warm, less saline water.

denudation broadly, the uncovering of deeper rocks by any natural agency, i.e. by any agent of EROSION as well as by WEATHERING and MASS MOVEMENT, and therefore not to be confused with the term erosion, which excludes weathering and mass movement. Denudation is also applied generally to the lowering of a land surface by erosion and the removal of rock waste, and is thus synonymous with DEGRADATION. Rocks vary in resistance to denudation, hence the term differential denudation is applied to areas where this is apparent. CIRCUMDENUDATION.

dependency, dependent territory a territory relying on or subject to the control of another country of which it does not form an integral part.

dependency ratio in population, the ratio of the number of people who cannot be gainfully employed in a population (the dependants) to the number who are actively or potentially active (the employed or employable). The dependent population is sometimes classified as those in the age groups 0 to 14 and 65 years of age and over.

dependent variable a VARIABLE which is to be explained or predicted. It is dependent on one or more other variables which may control it or relate to it. Thus it is not under the control of the experimenter; and it will be affected by other variables which are being manipulated. INDEPENDENT VARIABLE.

depopulation a marked reduction in the number of inhabitants of an area. POPULATION-2.

deposit material laid down, a natural accumulation, a SEDIMENT, DEPOSITION.

deposition the action of laying down of material, especially of the debris transported mechanically by wind, running water, tides and currents in the ocean and seas; of the materials transported in solution, subject to evaporation and chemical precipitation (e.g. ROCK SALT) or to the intervention of living organisms (e.g. CORAL); or of organic matter, mainly the remains of vegetation (e.g. PEAT). Deposition is thus the opposite of DENUDATION, the two processes together acting on the earth's crust at or very near its surface.

depressed area an area in economic decline, with a high level of unemployment over a long period of time. DEVELOPMENT AREA.

depression 1. in general, the process of sinking, the action of pressing down, or the fact of being pressed down, the condition of being lowered in position, of being less active than usual **2.** any hollow or relatively sunken area, especially one enclosed by higher land, without an outlet for surface drainage **3.** in meteorology, specifically, a region of the atmosphere where the atmospheric pressure is lower than that of its surroundings, i.e. a low pressure system, a 'low' or 'disturbance' in midlatitudes and high latitudes, replacing the term CYCLONE (WARM FRONT). A deep depression is one in which the pressure at the centre is considerably lower than that at the edges; a shallow depression is one in which there is little difference between those two pressures. Figs 14, 15.

deprivation the state of being prevented from using, of being taken away from, of lacking something necessary or desirable. MULTIPLE DEPRIVATION, POVERTY CYCLE.

deranged drainage a confused DRAINAGE-2 system which produces a mosaic of small lakes, streams, marshes, small islands (as in Finland) caused by the haphazard distribution of (glacial) DRIFT-1.

derelict land land damaged by some process (e.g. by extractive or other industry) and/or neglect, abandoned and left to fall to ruin, incapable of being used in its present condition.

desalination, desalinization the process of removing dissolved salts from water, especially from sea water, or from the soil.

desegregation the process of abandoning the practice of SEGREGATION, e.g. of bringing to an end the provision of separate facilities, such as educational facilities, for different ethnic or social groups.

desert a region in which evaporation exceeds precipitation, from whatever cause, so that the moisture present is insufficient to support any but the scantiest vegetation. ARID, COLD-WATER DESERT.

desertification the spread of land degradation in arid, semi-arid and dry subhumid areas, leading to the outward spread of DESERT fringes, brought about by climatic variations and the activities of people and their livestock.

desert pavement, desert mosaic in a hot DESERT, an exposure of bedrock or of pebbles, closely packed after the removal of finer rock material, polished or smoothed by blown sand so that eventually the upper surfaces of the bedrock or pebbles are ground flat. The pebbles are often bonded together by salts drawn to the surface in solution by CAPILLARITY and precipitated by EVAPORATION, which act as a CEMENT-2.

desert soils SOILS of ARID regions where there is a net deficiency of rainfall (commonly areas with rainfall under 250 mm: 10 in), hence a lack of vegetation and a thin or discontinuous organic layer. There is commonly a surface layer of pebbles, the leached layer being only about 15 cm

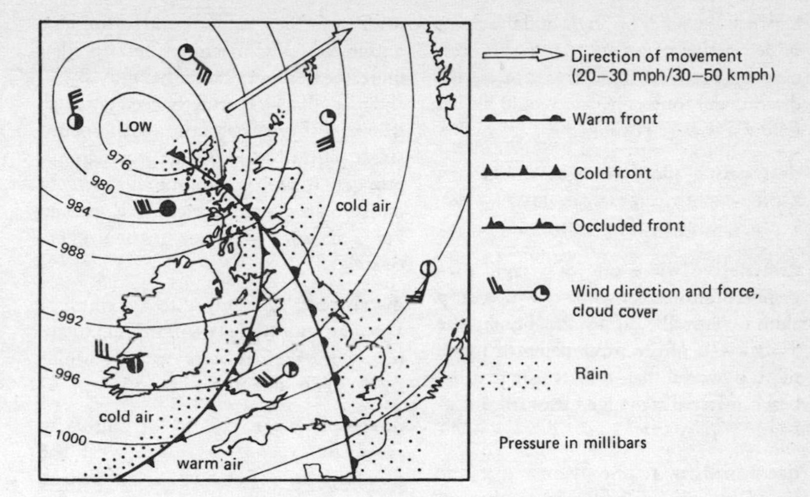

Fig 14 A depression as shown on a weather chart

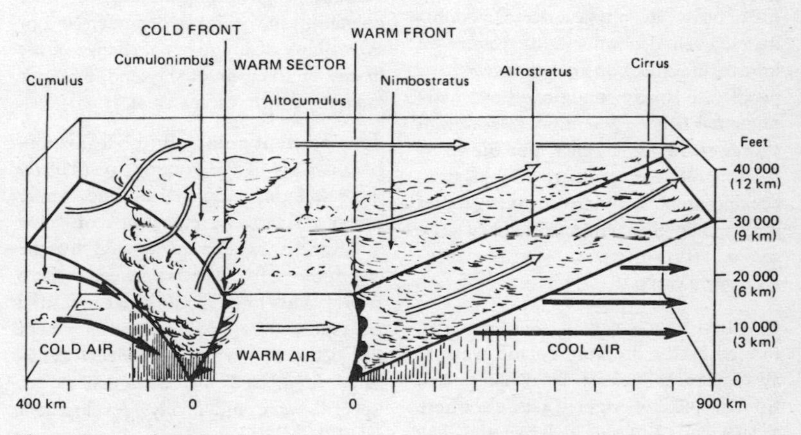

Fig 15 Section through a depression

(6 in) thick, underlain by a CARBONATE layer. SOIL CLASSIFICATION.

design disadvantagement the disadvantagement originating from buildings, the design of which fails to meet the needs of people who cannot avoid living or working in them, to such an extent that it contributes to social malaise. The term has been applied particularly to large (especially HIGH RISE) blocks of low-rent flats in densely built-up urban areas in Britain.

desire line a straight line drawn on a map

between the point of origin and the point of destination of a TRIP, i.e. the shortest distance between these two points, indicating the route a person would like to follow if it were possible.

destructive plate margin COLLISION ZONE, OCEANIC TRENCH, PLATE TECTONICS, SUBDUCTION ZONE.

destructive wave one of a rapid succession of strong storm waves which drop almost vertically on to the beach, the BACKWASH of the wave being so much more powerful than the SWASH that the beach material is dragged seaward. CONSTRUCTIVE WAVE.

determinism **1.** the theory that the world, or nature, or event, or human action is subject to causal law **2.** in geography, the belief that the ENVIRONMENT (particularly its physical factors) dominates, even determines, the pattern of human life and human behaviour, that people are largely conditioned by environmental factors. Scientific determinism expresses the same belief, but the justification for it is statistically based, proceeding from statistical analysis of data sets, rather than from individual case studies. ENVIRONMENTALISM, POSSIBILISM, PROBABILISM.

detritus **1.** fragmented rock material, formed by the breaking up and wearing away of rocks, that has been transported from the place of origin to a site elsewhere **2.** an accumulation of such material. The term DEBRIS has now generally superseded detritus.

developed *adj.* applied particularly to a country or a region frequently in connexion with the economy of the area, implying that the area is culturally and socially advanced, and that full use is being made of the natural and economic resources, skills, machinery etc. present there, the necessary capital being available, and that what had formerly been potential is being realized. DEVELOPING, DEVELOPMENT, THIRD WORLD. The gross national product (GNP) is commonly used to measure the degree of such development. For a broader view see UNDERDEVELOPMENT.

developing *adj.* applied to a country or a region, formerly UNDERDEVELOPED, now in the process of becoming DEVELOPED. BRANDT REPORT.

development **1.** the act of causing to grow, to expand, to realize what had formerly been potential **2.** in Britain, according to the Town and Country Planning Act 1971, the carrying out of building, engineering, mining or other operations in, on, over or under land or the making of any material change of use in any building or of the land. BRANDT REPORT, UNDERDEVELOPMENT.

development area in British legislation, certain parts of the country, particularly those suffering industrial decline, where DEVELOPMENT-2, especially of new industry, is to be encouraged by the government. In the years between the two World Wars some of the older industrial areas suffered from serious unemployment and decay and were successively designated 'depressed' or 'distressed' areas, 'special' areas, and finally 'development areas'.

Devensian in Britain, the final glacial stage of the Pleistocene (GEOLOGICAL TIMESCALE), characterized by fluctuating advances of the ice interspersed with warmer conditions, when birch and conifers became temporarily, and early human beings became strongly, established. In the

Devensian the extent and duration of the ice-cover varied from one area to another: nearly all of southern England and East Anglia was ice-free; but western Scotland lay permanently under ice for most of this stage. FLANDRIAN.

Devonian *adj.* of the fourth geological period of the PALAEOZOIC era (GEOLOGICAL TIMESCALE) when SANDSTONES, GRITS, SLATES and LIMESTONE were laid down in the sea (as apparent in south-western England) and red and brown sandstones, CONGLOMERATES, MARLS-3 and LIMESTONE (the OLD RED SANDSTONE) in lakes, and the characteristic life forms were ferns and lower fishes.

dew droplets of water deposited on any cool surface by CONDENSATION of WATER VAPOUR in the atmosphere, especially at night after a hot day. DEW-POINT, PRECIPITATION-1.

dew-point the temperature at which air, on cooling, becomes saturated with WATER VAPOUR, and below which CONDENSATION begins and DEW forms.

dew pond a shallow artificial POND-1, lined with a mixture of kneaded CLAY and straw and sand (sometimes), made especially on the chalk downlands of southern England, the lining preventing water from percolating downwards. It was long believed that these ponds were fed with dew, hence the name, but dew contributes very little to the water held: some of it comes from the condensation of sea mist, but most is derived from PRECIPITATION-1.

D horizon C HORIZON, SOIL HORIZON.

diabatic *adj.* of the thermodynamic process in which loss or gain of heat occurs (e.g. in an AIR MASS), the opposite of ADIABATIC.

diagnostic horizon a strictly defined SOIL HORIZON which clearly shows the soil-forming processes at work in an area, and is used as a basis for classifying soils in a SOIL CLASSIFICATION.

dialectic 1. the art of logical disputation, of critically examining the truth of an opinion or theory by question and answer **2.** in Hegelian philosophy, broadly, the logical subjective development in thought from thesis through antithesis to synthesis (DIALECTICAL MATERIALISM), or logical objective development in history by the continuous reconciliation, the unification, of parts, of opposites. Marx saw this process at work in his interpretation of the historical succession of MODES OF PRODUCTION. In DIALECTICAL MATERIALISM dialectic is not only equated with the way reality changes, it is declared to be the method of discovering the 'laws of motion', a method, it is asserted, which is applicable to all scientific disciplines. HISTORICAL MATERIALISM, MARXISM.

dialectical materialism the philosophy underlying Marxist theory, first formulated by Engels (MARXISM), based broadly on a modification of the standard theory of MATERIALISM-1 combined with a development of Hegelian philosophy (DIALECTIC-2), i.e. that everything that exists can be shown to derive ultimately from matter and that (using Hegelian philosophy) the development of nature, society and thought occurs by conflict between an original direction (the thesis), its direct opposite (the antithesis) and ultimately by synthesis (the reconciliation and unification of parts of these two extremes). Dialectical materialism maintains that dialectic is not only the way reality changes, it is the method to be used in discovering the 'laws of motion', a method applicable to all scientific disciplines. HISTORICAL MATERIALISM, MARXISM.

diamond a precious stone, the CRYSTALLINE form of pure CARBON, the hardest mineral (HARDNESS), usually occurring embedded in PIPES-3 in IGNEOUS ROCK or washed out and redeposited in PLACERS, measured in CARATS. The better quality diamonds are used as gemstones; the poorer in industry for cutting and as abrasives.

diaspora the dispersal of people from their perceived original homeland.

diastrophic eustatism, deformational eustatism a global change of ocean level (NEGATIVE MOVEMENT, POSITIVE MOVEMENT) due to a variation in the capacity of ocean beds, caused by filling in by sedimentation or movements of the ocean floors, a change which often leads to REJUVENATION. EUSTATISM, GLACIO-EUSTATISM.

diatom a subdivision of ALGAE, microscopic, unicellular, yellow-brown, with SILICA in the cell walls, occurring singly or in colonies which, with other divisions of algae, form part of marine and freshwater PLANKTON. The deposited remains of dead diatoms of the past appear today as diatomaceous earth (diatomite) or, with other decomposed organisms, as PETROLEUM. DIATOM OOZE.

diatom ooze OOZE consisting of the siliceous skeletons of DIATOMS deposited in the ABYSSAL zone of cold ocean water, occurring particularly in a belt around the earth in the Southern Ocean in latitudes 50°S to 60°S and in the North Pacific Ocean.

diffusion 1. the spreading out, the propagation, the dissemination through time of a PHENOMENON or phenomena (e.g. plants, animals, ideas, CULTURE-1, languages, knowledge, INNOVATION, techniques) over an ever-extending surface or over SPACE-2 from a single source (termed mononuclear diffusion) or from many sources (polynuclear diffusion). BARRIERS AND DIFFUSION WAVES, DISTANCE-DECAY PHENOMENON **2.** in meteorology, the seemingly random mixing of air bodies brought about by a slow process of mixing (termed molecular diffusion) or by TURBULENCE (eddy diffusion). The term is also applied to similar processes in liquids and light.

diffusion wave, innovation wave the movement of DIFFUSION-1 of INNOVATION, termed the innovation wave by T. Hägerstrand, who identified four stages in its progress, i.e. the primary (the beginning when the centres adopting the innovation are established and the contrast between them and remote areas is great); the diffusion (the start of the diffusion process characterized by the creation of new, rapidly-expanding innovation centres distant from the source, and a dimming of the contrast seen in the primary stage); the condensing (marked by a relative increase in the number of acceptances, equal in all locations irrespective of distance from the innovation source); and finally the saturation stage (characterized by a slowing down and eventual ending of the process, with apparent overall acceptance without regional variation). The wave will speed up or slow down according to the nature of the medium through which it moves; and if it meets another diffusion wave coming from another direction from another centre it may completely lose its identity.

dike, dyke (the spelling dyke is common but etymologically incorrect) **1.** a ditch, a wall, an embankment, a ridge **2.** in geology, an INTRUSION where the molten rock (MAGMA) has ascended through an approximately vertical fissure to solidify as a wall of rock often harder than the rocks of the surrounding strata. DIKE-SPRING, RING-DYKE. Fig 21.

dike- (dyke-) spring a SPRING-2 issuing along the line of a DIKE-2 where water from an AQUIFEROUS, PERMEABLE or PERVIOUS rock meets a dike of IMPERMEABLE rock which is penetrating the aquiferous surrounding strata.

dike (dyke) swarm in geology, a collection of DIKES-2 of the same age, usually with a common trend over a wide area, sometimes radiating from a common centre.

dilatation of rocks, the release of pressure effected within a rock mass when overlying layers are removed by DENUDATION, causing the rock to expand and split along expansion joints (dilatation joints) and concentric layers at right angles to the direction of the pressure release to split away from the upper surface, from which they are commonly removed by WEATHERING.

dip 1. the angle of maximum slope of an inclined surface **2.** in geology, true dip, the angle of maximum slope (i.e. maximum INCLINATION-2) of SEDIMENTARY ROCKS (or of rocks bedded with them) at a certain point. The term should not be applied to the INCLINATION-2 of land surfaces. APPARENT DIP, STRIKE, TRUE DIP. Figs 16, 42.

dip slip SLIP.

dip slope the surface slope of the ground where it inclines in approximately the same direction as the DIP-2 of the underlying rocks. CUESTA.

dip-stream a stream flowing roughly parallel to the DIP-2 of the underlying rocks.

Dirichlet polygon a polygon that contains within it areas which are nearer to the point around which they are constructed than to any other points, named after P. G. L. Dirichlet, 1805–59, German mathematician; also known as a Thiessen polygon or a first-order Brillouin region.

dirt band a dark band of ice, demarcated by light bands, formed within the ice of a GLACIER between the annual accumulation layers of FIRN. The almost bubble-free dirt band is formed when melt-water containing dirt is re-frozen. The light ice band, a mass of bubbles, may be the result of winter freezing of snow.

discharge of a stream the quantity of water passing through any cross section of a stream in a given unit of time. It is usually measured either in cubic feet per second (CUSEC) or cubic metres per second (cumec).

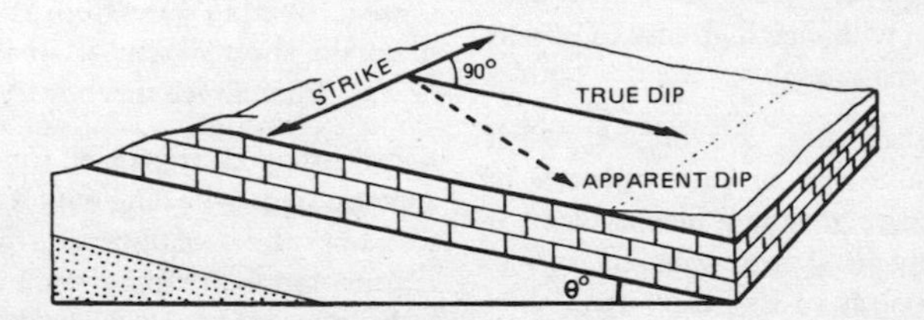

Fig 16 Dip

discordant *adj.* at variance, incongruous; not in accord, not harmoniously connected. CONCORDANT.

discordant coast a coast where the coastline cuts across the FOLDS-2 and FAULTS of the geological structure, i.e. across the 'grain' of the country; a transverse or ATLANTIC TYPE COASTLINE.

discordant drainage the condition of drainage when the surface drainage does not directly relate to the DIP-2 of the underlying strata.

discordant junction a river junction in which a tributary stream falls abruptly into the main stream, e.g. from a HANGING VALLEY. ACCORDANT JUNCTION.

disease ENDEMIC, EPIDEMIC, PANDEMIC.

diseconomies of scale an increase in unit costs arising from an increased scale of production. This rise in unit costs may be brought about by internal diseconomies (e.g. the need for a large administrative organization, the loss of contact between staff and management), or by the diseconomies experienced when an urban area grows so large that it becomes congested, transport costs rise and staff are no longer readily available. DISECONOMIES OF URBANIZATION, DISECONOMY, ECONOMIES OF SCALE.

diseconomies of urbanization the DISECONOMIES OF SCALE associated with large cities, with their high costs of labour, land, and transport.

diseconomy any unfavourable effect which results from an increased scale of production, e.g. diminishing returns or profitability. DISECONOMIES OF SCALE, DISECONOMIES OF URBANIZATION.

dispersed settlement a settlement pattern in which farmhouses and rural dwellings are scattered instead of being grouped together in a HAMLET, NUCLEATED SETTLEMENT, VILLAGE.

dissection the cutting of a land surface by EROSION, especially by eroding streams, into numerous valleys.

dissolved load the organic and inorganic material in SOLUTION-1 carried by a STREAM-1, as distinct from the BED LOAD and the SUSPENDED LOAD. The total amount of dissolved material in the water is usually assessed by evaporating a known volume of filtered water and weighing the dry residue.

distance-decay phenomenon the weakening, the fading, of process or pattern with increasing distance. It is apparent, for example, in transport flows in that as the distance between the point of origin and the point of destination increases the intensity of the flow tends to decrease.

distanciation TIME-SPACE DISTANCIATION.

distillation a process in which a LIQUID or a SOLID is subjected to EVAPORATION and CONDENSATION in order to purify it or separate it into smaller parts with different properties. GAS COAL.

distributary 1. a branch of a river which flows away from the main stream and does not return to it, as in a DELTA **2.** a branch canal distributing water from a main canal in an irrigation system **3.** an ice stream flowing from an ice sheet or ice cap.

distribution 1. in general, the action of apportioning, of dealing out, of allocating to distinct places, of dispersing to (or over) all parts of an area or space; or the condition of being so divided, allocated, dispersed **2.** the dispersal of COMMODITIES-1 among consumers **3.** the geographical range of an organism or group of organisms **4.** in statistics, a classification or arrangement, especially of statistical information. CUMULATIVE FREQUENCY DISTRIBUTION, NORMAL DISTRIBUTION, POISSON DISTRIBUTION.

disturbance in meteorology, a DEPRESSION-3 or LOW of no great intensity.

diurnal *adj.* in general, of, or belonging to, each day, completed once in one day.

diurnal range the difference between minimum and maximum values in 24 hours, e.g. as applied to air temperature.

diurnal rhythm, circadian rhythm the rhythmic physiological changes that, originating within an organism, occur in every 24 hours even when the organism is isolated from the daily rhythmic changes in its environment, e.g. sleep rhythm in animals (including humans), or leaf movements in plants.

divagating meander a MEANDER which is liable to variation from time to time because the surface on which it occurs approaches the condition of a PENEPLAIN.

divergence in general, the action of starting off from a point or source, and continuing in separate directions, with the result that the degree of separation increases with distance **1.** in climatology, a type of airflow in which in a certain area at a given altitude the outflow is greater than the inflow, resulting in a decrease in the air contained **2.** in oceanography, the movement of surface water away from a zone, brought about by wind-drift, resulting in the rise of water from the depths.

divided circle diagram popularly termed a pie diagram or pie graph, a diagram in which a circle, representing the total of the values, is divided into sectors, each sector being proportional to the value it represents.

division of labour INTERNATIONAL DIVISION OF LABOUR.

dock an enclosure or artificial basin, fitted with floodgates, in a harbour or river, in which vessels are loaded, unloaded, refitted, repaired. The pl. docks, denotes the dock basins with adjoining wharfs, warehouses, workshops and yards, offices. DRY (GRAVING) DOCK, FLOATING DOCK, TIDAL DOCK, WET DOCK.

dockyard an enclosure with docks and equipment for building and repairing ships, especially naval vessels.

doldrums the region of small pressure gradient, the belt of calms with high humidity and high temperatures occurring near the equator, approximately between 5°N and 5°S, especially over the eastern part of the oceans. ITCZ. Fig 5.

doline a shallow basin or funnel-shaped depression typical of KARST landscape. It usually has a flat floor, sometimes cultivated, linked to the underlying drainage system by a vertical shaft. The size and form vary, the diameter from a few metres to a kilometre, the depth from a few to several hundred metres. If formed mainly by direct solution of surface limestone, it is termed a solution doline; if by the collapse of a cave roof following subterranean solution, a collapse doline.

doline lake a body of freshwater occupying a DOLINE.

dolomite 1. a mineral consisting of equal molecules of calcium carbonate and magnesium carbonate, commonly occurring in evaporite deposits, e.g. from seawater; or as a replacement in LIMESTONE, some of the calcium having been replaced by magnesium; as a CEMENT; as a GANGUE mineral in HYDROTHERMAL deposits; and in carbonatites **2.** commonly applied to a rock consisting predominantly of that mineral, hence dolomitic limestone, a limestone with some dolomite. Dolomite rock is sometimes termed MAGNESIAN LIMESTONE. CALCIFICATION.

domain **1.** the estate or territory within defined limits over which control or influence is exerted **2.** the zone which immediately adjoins the CORE AREA of a CULTURE-1 (CULTURAL HEARTH) and into which the culture spreads. OUTLIER-2, SPHERE-5.

dome loosely applied to any dome-shaped (hemispherical) mass of rock or dome-shaped landform. More precisely applied to a structural feature where the underlying rocks form a dome, i.e. the strata dip away in all directions from a central, rounded area. BATHOLITH, DOME VOLCANO, LACCOLITH, OIL DOME, SALT DOME.

Domesday Book a documentary, detailed survey of England on a county basis compiled in 1086–7 on the orders of William I (the Conqueror), King of England, recording the extent, value, ownership of estates, census of householders, local customs, in two volumes, one covering Essex, Suffolk, Norfolk, the other the remainder of England apart from Northumberland, Durham, Cumberland and north Westmorland, which were excluded.

domestic animal a tame animal living under and dependent on human care.

domestic trade internal trade as opposed to international trade.

dome volcano a VOLCANO composed of highly viscous LAVA which, on eruption, congeals above and around the orifice instead of flowing away, the older lava sometimes being raised by pressure of the lava welling up from below.

dominant *adj.* controlling or ruling, most noticeable, commanding on account of strength or position.

dominant wave the largest, most powerful wave rolling on part of the coast.

dominant wind the WIND that blows with the most effect. It may, or may not, be the PREVAILING WIND.

dormant *adj.* sleeping, quiescent, applied specifically to a VOLCANO which has not erupted in historic time, but is not regarded as extinct.

dormitory town a town from which residents travel daily to work in an accessible nearby larger town or CONURBATION. COMMUTER.

dormitory village COMMUTER VILLAGE

dot map a map showing spatial distribution (commonly based on statistical data for an administrative unit) by the use of dots, usually of uniform size, each representing a specific number of the objects concerned. The value of the dot must be carefully chosen, bearing in mind the high and low quantities to be represented and their location. If statistical data only are available the dots have to be spaced evenly within the administrative unit; but with a knowledge of local conditions they can be placed more precisely to give a more realistic representation, less misleading than even spacing, but usually involving subjective judgement.

double tide a tidal regime in which there is a double high tide (the first falling a little before rising again to a second maximum, e.g. in Southampton Water, southern England) or a double low tide (termed a gulder near Portland, Dorset, England). It is due to the effects of the shape of the coast or of shallow water. TIDE.

down, downs, downland **1.** an open expanse of gently undulating, elevated land, usually of chalk and supporting PASTURE, typically the treeless CHALK uplands of south and southeastern England **2.** in Australia and New Zealand, midlatitude grasslands **3.** The Downs, the name given to part of the North Sea near the Goodwin Sands, off the east coast of Kent, England.

downsizing in industry, the reduction in number of employees.

downthrow in geology, the subsidence of rock strata on one side of a fault, the strata being lowered on the downthrow side. THROW OF A FAULT.

downtown (American) the main business district of a town or city. CENTRAL BUSINESS DISTRICT.

downward-transition region CORE-PERIPHERY MODEL.

down-warping a smooth, downward deformation or sagging of the earth's crust caused by the pressure of weight of a widespread and great mass of material, such as a continental ice sheet (e.g. as in the Great Lakes margin of the Canadian Shield) or sediments (e.g. as underneath the Mississippi delta). When the great weight is removed the crust recoils and in many cases large shallow lakes result. In North America down-warping and recoil contributed to the formation of the Great Lakes. GEOSYNCLINE.

downwind **1.** *adj.* situated to leeward. LEE **2.** *adv.* on the leeward side, in the same direction as the wind.

drainage **1.** the act of taking off excess water from the land by artificial channels **2.** the natural runoff of water from an area by streams, rivers etc. CONSEQUENT, OBSEQUENT, SUBSEQUENT DRAINAGE. The terms applied to the drainage pattern, system or network, i.e. to the arrangement of the main river and its tributaries, include ACCORDANT, ANTECEDENT, CENTRIPETAL, CONCORDANT, DENDRITIC, DERANGED, INCONSEQUENT, INLAND, INSEQUENT, PARALLEL, PINNATE, RADIAL, RECTANGULAR, RESEQUENT, SUPERIMPOSED, TRELLIS. Fig 17.

drainage area all the land with a common outlet for its surface water, synonymous with RIVER BASIN if the river flows into the ocean; but if several rivers flow into an inland sea the whole area draining to that sea may be included.

drainage basin the tract of land drained by a sole river system.

draw (American) a blind CREEK.

drift **1.** in geology, transported superficial deposits, especially those transported and deposited by ice, the two main types being stratified drift (FLUVIOGLACIAL DEPOSITION) and unstratified (BOULDER CLAY, TILL). In the British Geological Survey 'drift' maps cover all superficial deposits; the 'solid' edition maps cover the solid BEDROCK. **2.** slow movement, e.g. of surface waters in the ocean under the influence of prevailing winds, less distinct than a CURRENT **3.** the movement, and accumulation, of loose material such as snow (snowdrift) or sand (SAND DRIFT, SAND DUNE) caused by wind. CONTINENTAL DRIFT.

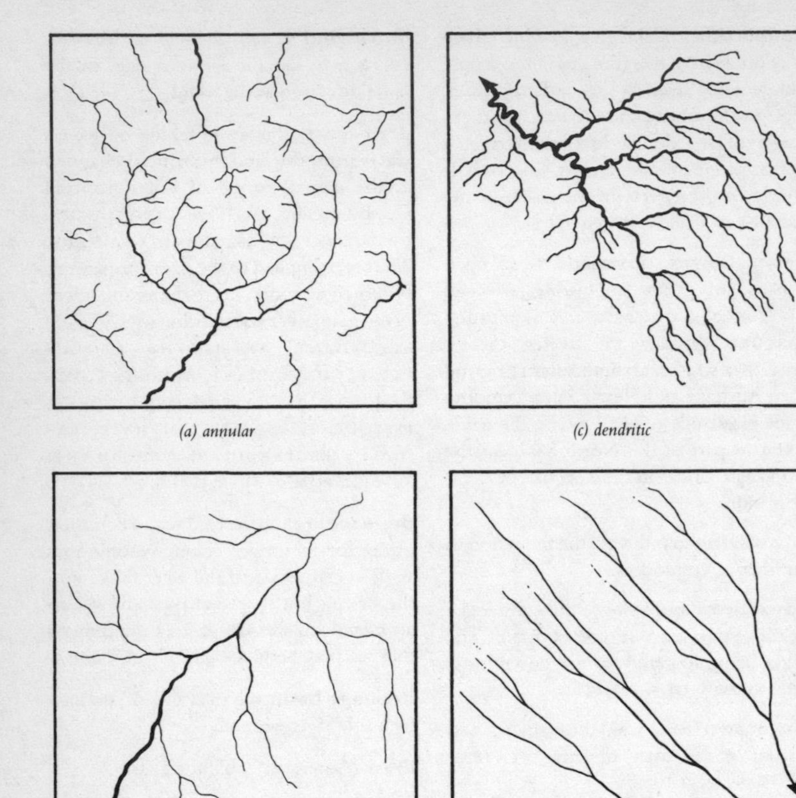

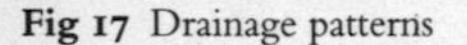
Fig 17 Drainage patterns

drift net a large fishing net, held down and open by weights at the bottom and floats at the top, that moves with the tide.

drizzle a very fine rainfall, with raindrops less than 0.5 mm (0.02 in) in diameter, falling continuously, especially associated with a WARM FRONT.

drought a prolonged, continuous period of dry weather, classified in British meteorology as ABSOLUTE DROUGHT, PARTIAL DROUGHT and DRY SPELL.

drove-road, drove-way a driftway, drift-way, an ancient road or track along which there is free right of way for cattle but which is not necessarily kept in order by any authority. Hence drover, one who drives droves of cattle, sheep etc. to a distant market, and is thus a dealer in cattle.

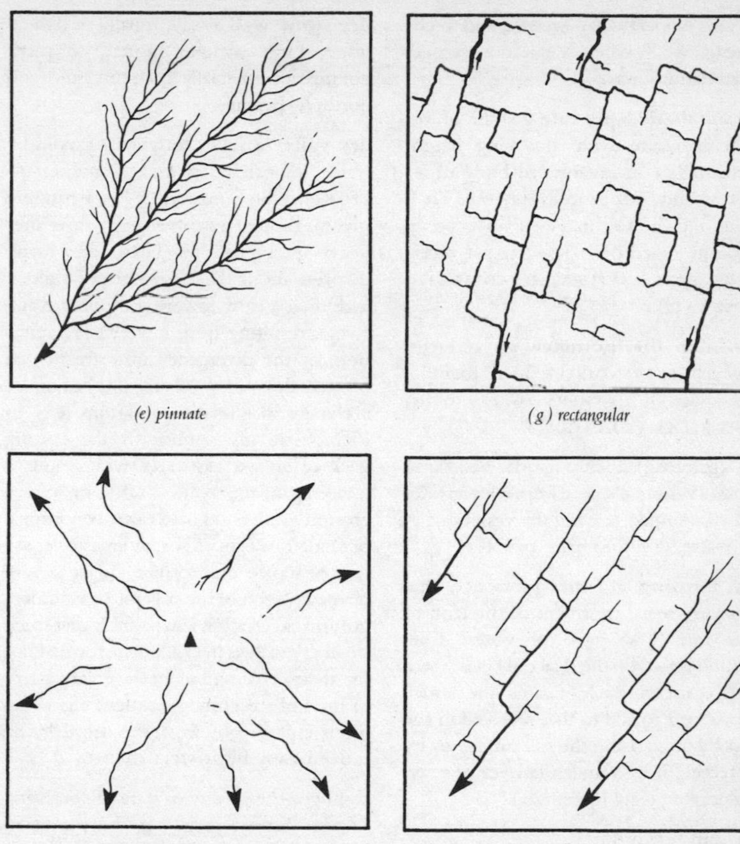
(e) *pinnate* (g) *rectangular* (f) *radial* (h) *trellis*

drowned valley a valley which was excavated in a land surface but owing to a change in sea level has been partly or wholly drowned by the sea. CONCORDANT COAST, FJORD, RIA.

drumlin a smooth, oval, low hill or mound composed mainly of BOULDER CLAY or glacial sands and gravels, occurring in a once-glaciated region, the long axis in line with the movement of the ice that deposited it. Rock drumlins are smoothed mounds of rock with or without their veneer of boulder clay. Drumlins often occur in groups (swarms) as a drumlin field or BASKET OF EGGS RELIEF. FALSE DRUMLIN.

dry *adj.* lacking moisture, specifically defined when applied to air (less than 60 per cent RELATIVE HUMIDITY) and to climate (generally when evaporation exceeds precipitation). ARID.

dry adiabatic lapse rate the rate of loss of temperature with increasing height occurring in an unsaturated body of air as it ascends adiabatically (ADIABATIC), about 1°C in 100 m (5.4°F in 1000 ft) of ascent. ENVIRONMENTAL LAPSE RATE, LAPSE RATE, SATURATED ADIABATIC LAPSE RATE.

dry-bulb thermometer an ordinary mercury THERMOMETER used together with a WET-BULB THERMOMETER to discover RELATIVE HUMIDITY.

dry (graving) dock a narrow basin into which a vessel passes and from which water is then pumped, leaving the vessel out of the water, dry, for repair. DOCK.

dry farming a farming practice that involves special treatment of the land to overcome a shortage of water. One method is to crop the land only every two years, conserving at least part of the rainfall of one year to add to that received in the next by pulverizing the soil surface or by protecting it by a mulch (a layer of straw or decaying plant leaves etc.).

dry gap WIND GAP.

drying oil an oil that has the property of drying and forming a thin elastic film on exposure to air, e.g. linseed oil, used industrially.

dry point settlement a settlement on a site not liable to flooding in a flood region, or on a patch of dry soil in a wet soil region. WET POINT SETTLEMENT.

dry spell **1.** in UK, any period of DROUGHT **2.** in USA, a period of 14 days without measurable precipitation. ABSOLUTE DROUGHT.

dry stone wall a wall, usually of natural stone, built without mortar, to mark boundaries, especially in southwestern and northern Britain.

dry valley a valley, originally carved by water (especially in CHALK and LIMESTONE), which no longer has a running stream, though a BOURNE may flow after heavy precipitation. There are many theories about the origin of dry valleys, including a slow lowering of the WATER TABLE resulting from lowered precipitation; or the divergence of a stream that formed the valley (RIVER CAPTURE); or a change in climatic conditions (e.g. in Pleistocene glaciations); or the cutting back of an ESCARPMENT with resultant lowering of the SPRING-LINE; or surface erosion under PERIGLACIAL conditions; or SPRING-SAPPING; or, in limestone, the disappearance of a former surface stream down a JOINT, or the collapse of an underground cavern. It is also possible that some small dry valleys in chalk were formed not by stream erosion but by the enlargement of lines of structural weakness, e.g. some joints enlarged by frost, the debris being moved away by SOLIFLUCTION.

dualism the quality or state of consisting of two distinct parts, e.g. the condition in the economy of a country where a relatively small group of well-educated, affluent, socially and economically advanced controlling elite live in the central city or in the larger towns, where the economy is dynamic and growing, industry (supported by large injections of capital) uses modern production techniques and management and is capital intensive, labour is specialized, commercial exchange is extensive and complex, the professions are gathered together and salaries are high, while the majority of the population (who are much poorer) live in the countryside where the economy is static, industry is labour intensive, techniques are traditional, trade with other areas is limited and services are inadequate. PLURALISM.

ductile *adj.* of metals, capable of being pressed or drawn into shape without the aid of heat. MALLEABLE.

dude ranch a RANCH organized for the entertainment of tourists, as a place for a holiday, mainly in the USA.

dumb-bell island an island consisting of two parts, often rocky, joined by a narrow isthmus, often of sand, which is never in any part of its length below high water mark. TOMBOLO.

dune a hill or ridge of sand piled up by the wind in dry regions (desert dunes) or along sandy coasts, often independently of any fixed surface feature which might form an obstacle. The form depends on the presence of such an obstacle (which may provide a nucleus), the type and quantity of sand, the characteristics of the land surface over which the sand is moved, the strength and direction of the wind, the presence or absence of ground water, and of vegetation which 'fixes' the sand. Coastal dunes are particularly affected by the presence or absence of vegetation and of welling-up ground water as well as by erosion by the sea. They are identified, in sequence from the sea inland, as FOREDUNES, MOBILE DUNES, STABILIZED DUNES. For other dune details see ADVANCED DUNE, ANTI-DUNE, ATTACHED DUNE, BARCHAN, HEAD DUNE, LATERAL DUNE, LONGITUDINAL DUNE, PARABOLIC DUNE, PLINTH, SEIF-DUNE, SLIP-FACE, TAIL DUNE, TRANSVERSE DUNE, WAKE DUNE.

duopoly the exclusive control of the supply of a product or service in a particular market by two suppliers, who thus dominate the market and between them control the price and scale of the supply. MONOPOLY, OLIGOPOLY, PERFECT COMPETITION.

durable goods goods that are not likely to wear out or decay for a long time, e.g. carpets, furniture, to buy which the consumer is prepared to travel some distance. CENTRAL PLACE THEORY, CONVENIENCE GOOD, ORDER OF GOODS.

duricrust a hard crust covering a relatively soft soil surface in semi-arid, flat areas with a short rainy season and a long, hot dry season. It consists of aluminous, calcareous, siliceous, ferruginous and magnesian materials, drawn to the surface by CAPILLARITY, which brings to the upper soil during the dry season the minerals dissolved during the wet season. At depth it forms duripan (HARD PAN).

dust minute particles of any comminuted dry matter, so fine that they can float in air.

dust bowl a semi-arid tract of land from which the surface soil, exposed by the unwise removal of the covering grassy vegetation by ploughing or overgrazing, has been blown away by wind (DEFLATION-1). The term became widely used after two or three very dry years (especially 1934–5) in southwestern USA, when strong winds raised huge DUST-STORMS in areas where grassy vegetation, formerly protecting the soil, had been removed by ploughing and the land had been cultivated without the necessary protection of WINDBREAKS. SOIL EROSION.

dust-devil a local swirl of wind, laden with DUST, forming a fast-moving pillar of dust, varying in breadth and height, common in most arid lands, especially in hot deserts. It is created by extreme, localized heating of the land surface, leading to strong CONVECTION currents, which gather up the dust. WHIRLWIND.

dust-storm a broad, general term applied to a strong dust-laden wind in arid and semi-arid regions, arising when the air is very hot, excessively dry and accompanied by high electrical tension. The turbulent wind gathers dust from the dry surface and carries it to heights up to 3000 m (10 000 ft), sometimes producing a wall of dust, sometimes a dust-laden WHIRLWIND, larger than a DUST-DEVIL. DUST BOWL, SANDSTORM, SIMOOM.

dyke, dyke-phase, dyke-spring, dyke swarm DIKE, RING-DYKE. Fig 21.

dynamic *adj.* of or related to motion or force.

dynamic equilibrium a state in which balance is maintained despite continual change, e.g. on a slope where the rate of weathering of the rock is balanced by the rate of removal of the weathered rock material. STEADY STATE.

dynamic metamorphism, dynamometamorphism the alteration of pre-existing rocks by intense pressure associated with earth movements, usually on a relatively small scale and without a great rise in temperature, so that new, well-defined rock is formed. METAMORPHIC ROCK, METAMORPHISM, THERMAL (CONTACT) METAMORPHISM.

dynamic rejuvenation REJUVENATION caused by EPEIROGENIC uplift of a landmass with accompanying tilting and warping.

dyne an absolute unit of force: that force which, when applied to a mass of one GRAM, produces an acceleration of one centimetre per second per second. On SI 10^5 dynes = 1 NEWTON. At sea-level at 45°N and 45°S a mass of 1 gram is subjected to a gravitational force of 980.616 dynes.

dystrophic *adj.* applied to a body of freshwater poor in plant nutrients and low in calcium, occurring typically in acid peat areas, the bed of the water being covered with undecomposed plant remains. EUTROPHIC.

E

eagre, egre a tidal wave or BORE.

earth 1. the PLANET on which we live, a flattened sphere (OBLATE SPHEROID) in orbit round the sun, fifth in size and third in order from the sun of the nine planets of the SOLAR SYSTEM. The polar diameter is 12 712 km (7899 mi), the equatorial diameter 12 755 km (7926 mi); the polar circumference 40 008 km (24 860 mi), the equatorial circumference 40 076 km (24 902 mi). It is generally agreed that the surface area is 510 100 448 sq km (196 949 980 sq mi), of which 361 059 266 sq km (139 405 122 sq mi) is water (70.78 per cent of the total surface) and 149 041 182 sq km (57 544 858 sq mi) is land (29.22 per cent of the surface); but some authorities give the surface area as 509 610 000 sq km (196 836 000 sq mi), of which 148 065 120 sq km (57 168 000 sq mi) is land. The mean density is 5.517; the mass is 5.882×10^{21} tonnes. ATMOSPHERE, AXIS OF THE EARTH, BARYSPHERE, CORE, CRUST, HYDROLOGICAL CYCLE, HYDROSPHERE, LITHOSPHERE, MANTLE, ORBIT OF THE EARTH, PLATE TECTONICS, ROTATION OF THE EARTH. TERRESTRIAL MAGNETISM **2.** the solid material of that planet, as distinct from air and water **3.** the disintegrated, loose material on the surface of it, the soil as distinct from the solid rock. Fig 18.

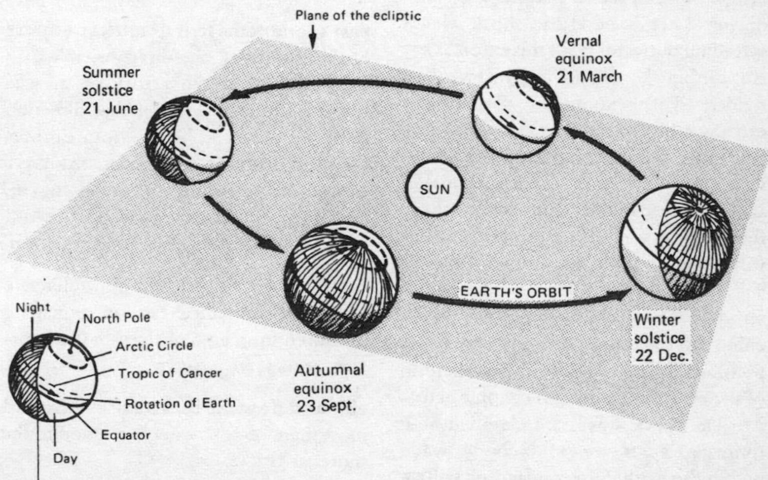

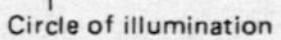

Fig 18 The earth's orbit, and seasons in the northern hemisphere

earthflow a slipping downwards of unconsolidated rock material on the earth's surface, due to its saturation by water and the influence of gravity, occurring on steep and shallow slopes. MASS MOVEMENT, MUDFLOW, SOLIFLUCTION.

earth-movement **1.** a movement of the earth's crust arising from disturbances in the earth's interior (COMPRESSION, DEPRESSION-1, FAULTING, FOLDING, TENSION, uplifting), including both the slow (secular) movements and the sudden (EARTHQUAKES and volcanic activity) **2.** a synonym for OROGENESIS.

earth pillar a high pillar of earthy material or soft rock, sandstone etc. capped by a stone or boulder which protects the underlying soft, easily eroded material. Earth pillars are common in dry regions which are subject to occasionally heavy downpours. DEMOISELLE.

earthquake a shaking of the ground caused by deep-seated disturbances, producing a series of elastic shock waves spreading outwards from the EPICENTRE. An earthquake usually originates from sudden adjustments in the crust of the earth, notably by movement along FAULTS (and thus of tectonic origin), or as a result of volcanic activity. Most severe earthquakes are associated with fault lines where there are no VOLCANOES to act as safety valves. Occasionally they occur in granitic and metamorphic rocks of continental SHIELDS. These may be due to the weakening of the crust arising from compression of plate edges; or the weakening of the crust from previous tectonic activity. The shock waves are classified as P (primary, PUSH WAVE), a body wave within the earth; S (secondary, or SHAKE WAVE); and L (surface, LONGITUDINAL WAVE). The degree of magnitude is usually measured on the RICHTER SCALE and of intensity (related to the effects of waves at the surface) on the MODIFIED MERCALLI SCALE. AFTERSHOCK, ASTHENOSPHERE, ELASTIC REBOUND, HOMOSEISMAL LINE, PLATE TECTONICS, SEISMIC FOCUS, SOIL LIQUEFACTION.

Earth Resources Technology Satellite ERTS. LANDSAT.

earth's crust CRUST.

earth's magnetic field TERRESTRIAL MAGNETISM.

Earth Summit, Rio the meeting of the Heads of Government in Rio de Janeiro, June 1992, at which contemporary environmental problems were discussed, including the need to reduce atmospheric POLLUTION; climatic change; SUSTAINABLE DEVELOPMENT; and the maintenance of BIODIVERSITY.

earth tremor a small, low intensity EARTHQUAKE.

east **1.** one of the four CARDINAL POINTS of the COMPASS, the direction at which the sun rises at the EQUINOX **2.** *adv.* towards the area lying in that direction from the observer, hence, from Europe, the NEAR EAST, MIDDLE EAST, FAR EAST **3.** *adj.* of, pertaining to, belonging to, coming from, or situated towards the east, e.g. of winds blowing from that direction.

easting the first part of a grid reference (GRID), the distance east on a map as measured from a point fixed in its south-west corner. NORTHING.

ebb and flow the backward and forward movement of tidal water in relation to the shore. SLACK-2.

ebb channel **1.** the channel in which the tide flows out most strongly in a river estuary **2.** the route taken by a TIDAL STREAM as it flows seaward after high tide, in some cases differing from the flood channel, the route taken by the flood tidal stream.

ebb tide the receding or outward movement of tidal water, after high tide and before low tide, in contrast to the FLOOD TIDE. SLACK-2.

echo sounder an instrument used especially for calculating the depth of water by measuring the time taken for a sound (sonic or ultra-sonic vibration) generated at the surface to return after being reflected from the sea floor (SOUNDING). It is also used to measure the thickness of ice and the depth of different densities of rock or of soil. SONAR.

eclipse the total or partial cutting off of light received by one celestial body from another by the interception of a third celestial body in passing between the other two. From the earth the eclipse of the sun (solar eclipse) occurs when the moon comes into line between the sun and the earth and casts a shadow on the earth; the eclipse of the moon (lunar eclipse) when the earth intercepts light from the sun to the moon and casts a shadow on the moon.

ecliptic the apparent path of the sun in the CELESTIAL SPHERE during the period of one year as seen from the earth, and relative to the fixed stars. The path makes a great circle, at an angle of 23°27' with the celestial equator; and it is divided into 12 imaginary sections, each identified by a sign of the zodiac. Fig 18.

ecocentrism the state or quality of being focused on the ELEMENTS-1 of the natural environment. ANTHROPOCENTRISM.

ecological balance the balance of nature, the balance maintained in a stable ECOSYSTEM by the gradual readjustments in the composition of a balanced community in response to natural SUCCESSION, changes in climate, or other influences. Such a delicate balance is easily upset by human activities, e.g. by the introduction or elimination of plants or animals, by pollution of the environment, by the destruction of habitats etc.

ecological factor any environmental factor which affects a living organism.

ecological fallacy the false assumption that characteristics or relationships observed in aggregated (AGGREGATE-1) data are also present in the individual data from which the aggregated data is produced. HISTORICAL FALLACY.

ecological validity a research finding held to be true in a range of natural settings or conditions, as distinct from one that is true in artificial settings or conditions only.

ecology the scientific study of the interrelationships between living organisms and the ENVIRONMENT (including the other living organisms present) in which they live. Without qualification the term ecology tends to be confined to plant ecology. Ecologists use many special terms in their studies, particularly in connexion with plant communities which have developed in harmony with the environment, the idea of development (CLIMAX, SUCCESSION) being fundamental. ECOSYSTEM, ECOTONE. Geography has been defined as HUMAN ECOLOGY, the interrelationship between people and place.

economic base theory a theory based on the assumption that urban and regional growth can be explained in terms of the numbers employed in the BASIC ACTIVITIES, responsive to external demand, and the NON-BASIC ACTIVITIES, meeting the internal demands of the city or region itself. LOWRY MODEL.

economic determinism in Marxism, the theory that the economic processes of a society have a determining effect on its other processes, particularly, for example, on the political processes. TECHNOLOGICAL DETERMINISM.

economic distance the distance a COMMODITY-1 may travel before its value is exceeded by the transport costs.

economic geography the branch of geography dealing with the interaction of geographical and economic conditions, with the production, spatial distribution, exchange and consumption of wealth, and with the study of the economic factors affecting the AREAL DIFFERENTIATION of the earth's surface.

economic good a good of economic value, i.e. it is useful, scarce and marketable. Economic goods are usually classified as CONSUMER GOODS and services, and PRODUCER GOODS and services. CAPITAL GOODS, COMMODITY, CONSUMER DURABLES, FREE GOODS.

economic growth the increase in real national income or (more commonly) in real national income per head over a long period of time.

economic man in CLASSICAL ECONOMIC THEORY, a person who is motivated solely by economic considerations, who manages personal income and expenditure strictly in accordance with personal, material interest, profit being the only objective. This concept assumes perfect knowledge of relevant circumstances and a perfect ability to use that knowledge in order to take the greatest advantage of it; and the gift of totally accurate prediction in order to achieve the lowest possible costs and the highest profits. BOUNDED RATIONALITY.

economic rent 1. the net surplus available to any factor of production (e.g. capital, labour, land) after the deduction (a) of all the costs, including interest on invested capital, involved in keeping it in its present use; and (b) the returns possible from an alternative use of the factor of production (termed opportunity costs). LAND RENT. This concept of economic rent is used particularly in agricultural geography to account for the farmer's decision to use land in one way rather than another (based on the assumption that the farmer will opt for the use yielding the higher economic rent) **2.** a variation of that application, used in studies of agricultural location, to account for the fact that the land is farmed at all. In such studies OPPORTUNITY COSTS are not taken into account, ignored on the assumption that only one use of the resource is possible, i.e. agriculture. The economic rent then becomes the net income earned by the farmer in excess of the net income which might be earned from producing at the margin. The margin is the point where the level of net income justifies the use of land for agriculture, and where a true economic rent (defined above) is non-existent. The position of the margin is usually governed by the level of transport costs involved in marketing, which increase with increasing distance from the market. David Ricardo, the nineteenth century classical economist, held this same concept of economic rent, but he considered variations in soil fertility resulted in variations of land use, resulting in differences in economic rent. **3.** a synonym for LAND RENT. RENT, VON THUNEN MODEL.

economies of scale the lowering of unit costs achieved by increasing the scale of production. The reduction in unit costs is brought about by internal economies, i.e. economies within the enterprise (e.g. greater specialization and division of labour, the spreading of research, development and other fixed costs over more production units) and from the external economies (economies of localization) which arise when firms in the same or similar industries are located close together, thereby benefiting from the availability of a skilled labour force, specialist services, supplies, infrastructure, marketing etc. (AGGLOMERATION-3), or when the growth of an entire industry reduces the costs in each individual firm. DISECONOMIES OF SCALE, ECONOMIES OF URBANIZATION.

economies of urbanization the ECONOMIES OF SCALE achieved by a wide range of industries from the circumstances of URBANIZATION-1, i.e. the well-developed physical structure and services, large labour force with diverse skills, large potential market. DISECONOMIES OF URBANIZATION.

economy 1. a SYSTEM-5 of production and distribution designed to meet the material needs of a country, region or society **2.** a part of such a system, e.g. agricultural economy.

eco-refugees people who leave their homeland to escape drought, DESERTIFICATION, soil erosion and other environmental problems.

ecosphere the BIOSPHERE and all the ECOLOGICAL FACTORS which affect organisms.

ecosystem ecological system, a SYSTEM-2 formed by the interaction of all living organisms (plants, animals, bacteria etc.) with each other and with the chemical and physical factors of the environment in which they live, all being linked by the transfer of energy and materials (FOOD CHAIN). The boundary of an ecosystem is difficult to define (the whole world may be considered as an ecosystem), but the term is usually applied to a small system where the net transfer of energy and materials across the boundary is low, e.g. a pond, a forest, a small oceanic island. An ecosystem is never totally self-contained or closed: solar energy received crosses the boundary, as does a foraging animal. The part of the world which forms the home for an ecosystem is termed an ecotope; but ecotope is sometimes used as a synonym for ecosystem.

ecotone a transitional zone between two HABITATS where different plant associations merge.

ecotope ECOSYSTEM.

ecotourism TOURISM planned to respect and safeguard the environment, based on the natural attractions of an area and aware of the concept of SUSTAINABLE DEVELOPMENT of fragile ecosystems.

ecumene, oecumene 1. the habitable world known to the ancient Greeks **2.** the part of the earth's surface suitable, through climatic conditions, for permanent human settlement. ANECUMENE.

ecumenopolis the city of the future, covering most of the habitable surface of the earth as a continuous system, forming a universal settlement, the limits determined by climatic constraints and the extent of fairly flat land. The term was introduced in 1961 by C. A. Doxiadis, who saw the natural hierarchy of large urban settlements as large CITY, METROPOLIS-2, CONURBATION, MEGALOPOLIS, ecumenopolis.

edaphic *adj.* related to, due to, dependent on, or having characteristics due to, the nature of the soil.

edaphic climax the vegetation CLIMAX produced by EDAPHIC FACTORS.

edaphic factors soil factors, the biological, chemical and physical properties of a soil. EDAPHIC CLIMAX.

edaphic formations vegetation formations classified according to the soil types that determine them, as distinct from CLIMATIC FORMATIONS.

eddy a swirling movement of fluid within a larger mass of fluid, in a direction contrary to that of the main flow, e.g. as in depressions or highs in the atmosphere, or in the water of a river as it encounters an obstruction in its flow. HYDRAULIC FORCE.

edge 1. loosely, a sharp ridge, especially one with an exposure of rock, a topographic term not generally used specifically in geographical writing, but commonly used in place-names to indicate an ARETE, ridge or mountain crest, e.g. Wenlock Edge in Shropshire, England **2.** in mathematics, a link, a route in a TOPOLOGICAL diagram. GRAPH-2.

EEC, European Economic Community, Common Market, European Union (EU) a group of European countries established 1 January 1958 on the basis of a treaty signed in Rome 25 March 1957. The aims included increased productivity, free mobility of labour, coordinated transport and commercial policies, and control of restrictive practices among members. The original six members, Belgium, France, Federal Republic of Germany, Italy, Luxembourg and Netherlands, were joined 1 January 1973 by the UK, Denmark (with Greenland) and the Irish Republic. Greenland, on gaining independence from Denmark, withdrew on 23 February 1982. Greece joined in 1981, Spain and Portugal in 1986, Sweden and Finland in 1995. Customs duties on trade between the six were phased out by 1 July 1968, between the nine by 1 July 1977. Some developing countries have associate status in the EEC, others have trading agreements with the Community. In 1992 the name was changed to the European Union (EU). The aims, operation and membership of the EU are detailed in the *Statesman's Year-Book*, Macmillan. CAP.

effective accessibility the extent to which a place or service is actually accessible, governed not only by the distance to be travelled but also by whether or not the means of transport, the time available, and social circumstances make an approach possible. ACCESSIBILITY, COST SPACE.

effective precipitation 1. that part of total precipitation which is of use to plants **2.** in hydrology, that part of total precipitation which flows into a stream channel.

effluent a pouring out, flowing away, hence a stream flowing out of a lake or out of a reservoir, or from land after irrigation; or the flow of waste liquid from sewage works or from a factory.

EFTA, European Free Trade Association a group consisting originally of seven European countries (sometimes known as the outer seven in contrast to the original six countries of the EEC) which linked themselves together effectively in 1960 for the purposes of trade, aiming to abolish tariffs on imports of goods originating in the group. The original members were the UK, Norway, Sweden, Denmark, Portugal, Austria and Switzerland, joined by Iceland 1970, Finland as an associate member 1961. The UK and Denmark left 1972 on joining the EEC.

egocentrism the state or quality of being egocentric, of perceiving the world with the self as centre. In the ordering of the world from the egocentric viewpoint (i.e. with the individual perceiving self as centre), the value of its components, the perceived objects, declines rapidly with increasing distance from the individual. ANTHROPOCENTRIC, ETHNOCENTRISM.

E horizon SOIL HORIZON.

ejido (Spanish) **1.** a system of land tenure reform under which HACIENDAS were transferred to communal ownership in Mexico **2.** an agrarian property, held and worked in common, and belonging to Mexican villagers.

elastic *adj.* of or pertaining to any substance that returns to its original shape or size after being deformed by stress so long as the stress does not exceed the elastic character of the substance and the substance itself is not over-fatigued. PLASTIC.

elastic rebound the recoil of rocks to a position near to that of the original after they have been forced apart by stress and the strain has relaxed, as in FAULTS and EARTHQUAKES.

E-layer HEAVISIDE LAYER.

elbow of capture of a river, RIVER CAPTURE.

electoral geography the branch of HUMAN GEOGRAPHY concerned with the study of election statistics and the geographical aspects of the organization, results and consequences of elections.

electric arc a luminous, continual electrical discharge producing very high temperatures (usually exceeding 3000°C: 5420°F), occurring when an electric current is carried by the vapour of the ELECTRODE (or by ionized gas) between two separated electrodes. ARC FURNACE.

electricity, electrical energy, electrical power POWER generated in dynamos in which energy is derived from turbines of two types: **1.** steam-turbines (producing thermal electricity) driven by heat from COAL (especially low-grade bituminous, lignite and brown coal), OIL, NATURAL GAS, PEAT, NUCLEAR ENERGY, GEOTHERMAL ENERGY (and, experimentally, SOLAR energy) **2.** hydraulic turbines (producing HYDROELECTRIC POWER) driven by WATER POWER, including tidal power (TIDAL POWER STATION). The output of electrical energy is commonly measured in kilowatt-hours (kWh).

electrode either of the two conductors (i.e. the anode or the cathode) by which an electric current is passed by an electrical circuit in an apparatus such as a discharge tube or an electrolytic cell. ELECTRIC ARC.

electromagnetic spectrum the range of wavelength (or of frequencies) of ENERGY radiated by ELECTROMAGNETIC WAVES. The spectrum is divided into bands on the basis of the type of wave, i.e., in order of decreasing wavelength, radiowaves (the longest), radar waves, infra-red radiation, visible light, ultra-violet radiation, x-rays and gamma rays. Fig 19.

electromagnetic wave a wave propagated through space or a medium by simultaneous periodic variation in the electric and magnetic field intensity at right angles to each other and to the direction of propagation. ELECTROMAGNETIC SPECTRUM, ENERGY.

electron ION.

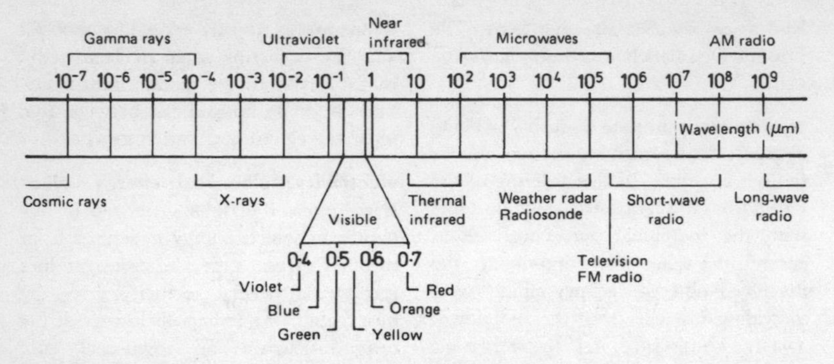

Fig 19 Electromagnetic spectrum showing bands and wavelengths

element 1. generally, a small quantity of something, present in a larger whole **2.** a basic component of a branch of study **3.** in ancient and medieval philosophy, one of what were considered to be the basic constituents of the universe, i.e. earth, air, fire, water **4.** rhetorically, 'to brave the elements', to go outside in unpleasant weather, e.g. in a heavy snowstorm **5.** in biology, the natural habitat of an organism **6.** in chemistry, a substance that has never been separated into simpler substances by ordinary chemical means because all its atoms have the same atomic number. There are over 100 such substances known, any of which, individually or together, are present in all matter, the first ninety-two (up to and including uranium) occurring naturally.

elevation 1. the vertical distance above a specific level, e.g. above sea-level **2.** the vertical angle between the horizontal and a high point, e.g. between the horizon and a star, or between a hill base and a hill top. ALTITUDE **3.** in architecture, the view of one of the sides of a building, or a drawing of this view.

ellipsoid a SPHEROID with a form that is regularly oval. GEOID, OBLATE.

el Niño effect a warm ocean current, originating in the warm equatorial current, which every seven to fourteen years temporarily replaces the cold current (Humboldt current) off the Peruvian coast. This occurs when the southeast trade winds in the Pacific lose their strength; it leads to a fall in the quantity of PLANKTON associated with the cold Humboldt current, and thus to a decrease in fish numbers. It also seriously upsets the tropical ATMOSPHERIC CIRCULATION, leading to major climatic anomalies. These self-reinforcing disturbances of ocean and atmosphere are termed the El Niño and Southern Oscillation (ENSO).

eluvial horizon the soil layer (broadly the A HORIZON) that is subject to ELUVIATION. HARD PAN, ILLUVIAL, ILLUVIATION, SOIL, SOIL HORIZON.

eluviation the process by which material (especially that consisting of BASES-2) is removed in SOLUTION-1 (LEACHING), or mechanically in water suspension, from the upper horizon or horizons of a SOIL by downward or lateral percolation of the water. ILLUVIATION, SUB-SURFACE WASH. In American usage eluviation is confined to the mechanical transport, leaching being the term applied to the process in which particles are moved in solution.

emergence the process of coming forth from concealment, hence applied particularly to **1.** the rise of the level of the land in relation to the sea, so that land formerly under the sea becomes dry, e.g. a RAISED BEACH **2.** the point at which an underground stream comes to the surface.

emigrant one who migrates (moves) voluntarily away from the land of birth to take up permanent residence in another country. EXILE, EXPATRIATE, IMMIGRANT, MIGRANT, MIGRATION, REFUGEE.

empirical *adj.* **1.** pertaining to, derived from, based on, or making use of, experience, trial and error, observation or experiment rather than theory or knowledge **2.** in mathematics, applied to a formula reached by inductive reasoning, not verified by deductive proof. DEDUCTION, INDUCTION.

empiricism **1.** the theory that all concepts are derived from experience and that all statements claiming to express knowledge depend on experience for their justification **2.** the philosophical system which considers true knowledge to be that which can be perceived and rejects that which cannot be verified (e.g. theoretical statements) **3.** the scientific method of proceeding by inductive reasoning (INDUCTION) from observation to the formulation of a general principle which is then checked by experiment. EMPIRICAL, LOGICAL POSITIVISM, POSITIVISM, SCIENTIFIC LAW.

empolder *verb* to reclaim by the creation of POLDERS.

enclave (French) **1.** an outlying part of a nation state lying within the territory of another nation state, as seen from the point of view of the territory in which it lies. EXCLAVE **2.** a small cultural or linguistic group surrounded by another cultural or linguistic group which is dominant.

enclosure INCLOSURE.

enculturation the process by which individuals are brought up to be members of a CULTURE or SOCIETY-3. ACCULTURATION.

endangered species plant and animal species with POPULATION-1 numbers so reduced (often by human activities) that their future existence seems uncertain. CITES.

endemic *adj.* restricted to a certain region or people, having originated there, applied especially to a disease that is normally confined and always likely to occur in certain areas, sometimes reaching EPIDEMIC proportions. PANDEMIC.

end moraine TERMINAL MORAINE.

endogenetic, endogenic *adj.* arising from within, having an internal cause or origin, e.g. applied to the geological processes which originate from within the earth (e.g. DIASTROPHISM, VULCANICITY) and landforms arising therefrom, in contrast to EXOGENETIC. Endogenetic is the usual English form, endogenic the American; but American authors sometimes differentiate endogenetic (applied to the product, e.g. the rocks formed) from endogenic (applied to the process).

endoreic *adj.* in-flowing, applied particularly to basins of inland drainage. AREIC, EXOREIC.

energy the capacity to do work possessed by a physical body or system of physical bodies, i.e. the capacity to move a force a certain distance, measured in JOULES. The two main types are KINETIC and POTENTIAL, other forms being mechanical (of moving bodies, stretched springs), chemical (energy stored in molecules of compounds), thermal (energy of heat, including geothermal, derived from the interior of the earth; and from the internal random movement of molecules), nuclear (of the nucleus of the atom) and radiant (e.g. solar), carried by ELECTROMAGNETIC WAVES, the only form of energy existing in free space, i.e. in the absence of matter. Each form of energy can be transferred to another form by suitable means, but the transformation is possible only in the presence of matter. ELECTRICITY, NATURAL RESOURCES, NUCLEAR ENERGY, POWER, RADIATION.

energy pyramid in ecology, a BAR GRAPH in the form of a pyramid showing the energy lost and retained in the different TROPHIC LEVELS-1 of a FOOD CHAIN. Each trophic level is represented by a horizontal bar representing the total energy received by the organisms in that level; and each bar is divided to show how much of that energy received is retained by and lost by the organisms in that level. Such a graph is particularly useful in providing a quantitative picture of an ECOSYSTEM.

Engels' Law the generalization propounded by Friedrich Engels, 1820–95, that as the income of an individual rises the proportion of the total spent on food decreases.

englacial *adj.* embedded in a GLACIER, as distinct from SUBGLACIAL, SUPERGLACIAL.

enhancement the processes and techniques used to sharpen the IMAGES-3 received by SENSORS in order to ease the interpretation of the PIXEL.

Enlightenment, the MODERNITY-2.

Enterprise Zone one of the zones, designated according to the UK Finance Act 1980, in economically derelict urban areas. Various concessions, designed to attract industry to such areas, included a ten-year exemption from rates, exemption from Development Land Tax, 100 per cent capital allowances, simplified planning controls, few government requests for statistical information.

entisols in SOIL CLASSIFICATION, USA, young MINERAL SOILS, lacking developed horizons, occurring in all climates.

entrepôt (French) anglicized **entrepot**, a place to which goods are brought to be stored temporarily while awaiting transfer to another country, and where they are not liable to customs duties. Commonly used as an *adj.*, e.g. entrepot port, entrepot trade.

entrepreneur a person who wholly or partly undertakes, manages and controls an enterprise and, in the case of a business, bears the financial risks involved.

entropy a measure of the amount of disorder in a SYSTEM-1,2,3. The more disordered the system the higher the entropy. NEGENTROPY.

environment that which surrounds, the sum total of the conditions of the surroundings within which an organism, or group, or an object, exists (including the NATURAL conditions, the natural as modified by human activity, and the ARTIFICIAL). The term is used broadly in geography, especially in human geography, where the emphasis is on the economic, cultural and social conditions of the surroundings. BEHAVIOURAL ENVIRONMENT, CONTROLLED ENVIRONMENT, CULTURE, ENVIRONMENTAL HAZARD,

ENVIRONMENTALISM, ENVIRONMENTAL STUDIES.

environmental determinism DETERMINISM.

environmental hazard a risk (usually to human beings) associated with the physical ENVIRONMENT. Such a risk may be natural (e.g. FLASH-FLOOD, LANDSLIDE, VOLCANIC ERUPTION) or produced by human activity (e.g. POLLUTION).

environmental impact the impression, particularly the undesirable or unpleasant impression, made on an ENVIRONMENT by the introduction of something alien to it, e.g. grazing animals in an area of sparse vegetation.

environmentalism the philosophical concept which stresses the influence of all the items and conditions of the ENVIRONMENT on the life and activities of people. In its extreme form it becomes environmental DETERMINISM.

environmental lapse rate static lapse rate, the actual rate of loss of temperature with increasing height at a specific location at a specific time, averaging some 0.6°C per 100 m (1°F per 300 ft). LAPSE RATE.

environmentally sensitive area ESA.

environmental studies a collective term for the studies which aim to make people aware of the conditions of the world they inhabit, and of the interrelationship of people with the physical, cultural and social items and conditions of their surroundings, i.e. the study of natural history, architecture, ecology, meteorology and many of the other studies commonly included in geography.

environs the area surrounding a place (not a person). ENVIRONMENT.

enzyme an organic CATALYST, one of a very large class of protein-containing compounds formed in and produced by living matter, which promote the chemical reactions on which life depends (e.g. digestion, reproduction). Each enzyme is usually responsible for one, or a few, of these reactions, the enzyme combining with a specific substance to bring about a chemical change in that substance without itself suffering permanent change. The process is usually dependent on temperature, pH, the presence of co-enzymes, and activators such as vitamins, metallic salts, etc.

eolith a crude stone tool showing some chipping believed to be the result of human work. In England and France flint was the stone most commonly used in the EOLITHIC (dawn of Stone Age) and in the PALAEOLITHIC (old Stone Age); in the NEOLITHIC (new Stone Age) stone implements were polished and refined.

Eolithic Age the earliest period of the STONE AGE when EOLITHS first came into use.

epeirogenetic, epeirogenic *adj.* of or pertaining to the formation of continents, applied to the type of mass earth movements which result in changes of level over large areas (e.g. continental uplift and depression, which may involve TILTING or WARPING but not intense FOLDING) in contrast to orogenic movement (OROGENESIS).

ephemeral *adj.* applied to a plant with a short life cycle (from germination to production of seed). It may produce several generations in a year, e.g. groundsel; or have a brief life when environmental conditions are suitable, e.g. after heavy rain in a hot desert region. ANNUAL, BIENNIAL, PERENNIAL.

epicentre the point on the earth's surface immediately above the centre of origin (the SEISMIC FOCUS) of an earthquake.

epidemic a disease that becomes widespread in a particular place at a particular time. ENDEMIC, PANDEMIC.

epidemiology the branch of medicine concerned with the study of the factors influencing the frequency and spread of diseases. It is concerned with the features and the prevalence of a disease in a population, rather than in an individual.

epilimnion the warmer top layer of water in a lake or the ocean, liable to be disturbed by wind and convection currents, lying above the THERMOCLINE. THERMAL STRATIFICATION.

epiphyte a plant which grows on the outside of another plant, using that plant just for support (i.e. not a PARASITE or a SAPROPHYTE), e.g. LIANA, LICHENS, MOSSES, orchids.

epistemology the branch of philosophy concerned with the study of the nature, origin, foundations, limits and validity of knowledge. IDEALISM-3, METAPHYSICS.

epoch GEOLOGICAL TIMESCALE.

equator an imaginary GREAT CIRCLE round the earth in a plane perpendicular to the earth's axis and equidistant between the north and south poles, thus dividing the earth into the northern and the southern hemispheres. Fig 27.

equatorial *adj.* or, near to, or pertaining to, the EQUATOR.

equatorial air mass a warm AIR MASS of high humidity, the source of which lies over the ocean in the zone of the EQUATORIAL CLIMATE.

equatorial belt the zone lying approximately between the latitudes 10°N and 10°S, the area of the EQUATORIAL CLIMATE.

equatorial climate the climate occurring in a belt on each side of the EQUATOR, near sea-level and between approximately 10°N and 10°S, characterized by constant high temperature (about 27°C: 80°F) and humidity with little range throughout day and night and the year, and approximately a 12-hour day, 12-hour night. It is subject to CONVECTIONAL RAIN, the maxima corresponding to the EQUINOXES.

equatorial current, equatorial counter-current the surface movement of ocean water in EQUATORIAL latitudes. In the northern hemisphere the north equatorial current moves towards the southwest or west; in the southern hemisphere the southern equatorial current moves towards the northwest or west; between the two the equatorial counter-current flows eastwards. EL NINO.

equatorial forest the luxuriant FOREST-1 mostly of evergreen hygrophilous (HYGROPHYTE) species, growing at lower altitudes in a belt from approximately 7°N to 7°S in areas of EQUATORIAL CLIMATE, particularly in the Amazon and Congo (Zaire) basins and also, with modifications, in parts of Indonesia, Malaysia, Sri Lanka and west Africa. The constant moisture and heat encourage rapid growth of tall, mostly HARDWOOD, trees, their spreading crowns forming a thick canopy. LIANAS and other woody climbers struggle up them to reach the light. In the modified areas, where tree growth is less dense, enough light penetrates to allow smaller plants to grow and form UNDERGROWTH. BUSH FALLOWING, RAIN FOREST, TIMBER.

equatorial front equatorial trough, ITCZ.

equifinality the end state, one of similarity, achieved by SYSTEMS-1,2,3 founded on different initial conditions and undergoing change. The term is applied especially to landforms, e.g. a PENEPLAIN, and in SYSTEMS ANALYSIS. MULTIFINALITY.

equilibrium a state of balance between opposing forces or effects. A body is said to be in a state of stable equilibrium if it returns to its original position after being moved by a small impulse; unstable if it continues to move away from its original position in the direction given to it by a small impulse; neutral if it comes immediately to rest and remains stationary in its new position after being moved by a small impulse. In geography the term equilibrium is usually applied to a static condition, i.e. one of no change; or a dynamic condition, i.e. one where a balance is maintained by continued adjustments in reaction of the opposing forces (LOWRY MODEL), e.g. in slopes where the rate of rock weathering and the rate of removal of the rock debris reach a state of balance, the angle of the slope continually accommodating to changing factors of weathering and removal. DYNAMIC EQUILIBRIUM, STEADY STATE.

equinox day and night of equal length. The moment or point when the sun crosses the EQUATOR during its apparent annual movement from north to south and is directly overhead at noon along the equator, i.e. 21 March and 22 September, respectively the spring (vernal) equinox and the autumnal equinox in the northern hemisphere. PRECESSION OF THE EQUINOXES. Fig 18.

era one of the major divisions of geological time. GEOLOGICAL TIMESCALE.

erg (north Africa: Arabic) the sandy desert of the Sahara, the KUM of central Asia. HAMADA, REG.

ericaceous soil an acid, lime-free soil, one on which plants of the genus *Erica*, family Ericaceae, flourish.

erosion the processes of the wearing away of the land surface by natural agents (running water, ice, wave action, wind, and including CORRASION and CORROSION) and the transport of the rock debris that results. This does not include the WEATHERING of rocks in situ or MASS MOVEMENT caused by the force of gravity. Erosion is used frequently, but incorrectly, as synonymous with DENUDATION. ACCELERATED EROSION, THERMAL EROSION.

erratic, erratic block a mass of rock or a boulder transported by a GLACIER or ice sheet (which usually has since disappeared) and deposited in an area remote from its place of origin. The erratics of VALLEY GLACIERS often perch precariously on the valley sides, and are thus termed perched blocks; far-travelled boulders are sometimes distinguished by the term 'exotic'. The former course of the glacier or ice sheet can be traced by tracking erratic blocks back to their sources.

ERTS Earth Resources Technology Satellite, LANDSAT.

eruption the process by which solid, liquid or gaseous material pours forth gently or explosively from the vent of a VOLCANO or from fissures in the earth's crust as a result of volcanic activity; and to the process by which hot water and steam pour forth from a GEYSER.

ESA environmentally sensitive area, an area in the EU with a fragile or rare HABITAT, and designated by the EU as being in need of special protection.

escalator region a REGION-I where occupational and CLASS mobility are higher than those in other regions.

escarpment 1. generally, any more or less continuous line of cliffs or steep slopes resulting from the differential EROSION of gently inclined strata or from faulting. FAULT **2.** specifically, the abrupt cliff face or steep slope of a CUESTA. The term escarpment should not be applied to the cuesta as a whole, and is preferable to the term SCARP which is sometimes applied to the steep face of a cuesta. Fig 42.

esker a long, continuous, sinuous, steep-sided ridge of glacial and fluvioglacial sands and gravels, deposited by the melt-water of a glacier or ice sheet. BEADED ESKER, KAME. Fig 20.

essential minerals the mineral constituents, not necessarily the major constituents, of an IGNEOUS or METAMORPHIC rock which determine its mineralogical classification and from which the rock is named. ACCESSORY MINERAL.

estancia a large farm in South America on which cattle are reared on a large scale. HACIENDA, LATIFUNDIA, RANCH.

estuarine *adj.* of or pertaining to an ESTUARY, applied especially to the deposits laid down in brackish water and to the environmental conditions of an estuary.

estuary the tidal mouth of a river where the saltwater of the tide meets the freshwater of the river current.

étang (French) a shallow pool or lake

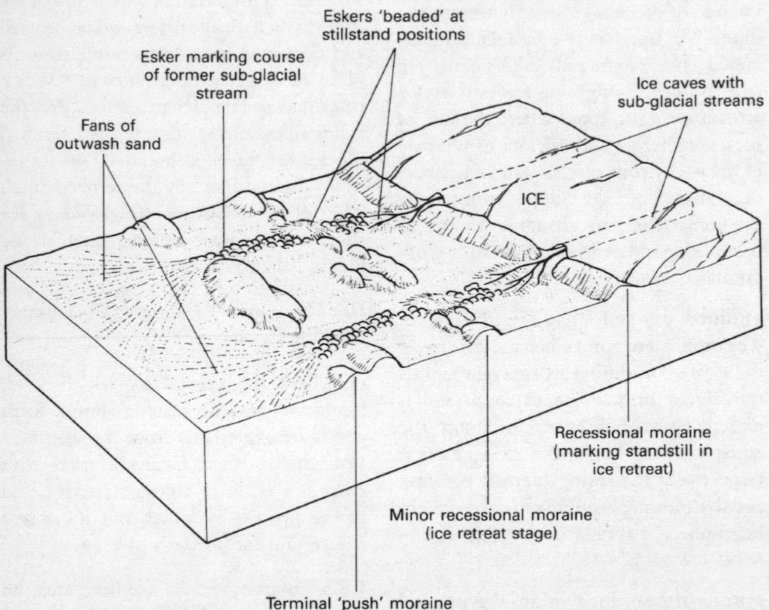

Fig 20 Eskers, moraines and outwash fans

caused by the ponding back of water draining from the land by beach material thrown up by the sea (an étang salé being one which communicates with the sea).

etesian wind (Turkish meltemi) a strong north to northwesterly wind blowing in Greece and other parts of the eastern Mediterranean, caused by a steep pressure gradient associated with the thermal low pressure lying over the Sahara. Strongest in the afternoon, dying out in the evening, it gives rise to rough seas, fog in coastal areas, and clouds of dust; and it may blow with such force and dryness in some localities that it prohibits tree growth.

ethnic *adj.* **1.** historically, pertaining to nations not Jewish or Christian, thus Gentile, heathen, pagan **2.** now, of or relating to people unified by common geographic origin, skin pigmentation, cultural, religious or linguistic ties, or deriving from or belonging to such ties. ETHNIC MINORITY.

ethnic cleansing a process in a multicultural state whereby one cultural group or ethnic community is expelled by force from territory shared with others by another more powerful group or community that wishes to create a homogeneous state/society, e.g. Bosnian Muslims by Bosnian Serbs in the civil conflict in the former Yugoslavian Republic in 1992.

ethnic minority, minority ethnic group a small ETHNIC-2 group, usually composed of IMMIGRANTS or the descendants of past immigrants, commonly living in large towns, sometimes in GHETTOES. Such a group is minor not only in the number of its members but also in the relationship of the group to the structures of political and economic power in the society in which it lives, which are controlled by bigger, more dominant groups. CULTURE CONTACT.

ethnobotany the branch of botany concerned with the scientific study or description of the relationship between people and plants, particularly the study of the knowledge and customs, and the understanding, of indigenous people of the plants in their own environment; and the uses to which plants and plant products are put by different cultural groups.

ethnocentrism the state or quality of being ethnocentric, i.e. regarding one's own ETHNIC group as being of superior, or supreme, importance, the human attributes of other groups decreasing with increased distance (physical, or the distance created by cultural differences, etc.) from one's own group. An extreme form of ethnocentrism is exhibited in racism. ANTHROPOCENTRIC, EGOCENTRISM.

etiology, aetiology in medicine, the science, doctrine, demonstration of the factors associated with the causation of disease. MEDICAL GEOGRAPHY.

euphotic zone the uppermost zone of ocean or lake depths where penetration by sunlight is sufficient for PHOTOSYNTHESIS to take place, rarely extending to depths below 100 m (55 fathoms: 325 ft). PELAGIC, PHOTIC.

European Community EEC.

European Free Trade Area EFTA.

eustatism, eustasy a world-wide simultaneous change of ocean level due to a rise or fall of the ocean, not to that of the land (ISOSTASY). This change of level may be due to GLACIO-EUSTATISM, with a consequent decrease or increase in the amount of ocean water, to ocean floor spreading (PLATE TECTONICS) or to changes in the capacity of ocean basins due, e.g., to infilling by sedimentation. DIASTROPHIC EUSTATISM, NEGATIVE MOVEMENT, POSITIVE MOVEMENT, REJUVENATION.

eutrophic *adj*. applied to a body of fresh water (e.g. a swamp or lake) which, having been over-supplied with organic or mineral nutrients, promotes excessive growth of ALGAE and other plants which draw on so much oxygen that little (or none in extreme cases) remains to support animal life, which is accordingly depleted or destroyed. DYSTROPHIC.

evaporation the process by which a SOLID OR LIQUID is converted to a gaseous state, to VAPOUR; or the action or process of driving off the liquid part of a substance in the form of vapour by means of heat. Water vapour in the atmosphere is the result of the evaporation of water from the earth's surface (HYDROLOGICAL CYCLE), a continual process dependent on air temperature, the quantity of water vapour already in the atmosphere, nature of the water or the land surface, and wind. The highest rate of evaporation occurs in hot DESERTS in conditions of great heat, dry atmosphere, lack of surface cover (plant or soil); the lowest rate in equatorial regions, with very high humidity (EQUATORIAL CLIMATE) and much surface cover. LATENT HEAT.

evaporite a SEDIMENTARY ROCK, such as gypsum, consisting of minerals which, having been precipitated from a solution, remain after EVAPORATION.

evapotranspiration the total return of water from the land to the atmosphere, i.e. the combined EVAPORATION from the soil surface and TRANSPIRATION from plants. POTENTIAL EVAPOTRANSPIRATION.

everglade a low marshy region, usually under water, with hummocks and small islands, and overgrown with tall grass, canes and trees sometimes hung with SPANISH MOSS. The term is specifically applied to a large area of this nature, the Everglades, a National Park in Florida, USA.

evergreen a plant that bears green leaves throughout the year, in contrast to a DECIDUOUS plant.

evolution 1. gradual change and development **2.** the gradual, cumulative change from a simple to a more complex form, e.g. as seen by Charles Darwin to be at work in the progressive diversification in the characteristics of organisms or populations in successive generations descended (according to his theory of evolution) from related ancestors, leading to the development of different species and sub-species. He based his theory of evolution on natural selection, the survival of the fittest. To summarize his theory, organisms in each generation produce many, variant offspring, but only the successful variants, i.e. those most fitted to their environmental conditions (ENVIRONMENT), survive and breed, transmitting their advantageous characteristics to their offspring. Thus when environmental conditions change, or organisms spread to new areas, new, strong, successful forms emerge, each well adapted to its particular environment. PUNCTUATED EQUILIBRIUM.

exclave a portion of an administrative or other kind of area, e.g. of a state, which is separated from the main part and surrounded by alien territory (and regarded in that territory as an ENCLAVE). If the portion is not physically separated, but is conveniently approachable only through alien territory, it is termed a pene-exclave. If, in practice, an exclave has ceased to be treated legally as such, it is in some cases termed a quasi-exclave. A temporary exclave is one created as a result of some territorial arrangement which has not been concluded, e.g. West Berlin, following the Second World War.

exfoliation (American spalling) onion weathering, the weathering process in which strains lead to the splitting off of the surface of rocks in scales or layers, common in hot DESERT and SEMI-ARID lands and in MONSOON lands with a marked dry season. It is due mainly to HYDRATION, resulting in the expansion of salt crystals in the pores of the rock surface. Ground water with its dissolved salts is drawn to the surface by CAPILLARITY where it is subjected to EVAPORATION, salt crystals being precipitated to form a film. HYDROLYSIS, SPHEROIDAL WEATHERING.

exhumation the action or process of bringing to the surface something buried in the ground, a term widely applied in geographical literature to previously existing surfaces uncovered by erosion, e.g. mountains, peneplained surfaces and plateaus buried by later deposits.

exile 1. banishment, expulsion from home or native land **2.** a person banished from, compelled to live away from, home or native land **3.** paradoxically, a voluntary exile, one who chooses to go into exile. EMIGRANT, EXPATRIATE, REFUGEE.

existentialism a body of philosophical doctrine (inaugurated by the Danish philosopher, S. A. Kierkegaard, 1813–55) concerned with the gulf between the existence of human beings (born with will and consciousness) and the kind of existence of natural objects (possessed of neither). Human beings are thus seen not to be part of an ordered, all-embracing, metaphysical scheme, but to exist in an alien world of objects from which they are estranged. Later existentialists, e.g. Jean-Paul Sartre, held that each individual, being free and responsible but not endowed at birth with character or goals, is self-creating in the situation and environment specific to him/her, through acts of will, of decision, and the self-development of his/her own 'essence'.

exogenetic, exogenic *adj.* arising from without, having an external cause or origin, hence in geology applied to the external processes (i.e. those at or near the surface of the earth), e.g. DENUDATION and DEPOSITION, and to the rocks and landforms arising therefrom. Exogenetic is the usual English form, exogenic the American. ENDOGENETIC, ENDOGENIC.

exonym the CONVENTIONAL NAME given by people of one linguistic group or nation to the place-names etc. of another.

exoreic *adj.* flowing to the outside, applied to normal drainage, i.e. flowing to the sea. AREIC, ENDOREIC.

exosphere the uppermost zone in the upper ATMOSPHERE-1 of the earth, in the IONOSPHERE, above c. 700 km (435 mi), from which neutral particles escape into space. The upper atmosphere may be defined as the atmosphere below the exosphere and above the TROPOPAUSE.

exotic *adj.* **1.** brought in from a foreign country or from a foreign language **2.** like, or imitating, something foreign.

expanded-foot glacier a GLACIER spreading out from the mouth of a valley to form a broad tongue of ice on the plain, a small PIEDMONT GLACIER.

expanded town in Britain, a town enlarged under the Town Development Act 1952 by accommodating the OVERSPILL population and industry from a major city. NEW TOWN.

expatriate one who voluntarily chooses to live (and in some cases work) away from his/her native country, but not necessarily permanently. EMIGRANT, EXILE, REFUGEE.

exponential curve a curve which may be assumed by a surface in accordance with a particular mathematical formula.

exponential growth GEOMETRIC PROGRESSION.

export something sent out from one country to another, the sender receiving in return money, goods or services in payment (IMPORT). Invisible exports consist of payments for services, which include interest on overseas investments and loans, earnings from transport, shipping, banking, insurance, and tourism. Visible exports are foodstuffs and raw materials, natural or partly processed, and manufactured goods. BALANCE OF PAYMENTS, BALANCE OF TRADE.

export base theory a theory that sees the development of a city or region as being linked solely to its export performance, i.e. to the magnitude of its exports to areas outside the city or region. BASIC ACTIVITY, NON-BASIC ACTIVITY.

exposure **1.** the state of being left uncovered, bare, without protection **2.** in geology, a place where the solid rocks reach the surface and are not obscured by soil etc., the act of uncovering being natural or artificial (a rock reaching the surface through a covering of soil produced by its own disintegration is best termed an OUTCROP) **3.** the position of a place with regard to compass direction or to climatic influences. ADRET, ASPECT.

extended family a FAMILY-3 group comprising not only parents and children (NUCLEAR FAMILY) but also blood relatives and relatives by marriage, all living close to one another.

extensive agriculture farming practised in large units in which the amount of capital and the labour employed is small in relation to the size of the unit. The work is sometimes highly mechanized, so that the yield per worker is high though the yield per ha (per acre) is low (due to low inputs), resulting in a high total output from the unit. The term is often, and incorrectly, applied to farming in any large unit. INTENSIVE AGRICULTURE.

external economies cost advantages which benefit a commercial activity but which are not produced within it, being derived from a source outside the activity, e.g. as in an AGGLOMERATION-4.

externality a concept concerned with the social and economic costs or benefits caused by the activity of an individual (or an institution) which do not enter the internal production costs of that activity but which affect the activity of another (or others) and over which the latter has/have no control. Thus the costs or benefits are externalized and fall on others. Externalities are often fortuitous and may be negative (creating costs) or positive (beneficial) to the recipient. AGGLOMERATION-4, EXTERNAL ECONOMIES, NEIGHBOURHOOD EFFECT.

exterritoriality the immunities granted to a diplomatic envoy and staff in accordance with international law. Not to be confused with EXTRATERRITORIALITY.

extinct volcano VOLCANO.

extraction the process of obtaining by pressure, distillation, evaporation, treatment with a solvent etc., e.g. of a metal from its ore.

extractive agriculture an agricultural practice which is more concerned with immediate short-term benefits than with conserving resources for future agricultural use.

extractive industry a PRIMARY INDUSTRY in which non-replaceable materials are removed in their natural state, e.g. mining, quarrying. Strictly, the term excludes forestry and agriculture, but some authors include forestry.

extrapolation 1. the estimation of values outside a range of known values based on the assumption that trends within the known existing range will be continued outside it **2.** drawing a conclusion about a future or hypothetical situation by using observed tendencies as a basis.

extraterrestrial *adj.* outside the earth or its atmosphere.

extraterritoriality the extension of jurisdiction beyond the borders of the state, involving the partial exemption from local law and jurisdiction enjoyed by diplomatic agents and others (such as troops) stationed in foreign countries, who thus remain under the laws of their own country, usually with the agreement of the foreign country in which they are living.

extreme climate a climate with a great range of temperature between the coldest and the warmest months.

extremes of temperature the highest and the lowest shade temperatures, or the highest mean and the lowest mean temperatures, recorded during any selected period (day, month, year) at a meteorological station.

extrusion the action of pushing out by mechanical force, emergence, e.g. the discharge of solid, liquid or gaseous material to the earth's surface during a volcanic ERUPTION. EXTRUSIVE ROCK.

extrusive rock VOLCANIC ROCK, an IGNEOUS ROCK resulting from the pouring forth or extrusion of molten material

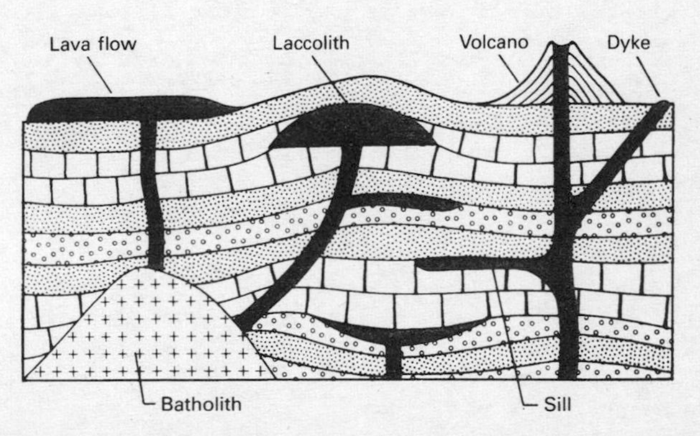

Fig 21 Extrusive and intrusive rocks: magma breaking through and cooling at the earth's surface forms extrusive rock; magma cooling in the crust forms intrusive rock (batholith, laccolith, sill)

(MAGMA) at the earth's surface as LAVA and its consolidation there. It usually has a glassy texture, and contains CRYSTALS smaller than those occurring in an INTRUSIVE ROCK. EXTRUSION, INTRUSION, PLUTONIC ROCK. Fig 21.

eye of a hurricane or other tropical revolving storm, the still, calm area, limited in extent, at the centre of the storm, where the atmospheric pressure is at its lowest.

eyot, ait a small island, a term surviving in many English place-names in its abbreviated form, -ey.

F

fabric structural material, especially the physical composition (textural and structural) of a compound, e.g. TILL.

facies 1. in botany, the general appearance and form of a plant, or the composition of a natural group **2.** in geology, the total character of any part of a FORMATION, shown by the FOSSILS it contains, the composition, colour, texture, form of STRATIFICATION and nature of the constituent rocks, or by other special features; hence used in STRATIGRAPHY in differentiating one rock STRATUM from another.

facilities the things (aids, equipment, structures, favourable conditions, opportunities) that make easier some action, specified activity or task.

factor 1. a circumstance, fact, agent or influence contributing to a result, effect or condition **2.** in arithmetic, a whole number which, when multiplied by one or more whole numbers, produces a given number (e.g. 2, 3, 4, 6 are all factors of 12) **3.** in algebra, an expression which will divide into a given expression **4.** in statistics, a quantity under examination in an experiment as a possible cause of variation.

factor analysis a branch of MULTIVARIATE ANALYSIS, a statistical technique which ignores the uniqueness of a number of variables (or test items) in a set of observations and aims to describe them in terms of a smaller number of more basic, hypothetical components. A factor which to varying degree underlies all items in a test is termed a general factor; one which underlies only a specific group of items is termed a group factor. A factor loading of an item on a factor is the correlation between the factor and the item.

factors of production the components needed in a production process, conventionally, capital, land, labour (including enterprise, e.g. of an ENTREPRENEUR).

factory farm a CAPITAL-INTENSIVE livestock farm in which the stock (e.g. pigs for meat, poultry for meat and eggs, cows for milk) is reared and tended under cover in carefully controlled conditions (ENVIRONMENTAL CONTROL), the food intake, rate of growth, egg production, milk-yield etc. being carefully monitored. The disposal of the effluent, e.g. from a pig unit, often presents difficulties. BATTERY SYSTEM.

fadama a floodplain in a wide, fairly flat, river valley, subject to annual inundation, common in the SAVANNA zones of Sudan and Guinea, supporting typical savanna vegetation (grasses, sedges, and the tree species *Mitragyna inermis* and *Borassus*). The term may also refer to an isolated depression on the low terrace or over-bank zone of a river course.

Fahrenheit scale a TEMPERATURE SCALE established 1715 by Daniel Gabriel Fahrenheit, German physicist, on which the ice point, the freezing point of pure water, is 32° (32°F) and the steam point, the boiling point of pure water at sea-level with a standard pressure of the atmosphere of 760 mm, is 212° (212°F). (0°F represents the melting-point of ice in a sal-ammoniac and water mixture, but 32°F represents the melting-point of ice in water.) The difference between ice point and steam point is thus 180°, and one Fahrenheit degree is 1/180 of the temperature interval between the two points; so 1° CELSIUS or CENTIGRADE equals 1.8° Fahrenheit. KELVIN SCALE, REAUMUR SCALE.

fall-line, fall zone a line or narrow zone where a number of nearly parallel rivers plunge over the edge of a plateau to the lowland below, marking the change in the character of their courses from the rocky channels with swift currents of the upper courses, to the more placid courses on the plain. In the USA applied particularly to the boundary between the ancient crystalline rocks of the Appalachian plateau and the younger, softer rocks of the Atlantic coastal plain.

fallow ploughed or cultivated land which is being allowed to rest, uncropped or partially cropped, for one or more seasons, or sometimes for a shorter period. BARE FALLOW, BUSH FALLOWING, GREEN FALLOW, LAND ROTATION, SHIFTING CULTIVATION, SWIDDEN FARMING.

false bedding, false-bedding LAMINAE, especially in sandstone, deposited under the influence of changing currents in shallow water areas, lying parallel to each other for short distances, but inclined at varying oblique angles to the general stratification. False bedding (a term synonymous with CURRENT BEDDING or cross bedding), is caused by swift local currents and swirling gusts of wind, thus studies of it provide much information regarding current direction and conditions of deposition.

false cirrus thick, grey CIRRUS cloud associated with the top of a thundercloud, and usually heralding bad weather.

false colour the use of a colour, arbitrarily selected, to portray a band (e.g. INFRARED RADIATION) on an image received by REMOTE SENSING. ELECTROMAGNETIC SPECTRUM.

false drumlin a formation that resembles a true DRUMLIN but is actually a rock mass overlain by a thin coating of DRIFT deposited by ice.

false origin a selected point in a GRID SYSTEM used on maps from which the position of any place can be expressed in terms of its COORDINATES from the selected point. The false origin differs from the TRUE ORIGIN (the intersection of the projection axes) in order to exclude negative values. NATIONAL GRID.

falsifiability a principle expounded by Karl Popper, 1902–94, Austrian-born, British philosopher, that a theory holds good until it is disproved and that falsification, not verification, is the proper objective of scientific procedures; that a theory which is not falsifiable is not scientific. Thus any scientific HYPOTHESIS must be expressed in terms that make it capable of being submitted to rigorous, sustained testing in an attempt to show it is wrong. For Popper falsifiability distinguishes science from pseudo-science. LOGICAL POSITIVISM.

family **1.** a group of people consisting of one parent/parents and a child/children, living together or not. **2.** a person's children **3.** a unit formed by people who are, or who are nearly, connected by blood or by affinity. EXTENDED FAMILY, NUCLEAR FAMILY **4.** a household denoting a group of people living in one dwelling, or with one head of household, including parents, children, servants **5.** persons descended or claiming descent from a common ancestor. CLASSIFICATION OF ORGANISMS.

family life cycle the progressive series of changes in size and composition which a NUCLEAR FAMILY (i.e. a married couple with dependent children) undergoes over time. The stages usually identified are marriage and the formation of the household, child rearing, child launching, contraction in size as children leave home, post-child (often associated with retirement). The patterns of the life cycle of other types of FAMILY differ with the composition of the family.

fan the alluvial (ALLUVIUM) or stony deposit of a stream where it issues from a ravine or canyon and drops its load on to a plain (ALLUVIAL FAN). Some DELTAS are fan-shaped (ARCUATE DELTA).

FAO Food and Agriculture Organization of the United Nations, a specialist organization of the United Nations established on 16 October 1945, with the aims of giving international support to national programmes designed to increase the efficiency of agriculture, forestry and fisheries, and of improving the conditions of people engaged in relevant activities.

Far East far to the east of Europe, i.e. east and southeast Asia, including China, Korea, Japan and, usually, Malaysia and the Indochinese peninsula. EAST.

farinaceous (Latin farina, flour) *adj.* rich in starch, the main nutritional constituent of CEREAL grains and edible roots, the common STAPLE foodstuffs.

farm any tract or tracts of land or of water, varying greatly in extent, worked as a unit, used for the cultivation of crops or for the rearing of LIVESTOCK or fish, under individual or collective management. Specific names such as DAIRY FARM, FISH FARM, FUR FARM indicate the purpose of the farm. AGRIBUSINESS, AGRICULTURE, BATTERY SYSTEM, COLLECTIVE FARMING, COOPERATIVE FARM, EJIDO, EXTENSIVE AGRICULTURE, FACTORY FARM, FARMER, FARMING, FAZENDA, FRAGMENTATION-2, HACIENDA, HUERTA, INFIELD-OUTFIELD, INTENSIVE AGRICULTURE, LATIFUNDIA, RANCH, SMALLHOLDING.

farmer a term used as loosely as is FARM, broadly one who, whether as tenant or landlord, is occupied in AGRICULTURE, AQUACULTURE or PISCICULTURE.

farming the activity of a FARMER. The main types of farming include CROP FARMING, DAIRY FARMING, MIXED FARMING, STOCK FARMING. Examples of some specific farming practices and activities are listed under FARM, but see also COMMERCIAL AGRICULTURE, PEASANT FARMING, SHIFTING CULTIVATION, SUBSISTENCE AGRICULTURE.

farmstead the land and buildings of a small farm.

fascism any political or social ideology of the extreme right that favours and encourages the gaining and retaining of power by the brutal use of force. MODERNISM.

fast ice sea ice, varying in width, firmly fixed along the coast, usually in the position of its original formation. It may stretch over 400 km (250 mi) from the coast.

fathom a nautical measurement of the depth of water, based on the span of the outstretched human arms, i.e. 1.829 m (6 ft); 100 fathoms = 1 cable.

fault in geology, a fracture or break in a series of rocks along which there has been vertical or lateral movement or both as a result of excessive strain (ELASTIC REBOUND). NORMAL FAULTS are those in which the rocks on one side have slipped down relative to the other; REVERSE FAULTS where they have been pushed up (a reverse fault of very low angle, the upper beds pushed far over the lower, is a THRUST FAULT); OVERTHRUST faults where the plane is near the horizontal. Faults may be nearly parallel to the DIP of the rocks (dip faults), to the STRIKE (longitudinal or STRIKE FAULTS, STRIKE-SLIP or TEAR FAULT), oblique (with planes between dip and strike); and there are groups of faults forming a complex fault or fault zone, STEP FAULTS, TROUGH FAULTS, ridge faults (forming a HORST). For some of the many other terms applied to faults see those that follow as well as COMPRESSION, FOOT WALL, HADE, HEAVE, LAG FAULT, NAPPE, THROW, TRANSFORM FAULT. Figs 22, 23.

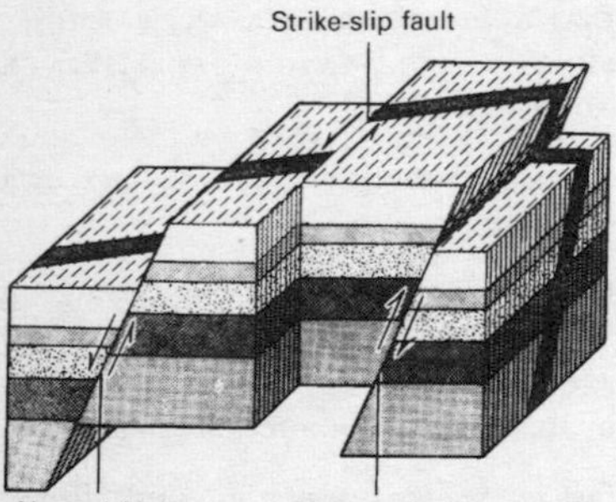

Fig 22 Diagram showing the direction of movement at some types of fault in the rocks of the earth's crust

fault apron an accumulation of rock debris at the base of a (fault) scarp resulting from rapid dissection, and formed by the coalescence of many ALLUVIAL CONES.

fault block a block of country, often of considerable size, bounded by FAULTS, in many cases resulting from vertical movements in the earth's crust. BASIN AND RANGE, HORST.

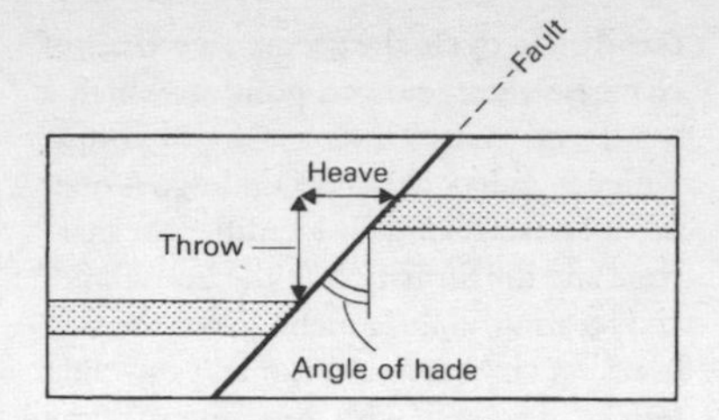

Fig 23 Terminology of a fault

fault breccia a rock composed of shattered angular fragments resulting from movement along a FAULT.

fault-line scarp a SCARP formed when, by faulting, a hard rock confronts a soft rock and the soft rock is subsequently eroded, leaving the hard rock standing as a cliff or scarp; but this scarp is not necessarily the original FOOTWALL of the FAULT, as it is in a true FAULT SCARP. OBSEQUENT FAULT-LINE SCARP, RESEQUENT FAULT-LINE SCARP, WATERFALL.

fault plane the surface of movement of a FAULT, where material may be shattered and broken off to form FAULT BRECCIA, or ground to ROCK FLOUR which may become consolidated.

fault scarp a steep face of rock resulting from the uplift along one side of a recently active FAULT. An early landform, it is rapidly affected by WEATHERING and EROSION, which may result in its disappearance, or its later development into a FAULT-LINE SCARP. ESCARPMENT, SCARP.

fault spring a SPRING-2 issuing along a FAULT where a PERMEABLE bed encounters IMPERMEABLE rock.

fauna 1. the animal life, considered collectively, of any given period, environment or region. BIOTA, FLORA **2.** the classification of the animals of a period, environment or region.

favela (Brazil: Portuguese) a shack, a slum. The plural, favelas, is applied to a SHANTY-TOWN, with shelters improvised from scrap materials, forming a SLUM settlement on the outskirts of a large town, e.g. of Rio de Janeiro.

fazenda (Brazil: Portuguese) a very large landed property in northeast Brazil originally granted by the King of Portugal to selected leaders when Portugal colonized Brazil. Fazendas became economically self-contained agricultural estates, with agricultural workers attached by various forms of tenure, e.g. SHARECROPPING. Today many fazendas are engaged in MONOCULTURE (e.g. SUGARCANE) and the sharecroppers have become wage-earners. LATIFUNDIA.

fecundity FERTILITY-2.

federal *adj.* **1.** of or pertaining to a form of government in which two or more states unite, but have some independence in internal affairs **2.** in USA, relating to the central government. FEDERAL DISTRICT.

federal district an area allocated as the seat of the capital of a country, e.g. District of Columbia (for Washington) in USA.

federal system of government, a political system which varies in detail from one country to another, but in which the main authoritative control is usually divided between a central or national government and a regional or state government, with local government being directly responsible to the regional or state government authority, not to the central or national authority. UNITARY SYSTEM.

feedback the action by which the output of a process is coupled to the input, i.e. the return to a system, process or device of part of its output. Normal (termed negative) feedback has a modifying effect in that it controls and corrects discrepancies in the working of the system, process or device, i.e. the response to a stimulus tends to counteract or inhibit the repetition of the stimulus (e.g. there is a lack of the feeling of thirst in an animal which has quenched its thirst: an animal in this situation will not drink any more even if water is available). Positive feedback, which is less common, does not have a modifying effect, but a snowballing effect in that it strengthens the stimulus (e.g. ever greater traffic flow on an ever-widened road: a growing volume of traffic which encourages the widening of a road, which in turn attracts a greater volume of traffic).

feldspar, felspar any of a wide-ranging group of crystalline minerals, white or pink in colour, consisting mainly of aluminosilicates of POTASSIUM, SODIUM, CALCIUM, barium. Usually classified as ALKALI or PLAGIOCLASE, feldspars are abundant in METAMORPHIC ROCKS and in ARKOSES, are used in the manufacture of porcelain and glass, and are a source of semi-precious stones, e.g. moonstone. Alkali feldspars are characteristic of all alkali igneous rocks, form constituents of such BASIC ROCKS as BASALT and GABBRO (in calcic form) and of sialic rocks (SIAL).

fell in northern England **1.** a wild elevated stretch of rough grazing or moorland, especially if interrupted by boulders or rock outcrops **2.** a summit, e.g. Bowfell, English Lake District.

felsenmeer (German) a boulder field, block field, rock-block field, rock river, an area on top of a flat-topped mountain or high plateau in a temperate climate, sometimes on lowland in much colder regions, covered with angular blocks of rock created in situ by frost action in JOINTS.

feminist geography a branch of HUMAN GEOGRAPHY that concentrates on the GENDER ROLE of women in society from the woman's perspective, particularly access to employment, social services etc. and the extent to which women are disadvantaged.

fen a lowland covered wholly or partly with shallow water with water-loving vegetation which decays to form PEAT. In contrast to a BOG, where the organic soil is very acid, the soil in a fen is alkaline, neutral or only slightly acid. ACID SOIL, ALKALINE SOIL, FEN PEAT, MARSH, SWAMP.

feng-shui, fong-choui, fung-shui (Chinese feng, wind; shui, water) in China, the spirit of wind and water of a locality, the good in a system of good and evil influences that inhabit natural features in the landscape, thus important in the siting of buildings and graves.

fenland **1.** a tract of land occupied by FEN **2.** with an initial capital letter, Fenland, a synonym for the FENS in eastern England, around the Wash.

fen peat ALKALINE or neutral black PEAT in which the plant structure is not apparent, cellulose is absent, but there is a high content of ash and proteins. It supports typical FEN vegetation in which sedges, grasses and rushes predominate. BOG PEAT.

Fens, Fenland in eastern England, a flat lowland below 15 m (50 ft) extending around the Wash in the lower basins of the rivers Witham, Welland, Nene and Great Ouse, consisting of naturally marshy land built up by the deposition of sediment by sea and rivers, forming alluvial flats near the coast and FEN PEAT further inland. Except for Wicken Fen, National Trust property, the area has been largely reclaimed (notably by Vermuyden and the fifth Earl of Bedford) and now provides land valuable for cultivation.

feral *adj.* like a wild beast, applied especially to an animal (once domesticated) or to a plant (once cultivated) which has now established itself in the wild.

fermentation a chemical change brought about by the action of ENZYMES on organic substances, with the release of energy, especially the ANAEROBIC breakdown by yeasts and bacteria of carbohydrates, yielding alcohols, acids and carbon dioxide. ANAEROBIC RESPIRATION.

Ferrel's Law a law postulated in 1856 by the American scientist W. Ferrel, who developed the concept of CORIOLIS FORCE. The law states that due to the force produced by the rotation of the earth, a body moving over its surface will be deflected to the right in the northern hemisphere, to the left in the southern, the force at the equator being zero, and increasing progressively with distance from the equator. The effects are especially noticeable in water and in air. BUYS BALLOT'S LAW.

ferric *adj.* pertaining to or containing IRON. OXIDATION.

ferricrete a DURICRUST in which the cementing agent is formed from iron oxides.

ferriferous *adj.* iron-bearing, applied especially to rocks. IRON.

ferrite any of the IRON ores.

ferromagnesian mineral a dark, dense

mineral, containing high proportions of IRON and MAGNESIUM.

ferromagnetic *adj.* possessing FERROMAGNETISM.

ferromagnetism the property possessed by iron, nickel and cobalt, as well as by some alloys, of holding magnetism in the absence of an external MAGNETIC FIELD and of responding characteristically to a magnetic field. Horseshoe and bar magnets are usually ferromagnetic. MAGNET.

ferruginous *adj.* of or containing IRON or iron RUST-1, or resembling the colour of rust, applied especially to reddish coloured rocks containing some iron.

ferry **1.** a boat (ferryboat) or an aircraft (air ferry) carrying people, goods, vehicles from one point to another on a regular route **2.** a place where people, goods, vehicles are carried, or the service carrying them.

ferry port a port serving ferries.

fertile *adj.* **1.** highly productive, e.g. a fertile soil **2.** capable of breeding or reproducing **3.** capable of developing, e.g. a fertile egg **4.** in demography, applied to a woman or man who has produced at least one live-born child.

Fertile Crescent the approximately crescent-shaped area of fertile land bounded by the rivers Tigris and Euphrates that was the site of the Sumerian, Babylonian, Assyrian, Phoenician and Hebrew civilizations, and where agriculture was practised with the help of elaborate irrigation and drainage systems. HYDRAULIC CIVILIZATION.

fertility **1.** the state or condition of being FERTILE **2.** in general, sometimes loosely applied to the potential capacity to reproduce (to produce live-born children), but fecundity is the more precise term for this **3.** in demography, actual reproduction performance (FERTILE-4). In measuring fertility in a human population a broad distinction is made between current fertility (measured by the number of births of a particular year) and period fertility or cohort fertility (the total number of children born to a COHORT throughout the reproductive lives of the members). REPRODUCTION RATE.

fertilizer a substance consisting of one or a mixture of chemical elements essential for plant growth (NITROGEN, PHOSPHORUS, POTASSIUM) added to the soil to enrich it or make good its deficiencies. Most plant fertilizers are now artificial, made in factories, but natural mineral fertilizers include POTASH salts, such as the famous Stassfurt deposits in Germany, NITRATE of soda from the deserts of Chile, PHOSPHATES of lime, notably from north America, and GUANO from the droppings of birds accumulated in arid lands, especially tropical islands. Natural organic fertilizers include fish meal, bonemeal, dried blood, dung, composted vegetable remains (COMPOST), SEAWEED.

fetch the distance of open water over which a wind blows and over which a sea WAVE-3, blown by the wind, travels, the length being an important factor in determining the height and energy of the wave and thus its effect on the beach and coast.

feud historically, an estate (LAND TENURE) held by a tenant (a vassal) on condition of services being rendered to an overlord. FEUDAL SYSTEM.

feudal system an economic, political and social system, a pre-capitalist MODE OF PRODUCTION common in western Europe between the ninth and sixteenth centuries, and elsewhere (e.g. in Japan to the nineteenth century), taking a wide variety of forms. Commonly the system involved a social hierarchy based on land held in FEUD, and the relationship of landlord to tenant, the former having jurisdiction over the latter. Neither labour nor the products of labour were COMMODITIES-1. The tenant, legally tied to the land, could own some of the MEANS OF PRODUCTION but the land and a set part of the product were the property of the landlord. For example, in England the MANOR was the economic unit. The lord of the manor held the land from the king in return for military service and homage. Tenants who occupied and cultivated the land held it from the lord in return for dues (services or the products of their labour). Bound in service to the lord or to the manor were the serfs, labourers 'attached to the soil', whose mobility was restricted and who were transferred from one lord to another if the land changed hands.

fiard FJARD.

field capacity in soil science, storage capacity, the amount of water held in a well-drained soil by CAPILLARITY after excess water has drained away by gravity and the rate of downward movement has materially decreased. INFILTRATION CAPACITY, WILTING POINT.

field system the medieval agricultural system in England and parts of western Europe. The arable land of the village consisted of unenclosed strips (STRIP CULTIVATION) held by different owners or cultivators, and was used as common pasture for a certain period in each year. Where the common arable was divided in two (one part cultivated, the other lying fallow) the system is known as Two-Field. More usual was the division into three (two cultivated, one part lying fallow), termed the Three-Field system. OPEN FIELD.

filling in meteorology, the condition in a low pressure system (DEPRESSION-3) when there is an increase of atmospheric pressure in the central area so that there is a filling-up, and dying away of the depression, in contrast to a DEEPENING.

financial exclusion the processes that prevent the access of poor and disadvantaged groups to the financial system. RED-LINE DISTRICT.

finger lake a long narrow lake occupying a U-shaped glacial valley.

finger millet a short-stemmed MILLET with the ear consisting of five spikes radiating from a central point, probably originating in the Indian subcontinent, now grown as a food grain there, in the drier parts of Sri Lanka, and in semi-arid areas of Africa, where it is often the staple food. The grain stores well for as long as five years, and is thus important as famine reserve food. Very little enters trade outside the growing areas.

fiord FJORD.

firn (French NEVE) snow which has been partly consolidated by alternate thawing and freezing but is not yet at the stage of being glacier-ice because, although the particles are partially joined together (which distinguishes it from snow), the small air spaces between the particles communicate with each other (which distinguishes it from ice).

First World the countries which have some form of capitalist, market-orientated economy, including the countries of northwestern Europe and of Australasia, and the USA, Canada, Japan. Spain,

Portugal, Greece, South Africa and Argentina are sometimes qualified as borderline. SECOND WORLD, THIRD WORLD.

firth (Scottish) loosely applied to an area of coastal water, e.g. an arm of the sea, the lower part of an estuary, a fjord, a strait.

fish a member of Pisces, a class of VERTEBRATE aquatic animals, many yielding nutritious flesh, some yielding oil for food or industrial use. Some (e.g. bream) live in freshwater, others in the ocean, where the PELAGIC, free-swimming species (e.g. herring) swim in shoals in surface waters, the demersal fish (e.g. plaice, sole, halibut) living near the sandy sea bed in shallow waters. The salmon, in nature, spends part of its life cycle in the ocean, part in a river. FISH FARM.

fish farm a FARM where fish are bred and reared for market, under carefully controlled conditions in tanks or in protected parts of coastal waters, lakes, rivers and their estuaries.

fishing port a PORT used by fishery vessels for anchorage, fuelling, repair etc. and for the landing of fish, where these are important among the other activities of the port.

fissile *adj.* able to be split, applied particularly to rocks which are easily split along their well-developed BEDDING PLANES, CLEAVAGE, LAMINAE or JOINTS, e.g. shale, slate.

fission the act or process of splitting into parts. In nuclear fission (NUCLEAR ENERGY) the atomic nucleus is split into roughly equal parts by bombardment with neutrons, with an overall loss in mass (FUSION), thereby releasing great energy.

fissure **1.** an extensive crack, a narrow opening cleft or fracture made by splitting, especially in a rock **2.** a linear volcanic VENT. FISSURE ERUPTION, VOLCANO.

fissure eruption linear eruption, a steady, non-explosive outpouring of LAVA, usually BASIC, from the depths of the earth to the earth's surface, occurring along a weak line (sometimes many kilometres in length) in the crust, often resulting in a BASALT plateau. CENTRAL ERUPTION, VOLCANO.

fixed capital the buildings and machinery of a firm. CAPITAL-3.

fixed costs **1.** costs that do not vary with volume of output within a certain range (unlike VARIABLE COSTS), i.e. capital investment in plant and machinery, rent, rates and certain other overhead costs incurred irrespective of the volume of production **2.** in spatial economic analysis, costs that are constant in space, e.g. the cost of finance, labour costs if fixed rates apply in the particular area, any components or materials sold at a uniform delivered price.

fjärd, fiard (Swedish) a term applied by some English authors to an inlet of the sea with low banks, the result of a rise in sea level (SUBMERGED COAST), in contrast with FJORD. This does not conform to the original Swedish application, which was simply to a large continuous area of water surrounded by SKERRYGUARD islands.

fjeld (Norwegian) widely applied in Norwegian to any upland rocky area, but restricted in English to an elevated, rocky plateau above the TREELINE, snow-covered in winter. FELL.

fjord, fiord (Norwegian) a term widely applied in Norway and Denmark to almost any sea inlet, but restricted in English to a deep narrow inlet in the coast with high rocky parallel sides smoothed by ice action,

HANGING VALLEYS and an irregular rocky floor, the floor being frequently deeper than the sea floor offshore, from which it is separated by a submerged sill near the entrance. It is caused by the submergence of a deep glacial valley. SUBMERGED COAST.

flagstone, flag a natural hard stone used for paving, capping walls etc., usually cut from SANDSTONE or some types of LIMESTONE which split conveniently and easily along BEDDING PLANES. FISSILE.

Flandrian the post-glacial stage of the QUATERNARY in northwest Europe (succeeding the DEVENSIAN) when the climate became less severe, woodland began to spread, and the sea level rose.

flash-flood a sudden violent FLOOD caused by exceptionally heavy rain in a normally dry valley in a semi-arid area, the torrential stream sometimes being laden with debris; or by the collapse or breach of a dam or sea wall. ENVIRONMENTAL HAZARD.

flat **1.** a stretch of level ground without marked hollows or elevations, or of country without hills **2.** any nearly level stretch of land within hilly country **3.** a low-lying tract of marshy land **4.** the low land through which a river flows **5.** mud-flat, a bank of mud over which the tide flows, exposed at low tide **6.** an alluvial deposit yielding tin or gold in a stream bed **7.** a horizontal deposit of ore in a BEDDING PLANE.

F layer in the ATMOSPHERE-1, a layer in the IONOSPHERE which reflects high frequency radio waves. It lies at a height of some 250 km (155 mi) above the earth's surface.

flexure **1.** the state or process of being bent **2.** the bending of strata etc. under pressure.

flint a heavy, very hard, grey or grey-black nodule or mass of a pure fine-grained SILICA, encrusted with white, occurring in the BEDDING PLANES or JOINTS in the Upper Chalk, and probably formed when silica-rich biogenic OOZE became mixed with chalk mud on the ocean bed, the silica being carried downwards in solution and precipitated round a core (e.g. the skeletal remains of a sponge, sea-urchin etc.), the flint nodule gradually enlarging within the chalk mud. Easily chipped to a cutting edge, flint was the main source of tools and weapons of the Stone Ages (EOLITH). It has long been used as a building stone; and when struck with steel it emits a spark, hence its use in a flintlock gun.

floating dock a large floating structure into which a vessel may pass and which can be used as a DRY DOCK. DOCK.

flocculation **1.** in soils, the process by which fine-grained particles come into contact and gather together into tufted masses to form CRUMBS, thereby improving the soil texture, especially the aggregation of colloidal (COLLOID) particles (clay) into tufted masses in the presence of an alkali (lime) **2.** the process of the aggregation of water-borne colloidal particles into small lumps which are able to settle out, occurring when a river carrying electrically changed colloidal clays in its fresh water meets and mixes with the sea, which carries electrically charged particles in solution in its salt water.

flood the overwhelming of usually dry land by a large amount of water that comes from an overflowing river or lake, an exceptionally high tide, melting snow, or sudden excessive rainfall.

flood basalt basaltic LAVA, very fluid at high temperatures, which has spread over a very extensive area as the result of a volcanic eruption or series of eruptions.

flood control 1. the regulation of excessive RUNOFF of water in order to prevent inundation of the land, e.g. by the building of river BARRAGES; the deepening of existing and the cutting of new channels to speed river flow; the making of temporary storage basins, such as TANKS; the conservation or planting of vegetation in a drainage area to slow down runoff **2.** the building of sea walls etc. to prevent inundation by exceptionally high tides.

flood hazard the dangerous chance of the inundation of usually dry land by water from an overflowing river or lake (e.g. caused by a break in a DAM), melting snow, sudden excessive rainfall, or an exceptionally high tide.

floodplain the relatively level part of the valley bordering a river resulting from alluvium deposited by the river in times of flood. Fig 29.

flood stage of a river, the stage when a river begins to overflow its banks. BANKFULL STAGE, OVERBANK STAGE, STREAM STAGE.

flood tide, flood-tide the advancing or rising tide, starting at low tide, ending at high tide, in contrast to EBB TIDE.

flora 1. the plant life considered collectively of a region or age **2.** a list of plant species of a particular area arranged in families and genera (CLASSIFICATION OF ORGANISMS), with descriptions and a key to aid identification. BIOTA, FAUNA.

flour 1. the finely-ground meal (the edible part of coarsely-ground GRAIN-5) of CEREALS such as WHEAT, BARLEY, RYE **2.** the finely-ground edible parts of other food crops, e.g. of POTATOES or CASSAVA. ROCK FLOUR.

flow 1. of a fluid, or something that behaves comparably, the smooth movement with a continuous change in shape, e.g. of lava, or of a stream confined between banks, or of grain, under the influence of gravity; or of water gushing forth from a spring; of air; of solid rocks under stress, without fracturing (i.e. plastic flow); of ice in a glacier **2.** a steady movement of ideas, goods etc. (e.g. of international trade).

flow diagram, flow chart a diagram showing a sequence of interconnected events, actions or items to indicate the progressive development or evolution of a theme, product, or other objective.

flow resources NATURAL RESOURCES.

fluid matter that does not have a fixed shape, i.e. a GAS or a LIQUID, being able to flow along a channel or tube. SOLID, VISCOSITY, VISCOUS.

fluidization a process in IGNEOUS activity by which hot gases, moving through fine-grained material, cause it to flow, thereby cracking and enlarging existing cracks in the rock by physical and chemical action. In some cases the enlargement of the crack is cylindrical, so that a PIPE forms.

flume 1. an artificial channel made to carry water over some distance for irrigation or for industrial use, e.g. for power, for transport (e.g. of logs), or in PLACER mining **2.** (American) a ravine or gorge with a stream flowing through it.

flush a sudden growth in the volume of a stream, creating a rush of water that does not quite overflow the banks.

fluting and grooving small ridges and

depressions caused by differential erosion, especially by wave action, on an exposed rock-surface with marked joining. JOINT.

fluvial, fluviatile *adj.* **1.** of or pertaining to a river **2.** found or living in a river. It is usual to apply fluvial to the action of the river (flow and erosive activity); and fluviatile to the deposits laid down by the river, or to the flora and fauna of a river.

fluvioglacial *adj.* produced by or due to the action of streams of water derived from the melting of the ice of the SNOUT of a GLACIER or the margin of an ICE SHEET, applied particularly to the deposits of an OUTWASH FAN. Some glaciologists prefer the term glaciofluvial because the action of the glacier precedes the flow of the streams.

fluvioglacial deposition the laying down of stratified DRIFT-1 by melt-water, especially in an OUTWASH APRON, a VARVE, or a PROGLACIAL LAKE. Stratified drift consists of rounded, washed and sorted sand and gravel, unlike TILL, which is angular or subangular and not sorted.

fodder 1. loosely, any food for cattle, sheep, horses **2.** specifically, dried hay, straw etc. used as food for some livestock, e.g. cattle, sheep, horses.

fog an opaque CLOUD in the ground surface layers of the atmosphere, consisting of condensed water vapour with smoke and dust particles held in suspension, that obscures vision for any distance up to one kilometre (International Meteorological Code). ADVECTION, ARCTIC SMOKE, COLD-WATER DESERT, FRONTAL FOG, HAZE, HILL FOG, ICE FOG, INVERSION, MIST, RADIATION FOG, SMOG, STEAM FOG.

fog drip, fog precipitation moisture deposited from fog especially in COLD-WATER DESERT coastal areas, where low CLOUD or FOG encounters trees, the water droplets saturate the branches, twigs, leaves, and then drip to the ground.

foggara in northwest Africa, especially in Mauritania, a gently inclined underground water channel bringing water for irrigation from AQUIFERS near the foot of mountains and carrying it, in some cases over long distances, to dry areas. A foggara is comparable with a KAREZ of Baluchistan, or a QANAT of Iran.

föhn (German) a warm dry wind blowing down slopes on the leeward side of a ridge of mountains, the warmth being due to the compression of the descending air, as in CHINOOK, NOR'WESTER (New Zealand), SAMUN, SANTA ANA. ADIABATIC.

fold 1. in farming, a pen or enclosure, especially a temporary one, often made with movable hurdles, for sheep or other domestic animals **2.** the condition when one part of something is made to lie on another, hence, in geology, used as a general term for bending (FLEXURE) in the rocks (STRATA) of the earth's crust resulting from COMPRESSION. The fold may be a simple arch or upfold (ANTICLINE) or a hollow or downfold (SYNCLINE), either of which may be SYMMETRICAL or ASYMMETRICAL, the axis of an asymmetrical anticline being so slanted that an OVERFOLD or RECUMBENT FOLD may result. A very large anticline is a GEANTICLINE or geoanticline; a big syncline a GEOSYNCLINE. A complex anticline is an ANTICLINORIUM, a complex syncline a SYNCLINORIUM. A fold with only one limb is a MONOCLINE (English usage); and repeated tight folds form ISOCLINAL strata (ISOCLINE). BREACHED

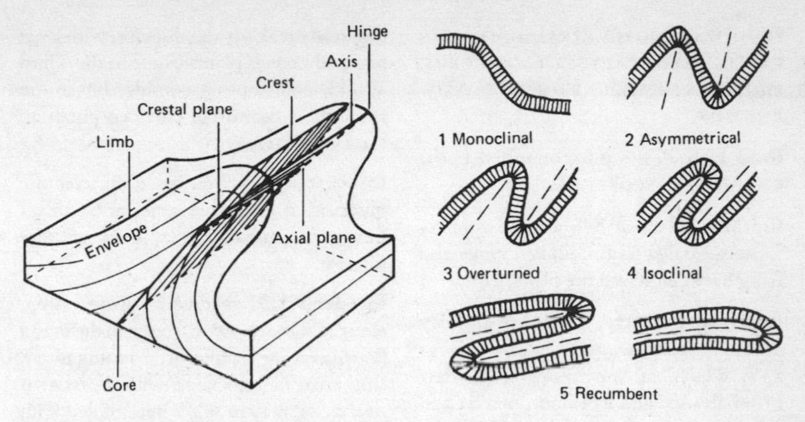

Fig 24 Types of fold, and terminology

ANTICLINE, CENTROCLINE, HETEROCLINAL FOLD, NAPPE, OROGENY, PERICLINE. Fig 24.

fold mountain a mountain resulting from the folding (flexuring) of the earth's crust, in contrast to a BLOCK MOUNTAIN. COLLISION ZONE, YOUNG FOLD MOUNTAIN.

foliation (Latin folia, leaves) **1.** the more or less parallel, wavy layers or bands of minerals found in some METAMORPHIC ROCKS, e.g. SCHISTS, GNEISS (METAMORPHISM) or in some IGNEOUS ROCKS which flowed as they cooled **2.** the wavy bands of ice seen in the depths of a GLACIER.

Food and Agriculture Organization FAO.

food chain a chain of organisms, existing in any natural community, along which energy passes, the organisms in each link obtaining energy by preying on or parasitizing those in the preceding link, and being so treated in turn by those in the succeeding link. At the first energy level, the first TROPHIC LEVEL-1, there is nearly always a green PLANT-1 or other organism capable of producing its own organic substance from inorganic compounds, using energy from the sun and making energy available to the consumer levels which follow it. At the second level are the primary consumers, the HERBIVORES, which supply energy to the secondary consumers, the smaller followed by the larger CARNIVORES. At each level much of the energy obtained is lost in RESPIRATION, so that progressively fewer organisms can be supported. Thus in a balanced community the BIOMASS of each trophic level is always greater than that of the succeeding level. OMNIVORES (diversivores) will eat almost anything; BACTERIA, fungi and various microorganisms are consumers operating at all levels. All organisms, including predators, die naturally if not eaten, their dead tissues being decomposed by microorganisms (decomposers) which release CARBON DIOXIDE, AMMONIA and mineral salts to the environment. The sequence up to the death of an organism is termed the predator chain, the energy flow after death being termed the detritus chain. Any natural community will have many interlinked food chains, making up a food cycle or

FOOD WEB. BIOTIC PYRAMID, CARBON CYCLE, NITROGEN CYCLE, PHOSPHORUS, PHOTOSYNTHESIS, PRIMARY PRODUCTION.

food web all the interconnected FOOD CHAINS in an ECOSYSTEM.

foothills a belt of hills aligned approximately parallel to a mountain range and lying between it and the plain below.

footloose industry, footloose activity a mobile enterprise, usually LIGHT INDUSTRY, which, by the nature of the raw materials used and its labour, market and transport needs, is not tied to a particular location. INDUSTRY.

foot wall the lower side of a FAULT.

forage crops in Britain, all CROPS-1 and grass (including grass from ROUGH GRAZING) grown on a farm specifically for grazing LIVESTOCK, but not PULSES or crops harvested as GRAIN-5.

ford a part of a river shallow enough to be crossed by wading, hence common in riverside place-names, e.g. Oxford.

Fordism a term connoting the system of fast mass production and the terms of employment pioneered by Henry Ford (1863–1947) in his vast Ford Motor Co. works in Detroit, Mich., USA. He aimed to produce low cost, low priced standardized motor vehicles at speed while sustaining good relations with his factory workers, who performed small, repetitive tasks (needing little or no skill) on a fast-moving assembly line for 8 hours a day. They were motivated by high wages related to their high productivity, and by the welfare services provided. Ford wanted them (and their families) to be able to afford the mass produced products of their labour. As time passed, the system began to break up and disperse, with large regional branch plants (some in the Third World) contributing considerably to the economic health of the corporation. POST-FORDISM.

forecast a prediction based on scientific observation, on experience, or by an estimate of probability, as in a weather forecast.

foredeep **1.** a relatively narrow, deep, elongated, steep-sided trough in the ocean floor, near or parallel to a mountainous land area, or associated with an ISLAND ARC **2.** such a trough when infilled with sediment, e.g. the Himalayan foredeep, now in the Indo-Gangetic plain. DEEP, OCEAN.

foredune a DUNE or line of dunes nearest to the sea in a system of dunes on a sandy shore, comparable with an ADVANCED DUNE in a hot desert.

foreland land or territory lying in front, e.g. **1.** a low PROMONTORY such as a CAPE or headland jutting seawards, e.g. North Foreland, Kent, England **2.** a stable continental mass on the margin of a GEOSYNCLINE **3.** in the study of ports, the seaward trading area connected with the port through maritime organization.

foreshore a term loosely applied to the part of the SEASHORE lying between the lowest LOW WATER line and the average HIGH WATER line. BACKSHORE.

forest **1.** an extensive, continuous area of land dominated by trees (sometimes specified as large trees in order to contrast forest with WOODLAND), sometimes including patches of pasture, economically exploited or not and (FAO qualification) capable of producing wood, or of influencing local climate or the water regime, or providing shelter for livestock and wildlife. BOREAL, CONIFEROUS, DECIDUOUS, EQUATORIAL, GALLERY, MONSOON, MOSSY, RAIN and TROPICAL FOREST and FORESTRY, NATURAL RESOURCES, PATCH CUTTING, SUSTAINED YIELD FORESTRY, TIMBER **2.** historically in medieval England, a legal term, an area of land, not necessarily tree-covered, set aside for a royal hunting ground (the hunting rights being the property of the Crown), subject to the strict Forest Laws and outside the Common Law, surviving in such place-names as Dartmoor Forest, Bowland Forest. PARK-1 **3.** the association with hunting similarly survives in the term deer forest, an unenclosed hunting ground, devoid wholly or partly of sheep, cattle, crops, used for deer stalking.

forestry the science and art of planting, cultivating and managing FORESTS-1, usually conveying the idea of management with a view to economic development, whereas WOODLAND management may not imply that. PATCH CUTTING, RESOURCE CONSERVATION, RESOURCE MANAGEMENT, SUSTAINED YIELD FORESTRY, TIMBER.

formation 1. in ecology, the largest natural vegetation type, the plants of a land BIOME **2.** in geology, sometimes termed a stage, the fourth division in the hierarchy of rocks in GEOLOGICAL TIME, a subdivision of a series, corresponding in the time interval to an AGE (GEOLOGICAL TIMESCALE). It consists of a STRATUM or series of strata which are similar in character (homogeneous), having some property which makes them distinctive.

form-line a CONTOUR based on general observations and sketched in, rather than one based on instrumental survey and determined by exact measurement.

fossil 1. specifically, the remains, or an impression left by the remains, or the traces, of a plant or animal, preserved in the earth's crust by natural processes, usually in SEDIMENTARY ROCKS, sometimes in metamorphic if the METAMORPHISM has not been too violent. The hard parts of the organism may be preserved intact, or they may be replaced by another mineral (CALCIFICATION, MINERALIZATION-2, PETRIFICATION). There may be internal or external casts of the organism in the rock, or impressions of soft parts, or carbon residues of the decomposed organism, or traces of plant and animal activities (e.g. tracks, burrows of animals, root passages of plants); or the organism may be preserved intact, e.g. mammoths in frozen ground in Siberia, insects in amber (fossil resin). PALAEONTOLOGY **2.** loosely applied to any feature of geological age buried in the earth's surface naturally or by geological agencies, e.g. FOSSIL FUEL.

fossil fuel combustible material derived from the fossilized remains of plants and animals, i.e. BIOFUELS, COAL, NATURAL GAS, OIL.

fossil water connate water, water trapped in SEDIMENTARY ROCKS during their deposition.

Fourth World THIRD WORLD.

fractal (Latin fractus, broken) *n.* and *adj.* a phenomenon composed of geometric patterns that are self-similar, repeated at every scale. Fractal geometry is concerned with the geometry of irregular objects, i.e. with natural or artificial forms that are not composed of straight lines (as in Euclidean geometry, which is based on straight lines). Attempts have been made to associate fractals with patterns of urban settlement.

fracture 1. a split, break or state of

being broken **2.** in geology, a clean break in STRATUM brought about by DEFORMATION under the strain of COMPRESSION or TENSION.

fractus cloud, fracto- ragged, shredded CLOUD indicating strong winds and stormy conditions in the upper atmosphere. The prefix fracto- is used to indicate a similar ragged condition in other cloud forms, e.g. fractocumulus, fractonimbus, fractostratus, etc.

fragmentation 1. the breaking of a whole into separate, isolated parts **2.** of a farm, the division of the land of a FARM into separate, isolated parts. If the farm remains a single holding it may become difficult to work as a unit if the land is divided (e.g. if the construction of a motorway cuts through the farm). If the fragmentation is the result of the law of equal inheritance, the land being parcelled out to family members, a number of small farm units of doubtful economic viability may be created. LAND REFORM.

free *adj.* **1.** unhampered, unimpeded, at liberty, without restrictions **2.** in chemistry, not combined, e.g. FREE OXYGEN.

free face a vertical facet cut in bare rock, part of the slope profile of a hillside, below the WAXING SLOPE, above the CONSTANT SLOPE, too steep for debris to rest on it (the debris falls downwards to form SCREE). SLOPE.

free goods goods (GOOD) which are so abundant that at any given time and place they are obtainable without cost or, being scarce enough to warrant a price, they are impossible to sell. CAPITAL GOODS, COMMODITY, CONSUMER DURABLES, CONSUMER GOODS, ECONOMIC GOODS, PRODUCER GOODS.

free oxygen gaseous OXYGEN or oxygen dissolved in a medium such as water. ANAEROBE, ANAEROBIC.

free port, free city an enclosed guarded city, usually a PORT or part of a port, where goods may be loaded or unloaded free from customs dues, most customs regulations and similar restrictions. It usually develops a large ENTREPOT trade.

freestone 1. a rock that can easily be cut in any direction, thus a fine-grained rock of uniform particles occurring in thick beds, e.g. sandstone, limestone **2.** in botany, a fruit in which the stone does not stick to the flesh; or the stone of such fruit.

free trade the commercial interchange between nation states or an economic group of nations and others, in which there are no tariff barriers or other protective measures such as quota restrictions.

freeze-thaw a form of WEATHERING in PERIGLACIAL areas where the temperature hovers around freezing point, below which frost breaks up the rock and above which the ice melts, so that water flows and carries away the rock fragments.

freezing point the temperature at which a substance in the liquid state changes to the solid state, e.g. at which pure water changes to ice, 0°C (32°F). FROST, ICE POINT.

F region the outer region of the IONOSPHERE, lying 200 to 400 km (125 to 250 mi) above the earth's surface.

freight 1. the hire of a vessel or vehicle to carry goods **2.** the service of carrying goods, originally by ship but now extended to land vehicles and aircraft **3.** the goods so carried **4.** the payment for that service, sometimes being specifically applied to goods carried in bulk, more slowly than the normal service and at lower cost.

freighter a ship or aircraft that carries cargo, i.e. FREIGHT-3.

freight rates the money paid to a carrier for the loading, transporting and unloading of goods.

frequency **1.** in general, the condition or quality of repeatedly occurring **2.** the number of times something, some event, or some process occurs in a given unit of time **3.** in statistics, the number of members of a POPULATION-4, or a number of occurrences of a given type of event, falling into a single class in a statistical survey of the variation of specified characteristics.

frequency curve a graph of a continuous FREQUENCY DISTRIBUTION. The class interval is shown on the horizontal axis in order of increasing value from the intersection of the two axes, the class frequency on the vertical axis in order of increasing frequency from the point of intersection. The data may group around the values in the centre of a range of values (CENTRAL TENDENCY), with a few low values (to the left) and about an equal number of high values (to the right). This forms a symmetrical curve, the curve of the theoretical normal distribution. If the curve is asymmetrical it is said to be skewed. If most of the data group around relatively low values and there is a long tail of very high values, the slope on the right flank of the curve becomes shallower and longer than that on the left, and the curve is said to be positively skewed. But if most of the data group around relatively high values, so that the slope on the left flank of the curve is shallower and longer than that on the right, it is said to be negatively skewed. If the curve of a positively skewed frequency distribution lacks a short tail and ends abruptly at the peak, it is termed a truncated positively skewed frequency distribution curve (POISSON DISTRIBUTION).

frequency distribution in statistics, the organization of statistical data brought about by separating the range of values covered by a set of observations on a single variable into CLASS INTERVALS, arranging the class intervals consecutively in increasing or decreasing order, and tabulating the number of observations (the frequencies) falling in each.

freshwater water that is not salt, registering less than 0.2 per cent salinity.

friable *adj.* easily crumbled, hence applied to rocks having that characteristic and to soils with a good CRUMB STRUCTURE.

friction the force which offers resistance to the relative movement of one surface sliding or rolling over another with which it is in contact.

friction of distance, friction of space the retarding effect of distance on human interactions, hindering perfect or immediate ACCESSIBILITY. Efficient transport partly offsets this friction; but site rentals and transport costs represent the charge for the friction remaining.

fringing reef CORAL REEF.

front the surface of separation between two large or small AIR MASSES of markedly different temperature and humidity, termed a frontal zone if the air masses are widely separated. ANAFRONT, COLD FRONT, KATAFRONT, OCCLUDED FRONT, OCCLUSION, WARM FRONT.

frontal *adj.* **1.** relating to a meteorological FRONT **2.** in, at, or relating to the forward part.

frontal fog FOG formed when very fine warm drizzle resulting from the passing of the WARM FRONT of a DEPRESSION encounters cold air near the ground, which becomes saturated.

frontal rain CYCLONIC RAIN.

frontier 1. that part of a country or other political unit which fronts or faces another country or other political unit, sometimes applied to the actual BOUNDARY, sometimes to a zone. MARCH **2.** in USA, the border or advance area of settlement, carrying few people.

frontogenesis the processes in the atmosphere that lead to the formation or intensification of FRONTS.

frost 1. the temperature of the atmosphere, at about 1.2 m (4 ft) above the ground surface, at or below the FREEZING POINT of water, expressed in the number of degrees C or F that the temperature falls below that freezing point. AIR FROST, GROUND FROST **2.** frozen dew, fog or water vapour, termed HOAR FROST or RIME, appearing as white ice-crystals on exposed surfaces. BLACK FROST, GLAZE **3.** a powerful, MECHANICAL WEATHERING, disintegrating agent effective when water that has penetrated cracks and gaps in rocks and soil freezes, expands, and splits apart the rock or soil particles, e.g. in jointed rocks in regular beds, shattered by frost along lines of weakness to produce rectangular blocks, a process known as BLOCK DISINTEGRATION.

frost action CONGELIFRACTION, CONGELITURBATION, FREEZE-THAW, FROST HEAVE, FROST THRUST.

frost heave the raising of the soil surface by FROST-1 formed in it. CONGELITURBATION, FROST THRUST.

frost line a boundary indicating the incidence or the seasonal limit of FROST-1, revealing areas **1.** that never experience temperatures below the FREEZING POINT of water, e.g. the altitudinal limit in tropical countries below which frost never occurs **2.** with a mean minimum temperature or the lowest mean monthly temperature higher than freezing point, 0°C (32°F) **3.** that do not experience frost at certain times of the year, e.g. with a frost-free period of a specific number of days, important to certain crops, such as maize or cotton, which need a frost-free growing season **4.** lacking a month with a mean temperature above 0°C (32°F), the boundary of a frost climate, a climate characterized by perpetual frost.

frost pocket a small, low-lying area into which cold air rolls down by gravity from a hillslope, the cold air having been created by swift radiation of heat from the hillslope (KATABATIC WIND). The cold air is trapped in the pocket and may stay at FREEZING POINT after the temperature of the air over the hillslope has risen. Thus fruit-growers, anxious to avoid frost damage to blossom, do not plant in frost pockets.

frost smoke fog-like clouds formed when cold air comes into contact with relatively warm sea-water. ARCTIC SMOKE.

frost thrust the sideways pressure of freezing ground water. FROST HEAVE.

full 1. the rounded part of a sandbank **2.** a ridge of sand or shingle created by wave action, usually lying almost parallel to the coastline. Depressions between fulls are known as SWALES, slashes or furrows.

fumarole (Italian) a vent in the ground in volcanic regions emitting steam and other gases (e.g. ammonium chloride, carbon dioxide, sulphur dioxide) in the form of powerful jets, a manifestation of dying or extinct volcanic activity. SOLFATARA.

functional analysis an analysis of phenomena on the basis of the part they play within a particular, wider organization, e.g. of rivers in terms of their role in DENUDATION, of towns in terms of the part they play within an economy.

functional classification of towns a classification of cities and towns based on their characteristic economic activities.

functionalism **1.** broadly, a concept of a culture or a society regarded as an entity, a system comprising smaller, differentiated and interdependent systems, all parts of which function to maintain one another and the whole, any change occurring in one part leading to readjustment in the others **2.** in design, the theory that the function of the thing to be designed should dictate the form of the design, extended in the 1920s and 1930s to the conviction that the form most closely concerned with function was the most beautiful, the most desirable.

fungus pl. fungi, a parasitic or saprophytic plant, lacking chlorophyll, having a worldwide distribution. Some are important in the breakdown of the remains of dead plants and animals, especially in highly acid soils where they largely take over the role of bacteria; some, e.g. the mushroom, are valued for food; some for the organic fermentation they produce, e.g. some fungi are the source of certain antibiotics such as penicillin.

funnel cloud a TUBA CLOUD, a spinning cone of dark grey cloud descending from the base of a low-lying CUMULONIMBUS cloud until it reaches the surface of the sea, where it may become part of a WATERSPOUT. It may also be associated with a TORNADO-2.

fur farm a FARM concerned with the rearing of certain fur-bearing animals under controlled conditions with the objective of producing pelts for sale.

furrow a narrow trench made in the earth by a plough, into which seed is sown. The ridge of earth turned up by the plough in making the furrow is known as the furrow slice.

fusion a melting or blending together, e.g. of metals, to form a fused mass, the components becoming indistinguishable. In nuclear fusion (NUCLEAR ENERGY) light atomic nuclei are exposed to extreme temperature and extreme pressure to form heavier atomic nuclei, without loss in mass, resulting in a great output of energy. LATENT HEAT.

G

G7 Group Seven, the seven leading world industrialized countries (Canada, France, Germany, Italy, Japan, UK, USA), that operate collectively in the world economy to coordinate economic exchange. They support the IMF with loans.

gabbro a coarsely crystalline BASIC, PLUTONIC ROCK, dark in colour, consisting essentially of calcium-rich PLAGIOCLASE FELDSPAR and a FERROMAGNESIAN MINERAL and ACCESSORY MINERALS, commonly occurring, e.g. as a LOPOLITH. GNEISS.

Gaia hypothesis the proposition put forward by James Lovelock, scientist, 1979, that all the processes in the atmosphere, hydrosphere, lithosphere and biosphere of the EARTH-I are interrelated and self-regulating in order to sustain the planet and its phenomena; and that organic life in all its forms is an indispensable element in the interaction. He regarded his theory as scientific and testable, not mystical or teleological as the title 'Gaia' (the Greek earth-goddess) might suggest. TELEOLOGY.

galaxy one of the great number of systems in the universe, consisting of stars (individual and in clusters), nebular and interstellar particles etc., and classified by shape (amorphous, irregular, ellipsoidal, elliptical, spheroidal, spherical, spiral). The earth lies in a spiral galaxy, the Milky Way.

gale a strong wind. BEAUFORT SCALE.

gallery forest (Italian and Spanish galleria, a tunnel) the forest fringing both river banks in otherwise open country. The foliage may meet in an arch over narrow streams, hence the association with a tunnel.

game theory the application of mathematical logic to the strategy, tactics and fluctuating odds in situations where two or more people (the players) are totally or partly in conflict, aware of the costs and benefits of all the potential results but unaware of the choices of the other(s); and where each opponent is free to adopt several courses of action in order to select the OPTIMUM strategy and achieve the desired goal of minimizing the maximum loss the other(s) can impose on him/her (the MINIMAX solution). HOTELLING MODEL.

gangue the worthless mineral material, e.g. QUARTZ, surrounding or accompanying a metallic ore in a LODE or a VEIN, of little value in relation to the ore itself. The term expresses a judgement of economic value: mineral material regarded as gangue at one place or at a certain period of time may be valued highly in another place or time where or when economic conditions are different; or it may be of higher value in places where the concentration of the mineral is greater.

ganister, gannister a fine-grained, hard, very pure ARENACEOUS-I, SILICEOUS-I rock underlying some coal-seams in the Lower COAL MEASURES, used in making heat-refractory bricks in furnace linings.

gap a break in a ridge or belt of hills. If the break is just an indent and no stream is present, it is termed a dry gap or WIND-GAP; if the cutting is deep and a stream flows through it, it is termed a WATER-GAP.

gap town a town situated in or at the entrance to a GAP, a natural centre for lines of communication, holding a commanding position.

garden city **1.** specifically, based on the definition of Ebenezer Howard who used the term in 1898 (apparently unaware of its use by A. T. Stewart in Long Island, USA, in 1869): a planned, self-contained, industrial-residential settlement, rural in character, with open spaces and trees, catering for inhabitants of various grades of society, providing properly planned facilities for cultural and recreational opportunities, with properly planned facilities for industry, and with a relatively low housing density (10 units per acre, or 35 persons per acre), the whole of the land being in public ownership or held in trust for the community, being of a size that makes possible a full measure of social life, but not larger, and surrounded by a rural belt. Such a settlement is exemplified in England by Letchworth (1903) and Welwyn Garden City (1920). **2.** now loosely applied to an open density, planned, industrial-residential settlement, similar to Howard's specification, but not necessarily so tightly restricted, e.g. with regard to land ownership, or an encircling green belt.

garden suburb a suburb with a well-planned, open layout (not to be confused with a GARDEN CITY), e.g. Hampstead Garden Suburb, northwest London.

garigue, garrigue (French) scrub vegetation, especially of aromatic XEROPHYTES and stunted evergreen oak, occurring on LIMESTONE in drier areas of Mediterranean climate. It is ANTHROPOGENIC-2 in origin, resulting from the effects of human settlements, fire and the browsing of DOMESTIC and FERAL goats on the earlier vegetation. MAQUIS.

garúa (South America) a heavy, dense mist or drizzle occurring in winter on the Pacific slopes of the Coastal Range or the lower foothills of the Andes, soaking the land and promoting the growth of quick-flowering plants and grasses. LOMA.

gas **1.** a fluid substance with neither definite volume nor definite shape, having a fixed mass but no fixed volume, the volume changing with temperature and pressure, filling and taking the shape of its container. FLUID, LIQUID, SOLID **2.** a fluid substance at a temperature above its CRITICAL TEMPERATURE-1. VAPOUR **3.** a substance or mixture of substances in a gaseous state used for some specific purpose, e.g. provision of light and heat, an anaesthetic etc. COAL GAS, NATURAL GAS.

gas coal a COAL, usually bituminous, used for making GAS-3 by DISTILLATION.

gash breccia a rock consisting of angular fragments of DOLOMITE and LIMESTONE in a MATRIX-3 of CALCITE and CLAY, peculiar to the CARBONIFEROUS LIMESTONE of southwest Wales.

gatekeepers senior employees in banks and building societies (sometimes in estate agents' offices) who have the power to advance financial loans to prospective home buyers. RED-LINE DISTRICT.

gateway city a settlement, usually occupying a favourable, commanding site, which acts as a link between two areas and in many cases becomes a PRIMATE CITY.

GATT General Agreement on Tariffs and Trade, a treaty signed in 1947 by 23 nations. By 1981 there were 85 contracting parties, by 1995 there were 100, with others participating, aiming to reduce barriers to international trade, to overcome trade problems, and to expand world trade by negotiation and consultation. It was succeeded by the WTO (World Trade Organization) in 1995, and now includes bodies dealing with intellectual property rights and services.

GDP gross domestic product, the total market value of commodities (goods and services) produced in a country in a given period of time, usually a year. No allowance is made for capital consumption and depreciation; the use of market prices ensures the value of indirect taxes and subsidies are incorporated. The value of intermediate goods used in the production of other goods is excluded, being incorporated in the market price of the final goods. The distinction between intermediate and final products is an arbitrary one, varying from one country (or one economic statistician) to another. COMMODITY, GNP.

geanticline, geoanticline a very large complex ANTICLINE, extending over many kilometres, commonly caused by compression of the sedimentary rocks within a GEOSYNCLINE.

gelifluction in PERMAFROST areas, the slow, spontaneous downward movement of soil and ground ice in periglacial areas, facilitated by the melting of ice in spring. MASS MOVEMENT, SOLIFLUCTION.

Gemeinshcaft, Gesellschaft (German, community, society) terms used to contrast one with the other, Gemeinschaft being the long-lasting, intimate social relationship between individuals, based on affection, kinship or membership of a close community such as a family, or a group of friends, or a religious sect; Gesellschaft being a loose, rational association of isolated individuals concerned with personal self-interest, each of whom enters the association for limited purposes and with limited commitments. Gemeinschaft is considered to be typical of pre-industrial, pre-capitalist, pre-urban society; Gesellschaft to be typical of industrial, capitalist, urban society.

gender male or female. A distinction is often made in the use of the terms gender and sex (male or female), gender relating to what may be perceived as socially produced differences between male and female, sex to the physical differences.

gender roles the differing parts played by males and females in the social and economic life of a society, often determined by custom and tradition. FEMINIST GEOGRAPHY.

gene the basic unit of the material of inheritance, part of a chromosome, passed on from parent to offspring and responsible for controlling the processes of growth, development and reproduction which distinguish each species. BIODIVERSITY, GENOME PROJECT.

General Agreement on Tariffs and Trade GATT.

general circulation of the atmosphere, ATMOSPHERIC CIRCULATION.

general land use model a hierarchical model devised to ease the interpretation of land use maps, consisting of three stable 'scape' territories (wildscape, farmscape, townscape) with intervening less-stable territories (MARGINAL FRINGE and RURBAN FRINGE).

general systems theory a concept developed by Ludwig von Bertalanffy in the 1940s–1950s as a framework for a science of 'wholeness' and organization. By making theoretical generalizations about the properties common to different types of SYSTEM-1,2,3,4,5 an attempt is made to discover ISOMORPHISMS between systems, to liken a system drawn from one discipline to that of another or others, and thus to produce a doctrine of 'wholeness'. This approach emphasizes the relationship between form and process and draws attention to the multivariate character of phenomena; so it has come to be applied to geographical systems where isomorphism may often be found, e.g. in systems displaying ALLOMETRIC GROWTH, ENTROPY, or the hierarchical structure of the processes involved in the filling of space. SYSTEMS ANALYSIS.

genetically modified organism GENETIC ENGINEERING.

genetic engineering the technology involved in manipulating the molecular building blocks of organisms (the GENES). Organisms treated in this way are termed genetically modified organisms (GMOs). Very broadly, genetic engineering in plants isolates chromosomal DNA with specific, desirable characteristics and transfers it into other species, e.g. the introduction into a plant of bacteria that produce insecticides. BIO-CATALYST, BIOTECHNOLOGY.

genome project the attempt to map and eventually to sequence the human genome, i.e. all the genetic information, contained in the chemical structure of the chromosomes, that describes individual human beings. GENE.

gentrification a process in which wealthier people move into, renovate and restore run-down housing in an inner city or other neglected area, housing formerly inhabited by poorer people, the tenure usually shifting from private rented to owner occupation. It occurs particularly in inner city areas as a result of the wishes of the wealthier people to have easy access to their jobs and the recreational facilities (e.g. theatres) in the central area. Once started in a district, the process spreads rapidly until most of the poorer inhabitants are displaced, so that the social character of, and the value of the property in, the neighbourhood markedly changes.

genus (pl. genera) one of the groups used in the CLASSIFICATION OF ORGANISMS, consisting of a number of similar SPECIES-1, sometimes of only one species, a group of similar genera constituting a family. BINOMIAL NOMENCLATURE.

geo, gio (Norse gya, a creek) in northern Scotland and Faeroes **1.** a long narrow inlet of the sea, penetrating cliffs, especially in areas of well-jointed OLD RED SANDSTONE **2.** in geomorphology, a coastal cleft, often marking JOINTS, FAULTS or DIKES from which material has been removed by wave action.

geode a hollow, rounded, rock nodule, its interior walls lined with inward-pointing crystals, common in SEDIMENTARY ROCK, especially in LIMESTONE.

geodetic distance the shortest distance ('as the crow flies' distance) between two places on the earth's surface.

geographic, geographical *adj.* of, pertaining to, or relating to GEOGRAPHY. On the whole, geographic is used more frequently by American authors than by British.

geographical imagination the personal concept of the world, local and global, that encompasses place, landscape, time and space, and people, is formulated by maps, the news media (especially television), the visual arts, literature in all its forms, travel brochures, personal travel experience, and local and international social contacts etc. It varies with experience and with the passage of time. COGNITIVE MAP, ICONOGRAPHY, TIME GEOGRAPHY.

geographical inertia the tendency of a place with established installations and services to maintain its size and its importance as a focus of activity after the conditions originally influencing its development have appreciably altered, ceased to be relevant, or disappeared. GEOGRAPHICAL MOMENTUM, INDUSTRIAL INERTIA, INDUSTRIAL MOMENTUM.

geographical information system GIS.

geographical mile a NAUTICAL MILE, a mile theoretically equal to one minute (1') of LATITUDE but, owing to the fact that the earth is a sphere flattened at the poles (GEOID), a minute of latitude varies slightly in length. The geographical or nautical mile has therefore been standardized internationally as being equal to 1.852 kilometres (1 km = 0.539 international nautical mile). KNOT, MILE.

geographical momentum the tendency of a place with established installations and services not only to maintain (GEOGRAPHICAL INERTIA) but also to increase its size and its importance as a focus of activity after the conditions originally influencing its development have appreciably altered, ceased to be relevant, or disappeared. INDUSTRIAL INERTIA, INDUSTRIAL MOMENTUM.

geography the study that deals with the material and human phenomena in the space accessible to human beings and their instruments (SPACE-2,3,4), especially the patterns of, and variation in, their distribution in that space, on all scales, in the past or present.

Human geography is concerned with human activities (of individuals and of groups) and organization in so far as these relate to the interaction (past or present) of people with their physical environment and with the environments created by human beings themselves, and the consequences of these interrelationships.

Physical geography is concerned with the physical characteristics and processes of the ATMOSPHERE-1, BIOSPHERE, HYDROSPHERE and LITHOSPHERE.

geoid **1.** an earth-shaped body **2.** the earth in geometrical terms, i.e. an OBLATE spheroid or ellipsoid (a SPHERE, SPHEROID or ELLIPSOID flattened at the poles), regarded as a mean sea-level surface (i.e. supposing all mountains and ocean basins to be levelled to the mean sea-level surface), or as an undulating surface related to gravitational force (GRAVITY-2), higher than the actual surface of the spheroid under the continents, lower under the oceans.

geologic, geological *adj.* of, pertaining to, derived from GEOLOGY. Geological is the usual English usage.

geological inversion STRATA which have the normal sequence of layers, but in reverse order (i.e. upside down, as in the lower limb of an OVERFOLD), occurring in regions of intense folding (FOLD-2).

geological structure STRUCTURE-3.

geological time the chronology of the history of the earth revealed by its rocks. The hierarchy of time-periods shown in the GEOLOGICAL TIMESCALE is not always strictly observed: there are variations in the names and the periods used. An epoch (time) may be divided into AGES-2, during which a FORMATION (of rocks) may be laid down. *See table.*

geological timescale

Era	*Period (time) or System (rock)*	*Epoch (time) Series (rock)*	*Duration in million years (approx.)*	*Million years ago (approx.)*
		Holocene (Recent)		(12 000 years)
	Quaternary	Pleistocene (Glacial)	c. 1.5	
				c. 1.5
CAINOZOIC (CENOZOIC)	Tertiary	Pliocene	5.5	
				7
		Miocene	19	
				26
		Oligocene	12	
				38
		Eocene	16	
				54
		Palaeocene	11	
				65
MESOZOIC (SECONDARY)	Cretaceous		71	
				136
	Jurassic		54	
				190
	Triassic		35	
				225
PALAEOZOIC (PRIMARY)	Permian		55	
				280
	Carboniferous	*American* Pennsylvanian Mississippian	65	
				345
	Devonian		50	
				395
	Silurian		35	
				430
	Ordovician		70	
				500
	Cambrian		70	
				570
PRECAMBRIAN	PROTEROZOIC			
	ARCHEOZOIC			

geology the scientific study of the origin and nature of the earth's crust, the rocks of which it is composed, and the history of its development and changes. The study concerned with landforms is GEOMORPHOLOGY, GEOLOGICAL TIMESCALE.

geometric progression a series of numbers in which each is multiplied by a fixed number to produce the next, e.g. 1, 3, 9, 27. ARITHMETIC PROGRESSION.

geomorphology the study concerned with LANDFORMS, especially the genesis, evolution and processes involved in the formation of the surface forms of the earth.

geophysics an interdisciplinary scientific study concerned with the structure, physical phenomena and evolution of the earth's interior and crust.

geopolitics (German Geopolitik) **1.** the

study of STATES-2 or NATIONS viewed as organic entities in space, and as such subject to biological laws of growth and decline (in territorial extent and political influence, as well as in economic and social terms), being, like other organisms, engaged in a perpetual struggle for survival, for control over the space occupied. F. Ratzel is credited with the introduction of this concept in the late nineteenth century; it was taken up by the English geographer, Halford Mackinder; by the Swedish political scientist, Rudolf Kjellen; and later by the German geographer, K. Haushofer. But it became particularly prominent (and disreputable) before and during the Second World War when adopted by the Nazi Party in Germany to advance its theory of race superiority and to justify the Third Reich's demand for lebensraum **2.** the influence, or the study of the influence, of spatial aspects on the political nature, history, institutions, etc. of states or nations, and especially on their political and economic relations with other states.

geosophy the study of the nature and expression of geographical ideas.

geostationary satellite in REMOTE SENSING a SATELLITE-1 with an orbital speed synchronous with that of the rotation of the earth, much used in communication systems.

geostrophic flow, geostrophic wind the concept of a wind blowing parallel to the ISOBARS as a result of the force exerted by the horizontal ATMOSPHERIC PRESSURE gradient in one direction, balanced by the deflection of the CORIOLIS FORCE in the opposite direction. It is only in the upper atmosphere that winds come near to perfect geostrophic flow, because FRICTION near the earth's surface upsets the balance below 455 m (1500 ft) above sea-level, so that at these lower levels winds blow at an oblique angle to the isobars towards low pressure. FERREL'S LAW.

geosyncline a very large linear depression or SYNCLINE or down-warping of the earth's crust, filled (especially in the central zone) with a deep layer of sediments derived from the land masses on each side and deposited on the floor of the depression at approximately the same rate as it slowly, continuously subsided during a long period of geological time.

geothermal energy energy, in the form of heat derived from anomalies in the GEOTHERMAL GRADIENT, which gives rise in nature to the delivery of hot water and steam to the earth's surface by THERMAL SPRINGS and GEYSERS (HYDROTHERMAL). The hot water and steam can be used in generating ELECTRICITY, e.g. as in Iceland and New Zealand. In addition the heat of the rocks in the earth's crust can be used to warm cold water pumped down from the surface through a borehole. The water is passed over the HOT ROCK and rises to the surface by another borehole, where its acquired heat can be tapped locally.

geothermal gradient the rise in temperature in the rocks of the earth's crust with increasing depth. There are indications that the rate is not constant but speeds up with increasing depth. Estimates range from 1°C in 28 m (1°F in some 51 ft) to 1°C in 40 m (1°F in about 73 ft), but an average in the SIAL could be about 1°C in about 28.6 m (1°F in about 52 ft). The temperature of the base of the sial may be some 986°C (about 1806°F). GEOTHERMAL ENERGY, HOT ROCK.

gerrymander the drawing of the boundaries of electoral districts in such a way that one political party is given an advantage.

Gestalt, gestalt theory a shape, pattern, configuration, structure which can be perceived only as a whole or unity, with qualities different from those of its components: it cannot be expressed as the sum of its parts, because the parts acquire certain characteristics produced by their inclusion in the whole, and the whole has some characteristics belonging to none of the parts, giving it an additional, often indefinable, quality (e.g. a melody cannot be expressed as the sum of its components, the notes). Hence the Gestalt theory of the Gestalt school of psychology, that human perceptions, reactions, responses are all Gestalten, that enquiries about them should examine the whole to discover what are the components, rather than start with the components and try to synthesize them to form the whole. BEHAVIOURAL ENVIRONMENT, PHENOMENAL ENVIRONMENT.

geyser a hot spring that intermittently, sometimes at regular intervals, throws up a jet of hot water, steam etc. in areas that are or were volcanic. Famous examples are in the volcanic districts of Iceland and the USA. The original 'geyser', the Giant Geyser in Iceland, erupts on occasions to 55 m (180 ft), the Giant Geyser in Yellowstone National Park, USA, to 60 m (200 ft) at irregular intervals. The Waimangu Geyser, New Zealand, is said to have erupted to 305 m (1000 ft) in 1909. GEOTHERMAL ENERGY.

ghetto **1.** historically, the part of a town or city in which Jews were required to live in parts of medieval Europe, e.g. Poland, Italy **2.** the part of a town or city where members of a minority group live on account of economic or social pressure. SEGREGATION.

ghibli KHAMSIN.

ghost town a TOWN, usually once a mining town, now wholly abandoned, or inhabited by a few people if the town is now a tourist attraction, e.g. in Klondike, USA.

gibber (Australian: aboriginal) a large stone or boulder. A gibber plain is a desert area strewn with pebbles or boulders, a type of gravel desert.

gibbous *adj.* MOON.

gilgai (Australian) a low, rounded mound on the surface of the plain in parts of the Murrumbidgee basin, New South Wales, formed when fragments of the A HORIZON falls down cracks in the B HORIZON. When the soil becomes saturated the B horizon expands, and the A horizon is pushed up in a series of mounds, continually eroded until the B horizon is exposed at the surface.

Gini coefficient, Gini index a measure of concentration, calculated by taking the ratio between the area contained by the diagonal and the LORENZ CURVE and the area in the whole triangle below the diagonal. The value of the index increases with the increasing size of the area between the curve and the diagonal. The index is frequently used as a measure of inequality in the welfare approach (WELFARE GEOGRAPHY).

GIS Geographical Information System, a readily accessible collection of geographical data stored in digital form in a computer, easily analysed and constantly updated with new data (especially that provided by REMOTE SENSING).

glacial *adj.* **1.** icy, frozen, used very loosely in connexion with a glacier, ice age, lake etc. **2.** strictly, of or pertaining to a GLACIER.

glacial age, epoch, era, period a period of GEOLOGICAL TIME when much of the earth's surface was covered by GLACIERS. ICE AGE.

glacial boulder a BOULDER which has been carried by a glacier. ERRATIC.

glacial diffluence the branching of a VALLEY GLACIER, usually due to an obstruction in the channel which causes a build-up of ice in the main glacier and an overflow (often over a COL) into an adjoining valley.

glacial divergence the obstruction of an established drainage pattern by an advancing glacier or ice sheet, leading to a change of course in the drainage.

glacial drainage channel a stream bed cut by the melt-water of a glacier.

glacial drift DRIFT.

glacial lake a body of water contained by the wall of a valley and the margin of a glacier.

glacial outburst JOKULHLAUP.

glaciated valley U-SHAPED VALLEY.

glaciation 1. the occupation of an area by an ICE SHEET or by GLACIERS **2.** the action of ice on rocks over which it has passed **3.** the time when such took place (e.g. first and second glaciations).

glacier originally a river of ice moving down a valley (now termed a VALLEY GLACIER, ALPINE GLACIER or mountain glacier). Now applied to a mass of snow and ice formed by the consolidation (under pressure) of snow falling on high ground (FIRN, NEVE), moving outward and downward from the zone of accumulation to lower ground (or, if afloat, continuously spreading), in some cases carving a broad, steep-sided valley on its way. It may be active, moving quickly and carrying much material; or passive (the rates of accumulation and ablation being low, the slope gentle), moving little if at all, and carrying little or no material. ABLATION, ALIMENTATION, ICE CAP, ICE SHEET, and glacier qualified by the terms ACTIVE, APPOSED, CIRQUE, COLD, EXPANDED-FOOT, HANGING, MOULIN, PASSIVE, PIEDMONT, TIDAL, WALL-SIDED, WARM.

glacierization the occupation of an area by a glacier or ice sheet at the present day. GLACIATION.

glacier karst the KARST-like appearance of an ice mass caused by the action of melt-water lying under a covering of morainic and glaciofluvial material which, varying in thickness, leads to variations in the rate of ABLATION. As in true karst, underground streams, POLJES etc. are produced. GLACIOKARST.

glacier mice GLACIER MOUSE.

glacier milk a murky, greenish-white stream of melt-water containing ROCK FLOUR, issuing from a GLACIER SNOUT.

glacier mill MOULIN.

glacier mouse (Icelandic jökla mýs, glacier mice) a small, rounded, moss-covered stone occurring on some glaciers in Iceland.

glacier retreat the stage of a glacier when the ice is moving forward but the snout is receding, reached when the rate of ABLATION exceeds that of ACCUMULATION.

glacier snout the cavernous end of a VALLEY GLACIER.

glacier tongue the part of a glacier extending to the sea, and usually afloat.

glacio-eustatism a global change of ocean level caused by the abstraction of water by the growth of ICE SHEETS, or the return of water to the ocean when these melt. DIASTROPHIC EUSTATISM, EUSTATISM, GLACIO-ISOSTASY, NEGATIVE MOVEMENT, POSITIVE MOVEMENT, REJUVENATION.

glaciofluvial, glacifluvial FLUVIOGLACIAL.

glacio-isostasy the deformation of part of the earth's crust caused by the weight of an extensive ICE-SHEET. The great weight of the ice depresses the crust but the land surface recoils as the ice melts, in many cases leading to TILTING and WARPING-2. GLACIO-EUSTATISM, ISOSTASY.

glaciokarst, nival karst karstic landforms (KARST) in a limestone area resulting from the effects of GLACIATION and subsequent melting, the ice subjecting the area to differential erosion (ABRASION), and the melt-water from the ice and snow, charged with CARBON DIOXIDE, dissolving the CALCIUM CARBONATE in the limestone. GLACIER KARST.

glasshouse GREENHOUSE.

glaze, glazed frost a generally HOMOGENEOUS-2, transparent clear ice formed when **1.** raindrops or non-supercooled DRIZZLE droplets fall on an exposed surface with a temperature well below 0°C (32°F) (e.g. on a road, forming BLACK ICE) **2.** supercooled raindrops or drizzle droplets fall on objects with a surface temperature below or slightly above 0°F (32°F) (e.g. on tree branches) **3.** FROST recurs after some thawing. The weight of ice of a glazed frost can be very damaging, especially to the branches of trees. SUPERCOOLING.

glen (Scottish and Gaelic; Irish glean; Welsh glyn) a steep-sided, usually glaciated, valley characterized by a narrow floor, sometimes contrasted with a strath, a broad, open and cultivated valley.

gley, glei a sticky yellow and grey mottling in the soil, or the horizon thus mottled. It is caused by poor drainage, intermittent waterlogging reducing OXIDATION or causing the deoxidation of FERRIC compounds. The compounds which result are bluish-grey, mottled, sticky, clayey and compact. PODZOLIC SOILS, WATERLOGGED.

gleying the development of a GLEY structure in soil.

gley podzol PODZOLIC SOILS.

gley (glei) soils, gleysols soils subject to periodic or permanent waterlogging (WATERLOGGED), not artificially drained, and to some extent displaying the MOTTLED horizon below the surface layer typical of GLEYING. There may or may not be a humic, peaty topsoil. They are characterized by the extent to which humus, clay, sand, peat are present, and to the severity of waterlogging. Those with a peaty surface horizon overlying a clayey, MOTTLED impermeable horizon are termed humic gley soils or stagnohumic gley soils: they are excessively wet and typically non-calcareous. Those that lack the peaty surface horizon and have a subsurface horizon that is slightly less impermeable are termed stagnogley soils: in general these are non-alluvial and non-calcareous and are classified as loamy or clayey. Other gley soils lacking the peaty surface horizon are termed argillaceous (clayey) or sandy. PODZOLIC SOILS.

glint a steep cliff, steep terrace, steep edge of a plateau.

glint line an erosion escarpment, the conspicuous edge of a stretch of denuded rock, particularly the edge between an ancient SHIELD (such as the Laurentian shield) and younger rocks.

glint-line lake a lake, frequently one of a string, formed along a GLINT LINE.

global *adj.* **1.** of or pertaining to the earth as a whole **2.** spherical **3.** comprehensive, total, encompassing all or nearly all considerations, categories, items etc.

global city WORLD CITY.

globalization the process by which people, their ideas and their activities (economic, cultural, political) in previously relatively separated parts of the world become interconnected, integrated, drawn to the same SOCIAL SPACE at the same historical time. It does not imply homogenization, or the elimination of regional differences or of UNEVEN DEVELOPMENT. It leads to increased ecological interdependence; and the globalized world economy is based particularly on the transnational movement of information and the mobile FACTORS OF PRODUCTION (capital, labour, technology).

Global Positioning System GPS.

global warming the potential slight rise in the temperature of the earth's atmosphere arising from pollution due to natural causes (e.g. emissions from volcanic eruptions) and human activities (e.g. destruction of vegetation cover, burning of FOSSIL FUELS) that upset the natural balance of atmospheric gases (e.g. by adding excessive amounts of carbon dioxide), thus enhancing the GREENHOUSE EFFECT and possibly damaging the OZONE layer.

globate *adj.* shaped like a GLOBE.

globe 1. a spherical body **2.** the EARTH-1 **3.** a spherical model with a map of the earth (terrestrial globe) or the heavens (celestial globe) shown on it.

GMO genetically modified organism, a term applied in BIOTECHNOLOGY to a product resulting from GENETIC ENGINEERING.

GMT, Greenwich Mean Time the LOCAL TIME at the meridian of Greenwich, England (0° longitude, passing through the former Royal Observatory at Greenwich), the STANDARD TIME for the British Isles and the standard from which most of the countries of the world reckon their standard times. MEAN SOLAR TIME.

gneiss a foliated (FOLIATION-1), coarsely crystalline rock with characteristic alternating dark and light bands composed of dissimilar minerals (usually minerals in the MICA group alternating with bands of QUARTZ and FELDSPAR), having the composition of GRANITE but produced by the METAMORPHISM and recrystallization of both IGNEOUS rocks and sediments. Those resulting from the DYNAMIC METAMORPHISM of igneous rocks are termed orthogneiss; those from metamorphism of SEDIMENTARY ROCKS are termed paragneiss. Gneiss is the most common of crystalline rocks (the term covering a wide range of crystalline rocks), and gneisses make up a large part of the areas of ancient rocks, e.g. of the Laurentian shield.

gneissose structure a structure shown by coarsely crystalline rocks, characterized by discontinuous FOLIATION-1.

GNP gross national product, a measure of the total flow of output in an economy during any specified period of time (excluding goods or services used as inputs in the production of further goods or services), i.e. the GDP (gross domestic product) of a country combined with income derived from overseas investments less profits generated by production within the country but due to foreigners abroad.

gold an indestructible, yellow PRECIOUS METAL, almost resistant to chemical attack (NOBLE METAL). It occurs in alluvial deposits (PAN) or in quartz veins in a free state. It is sometimes associated with other metals in the quartz veins or in the other forms in which these occur, thus gold is sometimes mined directly or won as a byproduct from other metals. It is alloyed with other metals to achieve hardness or to change its natural colour. CARAT.

golf course, golf links terms used interchangeably, but strictly golf links (LINKS) should be restricted to a tract of land lying near the sea coast, on which the game of golf is played.

Gondwanaland the southern part of the great PRECAMBRIAN land mass, PANGAEA. Current scientific evidence suggests that it included today's Africa, Madagascar, Australia, part of South America, Antarctica and the Indian subcontinent, and began to break up in PALAEOZOIC times. CONTINENTAL DRIFT, LAURASIA, PLATE TECTONICS, TETHYS OCEAN.

good in economics, anything (morally good or morally bad, material or intangible) which is capable of satisfying a human want. CAPITAL GOODS, CONSUMER DURABLES, CONSUMER GOODS, ECONOMIC GOODS, FREE GOODS, PRODUCER GOODS.

GOOS the Global Ocean Observing System, backed by the UN Environmental Programme and designed to monitor the massive energy movements in the world OCEAN-1. The information gathered should lead eventually to improved forecasting of weather and of climatic change, including anticipation of the incidence of the EL NINO EFFECT, and of earthquakes.

gorge a rocky-walled, steep-sided deep, narrow river valley.

GPS Global Positioning System, a means of locating position on the earth's surface (land or water) by using a simple earth-based instrument that receives signals from two or three GEOSTATIONARY SATELLITES. The position of the receiving instrument is calculated by TRIANGULATION.

graben (German) a long, narrow, depressed tract of the earth's crust, a trough between parallel NORMAL FAULTS, the THROWS of which face in opposite directions. The term is sometimes applied to a RIFT VALLEY, but such usage is unacceptable to some authors on the grounds that a graben is a structural feature and not necessarily a valley.

gradation the bringing of the surface of the land to a common level by the processes of AGGRADATION and DEGRADATION.

graded river, graded stream a river which has by EROSION and DEPOSITION adjusted its channel and the slope of its bed so that the rate of its flow is exactly that needed for the carrying of the material supplied by the drainage basin, a delicately balanced condition which may be upset at any time.

graded sediments loose or cemented sediments with particles sorted by natural means, according to size. Various systems of grain size are used in the study of sediments and soils (SOIL TEXTURE, WENTWORTH SCALE), e.g.

Grade	*International diameter (mm)*	*American diameter (mm)*
stones	over 2.00	over 2.00
coarse sand	0.2–2.00	0.2–2.00
fine sand	0.02–0.2	0.05–0.2
silt	0.002–0.02	0.002–0.05
clay	below 0.002	below 0.002

gradient 1. the degree of slope of a land surface, stream, road, railway, away from the horizontal, expressed as a percentage, or an angular measurement from the horizontal, or as a proportion between its VERTICAL INTERVAL, reduced to unity, and its HORIZONTAL EQUIVALENT. SLOPE LENGTH **2.** the degree of variation in certain phenomena, e.g. atmospheric pressure, temperature, density etc. GEOTHERMAL GRADIENT.

graffito (Italian, pl. graffiti) **1.** unauthorized drawing or writing on a wall or other surface (the term originally implied scratching, as on the ancient walls of Pompeii and Rome etc.) **2.** an incised decoration, e.g. on a pot, produced by scratching through a superficial layer of plaster, glaze, etc. to reveal a GROUND-5 of different colour.

grain 1. the general trend of the geological structure, folds, faults etc. reflected in the dominant direction of mountain ranges, river valleys etc. **2.** the natural arrangement of strata in rock, coal etc. **3.** the direction and pattern in which wood fibres grow **4.** a very small hard particle, especially of a mineral **5.** the seed of a CEREAL; or harvested cereal crops in general. FLOUR.

gram 1. a PULSE, e.g. chick pea **2.** gram, gramme, gm, the unit of mass in the metric measurement system, one thousandth part of a KILOGRAM. KILO-.

granary 1. a storehouse for threshed CEREAL (grain) crops **2.** a region producing or exporting much grain.

Gran Chaco CHACO.

granite 1. specifically, a granular, coarsely-crystalline acid PLUTONIC ROCK, the most common and widespread of plutonic rocks, consisting essentially of ORTHOCLASE FELDSPAR with free QUARTZ and a MICA, formed by the slow cooling of a large INTRUSION of MAGMA or possibly by the transformation of pre-existing COUNTRY ROCK by a form of METASOMATISM **2.** loosely, any coarsely-crystalline, pale, IGNEOUS ROCK. Granite is used as a building stone, especially polished for facings, and as an AGGREGATE-4. ACID ROCK, BATHOLITH, GNEISS.

granular disintegration a type of MECHANICAL WEATHERING, the breaking down of porous rocks (POROSITY) into fragments to form a granular mass, caused either by the freezing and thawing of water absorbed in the rock pores (which dislodges rock particles), or by differential contraction and expansion caused by INSOLATION-1,2 (which results in the disintegration of the GRAINS-4).

grape the juicy edible green or purple fruit growing in bunches on a grapevine. Some grapes are used for dessert, others for wine, or for drying as currants, sultanas, raisins. Grape vines flourish in Mediterranean conditions and in mild midlatitude areas (including southern England) in the open air, or under glass in cooler areas. VINEYARD, VITICULTURE.

graph 1. a geometric representation of

a function, i.e. a diagram indicating the relationship of one variable quantity to one or more others by showing their values as distances from two axes (occasionally three) usually placed at right angles to one another, one of the VARIABLES to be plotted being scaled along each of the axes. By convention, in a two-dimensional graph showing a FREQUENCY DISTRIBUTION, the DEPENDENT VARIABLE is plotted along the vertical axis (i.e. the *y*-axis), representing the FREQUENCY of occurrence, the INDEPENDENT VARIABLE along the horizontal axis (i.e. the *x*-axis) **2.** a mathematical structure, a TOPOLOGICAL diagram, consisting of a set of objects (which may also be termed access points, junctions, nodes, points, terminals, or vertices, sing. vertex) some of which are connected to each other by edges (which may also be termed links or routes), thus establishing a binary relation between the objects (access points, junctions etc.). Such a graph, being topological, ignores the spacing of the objects and the lengths of the connecting edges (links, routes). It is used to represent a NETWORK and is basic to GRAPH THEORY. BAR GRAPH, DIVIDED CIRCLE DIAGRAM, HISTOGRAM.

graphicacy the state or condition of being able to form an idea of (to conceptualize), interpret and express relationships (e.g. two- or three-dimensional spatial relationships) that cannot be expressed in words and/or mathematical terms alone (i.e. not in the languages of LITERACY and/or NUMERACY alone) by the use of maps, diagrams, graphs (CARTOGRAPHY in its widest sense) and illustrative material such as photographs, supported by the skills of literacy and of the various branches of MATHEMATICS. ARTICULACY, INGRAPHICATE.

graph theory the branch of mathematics concerned with the properties of GRAPHS-2, i.e. with the vertices (access points, junctions, nodes, points, terminals) and edges (links, routes). Graph theory is used to describe NETWORKS, to indicate ACCESSIBILITY, etc.

grass a loose HERBACEOUS plant with shallow, fibrous roots, and with a world-wide distribution in areas with sufficient moisture. Among the grasses are CEREALS, reeds, bamboos, pasture grasses. Grasses provide valuable food for people and their livestock, through their starchy seeds, rich in protein, and their nutritious foliage. The leaves of some species are used in paper-making, the stems of others (BAMBOO) provide structural timber. SUGAR CANE.

grassland land covered with grass **1.** occurring naturally in areas with enough moisture to stimulate and maintain seasonal growth, e.g. tropical (SAVANNA), midlatitude (DOWNS, PAMPAS, PRAIRIE, PUSZTA, STEPPE, VELD), mountain (ALP) **2.** cultivated for pasture or hay, either permanent or short LEY (grown in rotation with arable crops), usually sown and fertilized.

gravel a loose, water-worn sediment in which small rounded stones, variously defined as 2 to 10 mm (0.08 to 0.4 in), or 2 to 20 mm (0.08 to 0.8 in), or 2 to 50 mm (0.08 to 2 in) in 'diameter', predominate. It is used especially in making CONCRETE. When consolidated it becomes a CONGLOMERATE. BOULDER, COBBLE, PEBBLE.

gravitation **1.** the act or process of moving under the force of GRAVITY-2 **2.** the attractive force between any two pieces of matter, between two bodies, that is independent of the chemical nature of those bodies, or the presence of bodies between them **3.** the law (NEWTON'S LAW OF UNIVERSAL GRAVITATION) stating that any two pieces of matter attract one another with a force which is directly proportional to the product of their masses, and inversely proportional to the square of the distance between them.

gravity **1.** weight, heaviness **2.** the gravitational force between the earth and a body on the earth's surface or in the earth's gravitational field. The force decreases with increasing distance from the earth, the force of gravity being inversely proportional to the square of the distance. GRAVITY MODEL **3.** *adj.* synonymous with gravitational.

gravity anomaly a deviation from the predicted value of the GRAVITY of rocks, termed positive anomaly if exceeding expectation, negative anomaly if below. The deviation may be due to the existence of rocks of unexpectedly different densities or magnetic characters in the earth's crust, so the detection of gravity anomaly is important in the study of the structure of the EARTH'S CRUST. ANOMALY, ISOSTASY.

gravity model a mathematical MODEL-2 inspired by Newton's Law of universal gravitation, i.e. the force of attraction of two masses, or bodies, is proportional to the product of their masses and inversely proportional to the square of the distance between them. By analogy with this physical law a mathematical gravity model may be used to predict the attraction of places, and thus the potential movement of people between them.

Thus, in a crude analogy, the force of attraction becomes the transactions (T) between two places, i and j (T_{ij}); the masses become the two places represented by the numbers of their inhabitants (P), P_i representing the population of one place, P_j the other; and the distance between them D_{ij}:

$$T_{ij} = \frac{P_i P_j}{D_{ij}^2}$$

Using this equation the relative transactions between towns A and B (T_{AB}) and A and C (T_{AC}) can be calculated:

$$T_{AB} = \frac{\textit{Population of } A \times \textit{population of } B}{\textit{Distance between } A \textit{ and } B^2}$$

$$T_{AC} = \frac{\textit{Population of } A \times \textit{population of } C}{\textit{Distance between } A \textit{ and } C^2}$$

This over-simplified version of a gravity model is usually much refined in practice. REILLY'S LAW OF RETAIL GRAVITATION.

gravity slope that part of the slope of a hillside which is steep in comparison with the gentler slope, the HALDENHANG, which occurs lower down.

gravity slumping, gravitational slumping, sliding the downward slipping of rock masses or sediments due to GRAVITY-2. MASS MOVEMENT, SLUMP.

gravity water, gravitational water water in the soil, moving by the force of GRAVITY-2 towards the WATER TABLE. Either term may appear as a synonym for VADOSE WATER.

grazing a PASTURE area where livestock feed on growing grass or other HERBS-1. A distinction is often made between land used for grazing and land which is precisely similar, but from which the herbage is cut for hay or drying. MEADOW, ROUGH GRAZING.

great circle any hypothetical circle on the earth's surface the plane of which passes through the earth's centre, and thus divides the earth into two HEMISPHERES. The EQUATOR is a great circle, and each MERIDIAN is half of a great circle. The number of great circles which can be drawn on a sphere is limitless. The shortest distance between any two points on the earth's surface is the arc of the great circle on which they lie, hence the use of the great circle routes by aircraft, e.g. over the north polar regions.

Great Ice Age ICE AGE.

green belt a tract of open country of varying width, and not necessarily continuous, with farmland, open recreational areas, woodland etc. surrounding a large built-up area and, by planning regulations, protected from further building in order to prevent URBAN SPRAWL and to provide an AMENITY.

green fallow partial fallow, fallow land from which a quickly maturing green crop (e.g. turnips, potatoes) is taken, the crop being planted in rows and weeded by hoeing. Bastard fallow (pin fallow) is a type of green fallow, the land being sown with a quickly maturing crop (e.g. vetches) in autumn, fed-off in spring. FALLOW.

green field, greenfield site a planning term for a plot of land, previously undeveloped, for which DEVELOPMENT-2 is proposed, or on which it is in progress.

greenhouse glasshouse, a building with glass or clear plastic roof and sides, sometimes heated artificially, used to provide the protection from air turbulence and warmth needed in some areas by certain growing plants. The glass or clear plastic allows incoming short-wave SOLAR RADIATION to enter the structure, but acts as a barrier to outgoing long-wave radiation from the earth. GREENHOUSE EFFECT.

greenhouse effect the phenomenon in which the ATMOSPHERE-1 near the earth's surface holds heat. Incoming short-wave SOLAR RADIATION (which includes visible light and heat) has no difficulty in passing through the atmosphere to the earth's surface. It is absorbed there by materials that re-radiate on longer wavelengths. But this outgoing returning long-wave RADIATION from the earth does have difficulty in passing through the atmosphere: it is inhibited especially in the presence of cloud, being absorbed and radiated back to earth by water vapour, particles and carbon dioxide. Thus the atmosphere near the earth's surface is warmed by long-wave radiation and, like the glass of a greenhouse, restrains the rising heat, an effect enhanced by cloud cover. ATMOSPHERIC WINDOW, GLOBAL WARMING.

green labour people who for the first time go outside the home to work for payment. COTTAGE INDUSTRY.

green lane an unmetalled road, still bearing a turf or grass surface, but constituting a right-of-way. Often identical with DRIFTWAY.

green manure a green crop, especially one rich in nitrogen, e.g. clover, which is, while still green, ploughed directly into the soil to increase soil fertility and improve structure. LEGUMINOSAE, PEA, RAPE.

green revolution an AGRICULTURAL REVOLUTION in less developed countries, especially in Asia in the 1960s and 1970s, which gave rise to increased food production, brought about mainly by the rapid improvement of yields of cereal crops (particularly of RICE, MAIZE and WHEAT) resulting from the introduction of new high-yielding varieties and the techniques, relying on an abundant water supply and the use of fertilizers, necessary to grow them.

green village a small nucleated settlement in which the homesteads surround an open space or village green.

Greenwich Mean Time GMT.

Greenwich meridian the prime MERIDIAN (0° longitude) passing through the former Royal Observatory at Greenwich, England, from which other meridians are calculated, expressed east or west, to 180° longitude.

grey box approach BLACK BOX APPROACH.

grey-brown (gray-brown) podzol (podzolic soil) an acidic soil, transitional between a PODZOL and BROWN FOREST SOIL, less leached and with a greater organic content than a podzol. It occurs mainly in western Europe (including the UK) and northeast USA, where humid conditions lead to some leaching, but less than in the climatic conditions resulting in a true podzol. It supports good quality grassland, and is suitable for mixed farming. The upper layer of the A HORIZON has a MULL-HUMUS form, and there is marked ILLUVIATION of clay in the B HORIZON. SOIL, SOIL HORIZON, SOIL PROFILE.

grey earth a soil in arid areas of mid-latitudes, deficient in organic matter and nitrogen, commonly salt encrusted, because the lack of rainfall discourages plant growth and LEACHING is rare.

grid 1. an arbitrary network drawn on a map so that an exact reference can be made to any point on the map. In reference grids for small areas the Cartesian coordinate system is commonly used, the location of a point being identified by its distance from two reference lines intersecting at right angles. In this system the grid usually consists of two sets of parallel lines at right angles, the lines being such a distance apart as to represent a fixed distance (e.g. 1 km, 10 km). The lines are numbered eastwards and northwards from a point fixed in the southwest corner of the whole area covered (FALSE ORIGIN); thus any point can be expressed exactly in terms of its EASTING and NORTHING, without the cumbersome use of LONGITUDE and LATITUDE. Grid north is the direction of the approximately north to south grid lines, but coincides with TRUE NORTH only along the meridian of origin. BEARING, NATIONAL GRID **2.** any uniform pattern laid on a surface for the purpose of mapping (and subsequently spatially analysing) data, or for calculating values at the NODE-2 of each unit of the pattern so that ISOPLETHS may be drawn **3.** a NETWORK-2, e.g. of electricity cables or gas pipelines for a power supply.

grike, gryke a deep cranny or fissure lying between ridges and traversing LIMESTONE pavements, caused by CARBONATION-SOLUTION along well-defined joints.

grit, gritstone an ambiguous term applied to **1.** small, usually angular, particles of rock **2.** coarse-grained SANDSTONE, or sandstone (coarse or fine) made up of angular grains **3.** sandstone of grains of conspicuously unequal size **4.** stratigraphically, in names of such FORMATIONS as Millstone Grit.

groin GROYNE.

gross domestic product GDP.

gross national product GNP.

gross reproduction rate REPRODUCTION RATE.

ground fog RADIATION FOG.

ground frost a FROST occurring on the surface of the ground or in the upper soil layer when the ground surface temperature drops to -1°C (30.2°F), a condition likely to have a marked effect on plant tissues. AIR FROST.

ground ice **1.** ice formed on the bed of a river, lake or shallow sea when the water as a whole remains unfrozen (also termed anchor ice or bottom ice) **2.** fossil ice, a mass of clear ice associated with PERMAFROST.

ground information in REMOTE SENSING, information concerning the state of the physical environment derived from aerial photographs and measurements (based on sample estimates) of soil moisture, temperature, biomass etc. Ideally this information is recorded at the time when a remote sensor, also collecting data, is passing over the area concerned.

ground-mass, groundmass the compact fine-grained mineral material of an IGNEOUS ROCK, in which the crystals are embedded. MATRIX-3.

ground moraine debris carried by the underside of a glacier or ice sheet and deposited as TILL. BOULDER CLAY, LODGEMENT TILL, MORAINE.

groundnut, monkey nut, peanut a leguminous plant (LEGUMINOSAE), native to South America, now an important food plant on light soils. The stalk bearing the flowers lengthens after germination and forces the seed pods into the soil, where they develop (underground). Harvesting is carried out by hand or machine. The nuts (seeds) are high in protein, oil and vitamins B and E. They are cooked, eaten raw or roasted, roasted and ground to form peanut butter. The oil extracted from them is valuable as a cooking oil, and in the making of margarine, soap and toilet preparations. After oil extraction the residue is made into oilcake for animal feed.

ground rent the rent paid by a leaseholder to the freeholder for the privilege of using or leasing the ground on which a building stands. LAND TENURE.

ground swell the deep, slow rolling of the sea (sometimes caused by a distant storm or a SEISMIC disturbance) which in passing into shallow water raises the height of the waves.

ground water, groundwater PHREATIC WATER, all the water derived from PERCOLATION of rainwater, from water trapped in a sediment at its time of deposition (CONNATE WATER) and from magmatic sources (JUVENILE WATER), lying under the surface of the ground above an IMPERMEABLE layer, but excluding underground streams. VADOSE WATER, WATER TABLE. Fig 41.

Group Seven G7.

growing season that part of the year when plant growth is active on account of favourable temperature, availability of moisture, and sufficient hours of daylight. The duration of the season usually decreases with increasing distance from the equator. The crucial temperature for many plants in midlatitudes is 5°C (42°F), below which vegetative growth does not take place and the plant is dormant; but the number of continuously frost-free days while the plant is growing to maturity is sometimes an overriding factor. COTTON, MAIZE.

growth area, growth centre, growth point in planning, a district in which economic growth starts or is deliberately concentrated as part of regional development strategy (REGIONAL PLANNING) and continues if stimulated, thereby benefiting the area and its surrounding region.

groyne, groin a construction of timber, concrete, or stone jutting seawards, usually at right angles to the coast, to combat LONGSHORE DRIFTING of sand and shingle and maintain a level BEACH.

guano the thick deposit formed by the excreta of seabirds valued, as a source of nitrogen and phosphates, for FERTILIZER, e.g. occurring on the many islands of the dry Chilean and Peruvian coasts where there is little or no rainfall to wash it away.

gulch in western USA, a narrow, deep, rocky ravine.

gulf **1.** a large inlet of the sea, cutting into the land more deeply than a bay. It is more enclosed by the coast than a bay is; and it may itself contain one or more bays **2.** a deep hollow, a chasm, an abyss, e.g. a steep-walled SINK HOLE with a flat alluvial floor in which an underground stream may sink or rise.

gully, gulley a narrow channel worn in the earth by the action of water, especially a miniature valley resulting from a heavy downpour of rain. GULLY EROSION.

gully erosion, gullying SOIL EROSION resulting from a sudden heavy downfall of rain which cuts channels (GULLY), especially in soil or soft rock.

gumland in New Zealand, land from which fossil kauri gum has been obtained, but commonly applied to land which may or may not have yielded gum, now covered with scrub.

gust a sudden, strong, brief rush of wind.

Gutenberg Discontinuity the discontinuity occurring between the lower surface area of the MANTLE and the CORE of the earth, where the material of the earth's interior is in transition between the solid and plastic state, and where the speed of the P-waves of EARTHQUAKES slackens and S-waves are absent. In the Gutenberg Channel, lying between 100 and 200 km (60 to 120 mi) beneath the LITHOSPHERE, at the upper surface of the mantle, the material becomes more plastic and both P-waves and S-waves exist but are slowing down. ASTHENOSPHERE, PUSH WAVE, SHAKE WAVE.

guyot tablemount, a topographical feature of the ocean bed, a flat-topped volcanic mountain, occurring especially on the floor of the Pacific ocean (as opposed to a SEAMOUNT with its pointed summit). It may rise to a great height, e.g. 3200 m (some 10 000 ft), but the summit usually lies well below the ocean surface, e.g. 800 m (2700 ft) below.

gymnosperm a plant producing seeds not enclosed in an ovary (e.g. conifers), and thus unlike an ANGIOSPERM. By the end of the CRETACEOUS period angiosperms had succeeded gymnosperms as dominant land plants.

gyrocompass a COMPASS that makes use of the properties of a continuously driven gyroscope (a fast-spinning wheel mounted in such a way that its plane of rotation can vary, but which, unless subject to a very great disturbing force, keeps the plane constant in space). In the gyrocompass the spinning axis of the continuously driven gyroscope can move only in the horizontal plane and, due to the rotation of the earth, this axis aligns itself with the axis of the earth, thus pointing TRUE NORTH (unlike a magnetic compass, which points to MAGNETIC NORTH).

H

habitat a place or a kind of place that provides a particular set of environmental conditions for the organism or organisms inhabiting it. Some geographers use the term as a synonym for ENVIRONMENT.

Habitat I, Habitat II United Nations conferences on human settlement, attended by world leaders, planners, architects, population experts etc., to discuss the problems of rapidly expanding and overcrowded cities.

habitus the individual's sense of 'home', of place in the world, based on geographical background, cultural origin, inheritance, experience and social networks carried through life, subject to refashioning viewed through the rosy spectacles of passing time and increasing distance.

hachure (French) short lines of shading on a map drawn at right angles to the CONTOURS to represent slopes, the lines being thicker and more closely spaced at areas of greatest slope, but not giving specific information about altitude.

hacienda (Spanish) in Spain and former Spanish colonies, a large agricultural estate with a dwelling house. The term is sometimes applied to a country house and its associated activities, e.g. in monoculture, cash crop farming, sheep or cattle ranching (and some would add mining or manufacturing conducted on the estate); sometimes it is restricted to the country house itself. FARMING, LATIFUNDIA.

hade the angle made by a mineral vein, lode or fault plane with the vertical, i.e. the slope of the FAULT as contrasted with the DIP of the beds. Fig 23.

Hadley cell a thermally driven vertical circulation cell (ATMOSPHERIC CELL-2) forming that part of the general circulation of the ATMOSPHERE-1 between the equator and approximately 30°N and 30°S. Warmed air rises from the equatorial area to high altitudes, spreads polewards and, cooled, descends at about 30°N and 30°S, flowing towards the equator in the lower levels of the atmosphere, as TRADE WINDS at the surface. The cell takes its name from G. Hadley who, in his explanation of trade winds in 1735, issued his modification of the original theory on the circulation of the atmosphere put forward by Edmund Halley in 1686. Fig 5.

haematite, hematite an important iron ore, grey, black or reddish in colour, abundant and widely distributed, valued for its lack of phosphorus, occurring in CRYSTALLINE, MASSIVE or granular (in grains) form, sometimes in kidney-shaped nodules (kidney ore), as a CEMENT-2 in SANDSTONE, as an ACCESSORY MINERAL in IGNEOUS ROCK, and in HYDROTHERMAL veins and replacement deposits. LIMONITE.

haff (German, pl. haffe) a coastal lagoon of fresh or brackish water, fed by a stream which is blocked by a NEHRUNG, through which it is linked to the sea by a channel, typical of the Baltic coast of Germany, Poland and Russia.

hail, hailstone a small ball or piece of ice with a concentric layered structure, the diameter usually ranging between 5 and 50 mm (0.2 to 2 in), falling separately or agglomerated from CUMULONIMBUS cloud at the passing of a COLD FRONT, sometimes damaging crops, trees, greenhouses and injuring livestock and people. Hailstones are caused by the fast ascent of moist air in which frozen droplets of ice are carried ever higher by the force of the updraught, their size growing as additional WATER VAPOUR freezes on them, until their weight overcomes the force of the ascending air current and they fall to earth. On the way they sometimes gather more ice from supercooled (SUPERCOOLING) water droplets in the moist air. PRECIPITATION-1.

hailstorm a violent fall of HAIL.

haldenhang wash slope, the part of a slope that is less steep than the part above it, occurring at the foot of a rock wall, usually beneath an accumulation of TALUS.

half-life in physics, the time needed for one half of the ATOMS of a sample of a radioactive element (RADIOACTIVITY) to disintegrate or change into the ISOTOPE of another element. The half-life of a radioactive element is known and constant, e.g. the half-life of carbon 14 is 5600 years (but see RADIOCARBON DATING); of radium 226 is 1620 years; of uranium 238, U^{238}, is 4.51×10^9 years, decaying to lead 206; of U^{235} is 7.13×10^{18}, decaying to lead 207; potassium 40 has a half-life of 47 000 mn years, decaying to strontium 87.

halo a ring or rings of diffused light seen from the earth to encircle the sun or moon in conditions of thin cloud, caused by the refraction of light by water-drops or ice crystals in cloud. It usually appears to be white, but if it is sharply defined, the innermost ring appears red, the outermost blue.

halo effect of a BOUNDARY, the detrimental effect of a boundary on locations close to it, making them unattractive to people, e.g. of a political boundary in disputed territory, resulting in the emigration of the population; but there may be a beneficial effect (INTERVENING OPPORTUNITY EFFECT).

halophyte a plant that tolerates soil impregnated with salt, e.g. the soil of a salt marsh, the seashore, a salt desert.

ham in southern England, a local term **1.** a plot of meadow land, especially a tract of rich pasture by a river **2.** (Old English ham, home) a settlement ranging in size from a single homestead to a town, common in place-names in the parts of England influenced by Anglo-Saxons, surviving in HAMLET.

hamada, hamāda, hammada (north Africa: Arabic) the rocky desert of the plateaus of the Sahara, stripped of sand and dust by air currents, the surface smoothed by ABRASION, ERG, REG.

hamlet a small group of dwellings in the English countryside, usually smaller than a village and lacking a church. HAM-2.

hanger a WOOD-2, most commonly of beech trees (hence beech-hanger), on the steep chalk slopes of southern England.

hanging glacier a truncated GLACIER projecting from a basin or shelf high on a mountainside, from which ice may break off and fall as an ice AVALANCHE.

hanging valley a tributary valley, the lower end of which is well above the bed of the main valley (and from which a WATERFALL may descend), commonly found where the main valley has been deepened by a GLACIER which has since disappeared. U-SHAPED VALLEY.

harbour **1.** an anchorage haven, stretch of water, close to and sheltered by the shore and protected from the sea and swell by artificial or natural walls which allow access for vessels through a narrow entrance (the harbour mouth). In the harbour vessels can lie at anchor, secured to buoys or alongside wharfs, piers etc. The facilities of a PORT may or may not be provided **2.** a term used in place-names in England, especially applied to farms in sheltered situations.

hardebank KIMBERLITE.

hardeveld (Afrikaans) VELD where a thin layer of soil is underlain by hard gravel or rock, making ploughing impossible.

hardness **1.** the quality of being hard, in solid objects generally, of being resistant to cutting, cracking or crushing **2.** in metals, resistance to indentation or deformation **3.** in minerals, resistance to ABRASION. The hardness of minerals is indicated on a scale devised by Friedrich Mohs (Mohs' Scale), a German mineralogist, 1773–1839, in which the softest (talc) was represented by 1 and the hardest (diamond) by 10 (subsequently revised, the diamond now being 15) **4.** of water, HARD WATER.

hardpan an indurated (INDURATION) or cemented layer of soil of varying character found at varying distances below, or sometimes at, the upper surface of the soil. It is usually formed from material carried down mechanically or in solution by rainwater percolating from the surface, and later deposited. The term is applied to the layer or the material, e.g. clay pan, a dense subsoil formed by the washing down of CLAY or syntheses of clay; duripan or silcrete, cemented SILICEOUS minerals; fragipan, an acid, cemented, platey layer between the parent material and the upper soil layers; iron pan, a layer of sand or fine gravel cemented with IRON oxides; laterite or plinthite, a layer cemented with FERRIC oxide; lime pan, a thick layer of CALCIUM CARBONATE; moorpan, compact redeposited HUMUS compounds. LATERITE.

hardware the physical equipment used for data collection and handling, including REMOTE SENSING equipment, computers etc. and their parts. SOFTWARE.

hard water water which does not easily form a lather with soap or detergent on account of the inhibiting effect of dissolved calcium, magnesium and iron compounds present in it, derived from the rocks over or through which the water has passed (e.g. water flowing over LIMESTONE). The greater the quantity of these salts in it, the harder the water.

hardwood **1.** any DECIDUOUS or evergreen BROADLEAVED tree, with vessels in its wood, that produces close-grained wood **2.** the TIMBER of such a tree. Most hardwoods, e.g. oak, beech, maple, walnut, grow in temperate regions. Other examples are the evergreen oak, native to the region of Mediterranean climate; the eucalyptus, native to eastern Australia; the hard, heavy teak of the regions of monsoon climate; and ebony and mahogany, growing in tropical forests. All are used mainly in furniture making. SOFTWOOD.

harmattan, hamattan a strong, dry wind blowing from a northeasterly or sometimes easterly direction over northwest Africa, from the Sahara to the northwest African coast, the southern limit averaging 5°N in January (mid-winter) and 18°N in July. Heavily dust-laden and parching in the interior, it helps to evaporate the high humidity of the Guinea Coast, and thus seems a relatively cool and healthy wind in that area, hence its local name there, the Doctor.

Hawaiian volcanic eruption a non-explosive volcanic eruption in which basic (BASE-2) and highly fluid LAVA flows over a large area and hardens to form a volcano shaped like a shield (SHIELD VOLCANO). PELEAN, STROMBOLIAN, VULCANIAN ERUPTION.

hay the stems and leaves of grasses, cut and dried, and used as FODDER.

hazard a risk, a chance associated with danger or with a damaging effect. ENVIRONMENTAL HAZARD, FLOOD HAZARD.

haze the condition in the atmosphere near the earth's surface when visibility is restricted to more than 1 km (0.6 mi) but less than 2 km (1.2 mi), commonly resulting from the suspension in the air of solid matter such as dust or smoke particles, or from shimmering caused by intense heat, which produces irregularities and changes in density in the layers of the atmosphere. FOG, MIST.

HDC highly developed country, a country which is well advanced in realizing its full potential. LDC, MDC, UNDERDEVELOPMENT.

head dune a sand DUNE formed on the windward side of an obstacle, where the air is stagnant.

headland a comparatively high promontory with a steep face, projecting into the sea or a lake.

headwall the steep wall that forms the back of a CIRQUE.

headward erosion the action of a stream in extending its valley upstream.

heartland, heart-land 1. the central part of a land mass inaccessible to a seapower. The term was introduced by Sir Halford Mackinder, 1904, with reference to the heart of the Euro-Asian continental mass, and subsequently developed by him (WORLD-ISLAND). He postulated his theory of world power, that the fertile, self-sufficient heartland of the world-island, being land based, held the key to potential world domination, rather than a vulnerable maritime-based state **2.** the heartland of the USA, usually defined as the area within the rectangle Boston–Washington–St Louis–Chicago.

heat balance the condition of equilibrium in which radiation reaching the earth and its atmosphere from the sun (INSOLATION) is approximately equalled by radiation and reflection (ALBEDO) from the earth. HEAT ENGINE.

heat engine a mechanical system set in motion by heat ENERGY, e.g. as in the ATMOSPHERE-1, where differences in net RADIATION balance (HEAT BALANCE) provide the thermal power, and the various movements in the atmosphere distribute the energy (the heat energy being converted to KINETIC ENERGY).

heath, heathland, heather moor an uncultivated, open tract of land, with poor acid soils, supporting shrubby plants of which the dominant are the common heather or ling, *Caluna vulgaris*, and species of heath, *Erica*. The term heath or heathland may persist in place-names where such vegetation has almost or completely disappeared.

heat island, heat-island the persistent warmth of the densely built-over part of a large town, the overlying air having an average temperature higher than that of the air overlying the more open surrounding area. This is due in part to the storage of solar heat in roads and the mass of building etc. and its slow release, in comparison with the heat stored and released more quickly from the surrounding, more open area; in part to the effect of buildings in reducing wind speed; in part to the heat derived from the internal heating of buildings, from industry, transport, power generation, human metabolism; in part to the blanket effect of polluted air overlying large towns. The effects of a heat island spread to areas lying at some distance from it.

heat wave, heave-wave a relatively long unbroken spell of abnormally hot weather.

heave of a fault the forward, lateral displacement of STRATA on an inclined NORMAL FAULT, expressed as the horizontal distance between the ends of the surfaces of the displaced strata. THROW OF A FAULT. Fig 23.

Heaviside Layer the E layer of the ATMOSPHERE-1, lying in the IONOSPHERE at about 100 to 120 km (60 to 72 mi), and reflecting medium and long radiowaves (ELECTROMAGNETIC SPECTRUM) back to the earth while allowing short radiowaves from the earth to penetrate until they reach the APPLETON LAYER. Oliver Heaviside, 1850–1915, suggested the existence of such a layer; but it is also known as the Kennelly-Heaviside Layer, Arthur Edwin Kennelly, 1861–1939, having made the same prediction.

heavy industry a SECONDARY INDUSTRY in which the weight of materials used per worker is high, the machinery used is bulky and the finished products have a low value in relation to their weight, e.g. ship building. BASIC INDUSTRY, LIGHT INDUSTRY.

heavy land, heavy soil land that is difficult to plough, especially in wet conditions, e.g. a thick CLAY soil.

hectare ha, a metric unit of area, equivalent to 10 000 sq m or 2.471 ACRES. 100 hectares equal 1 square KILOMETRE.

hecto- h, prefix, a hundred, attached to SI units to denote the unit × 10^2, e.g. hectolitre (100 litres). CENTI-, KILO-, MILLI-.

hegemony (originally used with reference to the states of ancient Greece) predominance, leadership **1.** the predominance of, the leadership exercised by, one state in relation to others in a group, e.g. in a union or confederacy of states **2.** in Marxism, the predominance of one SOCIAL CLASS over others.

hemisphere the half of a sphere formed on each side of a plane passing through its centre. The earth's surface is commonly considered to be bisected by the EQUATOR, producing the northern hemisphere and the southern hemisphere. The eastern hemisphere (Asia, Africa, Europe, Australia and New Zealand) is separated from the western hemisphere (the New World, i.e. North and South America) by the meridians 20°W and 160°E. GREAT CIRCLE.

herb 1. in botany, a non-woody vascular plant lacking parts that persist above the ground (ANNUAL, BIENNIAL, PERENNIAL) in the unfavourable season, thereby enhancing its chances of survival, as distinct from a SHRUB or a TREE **2.** a plant that is not necessarily a herb in the botanical sense, but is valued for its fragrant, medicinal or flavouring properties.

herbaceous *adj.* of, pertaining to, or resembling a HERB-1.

herbivore a plant-eating animal, a primary consumer in the FOOD CHAIN. CARNIVORE, OMNIVORE.

Hercynian *adj.* applied to the earth building movements and associated mountain remnants of Upper Carboniferous to Permian times (GEOLOGICAL TIMESCALE) in Europe, variously and confusingly named. Hercynian is applied by some authors to the whole mountain system of central Europe (also termed the ALTAIDES); others use it as a synonym for VARISCAN. (Others restrict the use of Variscan to the eastern sector of the Hercynian earth movements, applying ARMORICAN to the western, i.e. to earth movements in Brittany and southwestern Britain.) There are remnants of mountains of similar age in Asia (the Urals, T'ien Shan, Nan Shan), in North America (the Appalachians) and in South America (the foothills of the Andes).

hermeneutics the art, science or skill of interpretation, of the classification of meaning, of understanding the significance of human actions, statements, institutions, products.

heteroclinal fold a FOLD-2 of which one side slopes at an angle steeper than that of the other.

heterogeneous *adj.* **1.** diverse or dissimilar in kind or character **2.** composed of different or disparate ingredients or elements **3.** a rock composed of diverse materials **4.** rocks varying in nature or kind and adjacent to each other. HOMOGENEOUS.

heterotrophic *adj.* applied to an organism which needs an organic food supply from outside itself in order to produce its own constituents and (with a very few minor, PHOTOTROPHIC, exceptions) to obtain energy. All animals and fungi, most bacteria, and some flowering plants are heterotrophic.

heuristic *adj.* **1.** useful for discovery or for solving problems, e.g. applied to a method of education which encourages the student to gain knowledge by personal investigation; or to a procedure (heuristic method) for discovering an unknown goal by a progressive sequence of operations or investigations based on a known criterion, e.g. a person put against a wall in a totally dark room and asked to find the door will discover it by moving systematically in the same direction round the walls **2.** of practical, though perhaps unexplained, use in invention or discovery, e.g. in social science, applied to conceptual devices such as MODELS-2 and working hypotheses the aim of which is not to describe or explain the facts but to suggest or to eliminate possible explanations.

H horizon O HORIZON, SOIL HORIZON.

hide **1.** the raw or dressed skin of dead cattle, used especially in shoe making. The term usually appears linked to SKIN in trade statistics **2.** (Anglo-Saxon) an inexact measure of land in the DOMESDAY BOOK, reckoned at four YARDLANDS, but the term sometimes refers to an area of land capable of supporting one family.

hierarchy the organization of a SYSTEM-6 into successive ranks, in ascending or descending order.

high an ANTICYCLONE, an area of high ATMOSPHERIC PRESSURE.

high farming a term applied to the period in England in the 1850s and 1860s when the large landowners adopted new farming techniques and spent considerable amounts of money on underdraining their land, improving its fertility, raising the quality of their livestock and crops, and improving farm buildings.

highland, highlands in general, any tract of high or elevated land or the more mountainous parts of any country. Used specifically, usually in plural, in proper names, e.g. the Highlands of Scotland. LOWLAND.

high latitudes LATITUDE.

highly developed country HDC.

high plains PLAINS lying at an elevation above some 600 m (2000 ft).

high pressure HIGH.

high technology the application of the knowledge of and/or the use of advanced techniques, complex equipment, and materials drawn from any convenient source to a task or an industrial process, or to the solution of a problem arising from the interaction of people with their ENVIRONMENT. INTERMEDIATE TECHNOLOGY, LOW TECHNOLOGY, TECHNOLOGY-3.

high tide **1.** the TIDE at highest flood **2.** the level of the sea at, or the time of, the highest flood. FLOOD TIDE, LOW TIDE, NEAP TIDE, SPRING TIDE.

high water the state of the TIDE when the water is at its highest for any given tide. The level varies through the year so that the high water ordinary spring tide (HWOST) is higher than high water ordinary neap tide (HWONT), both of which may appear on maps, HIGH-WATER MARK, LOW WATER, NEAP TIDE, SPRING TIDE.

high-water mark the mark left by the TIDE at HIGH WATER, or at the highest level ever reached, similarly applied to the water level in a river or lake. On British Ordnance Survey maps the high-water mark indicates the high-water mark of medium tides (HWMMT).

hill loosely, a natural elevation of the earth's surface, not so high as a MOUNTAIN. It is sometimes defined as an elevation under 300 m (1000 ft), but such exactness is misleading: some ranges termed hills exceed 2680 m (8800 ft).

hill farming generally, farming in hill country, in Britain traditionally PASTORAL, and devoted to SHEEP rearing. There are EEC subsidies designed to encourage the continued farming of poor hill land. LESS FAVOURED LAND.

hill fog low CLOUD covering the higher levels of HILLS as seen by the observer standing below.

hill village a village, a hamlet, or even a small market town or other non-agricultural settlement occupying a site from which the ground slopes markedly downwards.

hinge of fold that part of a FOLD where the STRATA is under maximum STRESS.

hinterland **1.** originally the district behind a settlement on a coast, especially the area serving and served by a PORT-1 **2.** the area influenced by any settlement **3.** the sphere of influence of an establishment within a settlement.

histogram a two-dimensional graph showing a FREQUENCY DISTRIBUTION by means of rectangles, the widths of the rectangles being proportional to the CLASS INTERVALS or CATEGORIES (shown on the horizontal axis), the heights (on the vertical axis) showing the frequencies of occurrence in each CLASS. If the frequency of each class is expressed as a percentage of all classes (i.e. not by the absolute frequency of each class), the resultant diagram is termed a relative frequency histogram. BAR GRAPH, FREQUENCY-3, FREQUENCY CURVE.

historical fallacy the false assumption that a relationship observed in a CROSS-SECTIONAL STUDY will be present in a similar LONGITUDINAL STUDY, or vice versa. ECOLOGICAL FALLACY.

historical geography the geography, physical and human, real, perceived or theoretical, of the past.

historical materialism a materialist conception of history, the Marxist theory of history, seen as a natural process related to human material needs. Marx believed that this historical human evolution was the product of the class struggle arising between and amongst the exploiting and exploited classes throughout the succession of the MODES OF PRODUCTION, each mode of production determining the general character of the social, political, intellectual and spiritual processes in a society (social existence being determined by human consciousness, not vice versa). Marx and Engels came to modify their views that such economic determinism governed all aspects of historical development; and Marxists have since variously interpreted the theory and its concepts. DIALECTICAL MATERIALISM, MARXISM, MATERIALISM.

histosols in SOIL CLASSIFICATION, USA, an order of soils developed mainly by the accumulation of organic matter in a waterlogged area, and thus including BOG soils and soils rich in PEAT.

hoar frost, hoar-frost, hoarfrost a white deposit of ice with a crystalline appearance, formed directly from the cooling of water vapour on surfaces with a temperature below that of DEW-POINT when the dew-point is below 0°C (32°F). FROST, RIME.

hobby farmer a PART-TIME FARMER who owns the freehold of, or rents, a FARM, who may or may not use any building on it as a main residence, who does not rely on the farm output for a livelihood, whose main income comes from another source, e.g. from an urban occupation, and who farms mainly for pleasure.

hoe a projecting ridge of land, a height ending abruptly or steeply. The term is now obsolete except in place-names, e.g. Plymouth Hoe, Ivinghoe.

hogback, hog back, hogsback, hog's back an elongated narrow ridge shaped like the back of a hog (pig), the result of unequal erosion on alternating hard and soft layers of steeply inclined rocks. It differs from a CUESTA because both slopes of the ridge (i.e. the dip slope and the scarp slope) are steep and more nearly equal.

holding in agriculture, land held or occupied by legal right for purposes of farming. The term is sometimes used as a synonym for FARM, but several farms may be combined and worked as one holding. SMALL HOLDING.

holism *adj.* holistic, a philosophy so named by J. C. Smuts that there is a tendency in nature to produce 'wholes' (whole bodies or whole organisms) from an ordered grouping of unit structures, the whole being greater than the sum of the properties (PROPERTY-3) and relationships of its parts (GESTALT). This phenomenon can be recognized in other 'wholes', leading to the doctrine that a functioning whole (e.g. an organization, institution, society) affects its component parts, is inimical to analysis, and that therefore the parts should not be studied in isolation. It has also led to a theory of science which sees science not as a collection of disparate parts but as an integrated system.

Holocene the period in which we are living, sometimes termed Recent or Post-glacial, i.e. the most recent of the geological periods, the youngest of the Quaternary (GEOLOGICAL TIMESCALE), and the rocks of that period, dating from the end of the last ICE AGE.

holocoen in ecology, the whole environment, i.e. including all the living (BIOCOEN) and the non-living (ABIOCEN) components.

holotype in biology, the type specimen, i.e. the original specimen(s) forming the basis of identification of a new SPECIES. BINOMIAL NOMENCLATURE, CLASSIFICATION OF ORGANISMS.

homeostasis, homoiostasis the maintenance of a constant, balanced internal environment within a system (OPEN SYSTEM), person, group etc. STEADY STATE.

homestead **1.** a house or home, especially a farm with its associated buildings **2.** in USA, a plot of land adequate for the residence and maintenance of a family, given special and legal meaning, notably under the 1862 Homestead Act, a grant of 65 ha (160 acres) being regarded as sufficient to support a family, rising to some 255 ha (640 acres) in mountainous or semi-arid areas **3.** a rural settlement of dispersed farms.

homocline MONOCLINE.

homogeneous *adj.* **1.** similar in kind, character or nature **2.** having at all points the same composition and properties, the opposite of HETEROGENEOUS.

homoseismal line CO-SEISMAL LINE.

honeycomb weathering a type of WEATHERING which causes a rock surface to resemble that of a honeycomb. The current theory is that hard material filling some of the joints in the rock is left standing as softer material is worn down more speedily by wind and water. The same pattern is sometimes seen on a shore, where wind and water act together on small pools in the rocks, enlarging the hollows in which they lie to such an extent that they are eventually separated only by steep, sharp ridges.

honeypot a place with some special interest (e.g. scenic beauty, historical association) that is over-popular with visitors and becomes overcrowded at peak holiday times. In some sensitive areas (e.g. national parks in Britain) honeypot locations have been identified and enhanced by the provision of picnic sites, car parks, toilet and refreshment facilities, information desks etc. to concentrate visitors and traffic and divert them from fragile, unspoilt areas elsewhere.

hop a PERENNIAL HERBACEOUS climber native to Europe and western Asia, now cultivated there and in other midlatitude areas for the sake of its essential oils and soft resins used in flavouring beer in Europe since the Middle Ages.

horizon **1.** visible horizon (also termed apparent, geographical, natural, physical, or sensible, horizon), the line at which the earth or sea and the sky appear to meet when seen from any given viewpoint, excluding anything interrupting or obstructing the view. In clear visibility, to a person standing at an elevation of 3 m (10 ft) from the horizontal the horizon is nearly 6.5 km (just over 4 mi) distant; at 30 m (100 ft) it is 21 km (13 mi) away **2.** in geology, a plane in a series of geological

STRATA, or the level at which a particular FOSSIL occurs **3.** in soil science, a distinct soil layer with more or less well-defined characteristics produced by soil-forming processes. SOIL, SOIL HORIZON.

horizontal equivalent HE, the distance between two points on the land surface when projected on to a horizontal plane, as on a map. If two points on a hillside, say 75 m (245 ft) apart measured down the actual SLOPE LENGTH, are projected on to a map they will be a shorter distance apart in the horizontal plane: this is the HE or Horizontal Equivalent, say 60 m (195 ft) in the example. A rise of 1° in an HE of 30 m (100 ft) represents a VERTICAL INTERVAL of about 0.5 m (1.74 ft). GRADIENT.

horn a pyramidal peak in a mountain range occurring when several CIRQUES are formed back to back, thereby leaving a high central mass, an unreduced part of the original mountain range, with marked faces and sharp ridges (ARETE). The term is common in mountain names, e.g. the Matterhorn.

horse latitudes the belts of calms (CALM), the subtropical belts of high atmospheric pressure (ANTICYCLONE), moving north and south with the sun, lying north and south of the equator from about 30° to 35°N and 30° to 35°S (but interrupted by the land and sea pattern), regions of the descending air which flows towards the equator and the poles to produce calm, stable, dry weather conditions with light variable winds. The origin of the name is uncertain. Fig 5.

horseyculture the practice of pasturing horses and ponies for riding on relatively small plots of land, usually on town outskirts.

horst an elevated block of the earth's crust, usually with a level summit, standing prominently above parallel NORMAL FAULTS, formed either by the sinking of the crust on each side outside the faults, or by the uplift of a block between the faults. In some cases it is denuded to the extent that it is no longer upstanding. RIFT VALLEY.

horticulture originally the cultivation of a garden, now applied to the intensive cultivation of vegetables, fruit and flower crops on relatively small plots, in a MARKET GARDEN, GREENHOUSE or nursery.

Horton's Law the theory relating to streams and their order (STREAM ORDER) and the area of the basin they drain, i.e. that the number of streams of a given order decreases sharply with increasing order, and that the total length of a drainage net increases regularly with order. Generally, in most drainage basins the number of streams of a given order is about three times the number in the next higher order.

Hotelling model a MODEL-2 used by Harold Hotelling, economist, to account for the locations of two firms in competition with each other, based on the assumptions that these firms are producing identical goods, that their production costs are the same at all locations, that the price of the goods covers the cost of transport to the consumer, that the market to be satisfied is LINEAR-2, with customers evenly distributed along its length, and that the demand for the goods is inelastic.

hot rock a deep-seated rock in the earth's crust which has a temperature higher than might be expected from the normal GEOTHERMAL GRADIENT. GEOTHERMAL ENERGY.

hot spot PLUME.

hot spring THERMAL SPRING.

Hoyt's sector model SECTOR MODEL.

huerta (Spanish, derived from Latin hortus, a garden) a highly cultivated, irrigated area along the eastern coastlands of Spain which may yield several crops a year of vegetables, fruit etc. INTENSIVE AGRICULTURE.

human agency the capacity of human beings to act, in the light of their experience and creativity, e.g. to improve a soil; or to reproduce and reinforce SOCIAL STRUCTURES by repeatedly conforming to the rules, constraints and conventions of their social system; or to change their circumstances, or an even wider area of the social structure, e.g. by coming together as a group on the basis of some shared, experience (such as place of residence, social class, gender) and as members of that group pursuing an agreed goal.

human ecology **1.** the study of the interrelationships between human beings and the ENVIRONMENT (physical and social) in which they live. Some authors maintain that GEOGRAPHY is synonymous with human ecology **2.** in sociology, the study of the interrelationship between human beings.

human geography one of the two parts into which GEOGRAPHY is often separated, the other being PHYSICAL GEOGRAPHY. Human geography is concerned with the study of those features and phenomena in the space accessible to human beings which relate directly to or are due to people (as individuals or in groups), their activities and organization, past or present. It concentrates on the interrelationship of people (as individuals or in groups) with space, with their physical ('natural') environment and with their social (societal) environment, covering spatial and temporal distribution, the organization of society and social processes etc. on a local to global scale.

humanism any theory, or doctrine, or movement, which is concerned with the primacy of human beings and their interests (as distinct from the primacy of a divine being, of nature, of STRUCTURES-1, or of SYSTEMS); or with the human race in general (as distinct from an individual member of it); or with the studies of human culture and creativity, particularly those exemplified by the cultures of classical Greece and Rome, the prime concern of the humanists of the RENAISSANCE. HUMANISTIC.

humanistic *adj.* relating to, characteristic of, HUMANISM or of a HUMANIST, e.g. one of the humanists of the RENAISSANCE who were primarily concerned with the products of human effort as revealed in the history, language, literature and the arts of classical Greece and Rome.

humanistic geography humanist geography, an approach in HUMAN GEOGRAPHY which emphasizes the SUBJECTIVE as distinct from the OBJECTIVE in that it stresses the importance of perception, creativity (e.g. in literature and landscape painting), thinking and beliefs as well as human experience and values in formulating people's attitudes to their environment and in affecting their relationships with it. HUMAN AGENCY, HUMANISM, HUMANISTIC.

Humboldt current a cold current usually rich in PLANKTON and fish, flowing northwards off the Peruvian coast. It is intermittently temporarily replaced by the warm EL NINO.

humic acid a complex organic acid found

in soils and resulting from the partial decay of organic matter. HUMUS.

humic gley soil GLEY SOILS.

humidity the state of the ATMOSPHERE-1 with respect to its content of WATER VAPOUR, warm air being able to hold more water vapour than cold air. When the air is holding its maximum amount of water vapour it is said to be SATURATED. ABSOLUTE HUMIDITY, RELATIVE HUMIDITY, SPECIFIC HUMIDITY.

humid tropicality, humid tropics the climatic condition relative to a standard period of time (e.g. a month) in which the RELATIVE HUMIDITY exceeds 65 per cent, pressure 20 mb, mean temperature 20°C (68°F). To this is sometimes added rainfall exceeding or equalling evaporation for the period, approximating to 75 mm (3 in) per month. The term humid tropics is applied to tropical areas in which the condition of humid tropicality prevails for a minimum of nine months of the year. TROPICAL CLIMATE, TROPICS.

humification in soil science, the transformation of organic material into HUMUS by slow DECOMPOSITION-1 and OXIDATION.

humus loosely, organic matter (of vegetable or animal origin) in the soil, but the term is better restricted to decomposed, amorphous organic matter. HUMIC ACID, LITTER, MODER, MOR, MULL.

hundred in England, an old subdivision of a county or shire with its own court, still occasionally used. The origin of this subdivision is obscure: it may originally have been the area inhabited by 100 families, or equivalent to 100 HIDES, but these suppositions are doubtful. YARDLAND.

hundredweight cwt, a measure of weight equal in British measures to 112 lb (50.802 kg), in US to 100 lb (45.359 kg).

hurricane 1. in the BEAUFORT SCALE, wind of force 12, with velocity equivalent at a mean velocity exceeding 34 m/sec or 121 km (75 mi) per hour **2.** a violent cyclonic storm with torrential rain and thunderstorms and wind velocity over 117 km (73 mi) per hour (often exceeding 160 km: 100 mi per hour), originating in latitudes 5° to 20°N over the west Atlantic, moving west-northwest over the Caribbean Sea and the Gulf of Mexico to Florida, then northeast at about 30°N along the eastern coast of the USA. CYCLONE, TROPICAL REVOLVING STORM, TYPHOON.

HWMMT, HWONT, HWOST HIGH WATER.

hydration in the MECHANICAL WEATHERING of rocks, the process in which minerals combine with water and expand, thereby exerting pressure within the rock pores (EXFOLIATION). Minerals that have undergone hydration are very likely to be affected by CHEMICAL WEATHERING.

hydraulic *adj.* of or pertaining to water in motion or to the pressure exerted by water when carried through pipes, or to mechanical devices operated by moving fluids. WATER POWER.

hydraulic cement a CEMENT-1 capable of hardening under water. POZZOLANA.

hydraulic civilization a rural-urban agrarian civilization depending on big, productive waterworks for irrigation, flood control and power, as distinct from a rural-urban civilization depending on rainfall (termed non-hydraulic). FERTILE CRESCENT, HYDRAULIC HYPOTHESIS.

hydraulic force the eroding power of water on rocks, by turbulence, eddying, wave action. If the water carries a load of material the process of EROSION is termed ABRASION.

hydraulic hypothesis the proposition that the problems attached to the development of agriculture in broad, seasonally flooded river valleys could have been tackled in prehistoric times only by the integrated, collective efforts of many small groups, the members of which would have built and maintained large scale irrigation works. This large labour force would have needed a supervising authority which could ensure a fair distribution of water to the groups over space and time; and that this led to the emergence of URBAN centres. HYDRAULIC CIVILIZATION, URBAN HEARTH.

hydric *adj.* having abundant water. MESIC, XERIC.

hydrocarbon an organic compound, solid, liquid or gaseous, consisting primarily of CARBON and HYDROGEN, e.g. petroleum, coal, natural gas. ACID RAIN.

hydroelectric power HEP, electric POWER generated in dynamos moved by the energy of falling water. ELECTRICITY, WATER POWER.

hydrogen the lightest element, gaseous, inflammable, occurring in very many compounds, but not normally in the uncombined state on earth, used in many chemical processes, as a component in rocket fuel, in producing (by combustion with OXYGEN) high-temperature flames and (in liquid form) as a cooling agent. HYDROCARBON.

hydrogen ion the positive ION of HYDROGEN, affecting the properties of ACIDS. The hydrogen ion activity in a SOLUTION is used to assess the acidity and alkalinity of soils. PH.

hydrograph a graph indicating the rate of flow or the level of water in a STREAM or an AQUIFER measured at a given point during a selected period of time.

hydrography the science and art concerned with the study of all bodies of water on the earth's surface, and especially with charting oceans and seas, their beds and coastlines, and the tides, currents, winds, with particular regard to safe navigation. The terms hydrography and HYDROLOGY overlap, and the tendency now in Britain is to restrict hydrography to cover survey and mapping etc. with an emphasis on marine waters, and to apply hydrology to the study of the water of land areas.

hydrological cycle the continuous circulation of water from the earth's surface to the ATMOSPHERE-1 to the earth's surface, brought about by EVAPORATION from the surface water and the land and by EVAPOTRANSPIRATION from vegetation, giving rise to WATER VAPOUR in the atmosphere which condenses (CONDENSATION), forms CLOUDS, and returns to the earth's surface as PRECIPITATION, swelling the oceans, seas, lakes and rivers, or become GROUND WATER. HYDROSPHERE. Fig 25.

hydrology **1.** the scientific study of the occurrence, movement, properties and use of water and ice on or under the earth's surface, from its precipitation to its discharge to the sea or its return to the atmosphere, with a particular emphasis in Britain on inland water. HYDROGRAPHY **2.** specifically, in USA Geological Survey, the scientific study of underground water resources as distinct from hydrography, there defined as applying to surface water

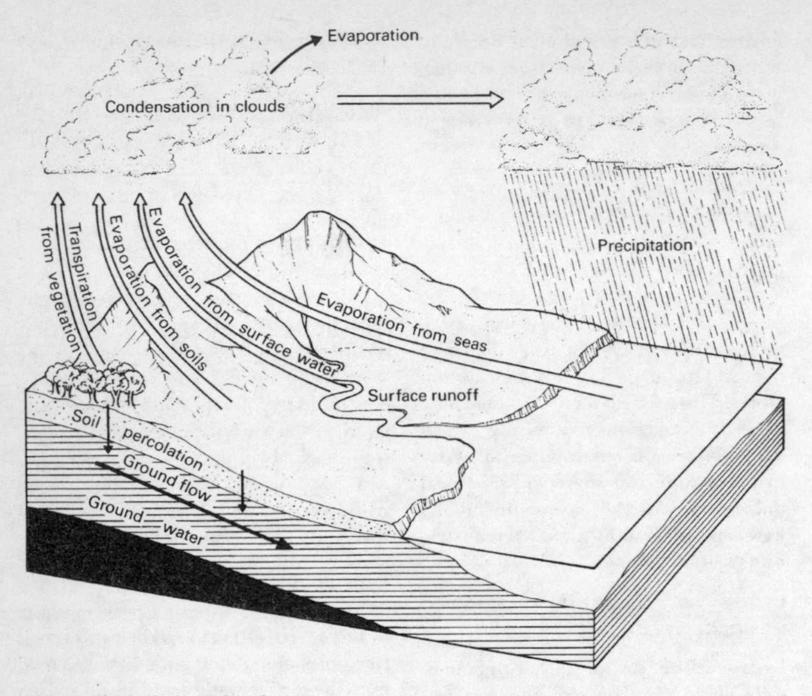

Fig 25 The hydrological cycle

supplies and resources. HYDROLOGICAL CYCLE.

hydrolysis a chemical reaction of water in which the reagent (i.e. the substances creating the chemical reaction of the water) is decomposed and HYDROGEN and hydroxyl (hydrogen, oxygen) and, commonly, other new compounds are added. This process is involved in **1.** the chemical WEATHERING of rocks, in which the salt constituents of the rock combine with water, and an ACID and a BASE-2 are formed. CORROSION **2.** the decomposition of organic compounds by interaction with water.

hydrophyte an aquatic (water) plant, one growing only in water or a saturated soil. HYGROPHYTE, MESOPHYTE, TROPOPHYTE, XEROPHYTE.

hydrosphere the water sphere, all the waters (liquid or solid) of the surface of the earth collectively, including soil and ground water, in comparison and contrast with the LITHOSPHERE and the ATMOSPHERE-1. The HYDROLOGICAL CYCLE blurs the distinction between the atmosphere and the hydrosphere.

hydrostatic pressure the pressure exerted by water at rest equally at any point within that body of water. ARTESIAN WELL.

hydrothermal *adj.* **1.** relating to hot water **2.** in geology, applied to the combined action of heat and water that brings about changes in the earth's crust, by making strong solutions, by altering processes in minerals (KAOLINIZATION) and by depositing minerals in VEINS and on the earth's surface around GEYSERS **3.** applied to the rocks (hydrothermal deposits), ore-deposits and springs (hydrothermal springs) so produced **4.** applied to vents in the OCEANIC CRUST through which gases and hot water jets rise. BLACK SMOKER, GEOTHERMAL ENERGY, HOT ROCK, METAMORPHISM, METASOMATISM, SINTER.

hygrometer an instrument used to determine the HUMIDITY or RELATIVE HUMIDITY of the atmosphere or a gas, some types making use of a human hair (which stretches and contracts), others a lithium-chloride strip (which has a varying resistance to moisture), the very slight reaction of the moisture-sensitive component being enlarged and registered on a graded scale.

hygrophyte a water-loving plant, one which thrives in moisture (e.g. a tree of the tropical RAIN FOREST) but is not an aquatic plant. Most of the coniferous forest trees of the west and southeast of North America, southern Chile, western Europe, parts of China and Japan, southeastern Australia and New Zealand are hygrophytes.

hygroscope an instrument that indicates, without actually measuring, HUMIDITY.

hygroscopic *adj.* the quality of having an affinity with water, applied to substances which absorb and retain moisture, e.g. salt particles in the atmosphere, which act as nuclei for the CONDENSATION of WATER VAPOUR.

hygroscopic moisture water in the soil held by such strong SURFACE TENSION that it is unavailable to plants. CAPILLARY MOISTURE, FIELD CAPACITY.

hypabyssal *adj.* half-abyssal, applied especially to an IGNEOUS ROCK or to an IGNEOUS INTRUSION which has risen towards the surface, but which has crystallized below it. They are thus intermediate in physical form between the rocks which are deep seated, having cooled from MAGMA deep in the earth's crust (PLUTONIC ROCK) and those which have been poured out and solidified on the surface (VOLCANIC ROCK). Hypabyssal rocks are found as DYKES, SILLS, small intrusions etc. ACID ROCK, BASIC ROCK, INTERMEDIATE ROCK, INTRUSIVE ROCK.

hypermarket a very big, self-service store which is larger than a SUPERMARKET, sells a wide variety of foodstuffs, household goods etc., has an extensive car park and is usually situated on the outskirts of a town.

hypolimnion the non-circulating coldest layer of water, with the least oxygen, in a deep LAKE or ocean, lying below the THERMOCLINE. THERMAL STRATIFICATION.

hypothesis pl. hypotheses **1.** an idea or proposition that is not the outcome of experience, but is formulated and used, as an untested assertion, to explain certain facts, or the relationships between two or more concepts (ALTERNATIVE HYPOTHESIS, NULL HYPOTHESIS) **2.** the primary assumption of an argument.

I

ice the solid form of water, formed in nature by the freezing of water (as in a river or sea ice), by the condensation of atmospheric WATER VAPOUR at temperatures below freezing point direct into ICE CRYSTALS, by the compaction of SNOW (with or without the movement of a GLACIER), by the seepage of water into snow masses and its subsequent freezing. The DENSITY-2 of ice is lower than that of water, thus ice floats.

ice age any period in the long course of geological time, from Precambrian time onward (GEOLOGICAL TIMESCALE), when extensive GLACIERS covered large parts of the land surface in the northern and southern hemispheres. The greatest and best known of these ice ages is the last or Great Ice Age, also known as the Glacial Epoch, which occurred in the PLEISTOCENE epoch when human beings had already appeared on the earth's surface. In Europe and North America there were at least four fluctuations in the Pleistocene glaciation, periods with extending ICE SHEETS being interrupted by warmer interglacial episodes. Many local names have been given to these periods of ice expansion. In the European Alps four were distinguished and named, from the oldest to the youngest: Günz, Mindel, Riss and Würm. In North America the four periods corresponding to those of Europe are, from the oldest, Nebraskan, Kansan, Illinoian, Wisconsin, with the Iowan as the earliest stage of the Wisconsin.

ice anvil the flattened head of an ANVIL CLOUD, formed by very small snow and ice crystals falling down below the spreading layer. INCUS CLOUD.

iceberg a large mass of land ice, often of great height, broken off from a GLACIER or from an ICE SHELF and floating in the sea, at the mercy of winds and currents. The ratio of ice below the water to that above is some three or four to one. Arctic icebergs have pinnacles in their superstructures; coming from Greenland, they are carried south by the East Greenland and Labrador currents to the Atlantic (a few passing through the narrow, shallow Bering Strait to the Pacific). The superstructures of the Antarctic icebergs, which come from the Ross ice shelf, are flat, and these icebergs float northwards to about 60°S in the Pacific.

ice cap a permanent covering of ice, a dome-shaped GLACIER, smaller than an ICE SHEET, covering a highland area or an island in high latitudes.

ice crystal a single ice particle with regular structure.

ice-dam lake, ice-dammed lake a lake formed by a barrier of ice stretching across a valley mouth.

ice field, ice-field, icefield 1. generally, an extensive area of LAND ICE **2.** specifically, a large continuous area of PACK ICE or sea ice, more than 8 km (5 mi) across, defined as large (over 20 km: 12.5 mi across), medium (15 to 20 km: 9 to 12.5 mi across), or small (10 to 15 km: 6 to 9 mi across).

ice-floe any separate piece of floating sea ice, level or hummocked, thinner than an ICEBERG, termed light (up to 1 m: 3.3 ft thick) or heavy (exceeding 1 m: 3.3 ft thick). Floes are described as vast (over 10 km: 6.2 mi across), large (between 1 and 10 km: 0.6 to 6.2 mi across), medium (200 to 1000 m: 655 to 3280 ft across), small (10 to 200 m: 33 to 655 ft across).

ice fog a formation of ICE CRYSTALS suspended in the air, so numerous that they restrict visibility at the earth's surface and reflect sunshine to produce optical phenomena such as haloes and luminous pillars. The sun shining above an ice fog can produce an effect so dazzling that it can be dangerous to the naked eye. STEAM FOG.

Icelandic low the mean atmospheric subpolar low pressure area (LOW) lying over the North Atlantic ocean between Iceland and Greenland, particularly in winter, comprising swiftly moving areas of low pressure interspersed with occasional periods of higher pressure. ALEUTIAN LOW.

ice pedestal SERAC.

ice point the melting point of pure ice at standard pressure. ABSOLUTE ZERO.

ice sheet a continuous mass of ice and snow of considerable thickness and covering a large area of rock or water, e.g. the ice masses occupying most of the Antarctic continent and Greenland at the present day. An ice sheet of under some 50 000 sq km (20 000 sq mi) and overlying rock is termed an ICE CAP. ICE AGE.

ice shelf a thick ICE SHEET of great extent and with a level or undulating surface, fed by snow and sometimes by glaciers, which has reached the sea and is floating, although parts may be aground.

ice wedge a mass of ice in the ground, shaped like a wedge with a point facing down, formed when melt-water, penetrating cracks in the ground in summer, freezes in winter under PERIGLACIAL conditions. Repeated melting and refreezing breaks up the material around the wedge, thereby increasing its width and penetration.

iconic model a simple scaled-down representation of reality in which the phenomena selected are shown in the form in which they exist, reduced to scale, e.g. people shown as reduced-to-scale people. ANALOGUE MODEL, SYMBOLIC MODEL.

iconography the study of the sources and meanings of images used in the visual arts and literature and by society, particularly as applied in HISTORICAL GEOGRAPHY and HUMANISTIC GEOGRAPHY. GEOGRAPHICAL IMAGINATION.

idealism **1.** the belief that a system or a standard conceived as perfect or nearly perfect (but which is unlikely to exist in the real world) exists **2.** in philosophy, one of the various approaches which, in general, maintain that ideas are the only real things, i.e. that the object of external perception consists (either in itself or as perceived) as an idea, a notion, that the only things which exist in reality are minds or mental states, or both. For example, the metaphysical doctrine (METAPHYSICS, ONTOLOGY) maintains that ultimate reality is either mental or spiritual; and the epistemological doctrine (EPISTEMOLOGY) holds that either the objects of perception or ideas are the only knowable entities. MATERIALISM, REALISM.

ideology **1.** the science of ideas, concerned with their origin and nature, particularly those springing from sensory stimulation **2.** ideal or abstract speculation **3.** a set of ideas adopted as a whole, held implicitly and maintained regardless of the course of events, used in support of, justifying, an economic, social or political theory, or the conduct of a class or group **4.** the way of thinking of an individual class or culture.

idiographic approach explanation concerned with individual cases or situations (explanation by case history) rather than with general types or theories (a law-seeking approach). NOMOTHETIC APPROACH.

igneous *adj.* of, pertaining to, containing, resembling, or emitting, fire.

igneous rock a rock which has originated from the cooling and solidification of MAGMA from the heated lower layers of the earth's crust. The chemical composition of igneous rocks depends on the nature of the magma. Arranged in order of their SILICA content they are classified as ACID ROCK (e.g. GRANITE), INTERMEDIATE ROCK (e.g. ANDESITE), BASIC ROCK (e.g. GABBRO), ULTRABASIC ROCK (e.g. PERIDOTITE). Their character is also affected by the mode of cooling. When solidified slowly at depth they are coarsely crystalline (e.g. GRANITE), and termed INTRUSIVE OR PLUTONIC; when rapidly, at the surface, they have fine crystals and are termed extrusive (EXTRUSIVE ROCK) or volcanic (VOLCANIC ROCK) (e.g. ANDESITE, BASALT); between the two are HYPABYSSAL. A few igneous rocks (e.g. TUFF) are formed from compacted or cemented fragments of pre-existing igneous rocks.

illuvial horizon the SOIL HORIZON which has received material in SOLUTION or SUSPENSION from the overlying soil layer(s). ARGILLIC HORIZON, ELUVIAL HORIZON, ELUVIATION, ILLUVIATION.

illuviation the process by which material removed in SUSPENSION or SOLUTION from the upper part of a soil (the ELUVIAL HORIZON) is washed down and deposited in the lower layers or ILLUVIAL HORIZONS (usually the B HORIZON). ARGILLIC HORIZON, HARD PAN.

ilmenite an oxide of IRON and titanium, a black, crystalline ore, occurring in detrital SAND, BASIC IGNEOUS ROCK and METAMORPHIC ROCK.

image **1.** a concept, a mental picture. COGNITIVE MAP, PERCEPTION **2.** something that represents or is taken to represent something else **3.** a counterpart, a copy (solid or optical, or the product of REMOTE SENSING) of an object, the same size as the object, or diminished or magnified, erect or inverted (e.g. in optics a copy produced by a mirror or lens).

IMF International Monetary Fund, an agency of the United Nations established in 1944 to promote international financial cooperation, expand world trade, eliminate foreign exchange restrictions. It does not advance loans. G7, WORLD BANK.

immature soil a young, imperfectly developed soil occurring on recently laid deposits or where EROSION keeps pace with the development of a SOIL PROFILE. AZONAL SOIL.

immigrant **1.** a person who voluntarily comes from the home country to settle in another, especially for the purpose of permanent residence **2.** a plant or animal which comes from its native habitat into another. EMIGRANT, EXILE, EXPATRIATE, MIGRANT, MIGRATION, REFUGEE.

imperialism 1. the making of an empire through the extension by one sovereign state of its authority over other territories, by military conquest or by political or economic means, thereby creating a relationship in which those territories contribute resources to, and become dependent on, the dominant sovereign state, usually to the economic benefit of the latter. COLONIALISM is a form of imperialism **2.** the policy or the doctrine of such an extension of authority.

impermeable *adj.* not PERMEABLE, not permitting the passage of fluids, especially of water. IMPERVIOUS ROCK.

impermeable rock a rock which does not allow water to soak into and through it because it is IMPERVIOUS or non-porous (or practically non-porous) or both. IMPERVIOUS ROCK, PERMEABLE ROCK, PERVIOUS, PERVIOUS ROCK, POROSITY.

impervious *adj.* not PERVIOUS, impenetrable, that which cannot be entered or passed through.

impervious rock rock through which water cannot pass freely, e.g. CLAY (POROSITY), unfissured GRANITE. Some authors use impervious as synonymous with impermeable, but some draw a certain distinction, outlined in PERMEABLE ROCK, PERVIOUS ROCK. (PERMEABILITY should not be confused with porosity.)

import something brought in, introduced from a foreign or external source, or from one use, connexion, or relation, into another, especially goods or merchandise brought into a country from a foreign source in the course of international trade. BALANCE OF TRADE, EXPORT.

import substitution the replacement of previously imported goods by home produced goods in order to improve the BALANCE OF PAYMENTS.

improved *adj.* made better in quality, more productive, more valuable.

improved land frequently used as a technical term, but not always with the same meaning. In the Agricultural Statistics of most countries it refers to farm land where by ploughing, cultivation, manuring, or some form of management, the natural condition of the land has been improved. In Britain it covers ploughland and grassland in fields, but not open moorland; similarly in the USA, land in farms but not open range land.

inceptisols in SOIL CLASSIFICATION, USA, an order of young soils with weakly developed horizons, occurring in variable climates, e.g. brown earths (BROWN FOREST SOIL) and TUNDRA SOILS.

incised meander a MEANDER deeply sunk into the general level of the surrounding country. This may be the result when a mature river, with extensive meanders, is rejuvenated by an uplift of the land and begins to cut down its bed. Most authors use incised and INTRENCHED (or entrenched) as synonymous; but some give a narrower meaning to intrenched, applying that term to a meander in which the valley sides are roughly of the same slope, and using the term ingrown meander when they are not of similar slope.

inclination 1. the angle of approach of one line or plane to another **2.** a slant, a slope, a deviation from the vertical or horizontal. DIP.

inclination-dip the angle of DIP of a FAULT, STRATUM or VEIN, measured from the horizontal. DIP SLOPE.

inclosure in England, the legal act, permitted by an Act of Parliament, either general or special, whereby open land or land formerly worked in common (OPEN FIELD) is cut up into individual fields surrounded by fences, walls, hedges etc. The spelling of the physical process is usually enclosure, and although that of the legal Act should be inclosure, the two spellings are commonly used interchangeably.

incompetent, incompetence (of rocks), incompetent bed COMPETENCE.

inconsequent drainage a drainage pattern not conditioned by (not consequent on) the present structure, being either ANTECEDENT or SUPERIMPOSED. ACCORDANT, DISCORDANT, INSEQUENT DRAINAGE.

incus cloud the ANVIL CLOUD spreading above a CUMULONIMBUS.

indenture a contract that binds one person to work for another for a prescribed period of time.

independent variable a VARIABLE which produces or causes (or is thought to produce or cause) changes in another variable (termed the DEPENDENT VARIABLE) and, in an experiment, is manipulated by the experimenter. REGRESSION ANALYSIS.

index **1.** an indicator, a sign **2.** in mathematics, a numerical ratio or other number deduced from observations and used as an indicator or measure of a process or condition, e.g. cost of living index.

Indian summer a period of calm, dry, mild weather, with cloudless sky but hazy atmospheric conditions, occurring fairly regularly in late autumn (fall) or early winter in the USA and the UK. The origin of the term is uncertain.

indicator plant, indicator species a plant or SPECIES-1 which shows by its presence in a locality the existence of a particular environmental factor or certain environmental conditions, e.g. the OLIVE, an indicator of the MEDITERRANEAN CLIMATE.

indigenous *adj.* originating in, native to, a particular area, not introduced, applied to **1.** plants, animals, human population **2.** a rock, mineral or ore in its place of origin, not carried in from somewhere else.

indirect contact space the part of an urban area perceived by an individual from secondhand contact, e.g. from information given by acquaintances, the mass media (newspapers, radio, television etc.), and so on. AWARENESS SPACE.

induction **1.** the drawing of a general conclusion from a number of known facts **2.** the conclusion reached in this way. DEDUCTION.

induration the hardening of a rock by heat, pressure or cementation, or of SOIL HORIZONS by chemical action (HARD PAN).

industrial archaeology the study of past industrial processes and methods, especially of the early period of the INDUSTRIAL REVOLUTION, based primarily on examination of the physical remains, e.g. of buildings (including housing associated with manufacturing), machinery, mines, means of transport, and their equipment.

industrial city the type of city commonly found in industrialized countries today, broadly conforming to the CONCENTRIC ZONE THEORY of Burgess. INDUSTRIAL

SOCIETY, POST-INDUSTRIAL CITY, PRE-INDUSTRIAL CITY.

industrial complex a large assemblage of manufacturing enterprises concentrated in a relatively restricted area (thus differing from an industrial region, where widespread industry covers an extensive area), served by good transport, commercial and financial facilities. It usually comprises one or more basic manufacturing industries combined with diverse other manufacturing enterprises, technically and economically interdependent (thus differing from an industrial centre, which is small, with less diversified manufacturing enterprises).

industrial crop, commercial crop a CROP-1 grown not for food but as a raw material for manufacturing industry, e.g. COTTON.

industrial geography a branch of ECONOMIC GEOGRAPHY concerned with manufacturing industry, particularly its location and spatial distribution.

industrial inertia the tendency of an industry or firm to remain in a location or site after the conditions originally influencing the choice of that location or site have altered, ceased to be relevant, or disappeared. GEOGRAPHICAL INERTIA, GEOGRAPHICAL MOMENTUM, INDUSTRIAL MOMENTUM.

industrialization **1.** the process of growth of large-scale machine production (MECHANIZATION) and the factory system **2.** the process of setting-up such organizations, especially in the introduction of MANUFACTURING INDUSTRY (SECONDARY INDUSTRY) in countries or regions where people are engaged mainly in agricultural activities (PRIMARY INDUSTRY). It is usually accompanied by the establishment of service industry (TERTIARY INDUSTRY) and by social change, e.g. in patterns of consumption, and in migration of people from rural areas to the growing urban settlements. AIC, COTTAGE INDUSTRY, DE-INDUSTRIALIZATION, NIC, UNDERDEVELOPMENT.

industrialized country generally, a country in which the contribution of manufacturing industry (SECONDARY INDUSTRY) and the service industry (TERTIARY INDUSTRY) to the economy is greater than that of PRIMARY INDUSTRY, e.g. of agriculture.

industrial linkage all the exchanges between an industrial enterprise and the factors, material and non-material, influencing its location, including, for example, the exchange of information with other firms engaged in similar work in the locality.

industrial momentum the tendency of an industry or firm in a given locality not only to maintain its activity (INDUSTRIAL INERTIA) but also to increase its importance after the conditions originally influencing its establishment in that locality have appreciably altered, ceased to be relevant, or disappeared. GEOGRAPHICAL INERTIA, GEOGRAPHICAL MOMENTUM.

industrial park a tract of land, in some cases park-like (PARK-2), planned and officially designated for the accommodation of clean, relatively small industrial enterprises. OFFICE PARK, PARK-6, RESEARCH AND SCIENCE PARK.

industrial region INDUSTRIAL COMPLEX.

industrial revolution the changes generated by the MECHANIZATION of MANUFACTURING INDUSTRY, i.e. the change from domestic industry to the factory system, which leads to the mass production of goods, with the consequent great changes in social, economic and technical structures. The term Industrial Revolution is commonly applied to the period in Britain in the eighteenth and early nineteenth centuries when the mechanization of the textile industry produced such changes.

industrial society POST-INDUSTRIAL SOCIETY.

industry any work performed for economic gain, but popularly applied especially to MANUFACTURING INDUSTRY (SECONDARY INDUSTRY). Industries are variously described as BASIC, EXTRACTIVE, FOOTLOOSE, HEAVY, LIGHT, LINKED, PRIMARY, SECONDARY, TERTIARY and QUATERNARY. See terms qualified by industrial; and INDUSTRIALIZATION. LOCALIZATION OF INDUSTRY, LOCATION OF INDUSTRY, MARKET ORIENTATION, RESOURCE ORIENTATION.

infield-outfield a system of farming which confines intensive cultivation, manuring etc. to enclosed fields near the farm (infield), the outer parts of the farm (outfield) being used for grazing or periodically cropped on a SHIFTING CULTIVATION basis.

infiltration the ABSORPTION and downward movement of PRECIPITATION in the REGOLITH. INFILTRATION CAPACITY.

infiltration capacity, infiltration rate the maximum rate at which water can be absorbed and seep downwards through the soil. Infiltration rate is now the preferred term. FIELD CAPACITY, OVERLAND FLOW.

inflation a general increase in prices produced by an increase in the proportion of currency and credit to the goods available.

influent stream a stream, common in CHALK and LIMESTONE country, that has its bed higher than the WATER TABLE, and which flows into a cave, etc. The bed is usually IMPERMEABLE, being lined with fine silt; but part of the water inevitably disappears down cracks or joints in the chalk and limestone.

informal sector unofficial employment, particularly in the service industry, hidden from the authorities, thus sometimes termed the black economy.

infra-red radiation, infrared radiation ELECTROMAGNETIC WAVES in the wavelength range of 0.7 to about 200 microns, invisible, perceived as heat, with wavelengths just a little longer than those of red light but less than short microwaves, used especially in some cooking devices and in photography with infra-red or colour infra-red film. Infra-red photographic devices used in REMOTE SENSING successfully distinguish features on the surface of the earth through darkness or cloud.

infrastructure the basic structure, the framework, the system which supports the operation of an organization (e.g. the power and water supplies, the transport and communications facilities, the drainage system), which makes economic development possible, the basic capital investment of a country or enterprise.

ingrown meander INCISED MEANDER.

initial advantage the advantage gained by a city, region or nation by being the first in some respect, e.g. in establishing a market area, or in adopting a new technique. Some authors maintain that urban centres which are first in establishing industry or in developing new techniques build up a self-generating lead over other centres in terms of the size of their population and in the volume and variety of their industries.

inland basin a DEPRESSION-2 entirely surrounded by higher land, with or without an outlet to the ocean. BASIN-8.

inland drainage INTERNAL DRAINAGE.

inland sea a large, isolated expanse of water, without a link to the open sea.

inlet **1.** a narrow opening by which the water of the sea, of a lake, or river, penetrates the land **2.** the passage between islands into a lagoon.

inlier an exposed rock formation entirely surrounded by geologically younger rocks.

inner city an undefined area with a wide range of economic and social problems, lying within a long-established, generally large, urban area. The term may be applied to such an undefined area, usually in economic decline, lying between the commercial centre of the city (CENTRAL BUSINESS DISTRICT) and its suburbs. That is a zone usually characterized by aged, run-down housing in multiple occupation by people with low incomes (especially new immigrants), people who stay for a relatively short period of time, and a dwindling number of aged 'local' people. GENTRIFICATION, INDUSTRIAL CITY, VILLAGE-2.

inner lead the calm water lying between the mainland and the skerries. SKERRY, SKERRY-GUARD.

innovation the making of changes, the introduction of practices, processes etc. which are new in a particular context.

innovation wave a concept of the DIFFUSION of INNOVATION as being analogous to the onward movement of an ocean wave.

inorganic *adj.* **1.** of or relating to substances not composed of plant or animal material. ORGANIC **2.** in chemistry, of or pertaining to substances that are not organic, hence inorganic chemistry, the branch of chemistry concerned with the scientific study of chemical elements and their compounds other than the CARBON compounds (but including the simpler compounds of carbon, e.g. CARBON DIOXIDE).

input-output analysis a method used in ECONOMICS to trace the connexions between products and services (the output) and the resources needed to produce them (the input). During any period of time the output of one sector of the productive system may become the input of another. The quantities, commonly expressed in money values, are displayed in a matrix as an input-output table, each sector of the productive system being assigned a row (showing the destination of the outputs) and a column (showing the provenance of the inputs). Outputs which are absorbed in the productive system are termed intermediate outputs, those which pass out of it into final demand are final outputs; inputs derived from the system are intermediate inputs, those that come in from outside (e.g. land, labour, capital) are primary inputs. For each sector the sum of the output entries (i.e. total revenue), displayed in the row, equals the sum of the input entries (i.e. total costs) displayed in the column. The interdependence of different sectors of the productive system is thus revealed and can be analysed; and the effects of changes in one sector on those in another traced.

inselberg (German) an isolated hill of

circumdenudation, e.g. a steep-sided, isolated residual hill, common in semi-arid and savanna lands, rising from a plain which is, in many cases, monotonously flat (PEDIMENT).

insequent drainage, insequent stream a DRAINAGE-2 pattern developed on the present land surface (especially on horizontal strata), bearing no direct relation to the underlying structure, and seemingly determined by accident. DENDRITIC DRAINAGE, INCONSEQUENT DRAINAGE.

inshore *adj.* an imprecise term, moving towards the shore (e.g. inshore breeze); or applied loosely to that which is close to the shore; or shorewards of a position in contrast to seawards of it.

in situ (Latin, in place, in position) *adv.* associated with the occurrence of a fossil, mineral, rock or soil in its original place of deposition or formation, e.g. a RESIDUAL DEPOSIT.

insolation **1.** exposure to the rays of the SUN **2.** SOLAR RADIATION received, applied especially to that reaching the surface of the earth, greatest at the equator for the year as a whole, but polewards decreasing, at first slowly, then more rapidly, then again slowly, the variation throughout the year being least at the equator, most at the poles. It is an important climatic factor and has a (sometimes disputed) role in atmospheric weathering. The interrelationship of relief and insolation is important in human geography (ADRET) **3.** the rate at which solar radiation reaches a specified area.

insular *adj.* of, inhabiting, situated on, forming, characteristic of, an island.

insular climate a climatic regime with little range of temperature between summer and winter, characteristic of many islands and some sheltered coastal areas.

integration in society, the process by which a sub-group, e.g. an ETHNIC MINORITY, adapts to, fits into, and participates fully in the social and economic structure in which it finds itself, while keeping its identity, its individuality and cultural distinction. ACCOMMODATION, ASSIMILATION, PLURAL SOCIETY.

intensive agriculture a farming system in which large amounts of capital and/or labour are applied to a relatively small area of land to achieve high yields per unit area. ESA, EXTENSIVE AGRICULTURE.

interaction in general, reciprocal action, the action or influence of forces, objects, or persons on each other.

interbedded *adj.* applied to a layer of rock deposited in sequence between two other BEDS-2.

interface **1.** the surface that constitutes the common boundary between two bodies, two systems, two spaces, two different contiguous parts of the same substance, or between phases in a heterogeneous system (e.g. the surface formed between a liquid and a solid), extended to cover other boundaries, especially if they are ill-defined **2.** the connexion between two pieces of equipment, by analogy extended to cover the intercommunication between different social groups.

interfluve the tract of land between two adjacent rivers, regardless of its character.

interglacial *adj.* of a period of time between two glacial periods; sometimes used as a noun, referring to an INTERGLACIAL PERIOD and/or to a deposit formed in that period.

interglacial period a period of time with a relatively warm climate, when the ice sheets retreated, occurring between two periods of glacial cold, e.g. as in the ICE AGE, which was not a period of unremitting cold.

interlocking spur one of the series of protrusions (SPUR) of land that, lying between bends in the winding course of a young river in its V-shaped valley, juts into a concave bend and interdigitates with its opposing neighbours lying upstream and downstream. Interlocking spurs thus obscure the upstream or downstream view of the river. They are caused initially by the stream's flowing swiftly round an obstacle in its course, undercutting the bank opposite the obstruction, and thereby making the concave bend. MEANDER.

intermediate rock an IGNEOUS ROCK classified chemically as being between acid and basic in its composition (ACID ROCK, BASIC ROCK), that is it has a SILICA-2 content lying between 52 and 66 per cent and no free QUARTZ. It may be plutonic (intrusive) or volcanic (extrusive). EXTRUSIVE ROCK, HYPABYSSAL, INTRUSIVE ROCK, LAVA, PLUTONIC ROCK, VOLCANIC ROCK.

intermediate technology a TECHNOLOGY-3 which aims to introduce to small groups of people in UNDERDEVELOPED lands those parts of advanced scientific knowledge and industrial processes which, combined with resources and materials readily available locally, will match their knowledge and skills, meet their needs, and improve their quality of life on a long-term basis. ALTERNATIVE TECHNOLOGY, HIGH TECHNOLOGY, LOW TECHNOLOGY.

intermittent saturation zone in soil science, a layer, lying below the surface soil, which may hold VADOSE WATER in a period of prolonged rainfall, but which quickly dries out in even a short-lived DROUGHT.

intermittent spring a spring that flows from time to time, usually depending on the height of the WATER TABLE (itself affected by fluctuations in precipitation).

intermittent stream a stream which does not flow continuously but dries up from time to time, e.g. a BOURNE.

intermonate *adj.* lying between mountains, e.g. the intermontane plateaus, the high plateaus, lying between the east and west ranges of the Andes.

internal migration the movement of people within a country, e.g. in search of employment.

international airport an airport with facilities suitable for handling international traffic and meeting the needs of international airlines.

international date line an imaginary line agreed internationally which follows the meridian of 180°, with some deviations to accommodate certain land areas. In crossing the line from west to east a day is repeated; in crossing it from east to west a whole day is lost. Thus an aircraft flying from Japan to Honolulu arrives at an earlier time on the same day; an aircraft leaving Honolulu late on Monday evening would not reach Japan until Wednesday morning, although the duration of the flight is only a few hours. Fig 43.

international division of labour 1. the separation of employment into parts on an international scale, a feature of the 'old' international division of labour (OIDL) (i.e. before c. 1940) when there was specialization on a territorial basis in the tasks performed to supply world markets.

To summarize, broadly, subject territories of a dominant country provided it with primary products (foodstuffs and other RAW MATERIALS) to augment the food supply of that dominant country and meet the needs of its manufacturing industry. The principal exports of the dominant country were manufactured goods, sold to the subject territories and to other countries. The subject territories bought these manufactures with earnings from the exports of their primary products. In the process the most powerful controlling powers became dominant in the international organization of production and trade, and in world markets **2.** in the 'new' international division of labour (NIDL) (i.e. after the mid-1940s) the production of manufactured goods is widespread throughout the world (not concentrated, as formerly, in the home territories of dominant countries), countries which were formerly predominant exporters of primary products having developed their own (in many cases substantial) manufacturing industry and, benefiting from improved transport and communications as well as from investment from external sources, entered international trade as exporters of manufactured goods. In many cases their production costs are lower than those current in the formerly dominant countries; their prices are highly competitive in world markets; and many of them specialize in particular products. Such developments have far-reaching social, economic and political effects; and are especially important to the MULTINATIONALS, with their world-wide interests and ability to finance technical advance. AIC, INDUSTRIALIZATION, NIC, UNDERDEVELOPMENT.

International Monetary Fund IMF.

internet the global communications computer network that by means of a common protocol links smaller computer networks and personal computers throughout the world. JANET.

interquartile range in statistics, a measure of dispersion (the spread of values) of a FREQUENCY DISTRIBUTION. A quartile is produced by splitting a distribution into four equal parts, the quartiles being those values of the VARIABLE below which lie 25 per cent, 50 per cent and 75 per cent of the distribution. The interquartile range is the distance between the 75 per cent (upper quartile) and the 25 per cent (lower quartile). It thus contains one half of the total frequency and provides a simple measure of dispersion. MEDIAN.

Intertropical Convergence Zone (ITCZ), Intertropical Front (ITF) ITCZ.

intervale (American) a tract of low-lying land, especially between hills or bordering a river.

interval scale in statistics, a MEASUREMENT scale which lacks an absolute zero, but in which the INTERVALS between the scale points are equal. A VARIABLE measured on this scale is termed an interval variable.

intervening location effect the effect on a place of lying in an intermediate position on a route which is well served for reasons other than those of benefiting the place itself. The result is that the place enjoys better services than its own characteristics would justify. ACCESSIBILITY, SHADOW EFFECT.

intervening opportunity effect of a boundary, the beneficial effect of a BOUNDARY on locations close to it, making such locations attractive to people, e.g. in giving shoppers from one side of the boundary access to goods sold at lower prices on the other. HALO EFFECT.

intrazonal soil a well-developed soil with the form and structure reflecting the influence of some local factor of relief, parent material or age, rather than of climate and vegetation. ANTHROPOMORPHIC SOIL, AZONAL SOIL, SOIL, SOIL CLASSIFICATION, ZONAL SOIL.

intrenched meander an INCISED MEANDER with steep, symmetrical valley sides, produced by swift, vertical erosion.

intrusion the forceful entry of a mass of molten rock (MAGMA) in the pre-existing rocks of the earth's crust, sometimes CONCORDANT as a sheet or SILL along BEDDING PLANES, or as a lens-shaped mass (LACCOLITH and PHACOLITH), sometimes DISCORDANT, i.e. across the beds (DYKE). HYPABYSSAL, INTRUSIVE ROCK.

intrusive rock PLUTONIC ROCK, an IGNEOUS ROCK with CRYSTALS larger than those occurring in an EXTRUSIVE ROCK, formed from the consolidation of molten material (MAGMA) which has penetrated or been forced into pre-existing solid rocks at depth in the earth's crust. EXTRUSION, HYPABYSSAL, INTRUSION, VOLCANIC ROCK.

invasion and succession in urban land use studies, the sequence of changes by which units of population or of land use replace those existing in another area, the invading population or land use type eventually achieving numerical superiority and controlling the area invaded, e.g. as in GENTRIFICATION. The term is used particularly to explain the outward growth in the CONCENTRIC ZONE GROWTH THEORY.

inversion the reversal of the normal or expected order of position. GEOLOGICAL INVERSION, INVERSION OF TEMPERATURE.

inversion of temperature a phenomenon in which there is in the air a temperature increase with increasing height instead of the normal decrease (LAPSE RATE). It can occur at high altitudes (e.g. when a cold air mass flows under a warm one, as at a COLD FRONT, or when a warm air mass flows over a cold one, as at a WARM FRONT, or when an OCCLUSION develops); or near the earth's surface (e.g. in temperate latitudes on a calm, clear night when radiation of heat from the ground at night is rapid, or when warm air flows over a cold surface; or in mountainous regions when radiation of heat from the upper slopes is rapid, and cold dense air behaves like cold water and flows down the valley). FRONT, RADIATION FOG.

invertebrate *adj.* lacking a backbone. VERTEBRATE.

inverted relief inversion of relief, a landscape displaying the effects of a long period of DENUDATION and EROSION, the upfolds in the strata (ANTICLINES) corresponding with low ground and the SYNCLINES with high ground, the opposite of UNINVERTED RELIEF.

invisible exports the income-earning items in the international trade of a country, representing not the sale or transfer of goods, but payments made by foreign countries to that country, for services provided (banking, insurance etc.), for shipping, air freight, for interest on investments; and also expenditure by tourists visiting the country (TOURISM) and remittances home by migrants. Invisible imports cover similar items, but for these payments are made by the country to foreign countries.

ion an ATOM or group of atoms with either an excess (cation, a positive ion) or a deficiency (anion, a negative ion) of electrons, which is therefore electrically unbalanced and electrically charged. (An electron is a fundamental particle, negatively charged, a constituent of the atom, orbiting in the nucleus.) An ion may be formed in a gas or in a solution and carry current through either. ACID, BASE, COLLOID, PH, HYDROGEN ION.

ionization the production of IONS, converting to ions, or being converted to ions, by the addition or removal of ELECTRONS from ATOMS, e.g. by addition to an ionizing solvent or by means of high-energy RADIATION, as in the IONOSPHERE.

ionosphere thermosphere, the outermost zone of the ATMOSPHERE-1, above the MESOPAUSE, at a height of about 65 km (40 mi), the lower level dropping to some 55 km (35 mi) in daylight, rising to some 105 km (65 mi) at night, so named because ULTRAVIOLET and X-RAYS radiated by the sun ionize its gases to a degree determined by the solar cycle, season and time of day. Particles arising from this IONIZATION concentrate in distinct layers (distinguished by the letters D, E, F2, F1) and refract radiowaves back to earth. APPLETON LAYER, HEAVISIDE LAYER, SUNSPOT. Fig 4.

iron a widely occurring heavy, METALLIC ELEMENT-6, the second most widespread MINERAL-1 (ALUMINIUM being the first), estimated to constitute chemically some 5 per cent of the earth's crust by weight. It does not occur NATIVE-1 (*adj.*) in nature, rusts readily in moist air (i.e. it combines with the oxygen of the atmosphere to form an oxide) and is chemically active, forming compounds. The chief ores are (a) haematite or red kidney ore and magnetite or magnetic iron ore which tend to occur in large masses associated with IGNEOUS or METAMORPHIC ROCKS; (b) the bedded ores of hydrated oxide of iron which include limonite and are usually very impure; (c) siderite; (d) sulphide ores of which iron pyrites and copper pyrites are the chief, although neither is an important source of metal, iron pyrites being more important as a source of sulphur and copper pyrites of copper. The impurities in the ore are important in processing, e.g. phosphoric iron ore needs special treatment, because PHOSPHORUS makes the iron brittle. PIG IRON, STEEL.

Iron Age the era in human development (succeeding the BRONZE AGE) when IRON was smelted and used for tools, utensils and weapons. It probably began among the Hittites c. 1400 BC, reaching southern Europe by c. 1000 BC and Britain by 500 BC.

Iron Curtain a term introduced by Winston Churchill in a speech in 1946 describing the divide between the then USSR and its associated communist states in eastern Europe on the one hand and the countries of western Europe on the other. The USSR, Poland, Czechoslovakia, Hungary, Romania, Bulgaria and East Germany were considered to be within the Iron Curtain, and sometimes Yugoslavia and Albania were included.

iron pan, ironpan HARD PAN.

irradiance the amount of radiant power (RADIANT FLUX) per unit area falling on a surface or object. RADIANCE.

irrigation the action of artificially supplying land with water to help the growth and productivity of plants. In addition to methods long used of bringing water by KAREZ, by CANAL-2 from a river or reservoir, and by BASIN IRRIGATION, more up-to-date methods include closed pipes (preventing loss by evaporation) and sprinklers producing artificial rain. PERENNIAL IRRIGATION.

isallobar a line drawn on a map to join places undergoing equal change in ATMOSPHERIC PRESSURE during a given period, plotted to indicate the development and progress of a PRESSURE SYSTEM.

isanomal, isanomalous line a line drawn on a map joining places with equal difference from the normal or average of any climatic element.

isarithm any line drawn on a map to link places having the same value or quantity.

island a naturally formed piece of land entirely surrounded by water, which is above water at high tide, a small island sometimes being termed an isle, very small an islet. By analogy applied to many other phenomena that are isolated, as an island, e.g. island site, a building site surrounded by roads; or a HEAT ISLAND.

island arc the disposition of an island chain in the form of an ARC, e.g. as in the Pacific ocean. DEEP, OCEANIC TRENCH, PLATE TECTONICS.

iso- (Greek, equal) a prefix (sometimes is- before a vowel) used in very many scientific terms, especially in geography, applying to various lines drawn on a map to link points having similar values or similar quantities, to which the general term applied is usually ISOPLETH.

Some standard isopleths (with the factor of similarity of their points specified) are: **isobar** barometric pressure; **isodapane** transport costs; **isohel** amounts of sunshine; **isohyet** amounts of rainfall; **isoneph** amount of cloudiness; **isoseismal** earthquake activity or intensity; **isoseismic** (CO-SEISMAL) or **isoseist** phase of earthquake wave at the same instant; **isotherm** temperature.

isocline, isoclinal folding a fold which is so intense that the two limbs now incline or dip in the same direction and to an equal amount, i.e. at approximately the same angle. Where such a fold coincides with a ridge or valley, the terms isoclinal ridge or isoclinal valley may be used. Fig 24.

isolated state term applied by Von Thünen to a notional state serving as the basis for his model (VON THUNEN MODEL), a state completely cut off from the rest of the world, dominated by a very large town which serves as the sole market and is situated at the centre of a broad, featureless, uniform plain (an ISOTROPIC SURFACE), bounded by an uncultivated wilderness which prevents communication between the state and the outside world. The plain has a uniformly fertile soil, is not crossed by a navigable river or canal, and the ease of movement over it is uniform. Production and transport costs on the plain are everywhere the same. The farmers provide the large town with agricultural produce in exchange for the manufactured goods produced in the town. They themselves haul their produce to market along a close, dense network of converging roads of equal quality, at a cost directly proportional to the distance covered. All the farmers wish to maximize their profits, so automatically adjust the output of crops to the needs of the central market.

isometric *adj.* having equal measure.

isomorphic *adj.* in general, and in biology **1.** of the same or an analogous form **2.** in botany, applied to ALGAE and some FUNGI that have alternating generations which are vegetatively identical **3.** in chemistry, mineralogy, having shape or structure similar to that of another, usually due to a similarity of composition **4.** in mathematics, applied to two data sets, or two theories, which are precisely equivalent in form and in the nature and product of their operations, the elements of one corresponding with those of the other. GENERAL SYSTEMS THEORY, ISOMORPHISM.

isomorphism the state or quality of being ISOMORPHIC, e.g. specifically in biology, the apparent similarity of individuals of different races or species; and in mathematics, a one-to-one correspondence between data sets.

isopleth **1.** common use in geography, ISO- **2.** a graph showing the frequency of any phenomenon as the function of two variables **3.** a straight line on a graph joining corresponding values of the variables when one of the variables has a constant value.

isostasy, isostatic theory, isostatic adjustment isostasy, the condition of relative equilibrium of the earth's crust, accounted for by the theory that the surface features of the earth have a tendency to reach a condition of gravitational equilibrium. Isostatic theory maintains that where equilibrium exists on the surface of the earth, equal mass must underlie equal surface area. Thus under an elevated plateau there would be rocks of low density, e.g. GRANITE; under ocean basins the rocks would be of high density, e.g. BASALT. The instability of continental margins where high mountains are found close to ocean deeps is explained by underground movement of magma to effect the necessary adjustment (isostatic adjustment). GLACIO-ISOSTASY, ISOSTATIC, PLATE TECTONICS.

isostatic *adj.* **1.** under equal pressure from all sides **2.** of, relating to, or characterized by, ISOSTASY.

isothermal *adj.* **1.** having the same temperature **2.** without change of temperature **3.** relating to or showing a change in pressure or in volume at a constant temperature.

isotope one of two or more forms of an ELEMENT-6, having the same atomic number as the other forms but identified by small differences in atomic weight and, usually, by minute differences (due to mass) in chemical and physical properties. An isotope is named by the mass number with the name or symbol of the element, e.g. carbon 14, or ^{14}C, a radioactive carbon. In natural RADIOACTIVITY one isotope changes very slowly but at a known rate into another that is more stable (HALF-LIFE), thus the proportion of one to the other present in a sample (e.g. of organic remains, or in a geological sample) which is being investigated gives a measure of the age of the material which can be expressed in years (RADIOMETRIC AGE). RADIOCARBON DATING.

isotropic *adj.* showing physical properties or actions equal in all directions. If the properties or actions shown are unequal, they are termed anisotropic.

isotropic surface a notional, unbounded, uniformly flat plain on which population density, purchasing power, transport costs, ACCESSIBILITY in all directions, etc. are kept uniform and unvarying. Christaller used this plain to show the theoretical distribution of CENTRAL PLACES in CENTRAL PLACE THEORY, specifying that CENTRAL GOODS should be bought from the nearest central place, all parts of the plain should be served by a central place (thus the COMPLEMENT-

ARY REGIONS should cover the whole of the plain), there should be minimum movement of consumers, and no central place should make excess profits. In order to fulfil all these specifications the plain has to be divided into hexagonal complementary regions. URBAN HINTERLAND.

isthmus a narrow strip of land, with water on each side, connecting two larger land masses, e.g. two continental land masses, or a mainland and a peninsula.

ITCZ, ITF Intertropical Convergence Zone, Intertropical Front (sometimes termed the equatorial front or equatorial trough), a broad trough of low pressure, a zone rather than a front, defined more sharply over land than over the ocean, where the tropical maritime air masses converge, i.e. where the northeast trade winds and the southeast trade winds meet, broadly in the region of the equator, but moving north and south according to season. The air masses may be almost stagnant, the winds light and variable, hence the old name of belt of calms or DOLDRUMS. The air is very unstable, a factor in the heavy CONVECTION RAIN of the equatorial belt; and in this area shallow, slow-moving DEPRESSIONS-3 develop which may stray from the zone towards the poles, intensify and become fast-moving tropical revolving storms (CYCLONE).

iteration in mathematics, the repeating of an operation on the product of the operation.

J

Janet Joint Academic Network, a computer network that links together most UK colleges and universities. Superjanet is the upgraded Janet; both link into the INTERNET.

jet stream a high altitude, fast-moving air current, a few thousand kilometres in length, a few hundred kilometres wide, a few kilometres in depth, more or less horizontal, flattened, tubular, occurring in the vicinity of the TROPOPAUSE, usually blowing more strongly in winter than in summer. POLAR FRONT JET STREAM, SUBTROPICAL JET STREAM, TROPICAL EASTERLY JET STREAM, WESTERLIES.

jetty a structure built to project from the shore into a sea, lake or river to break currents or waves, to shelter a HARBOUR-1, or to provide a landing stage.

joint a crack or fissure in a rock following a dominant direction along a line of weakness, usually transverse to the BEDDING, and produced by tearing apart under TENSION or shearing under COMPRESSION, but without dislocation (as in a FAULT). In stratified rocks (STRATIFICATION) the joints are usually at right angles to the bedding. If the BEDDING PLANES are well marked a number of parallel joints will divide up the rock bed into more or less regular blocks. There may be two or more sets of joints following different directions, in which case the most strongly marked set constitutes the master joints. A joint coinciding with a STRIKE is termed a strike-joint, with a DIP a dip-joint. EROSION and WEATHERING are helped by the weak surfaces of well-developed joints, which are also useful to the quarryman in stone extraction.

jökulhlaup (Icelandic) a glacial outburst, a sudden, sometimes catastrophic, flood of meltwater flowing from the underside of a glacier or ice sheet as a result of volcanic activity and geothermal heating which melts the ice.

joule j **1.** the unit of energy and work in SI, the work done when the point of application of a force of one NEWTON is displaced in the direction of the force through a distance of one metre **2.** a unit of heat, 1000 calories.

junction of rivers, ACCORDANT JUNCTION. DISCORDANT JUNCTION.

jungle a word brought home by the British from India where as jangala (Sanskrit) and jangal (Hindi and Marathi) it meant waste or uncultivated ground as opposed to cultivated land. Frequently such land was covered with scrub and tangled vegetation, including long grass, and so was the haunt of wild animals. It therefore has no precise meaning (and is best avoided in scientific literature, especially the term 'jungle-forests' applied to equatorial or hot wet forest).

Jurassic *adj.* of or relating to the middle geological period of the Mesozoic era (GEOLOGICAL TIMESCALE), when sediments of CLAYS and SANDS and CORAL REEFS were laid down in shallow seas, dinosaurs were at their peak, birds began to appear, and the flora included ferns and conifers. The rocks include CLAYS, LIMESTONES, SANDSTONES.

juvenile relief a landscape with steep-sided valleys characteristic of the early stages of a CYCLE OF EROSION.

juvenile water magmatic water, intratelluric water, water from great depths of the earth reaching the earth's surface for the first time, as a result of volcanic activity, i.e. not the METEORIC WATER which is already present in the ATMOSPHERE-1 and HYDROSPHERE.

K

K 1. KELVIN, the basic SI unit of temperature **2.** usually in italic, the symbol for a constant, e.g. in statistics **3.** K-VALUE.

kame an imprecise, unspecific term applied to any ridge or mound of poorly sorted water-laid materials (glacial sands and gravels) associated with former ice fronts. ESKER, MORAINE.

kame-and-kettle country an undulating landscape consisting of kame moraines and shallow depressions. KETTLE.

kame terrace a terrace formed of sand and gravel deposited by a stream of MELTWATER in the depressions between a GLACIER and the sides of its trough.

kaolin china clay, a fine, white CLAY occurring especially in pockets in GRANITE masses, resulting from the decomposition of FELDSPARS by HYDROLYSIS and by ascending heated gases (PNEUMATOLYSIS) and vapours (mainly CARBON DIOXIDE and superheated steam) from a deep-seated MAGMA. It is used in making ceramic ware (china, porcelain), paper, pharmaceuticals, rubber and cosmetics. The term is derived from the name of the mountain in China (Kaoling) from which it seems originally to have been obtained. SUPERHEATING.

karaburan black buran, a strong northeast wind laden with dust and sand blowing in daytime in the arid Tarim basin in central Asia, darkening the sky, changing river courses by depositing sand, and carrying dust particles great distances to settle as LOESS.

kārez term applied in Baluchistan to the qanat of Iran and the foggara of north Africa: an almost horizontal underground, hand-engineered irrigation channel or tunnel dug from the arid plains to tap water at the foot of a nearby hill range, the water flowing through by gravity.

karoo, karroo 1. a plateau in southern Africa between the Swartberge and Nuweveldberge, covered with semi-desert vegetation of small shrubs **2.** the vegetation in this region, extending into the Little Karoo, south of the Swartberge, and the Northern Karoo, Upper Karoo, or Karroid plateau, north of the Nuweveld range.

karre (German, usually in pl., karren) a channel or furrow varying in depth from a few millimetres to over a metre, and separated from others by ridges, caused by solution on limestone surfaces. CLINT, KARST, LAPIE.

karrenfeld a surface cut with and dominated by karren (KARRE).

karst originally the barren limestone plateau of Istria, between Carniola and the Adriatic, where nearly all the natural drainage is underground and where there are bare, limestone ridges, caverns, sinks and underground drainage caused by rainwater which, charged with CARBON DIOXIDE from the atmosphere, dissolves the CALCIUM CARBONATE in the porous

LIMESTONE, producing an uneven landscape. Karst is now applied to any area of similar limestone or dolomite country. COCKPIT-2, DOLOMITE.

katabatic wind a drainage wind, a cold wind that blows down a valley or slope, especially at night, when dense cold air, cooled by radiation at higher levels, flows downhill by gravity, behaving much like a stream of water. ANABATIC, FROST POCKET.

katafront a COLD FRONT in which warm air flows down over a wedge of cold air. FRONT.

kegelkarst (German) the term now applied internationally to several types of tropical humid KARST, sometimes termed cone karst in translation. COCKPIT-2.

kelvin K, the basic SI unit of temperature, defined from the triple point of water (the point at which water, ice and water vapour are in equilibrium), valued at 273.16K, ice point (ABSOLUTE ZERO) being 273.15K. The value of the degree in the KELVIN SCALE is the same as that of the degree in CENTRIGRADE. A temperature expressed in K (the symbol °K and the term degree kelvin have been superseded by K or kelvin) is equal to the temperature in °C less 273.15°C (commonly rounded to 273°C); and to express °C in K it is necessary only to add 273°C to the Centigrade value (e.g. −3°C = 270K).

Kelvin (K) scale a temperature scale with 1K (one KELVIN) equal to 1°C (CELSIUS, CENTIGRADE) but with an ABSOLUTE ZERO temperature calculated to be −273.15°c or −459.4°F (FAHRENHEIT). The advantage of the Kelvin scale is that it has no negative quantities. It is thus especially valuable if one is dealing with very low temperatures.

Kelvin wave a tidal system in an approximately rectangular sea area, the tidal range being greater on the right of the direction of a PROGRESSIVE WAVE, decreasing on the left, e.g. in the English Channel the tidal range on the south coast of England is less than that on the north coast of France. AMPHIDROMIC POINT, AMPHIDROMIC SYSTEM, OSCILLATORY WAVE THEORY OF TIDES.

kettle, kettle-hole, kettle-lake kettle seems originally to have been a local term applied to a pothole in a river, then, as a 'giant's kettle', to a pothole formed by whirling stones in a stream under a glacier. Later it came to be applied to a circular hollow in a stretch of glacial sands, gravels and clays, caused by the former presence of a great detached block of ice which had eventually melted. Such hollows became filled with water to form kettle-lakes; and the drifts in which they occurred became known as kettle-drift or kettle-moraine.

kettle-drift a mound or ridge of gravelly DRIFT-1 formed by water at or beyond the margin of the ice.

kettle-moraine a TERMINAL MORAINE pitted with many kettle-holes. KETTLE.

key, kay, cay a low sand and coral island, sandbank or reef, lying a little above high tide, dry at low tide, a term used especially in the West Indies and Florida, sometimes appearing in place-names, e.g. Key West.

Keynesianism the economic theory of John Maynard Keynes, 1883–1946, in part of which he states that a condition of unemployment may continue for long periods, even indefinitely, unless a government steps in to remedy it by spending to supplement a deficient private sector demand. AGGREGATE DEMAND, MONETARISM.

key village, king village a minor centre with facilities (e.g. a primary school, a village hall) to serve even smaller villages and hamlets, but reliant on major centres for other facilities.

khamsin a hot, dry, often dust-laden southerly wind known also as ghibli, samiel or LEVECHE, which blows intermittently for some fifty days in March, April, May from the deserts of the south across Egypt and the southeast Mediterranean, commonly after a HEAT WAVE, and often becoming humid in passing over the Mediterranean sea. The SIROCCO is similar, but warmer.

kibbutz, kibutz a form of COLLECTIVE or communal rural settlement in modern Israel, with egalitarian communal ownership of the land and collective economic and social organization. Originally the kibbutzim were devoted solely to farming, but many now produce manufactures. LAND TENURE.

kidney iron ore HAEMATITE.

kilo k, prefix, a thousand, attached to SI units to denote the unit $\times 10^3$. CENTI-, HECTO-, MILLI-.

kilogram, kilogramme kg, the basic SI unit of weight, 1000 GRAMS (2.2046 lb), being defined as the mass of a standard piece of platinum-iridium alloy kept in the Bureau International des Poids et Mesures at Sèvres, near Paris.

kilometre km, an SI unit of length, 1000 METRES (0.62 mile, or 3280.84 feet). In measures of area one square kilometre equals 100 HECTARES each of 10 000 square metres.

kimberlite a DIAMOND-bearing ULTRABASIC IGNEOUS ROCK containing MICA and other silicates, filling volcanic PIPES-3 in the Kimberley district of South Africa, termed hardebank at depth, changing near the surface to the softer blue ground which oxidizes at the surface to yellow ground.

kinetic energy the energy of a moving mass associated with its speed and equal to half the product of the mass and the square of its velocity. THERMAL CONDUCTION.

knickpoint, nickpoint a break of slope, particularly one in the long PROFILE of a river valley, occurring when a relative lowering of the sea level (NEGATIVE MOVEMENT) compels the river to regrade its course to the new sea level. In doing this the river makes a new curve which, by headward EROSION, cuts across an earlier one. The junction is marked by a break of slope (the knickpoint, or rejuvenation head) which, owing to continued erosion, moves progressively upstream. REJUVENATION.

knoll, knowe, know, knowle a more or less rounded small hill.

knot 1. a unit of speed, one nautical mile (standardized at 1.852 km) per hour, derived from the original use of pieces of knotted string fastened to the logline trailed from sailing vessels, the number of knots being measured against a period of time indicated by a sand-glass. It is tautological to refer to so many 'knots per hour'. MILE **2.** a complex of mountains, especially one where several ranges meet and the arrangement is irregular, e.g. the Pamir Knot.

kolkhoz (Russian contraction of kollektivnoe khoziaistvo) a collective farm in the former USSR. COLLECTIVE FARMING, SOVKHOZ.

Kondratieff cycle a theory of cyclic growth, boom, bust, spanning some fifty to sixty years and occurring in capitalist economies, postulated by Kondratieff (Russian economist) in 1925. He identified an innovation in new technology developed in times of depression as a major trigger for each new cycle, e.g. the first (eighteenth century to 1842) being steam power; the second (1842–97), railways and Bessemer steel; the third (1898–1939), electricity, chemicals and motor vehicles. Other economists have added a fourth (1946–73), air transport, electrical, electronic and oil-based industries; and, at present, a fifth, microprocessors and genetic engineering.

Köppen's climatic classification a classification devised by Vladimir Peter Köppen, 1846–1940, who based his climatic classification on the climatic needs of certain types of vegetation, and identified five major groups, A to E, to which he added H, the mountain zone. The major groups are: A, tropical zone (12 months with a temperature exceeding 20°C); B, subtropical zone (4 to 11 months with temperature exceeding 20°C, and 1 to 8 months ranging between 10°C and 20°C); C, temperate zone (4 to 12 months with temperature between 10°C and 20°C); D, cold zone (1 to 4 months with temperature between 10°C and 20°C, and 8 to 11 months below 10°C); E, polar zone (12 months with temperature below 10°C) (10°C = 50°F; 20°C = 68°F). These major climatic groups were subdivided to take account of refinements of rainfall and temperature characteristics, expressed by lower case letters. THORNTHWAITE'S CLIMATIC CLASSIFICATION.

kum any of the sandy deserts of central Asia, equivalent to the Saharan ERG.

kurtosis in statistics, a shape characteristic of a FREQUENCY DISTRIBUTION that reflects the pointedness of the peak and the length of the tails. FREQUENCY CURVE.

K-value in CENTRAL PLACE THEORY, a value given to a CENTRAL PLACE, to describe its place in, and the nature of, the hierarchy (CENTRAL PLACE HIERARCHY). The K-value expresses the total number of central places at a certain level in the central place hierarchy served by a central place at the next highest order in the system. The value includes the higher order place itself, e.g. in a K-3 hierarchy, the higher order place serves two adjacent lower order places (i.e. two places and the place itself). ADMINISTRATIVE PRINCIPLE, MARKETING PRINCIPLE, TRAFFIC (TRANSPORTATION) PRINCIPLE.

kyle (Scottish) a narrow channel or strait between an island and the mainland, or between two islands.

L

labelling a social process by which individuals or groups classify and categorize social behaviour in others, e.g. a particular group or area may be reputed to have socially undesirable characteristics (e.g. being characterized by criminality) and is stigmatized, given a disparaging name. People who are not 'socially undesirable' but who live in such a stigmatized area may have difficulty in finding work or obtaining credit. It has been suggested that labelling affects the behaviour of the labelled, in that people who are labelled come to see themselves in terms of the label, and behave accordingly.

labour **1.** workforce **2.** in economics, work as a factor of production. FACTORS OF PRODUCTION.

labour-extensive *adj.* needing a very small work force to achieve a very high output. CAPITAL-INTENSIVE, LABOUR-INTENSIVE.

labour-intensive *adj.* needing the efforts of a large work force for increased productivity or higher earnings, as opposed to CAPITAL-INTENSIVE. LABOUR-EXTENSIVE.

labour power in Marxism, the ability to work, or the commodity that workers sell, the exchange value being determined by the socially necessary labour (LABOUR THEORY OF VALUE-2) needed for subsistence, i.e. the cost of production and reproduction of labour itself. CONSTANT CAPITAL, MEANS OF PRODUCTION.

labour theory of value **1.** in CLASSICAL ECONOMIC THEORY, a theory proclaimed by Adam Smith and David Ricardo, that any two products will exchange one with another in proportion to the amounts of labour needed to make them, i.e. that VALUE-1 is the product of the expenditure of labour. The part played by capital in production is explained by assuming that the amount of capital used per unit of labour in making every product is constant, or by treating capital equipment as stored-up labour. **2.** in Marxian economics the theory is similar, one of the main tenets of Marxism being that value can be created only by the expenditure of human labour. Thus the price of a commodity should be the amount of labour time needed to produce it under normal conditions. The labour so required is termed socially necessary labour. CONSTANT CAPITAL, LABOUR POWER, VARIABLE CAPITAL **3.** in NEOCLASSICAL ECONOMICS it is asserted that capital and land as well as labour contribute to the production process, so that they also are entitled to a return and should be reflected in the price of a commodity.

laccolith a mass of IGNEOUS ROCK intruded along the BEDDING PLANES of SEDIMENTARY ROCKS, like a SILL but swelling out to form a lens-shaped mass, the flat base being concordant with the strata into which it is intruded, the upper surface swelling out in the shape of a dome causing the overlying strata to arch over it. Sometimes a laccolith is more complex, with several masses one above the other, a section through the whole looking like a cedar tree with spreading branches; hence the term cedar-tree laccolith. PHACOLITH. Fig 21.

lacustrine *adj.* of or pertaining to a LAKE, hence applied, e.g. to deposits laid down in a lake; or to terraces on lake margins left when the area of the lake diminishes.

lacustrine delta a DELTA spreading into a LAKE, built by a stream flowing into the lake.

ladang (Indonesia) SHIFTING CULTIVATION in the Malay archipelago and, particulary, in Indonesia.

LAFTA Latin American Free Trade Association, headquarters Montevideo, an organization of some South American states, established in February 1961 by Argentina, Brazil, Chile, Mexico, Paraguay, Peru and Uruguay, with Colombia and Ecuador (October 1961) and Venezuela (September 1966) associated by treaty, to promote economic cooperation. It has two subgroups: the Andean Group, established May 1969, comprising Bolivia, Chile, Colombia, Ecuador and Peru (Venezuela, expressing interest, has not yet signed); and the River Plate Association, comprising Argentina, Brazil, Paraguay and Uruguay.

lag fault a low-angled FAULT resulting from movement in a series of rocks, those nearer the top moving less than (lagging behind) those nearer the bottom of the series.

lagoon 1. a shallow area of salt or brackish coastal water completely or partly separated from the open sea by some more or less effective obstacle, such as a low sandbank (BARRIER BEACH, BARRIER ISLAND) or a CORAL REEF **2.** the sheet of water enclosed in an ATOLL.

lahar (Indonesia) a flow of mud arising from volcanic activity, the fine-grained volcanic material being impregnated with water derived from heavy rainfall during the eruption; or from the sudden emptying of a CRATER LAKE; or from melted snow (e.g. during the eruption of a snow-capped volcano).

laissez-faire, laisser-faire (French, let act, i.e. let things alone) a term originated by the physiocrats, the philosophy or the practice of the avoidance of planning, particularly, in economic affairs, as expressed in the avoidance of government control. The doctrine is based on the theory that general good and harmony will ensue if each individual is allowed to work for his or her economic advantage in a freely competitive economic system, a theory supported by Adam Smith, David Ricardo and others. CLASSICAL ECONOMIC THEORY.

lake a broad, general term applied to an accumulation of water lying in a depression in the earth's surface, normally to a sheet of water of considerable size, but sometimes to a small artificial ornamental feature, e.g. in a PARK-2,3; if very large, natural and saline, the term sea is commonly used (e.g. Caspian Sea, Sea of Aral, Dead Sea), if very small and natural, POND or POOL are used. An inflowing and/or outflowing river may or may not be present; and a lake may not necessarily be a permanent body of water (e.g. Lake Eyre, PLAYA). The CLASSIFICATION OF LAKES is usually based on the origin of the depression which they occupy (e.g. BARRIER LAKE, CRATER LAKE, GLACIAL LAKE, TROUGH LAKE). The term is also applied to a fairly large accumulation of ASPHALT, a viscous substance, lying in a depression in the earth's surface. ICE-DAM LAKE, LACUSTRINE, PATERNOSTER LAKES.

lake-dwelling a dwelling built on piles driven into a marsh or the bed of a shallow lake, common in some tropical areas today, and characteristic of certain periods of NEOLITHIC times in Switzerland, France and central Europe, to which the terms Lacustrine period or Lacustrine civilization are applied.

lamina pl. laminae, any thin plate, scale or layer. LAMINATION.

laminar flow **1.** non-turbulent flow of a FLUID closely following the streamlined surface of a solid object in the fluid (TURBULENCE) so that adjoining levels do not mix. **2.** the movement of a GLACIER along a slope caused by the thrust of the weight of solid ice in the upper part, in some cases so powerful that the SNOUT of the glacier moves uphill.

lamination in geology, STRATIFICATION on the finest scale, the usual definition being that each layer should be under 1 cm (0.39 in) in thickness, typically occurring in fine-grained SANDSTONES and SHALES.

land breeze a cool breeze (BEAUFORT SCALE) that blows at night from the land to the sea in coastal regions (or from the land surrounding a large lake to the lake), due to the differential heating of land and water. At night air cooled relatively quickly by radiation over the land descends, the atmospheric pressure over the land is slightly higher than that over the water, and the air over the land flows away from the land towards the warmer water. Land breezes are particularly likely to occur in equatorial latitudes and in other areas where temperature changes are regular in calm, settled weather. SEA BREEZE. Fig 26.

land bridge **1.** in geology, a land link between continents **2.** an ancient route used by migrating land animals **3.** an overland route lying between and connecting two sea routes.

land capability the potential usefulness of land for agriculture (including forestry) based on environmental factors, e.g. soil and climatic factors. LAND CLASSIFICATION.

land classification a systematic classification of land, usually devised for a specific purpose, e.g. as a basis for land use planning and the conservation of land

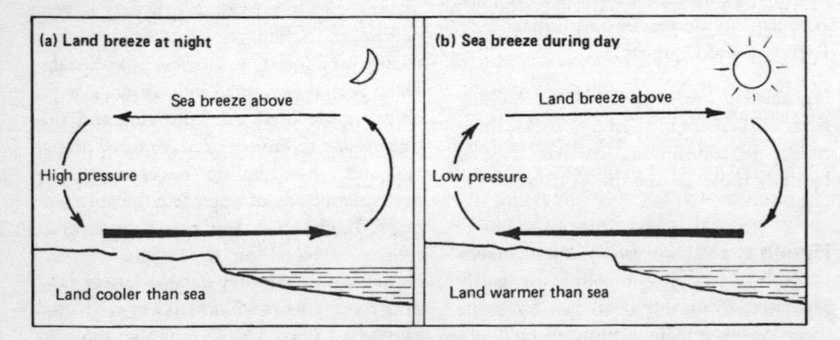

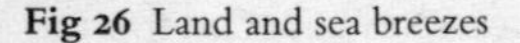
Fig 26 Land and sea breezes

resources, designed in most cases to indicate the quality, the relative fertility of the land for different types of farming or for some other land use. In some cases 'land capability' or 'potential land use' classes are favoured; but potential is a matter of judgement and may be radically changed by research findings and technological progress.

land degradation DESERTIFICATION.

landform the shape, form, nature of a specific physical feature of the earth's surface (e.g. a hill, a plateau) produced by the natural processes of DENUDATION and DEPOSITION (including WEATHERING, GLACIATION-1 etc.) and by TECTONIC processes. GEOMORPHOLOGY.

land ice ice formed from fresh water lying inland.

landlocked *adj.* applied to an area (particularly a state) which lacks a sea coast and thus does not have direct access to the sea.

landmass a very large area of continental crust (PLATE TECTONICS) lying above sea-level.

land reclamation a term applied broadly to cover not only the winning back, the recovering, of land that has been spoilt for agricultural use, but also the improvement of land so that it can be made useful, or more useful, for economic (including agricultural) or social purposes. Some of the types of land and the techniques employed are: land under water or waterlogged (by drainage or by the filling-in of a water-filled depression); arid land (by irrigation and, if saline, by chemical treatment); unstable slopes and loose soil (by planting of vegetation cover); land subject to water erosion (by planting of vegetation cover, by terracing, by embankment); land subject to wind erosion (by planting of vegetation cover including shelter belts of trees); land spoiled by quarrying, mining or industrial activity (by filling-in of quarries and pits, levelling, planting of spoil tips, restoration of soil profile); land impregnated with salt or industrial effluent (by chemical treatment); land covered with undesirable trees and/or scrub (by clearance).

land reform changes in a system of LAND TENURE, commonly brought about by government intervention and usually aimed at removing what is considered to be unfairness in the system, or at improving agricultural efficiency, etc., e.g. by the breaking up of big estates and the redistribution of the land as small holdings to farmers who become owner-occupiers; or by the consolidation of fragmented holdings (FRAGMENTATION-2) to form larger, more efficient, farming units. AGRARIAN REFORM.

land rent the concept of economic rent developed by J. H. Von Thünen for his model (VON THUNEN MODEL). He defined it as that part of the total (gross) product of land which remains as a surplus after the deduction of all costs, including interest on invested capital, i.e. the net profit earned by a farmer from his/her chosen productive system, opportunity costs (ECONOMIC RENT) being ignored. The net profit is governed by production costs and market price per unit of product, transport rate per distance unit for each product, the yield per unit of land, and the distance from the point of origin of the product to the market centre. RENT.

land rotation a regular system of land management in which land is cultivated for a few years and then allowed to rest, perhaps for a considerable period, usually by simply allowing scrub or bush to grow up over it. In due course it is cleared and cultivated again, the farms on which, or settlements from which, cultivation takes place being fixed. This is a type of SHIFTING CULTIVATION common in Africa. Land rotation should not be confused with the ROTATION OF CROPS.

LANDSAT one of the US satellites orbiting the earth without a crew, at a height of over 12 km (7.5 mi) and by REMOTE SENSING surveying the natural resources, land use, environmental conditions of the earth (e.g. crop disease, water pollution etc.). LANDSAT 1 was launched in 1972, LANDSAT 2 in 1975, LANDSAT 3 in 1978 followed by LANDSAT 4 and 5; LANDSAT 6 (launched Oct. 1993) was lost in space. LANDSAT was designed to pass round the earth in a sun-synchronous, near-polar orbit, completing 14 orbits a day, and achieving global coverage every 18 days, recording images by means of television-like cameras and four-channel multi-spectral line-scanner devices, each system operating in various bands of the green, red and two near-infra-red wavebands. NASA, SEASAT 1, SKYLAB, SPACE SHUTTLE.

landscape (Dutch landschap, the representation in painting of inland natural scenery) **1.** still used in the Dutch sense, e.g. a landscape by Constable **2.** the scenery itself **3.** an area of the earth's surface characterized by a certain type of scenery, comprising a distinct association of physical and cultural forms. From this came the separation of NATURAL LANDSCAPE from CULTURAL LANDSCAPE.

landslide, landslip 1. the sliding down under the force of gravity of a mass of land on a mountain or hillside **2.** the part which has so fallen. ENVIRONMENTAL HAZARD, MASS MOVEMENT, SLIP-2.

land tenure the rules and regulations governing the rights of holding, disposing and using land, i.e. the conditions on which land is held (TENURE), varying with the social and economic organization of the country concerned. In England and Wales the one who holds the land in perpetuity is known as the freeholder. The freeholder may or may not lease the land to another, the leaseholder (the tenant) for a certain period, usually on payment of rent. The leaseholder may, unless the terms of the lease prevent it, grant an underlease.

In other systems the tenant may, instead of paying money rent, pay for the right to use land by providing labour on the land which the owner has kept for personal use, or by giving the owner a share of the crop produced on his/her (the tenant's) holding (SHARE CROPPING). In collectivist systems the land is owned by a collective interest, e.g. the state (COLLECTIVE FARMING, STATE FARMING), or a small community (KIBBUTZ) etc. In SHIFTING CULTIVATION an individual farmer or a group of farmers establishes a right merely by using the land. LAND REFORM.

land use, land utilization, land utilisation terms commonly applied interchangeably to the use made by human beings of the surface of the land. A land use survey, though literally surveying and mapping the use of the land surface, usually also includes in sparsely populated countries the natural and semi-natural vegetation.

land use planning the demarcation of land for specific uses, usually (but not necessarily) over an extensive area, based on environmental, social and economic criteria, which takes into account present and possible future needs. PLANNED ECONOMY, PLANNING, REGIONAL PLANNING.

Land Utilisation Survey of Britain the first survey, started in 1930, fieldwork carried out mainly in 1931–3, was a voluntary organization established by L. Dudley Stamp with E. C. Willatts as organizing secretary. With the help of volunteers from educational institutions, the use of every acre of land in England, Wales, Scotland, the Isle of Man and the Channel Islands (Northern Ireland was separately surveyed later) was recorded on over 15 000 6-inch (1:10 560) field sheets, the results being reduced to the scale of 1 in to 1 mi and published in 150 sheets for all of England and Wales and for the more populous parts of Scotland. The published maps were accompanied by county monographs under the title *The Land of Britain*. The whole work is summarized in L. Dudley Stamp, *The Land of Britain: Its Use and Misuse*, Longman.

Basically the land use was classified in six categories, with some subdivisions: arable (brown), permanent pasture (light green), rough pasture (yellow), woodland (dark green, with subdivisions), gardens, orchards, nurseries (purple, with subdivisions), and land agriculturally unproductive (red). The picture shows Britain at a time of agricultural depression.

The Second Land Utilisation Survey of Britain, far more detailed than the original Survey, was started in 1960 under the direction of Alice Coleman of King's College, London. The 52 categories identified in the fieldwork were recorded on the scale of 1:10 000, the results being published on the scale of 1:25 000.

The third survey (Land Use – UK) was led by Rex Walford for the Geographical Association. Carried out by schoolchildren it is based on representative samples: 500 in rural, 500 in urban one-kilometre squares, plus 600 random squares. The results became available in 1997.

lapié (French) an exposed limestone surface in a KARST region, with etching, pitting, grooving, fluting, caused by CARBONATION, mainly, some authors maintain, from small free-flowing streams; equivalent to German karren (KARRE). There are also very small forms, micro-lapiés, German rillensteine, perhaps best termed rock-rills. CLINT.

lapse rate the rate of decrease of air temperature normally occurring with height (vertical temperature gradient), but varying with time and place. The average lapse rate in the atmosphere, termed ENVIRONMENTAL LAPSE RATE, is 0.6°C per 100 m (about 3.5°F per 1000 ft, or sometimes stated as 1°F per 300 ft). This continues up to the TROPOPAUSE unless an INVERSION OF TEMPERATURE occurs, causing an increase of temperature with height (termed an inverted lapse rate and indicated by a minus sign). ADIABATIC LAPSE RATE, DRY ADIABATIC LAPSE RATE, ENVIRONMENTAL LAPSE RATE, SATURATED ADIABATIC LAPSE RATE.

latent heat 1. the amount of heat energy (thermal energy) needed to bring about an isothermal chemical or physical change of a body without making it hotter, e.g. from a SOLID to a LIQUID (the latent heat of FUSION) or from a liquid to a GAS-1 (the latent heat of EVAPORATION or vaporization) or from a solid to a gaseous state (the latent heat of SUBLIMATION) **2.** the thermal energy released in such a process. SENSIBLE TEMPERATURE.

lateral dune a small DUNE lying beside a major dune within a dune pattern caused by an obstacle in a desert.

lateral erosion the wearing away of its banks by a stream. BANK CAVING, MEANDER.

lateral moraine the rock debris from valley slopes that lies on the surface of a GLACIER, making a low ridge along each side. It may form an embankment along the valley wall as the glacier melts. MORAINE.

laterite a subsoil product of WEATHERING in humid tropical areas (HUMID TROPICALITY) where there are alternating wet and dry seasons which lead to the formation of a mottled red-yellow and grey mass, sufficiently soft to be cut out with a spade but hardening on exposure to the atmosphere. In humid tropical conditions the soft grey clayey or sandy matter is leached of SILICA and ALKALI-1, leaving a concentration of sesquioxides of ALUMINIUM and IRON. When cut and exposed to the air a sponge-like red rock is formed, hard enough to be used for building, especially as foundations for light structures, for paths or secondary roads. The process of formation is called laterization. Sometimes the rock is sufficiently rich in iron to be usable as an iron ore. In other cases it is rich in alumina and grades into BAUXITE, the chief ore from which ALUMINIUM is extracted.

laterite soil, lateritic soil a ZONAL SOIL formed on LATERITE, porous and leached, and of little agricultural value.

latex the milky fluid exuding from the cut surface of some flowering plants and trees, coagulating on exposure to the air. Some of these fluids are commercially useful, providing, e.g., the raw material of RUBBER from *Hevea* species, or of chewing gum (chicle gum) from the sapodilla tree.

latifundia (pl. of Latin latifundium, a large estate) originally large landed properties in South America cultivated by peons (agricultural labourers) for the Spanish Crown, now applied to the extensively farmed large estates or ranches in Spain and South America in contrast to the intensively farmed HUERTAS. In some cases very small holdings on the estate are leased to tenants (LAND TENURE), the rest of the land being farmed by the landlord, who employs day labourers. The comparable Italian term is latifondo, but this is applied to an agricultural area with extensive cereal cultivation and grazing which includes large estates and peasant holdings. EXTENSIVE AGRICULTURE, FARMING, FAZENDA, INTENSIVE AGRICULTURE.

Latin America those countries of the NEW WORLD which were discovered, explored, or conquered by the Spaniards (or the Portuguese in the case of Brazil), i.e. that part of the New World where Spanish is spoken together with Portuguese-speaking Brazil, comprising the whole of mainland South America (except Guyana, French Guiana, Surinam), all the countries of central America (except Belize), as well as Mexico and the islands of the West Indies, Cuba and Dominica, where Spanish is spoken.

Latin American Free Trade Association LAFTA.

latitude the angular distance of any point on the earth's surface north or south of the equator, as measured from the centre of the earth, in degrees, minutes and seconds. The equator itself is 0°, the NORTH POLE is 90°N, the SOUTH POLE is 90°S. Low latitudes are broadly those between the TROPIC OF CANCER (23°30'N) and the TROPIC OF CAPRICORN (23°30'S). Midlatitudes extend from the Tropics to the ARCTIC CIRCLE (66°30'N) and ANT-

ARCTIC CIRCLE (66°30'S). High latitudes lie within the Arctic and Antarctic circles, i.e. from the Arctic circle to the North Pole, from the Antarctic circle to the South Pole. PARALLEL OF LATITUDE. Fig 27.

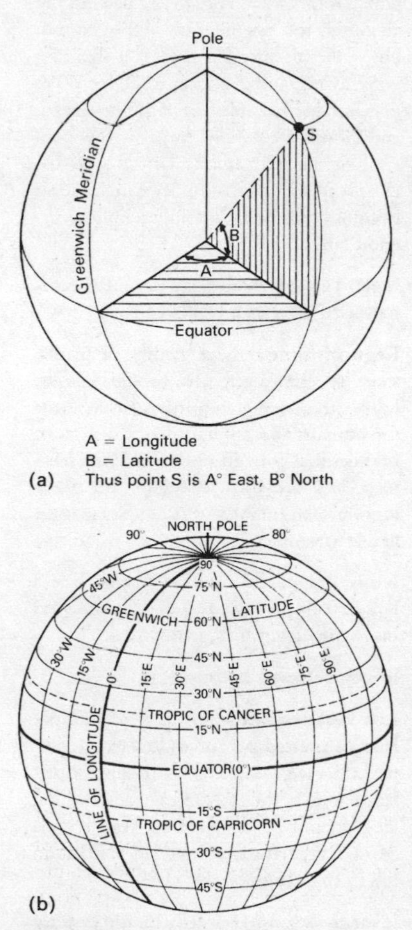

Fig 27 Latitude and longitude
(a) the basis
(b) some parallels of latitude, and some meridians

latosol a soil with thin A_0 and A_1 layers over reddish or red deeply weathered material which is low in silica and high in sesquioxides. A term not widely accepted outside the USA. LATERITE.

Laurasia the northern part of the great PRECAMBRIAN landmass, PANGAEA. Current scientific evidence suggests that it included today's North America, Europe, Asia (the Indian subcontinent excluded) and the Arctic. CONTINENTAL DRIFT, GONDWANALAND, PLATE TECTONICS, TETHYS OCEAN.

lava molten rock (MAGMA) that issues from a volcanic VENT or FISSURE to the surface of the ground, where it solidifies. Chemically it is divided into acid (with excess silica so that some crystallizes out as quartz), intermediate, basic, ultrabasic (ACID LAVA, BASIC LAVA, INTERMEDIATE ROCK, ULTRABASIC ROCK). Some lavas cool quickly and are glassy, others more slowly, giving crystals of minerals time to grow, so that they are CRYSTALLINE. AA, BASALTIC LAVA, PILLOW LAVA, VOLCANO, VOLCANIC ROCK. Fig 21.

law of superposition the assumption that in a sequence of stratified or sedimentary rocks (and in extrusive igneous rocks) the oldest bed is at the bottom, unless the sequence has been reversed by intense folding, faulting or other disturbance.

law of unequal slopes the assumption that if the opposing slopes of a ridge are dissimilar, one being steep, the other gently sloping, the steep one will be eroded more quickly than the other, so that the ridge retreats on that side, e.g. as in an ESCARPMENT.

LDC less developed country. HDC, MDC, NIC, UNDERDEVELOPMENT.

leaching the process whereby percolating

water removes materials from the upper layers of a rock, soil or ore, and carries them away in SOLUTION or SUSPENSION. It brings about the secondary enrichment of ores and POROSITY in LIMESTONE, and is especially important in the formation of soils in that it removes soluble salts from the A HORIZON to the zone of accumulation, the B HORIZON.

lead a blue-white, tarnishing to dark grey, soft, dense, metallic ELEMENT-6, not occurring native in nature, most being obtained from an ore in its sulphide form, galena, which often occurs with zinc sulphide in veins in igneous rocks or as irregular masses in some limestones; also in an oxidized form, cerussite. It is used in pipework (a declining use), protective coverings and linings, or as a shield against radioactivity. Some compounds are used in glass making and other manufactures, in accumulators; and as pigments.

leasehold, leasehold ownership, leaseholder control or ownership of rights to land or buildings over a certain period. LAND TENURE.

least cost location the siting of industry in a place where the costs of transport (assessed on weight and distance) and labour, and the advantages of industrial AGGLOMERATION-4 or DEGLOMERATION are most favourable in economic terms. COST SURFACE, WEBERIAN ANALYSIS.

least squares method REGRESSION ANALYSIS.

lee, lee side, leeward the sheltered side, i.e. the side opposite to that against which the wind blows (WINDWARD) or oncoming ice impinges (STOSS END); but see LEE-SHORE.

lee-shore the shore towards which the wind is blowing, thus dangerous to shipping. LEE.

lee wave, lee-wave a wave formation in an air-stream caused by a relief barrier which forces air to rise. In slight wind the air closely follows the form of the ground; but if the air speed is greater a standing wave forms; and if greater still a lee wave develops in the stable part of the air-stream and clouds frequently form in the cool crest of the lee wave (BANNER CLOUD, LENTICULAR CLOUD) under which the air becomes turbulent, a revolving phenomenon termed a rotor.

legend an explanation, a key to, the symbols used on a map or diagram.

Leguminosae a large family of plants, many species (e.g. beans, bersim, clover, lubia, peas) being important food plants for humans and animals. Most bear root nodules that contain nitrogen-fixing bacteria, and are thus useful in soil management. NITROGEN CYCLE, NITROGEN FIXATION.

lenticular cloud a cloud shaped like a lens, usually produced by an eddying wind in mountains or hills. LEE WAVE.

less developed country LDC.

less favoured land an area of infertile land of limited potential, low economic performance, and a low, dwindling population, one of the categories of agricultural land identified under EEC regulations. Most of the HILL FARMING land of Britain lies in this category.

levante (Spanish) a usually humid easterly wind blowing over the western Mediterranean area, taking its name from the direction from which it blows. It is due to a DEPRESSION-3 in the region, which sometimes causes heavy rainfall, and particularly affects southeast Spain, the Balearic islands, Gibraltar and northern Algeria, especially between July and October. When moderate it causes a BANNER CLOUD to develop from the summit of the Rock of Gibraltar, when strong it produces dangerous currents and eddies in the sea on the LEE side of the Rock. When especially stormy it is termed llevantades. SOLANO is an alternative term.

leveche (Spanish) a hot, dry wind, sometimes dust-laden, originating in the Sahara and blowing towards Spain, due to the eastward movement of a DEPRESSION-3 in the western Mediterranean area. KHAMSIN, SIROCCO.

levee, levée a natural or artificial embankment of a river which confines the river within its channel and hinders or prevents flooding, applied especially to those of the lower Mississippi.

level a nearly horizontal tract of land, unbroken by hills and valleys, applied specifically to certain tracts of such land, e.g. the Bedford Level in Fenland, England. There, in a famous experiment, three stakes were driven into the ground so as to stand exactly the same height above a water surface, by which the curvature of the earth could be seen and measured.

levelling in surveying, the process of establishing the difference in height between successive pairs of points by means of sighting through various instruments that incorporate a telescope with spirit level, graduated measuring rod, etc.

level of living in WELFARE GEOGRAPHY, the level of well-being (SOCIAL WELL-BEING) of a group of people in a particular place at a particular time. QUALITY OF LIFE, STANDARD OF LIVING.

ley, lay (ley is the usual spelling) arable land put down to grass and/or clover for a period of years, a short ley applying to two to four years, a long ley to longer periods, even up to twenty years. The character of the ley may be indicated, e.g. grass-ley, clover-ley, etc.

ley farming a system of farming in which grass-leys or clover-leys are an essential part of the land management.

liana, liane a climbing woody plant, with roots in the ground, characteristic of tropical forests.

lichen a group of slow-growing dual organisms formed from the symbiotic (SYMBIOSIS) association of a fungus and a green or blue-green alga (ALGAE), size variable, flat and leaflike or upright and branched in form. Lichen are primary colonizers of bare areas, the dominant flora in mountainous and Arctic regions where few other organisms can survive; but they occur also elsewhere on tree trunks, walls, exposed rock surfaces. A valuable food for animals (e.g. reindeer) in Arctic regions, they are also a source of dyes (e.g. in providing the dye litmus, for litmus paper, which is turned red by the application of an acid or blue by an alkali). LICHENOMETRY.

lichenometry an imprecise method of measuring the passage of time (DATING), e.g. the length of time a stone has been lying in situ in a MORAINE, based on the rate of growth of LICHEN (the assumption being that the diameter of the largest lichen growing on the surface under investigation is proportional to the length of time that the surface has been exposed to colonization and growth). It is not a completely satisfactory method because the rate of lichen growth is greatly affected by climatic and other factors, e.g. atmospheric POLLUTION-1.

lido (Italian, a barrier of sand or silt in front of a lagoon) one of the best examples of such a barrier (BARRIER BEACH) is that protecting the lagoon of Venice. This has been converted into a bathing beach and resort. The term lido came to be applied to it, and later extended even to a fresh-water or artificial lake beach/resort. The term is best avoided in its original, Italian, sense.

life cycle the progressive series of changes undergone by an organism from fertilization (the union of gametes, the reproductive cells) to death or, in a lineal succession of organisms, to the death of that stage producing the gametes which begin an identical series of changes. In some organisms (e.g. parasitic worms) there is a succession of individuals, connected by sexual or asexual reproduction, in the complete cycle.

life expectancy the amount of time, calculated by actuaries, an individual person or a class of people is expected to live.

light **1.** the wave band of electromagnetic radiation (ELECTROMAGNETIC SPECTRUM) to which the retina of the eye is sensitive and which is interpreted by the brain **2.** the part of the electromagnetic spectrum which includes infra-red, visible and ultra-violet radiation.

light industry **1.** loosely applied to a SECONDARY INDUSTRY which does not come under the definition of HEAVY INDUSTRY **2.** more specifically, a secondary industry in which the weight of materials used per worker is low, and the finished products have a high value in relation to their weight **3.** in town planning, any industry which may be located in a residential area without detracting from its amenity. FOOTLOOSE INDUSTRY.

lightning an electrical discharge seen as a flash occurring within or between clouds, or between a cloud and the ground, and giving rise to THUNDER. The flash is seen before the thunder is heard because light travels faster than sound. The distance of the flash from the observer can be assessed roughly as 1.6 km (1 mi) per 5-second interval between flash and thunder. THUNDERSTORM.

lignite a low-grade COAL, generally post-CARBONIFEROUS in age, intermediate in properties between the PEATS and BITUMINOUS COAL. The term is sometimes used as a synonym for BROWN COAL, but lignite is darker in colour, the vegetable structure is not so apparent, the carbon content is higher, the moisture content lower than that of brown coal. Lignite is used mainly as a fuel to produce heat in thermal-electric generators.

limb of a fold the rock strata on one or the other side of the central line (the AXIS) of a FOLD. Fig 24.

lime a caustic, infusible solid consisting essentially of CALCIUM OXIDE (quicklime), obtained by heating CALCIUM CARBONATE. It is used in agriculture and metallurgy, in building materials, the treatment of sewage, and in various manufacturing processes.

limekiln a chamber, usually of brick, in which CHALK or LIMESTONE is heated to a high temperature to produce LIME.

limestone a broad, general term for a SEDIMENTARY ROCK consisting mainly of CALCIUM CARBONATE with varying amounts of other minerals. The many types are distinguished by qualifying adjectives, which may refer to texture (e.g. oolitic, pisolitic, crystalline etc., limestone of tiny rounded grains being OOLITE, of larger grains PISOLITE; CRYSTALLINE limestone being MARBLE), to mineral content (dolomitic), to origin (e.g. organic, coral, sedimentary, precipitated, shelly etc.), to geological age (CARBONIFEROUS, JURASSIC), and to other characteristics. The popular concept of a limestone is that it is relatively hard, hence such references as to 'chalk and limestone', despite the fact that CHALK itself is a limestone, but can be comparatively soft. CALCITE, CAMBRIAN, LIME.

limestone pavement a horizontal or slightly inclined bare limestone surface, broken by GRIKES and CLINTS, apparently coinciding with the major BEDDING PLANES exposed by glacial erosion. Limestone pavements are particularly vulnerable to erosion by PRECIPITATION in the grikes; and to the activities of people who want to use pieces in their gardens.

limnology the scientific study of freshwater lakes, ponds and streams, covering characteristic physical, chemical and biological conditions.

limon (French) a superficial fine-grained deposit widespread in northern France, spread like a blanket regardless of minor relief, and providing brown, loamy soils. It seems to have been deposited, as wind-sorted, wind-borne material, around the margins of retreating ice sheets in the Great Ice Age, and in that its origin may be compared with that of LOESS. Some authors distinguish the wholly wind-sorted, wind-borne limon from the wind-sorted, wind-borne limon which may have been reworked and redeposited by later stream action.

linear *adj.* **1.** of, or in, lines **2.** long, narrow and of generally uniform width **3.** involving one dimension only in a unit of measure **4.** in mathematics, having the property of being able to be shown as a straight line on a graph **5.** having or involving the property that a change in one quantity corresponds to or is accompanied by a directly proportional change in another quantity.

linear eruption a FISSURE ERUPTION.

linear growth arithmetic growth, simple interest, the growth of a value over a period based on a fixed percentage calculated on the original principal only, so that the sum added at the end of each successive period is constant. ARITHMETIC PROGRESSION.

linear model a MODEL-2 in which the DEPENDENT VARIABLE is assumed to be related in direct proportion to one or more INDEPENDENT VARIABLES. When the variables are shown against each other on a graph, the line of best fit (a line drawn to come as close as possible to the data points) can be drawn. The fitting is accomplished by using the least squares method (REGRESSION ANALYSIS, including linear regression analysis).

linear scale SCALE.

linear settlement, linear town an elongated settlement, especially an elongated urban settlement, developed alongside a major routeway on account of either the constriction of the terrain (e.g. valley slopes rising steeply from the routeway) or the advantages of the ease of transport provided by the routeway.

line of best fit REGRESSION ANALYSIS.

line squall a phenomenon associated with the passage of a COLD FRONT, in which violent storms with strong gusts of wind and heavy precipitation occur simultaneously along a line extending in some cases from 480 to 650 km (300 to 400 mi). SQUALL.

link in a NETWORK-2,3,4, a line, a route, an edge (GRAPH) between NODES-2.

linkage the contact and the flow of information between two individuals, the connexion between and within different types of activities or different functions. INDUSTRIAL LINKAGE.

linked industries SECONDARY INDUSTRIES where the final product, e.g. motor vehicles, depends on many separate preparatory processes and materials. INDUSTRY.

links (Scottish, always pl.) a narrow coastal strip with accumulations of blown sand giving rise to sand dunes, supporting coarse grass and low shrubs, so often used for the game of golf that the term golf links, or links, is now regarded as almost synonymous with GOLF COURSE.

liquid fluid matter having a definite volume but no definite shape, taking on the shape of its container but, unlike a GAS-1, not expanding to fill the containing vessel, i.e. it keeps its own volume at any given temperature. CONDENSATION, FLUID, GAS, SOLID.

listed building in planning system in Britain since 1945, a building with some particular quality (e.g. architectural, artistic, historical; or an association with technological development or famous people) that is protected, and may not be altered or demolished without planning permission.

lithification **1.** the process of forming into stone **2.** in geology, the result of the transformation of an accumulation of loose SEDIMENTS-2 into a rock mass.

lithomorphic soils one of the seven groups in the 1973 SOIL CLASSIFICATION of England and Wales. They include RANKERS and RENDZINAS, and are generally well-drained soils with an organic surface layer lying on bedrock or little altered parent material which underlies the soil layer at a depth of some 30 cm (12 in).

lithosphere **1.** the earth's CRUST, including the SIAL and SIMA layers above the MOHOROVICIC DISCONTINUITY **2.** the sial, sima and upper part of the MANTLE above the Gutenberg Channel (GUTENBERG DISCONTINUITY). ASTHENOSPHERE, ATMOSPHERE-1, BIOGEOSPHERE, GEOTHERMAL GRADIENT, HYDROSPHERE, PLATE TECTONICS.

litter in soil science the layer of leaves, twigs and other organic remains lying on the soil surface (especially on the forest floor), which may decompose to form HUMUS (L LAYER).

littoral *adj.* of, on, or along, the shore, whether of sea, lake or river.

littoral current a CURRENT in the zone of the SURF running parallel to the SEASHORE, caused by waves breaking obliquely to the shore. LONGSHORE CURRENT.

littoral deposit the sand, shells, shingle, deposited in the area between the marks of HIGH WATER and LOW WATER. Some authors include the offshore mud and all the shallow water marine deposits.

littoral drift the movement of material in a LITTORAL CURRENT.

littoral zone variously applied to **1.** the aquatic zone between the marks of HIGH WATER and LOW WATER **2.** in biology, the upper part of the BENTHIC DIVISION, from the water surface to a depth of about 200 m (110 fathoms: 655 ft) **3.** the part of the benthic division favourable to the growth of green plants.

livestock the DOMESTIC ANIMALS commonly kept on a subsistence or commercial basis to provide food (e.g. eggs, meat, milk) and other raw materials (e.g. bonemeal, hides, WOOL) for people **1.** in general, applied to cattle, sheep, pigs to which may be added poultry (chickens, ducks, turkeys) (often); and goats and horses (sometimes) **2.** FAO international statistics include cattle, sheep, pigs, poultry (chickens, ducks, turkeys), buffaloes, horses, mules, asses, camels (all, except the mules, asses and camels being slaughtered for meat). FAO also include indigenous animals as livestock, as providers of meat, i.e. beef and mutton as well as buffalo, goat and pig meat. But reindeer (important in semi-NOMADISM in Lapland), caribou and other deer and antelope, all providing meat, skins etc., are not included. LIVESTOCK FARMING.

livestock farming, stock farming, pastoral farming, pastoralism the farming activity based on the rearing of animals (LIVESTOCK) for eggs, hides and skins, meat, milk, WOOL etc. as distinct from CROP FARMING. The methods used and the size of the enterprise vary greatly, from NOMADISM to small scale farming (where, for example, a small herd of dairy cattle may be kept) to large scale RANCHING, or to the large scale, highly organized, scientific rearing of animals under cover in a completely controlled environment (the term PASTORAL being perhaps an inappropriate description of the last). BATTERY SYSTEM, FACTORY FARM, GRAZING, PASTURE.

llano (Spanish) the extensive plains in the basin of the Orinoco in South America. These are treeless, and the term came to be transferred, in international literature, to the type of vegetation they supported, i.e. tropical grassland, or SAVANNA.

L layer the organic LITTER lying on the land surface and not yet incorporated in the soil. O HORIZONS, SOIL HORIZON.

load of a river, load of a stream all the solid matter transported by a river or stream, by being rolled and bounced (SALTATION) along its bed (BED LOAD), or carried in SUSPENSION or in SOLUTION. COMPETENCE OF A STREAM.

loam an old term variously applied, now best restricted to a soil having clay and coarser materials in such proportion as to form a PERMEABLE, FRIABLE and easily worked mixture. This is attained, according to standards in the USA, when clay is 7 to 27 per cent, silt 28 to 50 per cent and sand less than 52 per cent (GRADED SEDIMENTS). Loam is termed clay loam if the clay content lies between 27 and 40 per cent and the sand between 20 and 45 per cent; silty loam if clay is below 30 per cent, sand between 20 and 50 per cent, silt between 72 and 80 per cent; sandy loam if clay is between 10 and 20 per cent, sand between 50 and 70 per cent (SOIL TEXTURE). The best agricultural soils are usually loams, hence tracts with loamy soils, or loam-terrains, have always attracted settlers.

local climate a general term applied to the climate of a small area, e.g. a valley with a particular aspect (thus larger than the area of a MICROCLIMATE), which differs in one or more elements from the climate of nearby areas, those within, say, a kilometre or less. The difference(s) may be caused by slight variations in slope, aspect, type of soil or vegetation, or the presence (or absence) of water or tall buildings. MACROCLIMATE, MESOCLIMATE.

localization of industry the concentration of an industry in a certain district or districts. Various means have been proposed to measure the degree of concentration, such as a LOCATIONAL (LOCALIZATIONAL) COEFFICIENT.

local time local solar time, APPARENT TIME, the time expressed with reference to the MERIDIAN of a given place, e.g. it is twelve noon local time when the sun's centre crosses that meridian and the shadows of vertical objects at the given place are at their shortest. It would be inconvenient for every place to use its local time, hence the STANDARD TIME adopted for use in most countries. MEAN SOLAR TIME, ROTATION OF THE EARTH.

location 1. geographical situation, a part of SPACE-2 or a point or position in space where objects, organisms and fields of force (e.g. MAGNETIC FIELD) may be found or events occur **2.** the action of placing or the condition of being placed **3.** the fact or condition of occupying a particular place **4.** a local position, or a position in a series or succession **5.** the action of finding the position of someone or something; or the ability to do this **6.** in Australia, a farm or STATION-3. PLACE, POSITION, SITUATION.

locational (localizational) coefficient a statistical measure expressing the relative amount of a function or of a population present in a part of a large region. In P. Sargent Florence's formula the workers are shown region by region as percentages of the total in all regions, and the coefficient is the sum (divided by 100) of the plus deviations of the regional percentages of workers in the particular industry from the corresponding regional percentages of workers in all industry. Uniform dispersal gives a coefficient of 0, extreme differentiation a coefficient of 1.

location of industry the areal distribution of industrial activities.

location quotient LQ, a measure used to show the degree of concentration of a particular activity or characteristic in an area compared with a specified norm, e.g. the degree to which a particular industry is concentrated in a particular part of the country. Using this example, the location quotient for an area is the ratio between the area's share of the national total of the activity under consideration and its (the area's) share of all activities. For example, if engineering is the particular activity, the location quotient for the area is calculated by dividing the percentage of all the engineering workers in the country employed in the area by the percentage of the total employed population working in the area. If the resultant location quotient exceeds 1.0 it indicates a 'surplus' in the share of a particular activity or characteristic, if less than 1.0 a shortfall. LOCATIONAL (LOCALIZATIONAL) COEFFICIENT, LOCALIZATION OF INDUSTRY.

loch (Scottish; Gaelic; Irish loch or lough) **1.** a lake **2.** an arm of the sea (a sea loch), especially if it is narrow and bounded by steep sides, or partly landlocked.

lock a stretch of canal or river confined within gates so that the water level can be controlled to lift or lower a vessel to allow it to pass from one REACH of navigable water to another at a different level.

lode a mineral vein or systems of veins in a rock.

lodgement till 1. in general, GROUND MORAINE **2.** more precisely, debris deposited under ice while the ice is moving, the sticky clay being lodged on the under surface of the ice and the longer axes of the larger stones being aligned with the direction of movement of the ice, often resulting in DRUMLINS. ABLATION TILL,

TILL (includes deformation till, deformed lodgement till).

loess, löss (German dialect lösz; spelling loëss is erroneous) originally applied to a fine-grained yellowish loam occurring in the valley of the Rhine and elsewhere in Germany, comparable with the LIMON of France. Internationally the term has come to be applied to the fine-grained AEOLIAN deposit, the PERMEABLE, wind-sorted and wind-deposited morainic material, unconsolidated and unstratified, laid down away from the margins of the great ice sheets of PLEISTOCENE times, and covering vast areas in central Asia, Europe, North America and elsewhere. It ranges from clay to sand (GRADED SEDIMENTS), is buff or brownish in colour, is usually calcareous and contains concretions of CALCIUM CARBONATE and in some cases iron oxide. It has been suggested that increased rainfall helped to wash the fine-grained material down from the air to the ground; and also that much may have been reworked and redeposited by later stream action. The soils derived from loess are of high quality, being fine in texture, well-drained, fertile, deep and easily worked.

logical positivism a body of philosophical thought developed from the late 1920s under the leadership of Schlick and Carnap in Vienna, the main aim of which was to create a comprehensive philosophy of science stemming from EMPIRICISM and proceeding by INDUCTION. Among its tenets, logical positivism maintained that traditional METAPHYSICS, consisting of propositions that could not be verified by empirical observation, was without meaning (thus religious and moral statements, being metaphysical, were considered by most logical positivists to be meaningless); and that philosophy generally consisted of an analysis of language, of the discovery of verbal forms of expression for complex ideas and propositions, helped by formal logic. Logical positivism differs from POSITIVISM in that it accepts that some statements are verifiable without recourse to experience. FALSIFIABILITY.

Lo-Lo lift-on/lift-off of CONTAINERS. RO-RO.

longitude the angular distance between the MERIDIAN passing through a given point and the prime, standard, initial or zero meridian (usually considered to be the meridian passing through the old Observatory at Greenwich, London, England, and numbered 0°). This angular distance, i.e. longitude, is measured in degrees, minutes and seconds east or west of the Greenwich meridian (0°) to 180°, the meridian 180°E thus coinciding with 180°W. All points through which a meridian passes have the same longitude. The alternative term for meridian is line of longitude; and all lines of longitude meet all PARALLELS OF LATITUDE at right angles. The actual distance represented by 1° of longitude becomes less as the meridians converge towards the poles. Thus at the equator the distance is roughly 111 km (69 mi), at latitude 45° it is 78.8 km (nearly 49 mi), and at the poles it is zero. Fifteen degrees of longitude are equivalent to a difference of one hour in LOCAL TIME. LATITUDE. Fig 27.

longitudinal *adj.* **1.** of or pertaining to LONGITUDE **2.** of or pertaining to length as a dimension **3.** running lengthwise, e.g. LONGITUDINAL COAST.

longitudinal coast a CONCORDANT COAST, one running broadly parallel to the main geological structure or fold lines. Commonly occurring around the Pacific ocean, it is also termed a Pacific coast, in contrast to a transverse or ATLANTIC COAST.

longitudinal dune a SAND DUNE with its crest running parallel to the direction of the prevailing wind. TRANSVERSE DUNE.

longitudinal profile of a river long profile, RIVER PROFILE.

longitudinal study, longitudinal analysis a study or analysis in which selected variables are studied in the same group or groups of subjects at intervals over a period of time, often over many years, as distinct from a CROSS-SECTIONAL STUDY or analysis, in which a cross section of the population is sampled, similar variables being studied at different ages, but on different subjects at each age. Thus in a cross-sectional study the subjects are approached only once. ECOLOGICAL FALLACY, HISTORICAL FALLACY.

longitudinal valley a VALLEY with a trend roughly parallel to the STRIKE of the rocks or the grain of the country.

longitudinal wave L-wave, a form of shock wave produced by an EARTHQUAKE passing over the land surface and responding to the character of the rocks encountered. SHAKE WAVE, TRANSVERSE WAVE, WAVE.

long profile of a river the PROFILE-3 of a river bed, from the source of the river to its mouth. RIVER PROFILE, TALWEG.

long-range weather forecast a weather forecast for a period usually longer than five days, based on the handling by computer of great quantities of data, and the use of a synoptic analogue (MODEL-2).

longshore current a current generated by tides, waves, winds, running generally parallel to the coast. LITTORAL CURRENT, LONGSHORE DRIFT.

longshore drift the movement of material along a beach (not only in the surf zone) by the action of waves that meet the shore at an angle. The SWASH of a BREAKER carries the material up the beach at an oblique angle, the BACKWASH pulls some back seawards at right angles, so that there is a gradual build-up of material along the beach, reinforced if there is a LONGSHORE CURRENT. GROYNE, LITTORAL DRIFT. Fig 28.

loop district, the loop originally that part of the DOWNTOWN business district of Chicago, enclosed by a loop of elevated railways; later, by analogy, applied in other towns to what came to be called the CBD, or CENTRAL BUSINESS DISTRICT.

lopolith a large INTRUSION of igneous rock CONCORDANT with the strata, allied to a LACCOLITH or PHACOLITH, but hav-

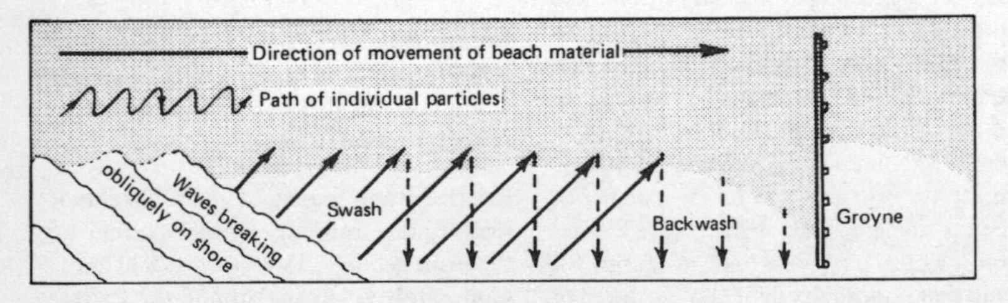

Fig 28 Longshore drift, showing wave action and the movement of beach materials

ing the form of a saucer-shaped basin. GABBRO.

Lorenz curve a curved line drawn on a graph to show, by its concavity, the extent to which a distribution (e.g. concentration of population) differs from a uniform distribution. Percentage values are used for each axis, the uniform distribution appearing as a straight line drawn across the graph at 45°. GINI COEFFICIENT.

low, atmospheric a low pressure system or DEPRESSION-3 in the atmosphere, indicated on a weather chart by closed isobars (ISO-), the values of which fall towards the centre. If the strength of the air flow around the system increases, the depression is said to intensify. ANTICYCLONE, CYCLONE.

low, on a beach SWALE.

lowland, lowlands an imprecise term applied broadly to relatively level land at a lower elevation than that of adjoining districts, often used to contrast with highland, and sometimes defined as lying at an elevation under 180 m (600 ft).

low latitudes LATITUDE.

low pressure system LOW, ATMOSPHERIC.

Lowry model a computer MODEL-2 devised by I. S. Lowry to portray the spatial organization of human activities in an urban area, so that the impact of public decisions on such an area can be evaluated and changes in urban form predicted in view of anticipated changes in key variables such as the pattern of employment, the efficiency of the transport system, or the growth of population. Using ECONOMIC BASE THEORY, the distribution of 'primary' and manufacturing employment (BASIC ACTIVITIES) in the area are plotted on one mile square cells (Lowry's measure) on the basis of POPULATION POTENTIAL-2; and service employment (e.g. retail activities) are plotted in proportion to the market or employment potential in each cell. The service activities naturally create further employment opportunities, located to take advantage of the market potentials. Constraints on land use in the cells are incorporated in the model (e.g. the minimum sizes of clusters of services, maximum housing densities etc.). By ITERATION more and more residents come into the cells, until the market potentials are so disturbed that the pattern of retail activities has to be modified. These activities are reallocated to the limits of the land use constraints, so that eventually a state of EQUILIBRIUM is reached, the final distribution of the population corresponding to that used initially to compute potentials.

low technology the application of the knowledge of and/or the use of simple methods, simple equipment and readily available inexpensive materials to a task or an industrial process, or to the solution of a problem arising from the interaction of people with their ENVIRONMENT. HIGH TECHNOLOGY, INTERMEDIATE TECHNOLOGY, TECHNOLOGY-3.

low tide **1.** the TIDE at lowest ebb **2.** the level of the sea or the time of the lowest ebb. EBB TIDE, HIGH TIDE, NEAP TIDE, SPRING TIDE.

low water **1.** LOW TIDE **2.** the low level of water in a lake or river. HIGH WATER.

lumber (American) TIMBER, especially that recently felled and roughly sawn into logs and planks.

lumbering (American) the act of **1.** felling and sawing of TIMBER and removing it from the area **2.** felling trees and sawing them into logs.

lumberjack (American) one who cuts TIMBER and prepares it for the sawmill or market.

lumberman (American) one employed in the felling of trees and in preparing them for market, especially as a manager.

lunar *adj.* of, relating to, similar to, the MOON.

lunar day the period of time in which the earth rotates once in relation to the MOON, i.e. 24 hours 50 minutes between two successive crossings of the same meridian, despite the fact that the earth rotates once in 24 hours. The discrepancy is due to the orbiting of the moon itself around the centre of gravity of the moon and the earth, causing it to cross each meridian 50 minutes later: hence the interval of some 12 hours 25 minutes between one high tide and the next, i.e. high tide on one day is 50 minutes later than the corresponding high tide of the day before.

lunar month the period from one new moon to the next, averaging 29 days 12 hours 44 minutes. MOON.

lynchet, linchet probably originally applied either to a strip of green land between two pieces of ploughed land, or (as in present use) to a narrow terrace on a hillside, especially in the chalk downlands of southern England. In most cases lynchets seem to mark old cultivation strips, perhaps of IRON AGE or earlier, usually lying parallel to the contours to make a level, well-drained strip of land, accidentally or intentionally preventing soil erosion; but sometimes running down the hillslope; and sometimes retained by a wall of stones. Their origin is much discussed. TERRACETTE.

lysimeter a simple device for measuring the percolation of water through the soil, and so of determining the soluble constituents removed in soil drainage. A container holding the material under investigation is fitted with instruments that measure the quantity or quality of the water that has passed through it, revealing what happens in the field.

M

macadam, macadam road, macadamized a road building method and its material. Early in the nineteenth century a Scottish road engineer, John Loudon McAdam, 1756–1836, discovered that certain types of stone, broken into angular pieces of nearly uniform size, would bind together under pressure to form a hard road surface. The use of such stone under the pressure of specially designed rollers became standard practice in road making. Macadam is the term applied to the broken stone, the roads being termed macadam roads or macadamized roads. The binding of the stone is helped if the pieces are coated with TAR, hence tarmacadam (tarmac), a later development.

macchia (Italian) MAQUIS.

machair (Scottish) whitish, shelly sand forming a low plain along the western shores of Scotland, and in the Hebrides in South Uist and Tiree, providing useful light arable soils (machair soils). The term is also sometimes applied to the calcareous vegetation growing on it.

mackerel sky rows of CIRROCUMULUS CLOUD or of small ALTOCUMULUS CLOUD that resemble the pattern on the skin of mackerel (a food fish of the North Atlantic). It occurs particularly in dry warm weather in summer when a thin sheet of cloud is broken by CONVECTION CURRENTS.

macro- (Greek) large, as opposed to MICRO-, small.

macroclimate the climate of a large region, in contrast to LOCAL CLIMATE, MICROCLIMATE.

macronutrient an ELEMENT-6 or combination of elements needed in relatively large quantities by an organism in order to maintain health, in contrast to a MICRONUTRIENT. The macronutrients needed in varying amount by plant crops are carbon, hydrogen, oxygen (from the air), and calcium, nitrogen, potassium, phosphorus, magnesium and sulphur (from the soil).

maelstrom (Dutch) a large dangerous whirlpool, a term derived from the name of a notorious whirlpool caused by tidal currents in the Lofoten Islands, off the west coast of Norway.

maestrale (Italian) MISTRAL (French).

magma extremely hot, viscous, molten rock material, charged with gas and volatile matter, lying under great pressure below the surface of the earth. It contains among a wide range of elements silica (ACID ROCK, ACID LAVA) and basic oxides (BASIC LAVA, BASIC ROCK). It is believed to have accumulated in magma basins, from which it may force its way to the surface; but it often solidifies underground to form INTRUSIVE PLUTONIC ROCKS. Less frequently it reaches the surface as LAVA from which EXTRUSIVE ROCKS, VOLCANIC ROCKS, are formed on solidification; or it stops short of the surface and solidifies in minor INTRUSIONS (e.g. dykes, DIKE), LACCOLITHS, SILLS as

HYPABYSSAL ROCK. When solidifying as a plutonic rock (such as GRANITE) in a great mass or BATHOLITH it heats and alters the surrounding rocks by what is known as CONTACT METAMORPHISM, converting sedimentary and other rocks into METAMORPHIC ROCKS through the distance from the molten mass known as the METAMORPHIC AUREOLE. The magma may dissolve and absorb some of the surrounding rocks (magmatic assimilation), and in particular penetrate fissures, joints, cracks and replace fragments of the adjoining rocks by its own mass (magmatic stoping). In the course of these changes, and in the slow cooling of the mass of the magma, a wide range of different igneous rocks may be formed, originating from one great magma basin (magmatic differentiation). The main mass eventually solidifies as a BATHOLITH or BOSS with associated smaller masses such as dykes (DIKE), LACCOLITHS, PHACOLITHS, SILLS. There may be a connexion in the earth's surface through a FISSURE or VOLCANIC NECK.

magmatic water JUVENILE WATER.

magnesian limestone a LIMESTONE containing 5 to 15 per cent MAGNESIUM carbonate (magnesite). As Magnesian Limestone (with initial capitals) it is the name given to the rocks of this character which constitute an important part of the PERMIAN rocks in northeast England.

magnesium a silver-white metallic element occurring naturally only in combination, especially as the carbonates magnesite and dolomite and as the chloride carnallite, present in seawater, in plants (e.g. in CHLOROPHYLL) and in animals (in bones). PEDALFER, PEDOCAL.

magnet an object producing a MAGNETIC FIELD, e.g. a piece of iron, cobalt, nickel or alloy exhibiting FERROMAGNETISM; or the magnetized needle of a COMPASS; or the soft iron core together with the surrounding coil through which an electric current is passed, of an electro-magnet.

magnetic *adj.* **1.** of or relating to a MAGNET or to magnetism **2.** having the magnetic north pole as a reference. MAGNETIC POLE **3.** producing a MAGNETIC FIELD or being capable of so doing.

magnetic anomaly a deviation from the predicted value of the earth's MAGNETIC FIELD, at a particular point on its surface, due to changes in the internal magnetic field (recorded in rocks at the time of their formation, PALAEOMAGNETISM) or to local concentrations or deficiencies of magnetic minerals. The symmetry of the deviations (MAGNETIC REVERSAL) revealed by the rocks on each side of an OCEANIC RIDGE supports the theory of sea floor spreading and PLATE TECTONICS.

magnetic bearing BEARING-2.

magnetic compass COMPASS.

magnetic declination, magnetic variation the angular distance between MAGNETIC NORTH and TRUE NORTH, expressed in degrees east and west of true (geographical) north, at any point on the earth's surface. It varies with place and time owing to irregularities in the earth's MAGNETIC FIELD. MAGNETIC POLE, TERRESTRIAL MAGNETISM.

magnetic field the space through which a magnetic force, created by an object with magnetic power at its centre, exists. It is formed by a permanent magnet or by a circuit carrying an electric current. FERROMAGNETISM, MAGNETIC, TERRESTRIAL MAGNETISM.

magnetic meridian any line joining the

north and south MAGNETIC POLES along which the free-swinging magnetic needle of a COMPASS aligns itself.

magnetic north the direction to which the magnetic needle of a COMPASS, swinging freely in a horizontal plane, points in its search for the magnetic north pole. BEARING-2, MAGNETIC DECLINATION, MAGNETIC MERIDIAN, MAGNETIC POLE, TRUE NORTH.

magnetic pole either of the two poles (the north or the south) of the earth's MAGNETIC FIELD, indicated by the magnetic needle of a COMPASS swinging freely in a horizontal plane. The precise locations vary, partly due to irregularities of the earth's magnetic field. The magnetic north pole moved 150 km (93 mi) northwards between 1984 and 1994. In 1994 the average position lay on the southwestern coast of Ellef Ringness Island, 1300 km (1186 mi) south of TRUE NORTH, situated in the Arctic ocean. MAGNETIC DECLINATION.

magnetic reversal a 180° swing round of direction of the earth's MAGNETIC FIELD (MAGNETIC ANOMALY), causing the MAGNETIC POLES to change position, so that the north magnetic pole becomes the south magnetic pole, and vice versa. This phenomenon, revealed by studies in PALAEOMAGNETISM, helped to confirm the theory of sea floor spreading and PLATE TECTONICS. The time period during which the reversal occurs is termed the magnetic interval, and the rocks formed at that time are termed a magnetic division.

magnetosphere the zone of influence of the earth's MAGNETIC FIELD, extending in the ATMOSPHERE-1 up to and including the EXOSPHERE and IONOSPHERE. SUNSPOT, TERRESTRIAL MAGNETISM.

maize a tall, coarse ANNUAL cereal, *Zea mais*, bearing large 'cobs' on which the edible kernels are carried. It is native to the Americas, where it is known as CORN, was brought to Europe by Columbus, taken to Africa by the Portuguese, where it spread to the areas with sufficient rainfall (and where, in central and southern Africa, it is known as mealies). It needs good, deep soil, plenty of moisture, at least 150 frost-free days, and summer warmth. New hybrid strains with short stalks ripen quickly, are pushing the limit of cultivation to cooler areas, and give higher yields. The grain does not make good bread, but maize is the staple grain for people in many maize-growing countries (especially in South America and Africa). The sweeter varieties grown for use as a vegetable are termed sweet corn. Maize grain, when very finely ground, produces cornflour, used in baking and confectionery; it also yields corn oil, used in cooking, and corn syrup, which is sweeter than sugar. But in Europe and North America the larger part of the crop is fed to animals, especially to cattle and pigs.

malleable *adj.* of metals, capable of taking a permanent change of shape by being beaten, pressed, rolled etc., i.e. by the application of STRESS-1. DUCTILE.

mallee (Australian: aboriginal term) a low, scrubby, EVERGREEN plant, *Eucalyptus dumosa*, growing in arid, subtropical parts of southern Australia.

mallee scrub a densely growing, low, EVERGREEN shrub formation of *Eucalyptus* species, growing in the arid, subtropical parts of southern Australia.

malt processed grain, especially of BARLEY, used in brewing and distilling. The grain is put in water to germinate, then heated and dried.

Malthusianism the body of doctrines derived from the writings of Thomas Robert Malthus, 1766–1834, a British economist and demographer, especially from *An Essay on the Principle of Population as it affects the Future Improvement of Society*. His thesis was that the human population, if unchecked, increases at a geometric rate (GEOMETRIC PROGRESSION) while the food and other resources needed for its subsistence increase only at an arithmetic rate (ARITHMETIC PROGRESSION); that population always grows to the limits of the means of its subsistence, checked only by war, famine, pestilence and the influence of miseries derived from a consequent low standard of living.

mammatus cloud a breast-shaped CLOUD (CLOUD FORM) usually occurring in the formation of thunderclouds, associated with CONVECTION.

managed fallow BUSH FALLOWING.

mandate, mandated territory a territory designated under Article 22 of the League of Nations. When the League of Nations was created after the First World War, in 1919, the former colonial possessions of Germany and Turkey (the losing countries) came under the control of the League, which allocated the administration of these territories (mandated territories) to certain powers, notably Great Britain and France (mandatory powers). The Class A mandates (Iraq, Palestine, Transjordan, Syria, Lebanon) were regarded as likely soon to be ready for independence and fairly soon became independent. Class B (Cameroons, Togoland, Tanganyika, Ruanda) were less developed; Class C (South-West Africa, New Guinea and certain Pacific islands, Samoa) were to be governed as part of the territory of the mandatory power. In 1945 the United Nations took over such responsibilities as remained, and the territories became TRUSTEESHIP TERRITORIES.

manganese Mn, a hard, brittle metal occurring in nature usually as an OXIDE, CARBONATE or SILICATE, found in a great variety of compounds, an essential MICRONUTRIENT for plants, and especially important in making certain types of steel.

manganese nodule a MANGANESE OXIDE concretion, with some iron, copper, cobalt and nickel, occurring in some clay rocks and on the floor of the ABYSSAL ZONE of the ocean (RED CLAY).

mangrove 1. a member of a number of genera and species of low trees and shrubs which grow and spread quickly on tidal mud in tropical areas, so that their dense root systems (which include adventitious, aerial roots which stretch out at a distance from the main stem, bend towards the ground, strike, and send up new trunks) are covered by salt or brackish water at each tide and effectively bind the mud. The roots under the mud have air supplied by aerial roots which rise above the surface **2.** a plant community dominated by such trees and shrubs. MANGROVE SWAMP.

mangrove swamp the association of low trees and shrubs covered by the collective term MANGROVE, growing with members of other families in tidal mud in DELTAS, ESTUARIES and along the coasts in TROPICAL regions. SWAMP.

manioc a shrub native to Brazil but widely cultivated as a staple food plant in tropical areas of South America and Africa, the source of cassava and tapioca, produced by subjecting the poisonous tuberous roots of manioc to heat and pressure, then drying. It is the non-poisonous granular substance produced that is known as cassava or tapioca (according to treatment), high in starch but low in protein. TAPIOCA STARCH.

man-made fibre SYNTHETIC FIBRE.

man-made soils the term applied by those who drew up the 1973 SOIL CLASSIFICATION of England and Wales to one of its major soil groups. Made by human endeavour, these are good deep soils, over 40 cm (16 in) thick, rich in HUMUS, with a surface layer which may or may not be totally artificial.

Mann-Whitney test in statistics, an hypothesis test used to analyse two unrelated samples of data in order to compare two population MEDIANS-2 (i.e. the samples are drawn from two different populations). The sample data must be ordinal (ORDINAL SCALE) and the sample sizes small. It is assumed that both population distributions have the same shape.

manor the extent of the land held by the lord of the manor together with, in feudal times in England (and to some extent in Wales), certain legal and administrative rights, dues and responsibilities (FEUDAL SYSTEM). The lord of the manor, holding land from the king, owned the soil of the whole, and kept for his own occupation and use or that of his servants the parts termed demesne lands. The remainder was worked by his tenants under various systems of payment (LAND TENURE). With INCLOSURE the lord of the manor retained the demesne lands and the ownership of the soil of the common grazing (COMMON), subject to COMMON RIGHTS. (The lordship of the manor is now in general little more than an archaic, empty title.) The affairs of the manor were controlled by manorial courts from the manor house or mansion of the lord.

mantle 1. of the earth, that part of the earth's interior lying between the CORE and the CRUST, consisting of ULTRABASIC ROCKS, probably solid unless the pressure of the overlying rocks is relieved, when they become viscous (MAGMA). The mantle is some 2900 km (some 1800 mi) thick, density about 3.0 to 3.4, the lower surface forming the GUTENBERG DISCONTINUITY, the upper the MOHOROVICIC DISCONTINUITY, the uppermost layer forming the ASTHENOSPHERE. Movements in the mantle, presumed to be convectional, are reflected in the structure of the earth's CRUST **2.** mantle rock, the accumulation of loose rock debris, consisting of weathered rock and soil, lying on the older solid BEDROCK and covering most of the land of the earth, alternatively termed REGOLITH.

manufacturing industry SECONDARY INDUSTRY, the making of articles or materials (now usually on a large or relatively large scale) by physical labour or mechanical power, in general terms the processing of raw materials and foodstuffs (or of semi-processed or recycled materials), the working-up of materials into a useful form. FOOTLOOSE INDUSTRY, HEAVY INDUSTRY, LIGHT INDUSTRY; and PRIMARY, TERTIARY, QUATERNARY INDUSTRY, MARKET ORIENTATION, MATERIAL ORIENTATION.

map a representation of the earth's surface or a part of it, or of the heavens, delineated on a flat sheet of paper or other material. Generally used loosely, thus diagrammatic maps are included. COGNITIVE MAP, MENTAL MAP, PLAN, TOPOLOGICAL DIAGRAM (MAP).

map projection 1. the method or methods and their results of representing part or the whole of the earth's surface on a plane surface **2.** the orderly arrangement of parallels and meridians which enables this to be done. It is impossible to map any part of a sphere on a plane, or a flat sheet, with complete accuracy. It is necessary therefore to choose the properties which are desirable for the purpose of the proposed map (correct area, shape, bearing, scale). True shape can be obtained for small areas only, and is incompatible with correct area. A map which shows correct areas cannot give true direction. No projection can give true distances over the whole surface.

Certain classes, e.g. perspective projections, can be obtained by geometrical construction: the zenithal or azimuthal (true direction from the centre); the gnomic (shortest distance between two points in a straight line); stereographic (preserves correct shape); and orthographic (produces effect of a globe). These are useful features but the modified, or non-perspective, projections are in greater use since the network of meridians and parallels can be calculated to meet particular requirements.

Two much-used classes are derived from the so-called developable surfaces, those of the cylinder and the cone (Mercator's projection is a modified cylindrical and Bonne's a modified conic). Conventional projections include the Mollweide (giving the whole surface within an ellipse) and various interrupted projections. The latter modify the central meridian or meridians to show conveniently the areas of most interest, with the result that the whole surface is not continuous (Goode's projection). The Peter's equal-area projection, introduced in 1973 in response to the misleading image of the TROPICS in Mercator's projection, shows each area of the earth at the same scale as any other. But the shapes of the land areas are greatly distorted, e.g. Africa appears twice as long as it is wide (in reality Africa east to west is as long as it is from north to south). Good atlases generally discuss in their introductory section the projections employed, and indicate the correction to be applied to distances. Projections frequently used for atlas maps are Mercator (bearings are straight lines and shapes correct, but areas in high latitudes are greatly exaggerated; also used for navigational charts); Mollweide (equal area but shapes greatly distorted on the margins); Bonne (modified conic, used for great continental areas); zenithal equidistant (correct direction and distance from the centre, used for polar regions). A number of projections have transverse and oblique forms, when in place of the equator as the axis, a suitable GREAT CIRCLE is used.

maquis, macquis, macchia, matorral (French maquis; Italian macchia; Spanish matorral) the low scrub vegetation of EVERGREEN shrubs and small trees characteristic of SILICEOUS soils in Mediterranean lands, the result not only of the summer aridity of the Mediterranean climate in association with the soils, but of human activities in the felling of the natural forest cover (including particularly the holm oak), of grazing animals (particularly goats) and of fire. GARIGUE.

marble a naturally occurring CALCIUM CARBONATE, a CRYSTALLINE LIMESTONE, veined or mottled by the presence of other crystallized minerals, produced by dynamic or thermal METAMORPHISM. It forms a hard rock that takes a high polish.

march, marches, marchlands (French marche) a boundary or frontier, the borderland, now generally applied to the borderland between two states (e.g. the Welsh Marches between Wales and

England), or a zone of land of debatable ownership between two states.

mare's tails wispy streaks of CIRRUS CLOUD, indicating strong winds in the upper ATMOSPHERE-1.

marginal fringe a zone of poor quality land where improved and unimproved patches are interspersed and where the area of the improved land expands or contracts according to fluctuations in farm prices. Thus neither improved farmland nor natural vegetation is dominant.

marginal land land which is just fertile enough to yield an average return from agricultural use, no more than is sufficient to cover the costs of production. It goes in or out of agricultural use according to fluctuations in economic conditions.

marine *adj.* **1.** of, relating to, found in, or produced by, the sea (examples given under RAISED BEACH) **2.** of or relating to shipping or navigation.

maritime air mass AIR MASS.

maritime climate a climate similar to an INSULAR OR OCEANIC CLIMATE, directly influenced by the proximity of the sea, which causes a comparatively cool summer and a comparatively mild winter on account of the different thermal capacities of land and water. There is thus a small daily and seasonal range of temperature, much cloud, fairly uniform precipitation, and humidity higher than that of a CONTINENTAL CLIMATE. This regime may be experienced on islands and in coastal regions in any latitude, but it is particularly prevalent in, and typical of, western mid-latitude coastal areas.

maritime polar air mass POLAR AIR MASS.

maritime tropical air mass TROPICAL AIR MASS.

market many applications, the most common in geographical writing being **1.** a congregation of sellers and buyers of goods **2.** a public place where goods are displayed and put on sale **3.** the demand for a COMMODITY-1 **4.** the outlet for a COMMODITY-1 **5.** the trade, or traffic, in a particular COMMODITY-1 **6.** the people concerned with buying and selling a particular COMMODITY-1 **7.** the class of persons to whom a particular COMMODITY-1 can readily be sold.

market economy, free economy an economic system which operates according to free market forces, i.e. in which there is free enterprise and competition, most of the production, distribution and exchange being in private hands, with government intervention kept to the minimum; and in which goods and services are allocated by the price mechanism, prices being determined by the free exchange of commodities for money (SUPPLY AND DEMAND). Today most countries with a market economy modify to some extent (by STATE INTERVENTION) the forces of the free market. SUBSISTENCE ECONOMY.

market gardening in Britain, the intensive cultivation of vegetable crops, soft fruit or flowers for sale. When organized on a large scale, with a marked concentration on one or two crops, it becomes comparable with TRUCK FARMING in the USA. HORTICULTURE.

marketing principle the main principle used by W. Christaller to account for the varying levels and distribution of CENTRAL PLACES in a CENTRAL PLACE SYSTEM. According to the marketing principle the supply of goods and services from a central place should be as close as possible to the place supplied. To achieve this a higher order central place will serve two centres of the next lower order (a K-3 hierarchy). See ADMINISTRATIVE PRINCIPLE, K-VALUE, TRAFFIC OR TRANSPORTATION PRINCIPLE. Fig 9(a).

market orientation the tendency of a firm or industry to be located close to its MARKET-3. MATERIAL ORIENTATION, RESOURCE ORIENTATION.

market town in Britain, a town with a legal right to hold a MARKET-2 on certain days, usually a settlement which developed at the intersection of routes. 'Market' commonly occurs in the name, e.g. Market Harborough.

marl 1. a calcareous CLAY OR MUDSTONE with an admixture of CALCIUM CARBONATE **2.** in agriculture, loosely applied to any friable clayey deposits.

marling the old practice of spreading MARL OR CLAY on light soil to improve its texture and its capacity to hold water.

marsh a wet area of mainly mineral (i.e. inorganic) soil commonly flooded periodically or at intervals and covered with water-loving vegetation. It differs from a SWAMP (where the summer water level is usually above the surface of the soil), a BOG (which has a mainly organic, acid peat soil), or a FEN (which has a purely organic soil, typically alkaline in reaction, though occasionally neutral or slightly acid). SALT MARSH.

Marshall Plan a popular name for the European Recovery Programme proposed by G. C. Marshall in the USA on 5 June 1947 that materials and financial aid should be provided for the countries of Europe by the USA on condition that the European nations took the initiative in cooperative action. It came into force in 1948, administered by OEEC, terminated 1952.

Marxism a doctrine based on the political, social and economic views of Karl Marx, 1818–83 (German economist, sociologist and philosopher), and Friedrich Engels, 1820–95 (German socialist philosopher), both of whom were profoundly influenced by German philosophers, particularly by G. W. F. Hegel, 1770–1831. Very broadly, the doctrine's philosophical bases are DIALECTICAL MATERIALISM and HISTORICAL MATERIALISM. The MODE OF PRODUCTION is viewed as the dominant factor governing economic and social interaction in society, the impetus for change in the mode of production being generated by class conflict, until ultimately the harmonious, classless, communist society (COMMUNISM-2) is to be achieved. In economic theory the Marxist view is especially concerned with the determination of the value at which a COMMODITY-1 can be exchanged for another; it broadly accepts David Ricardo's LABOUR THEORY OF VALUE-2. References that help to explain the doctrine will be found elsewhere in the *Dictionary*, labelled 'in Marxism'.

massif (French) a compact mass of mountain or highland with relatively uniform characteristics and well-defined boundaries, thus similar to the Massif Central of France.

massive *adj.* applied to a thick STRATUM of rock in which STRATIFICATION, jointing (JOINT), CLEAVAGE, FOLIATION-1 etc. are almost or completely absent.

mass media the systems or impersonal means used for the transmission of information and entertainment for the benefit of a large number of people, i.e. newspapers, radio, television, etc. POSTMODERNISM.

mass movement, mass wasting the spontaneous downward movement by GRAVITATION-1 of rock material, usually helped by rainfall or melt-water from snow or ice, classified as: slow, creeping, usually imperceptible except through extended observation (CREEP, GELIFLUCTION, ROCK CREEP, SOIL CREEP, SOLIFLUCTION); rapid (DEBRIS AVALANCHE, EARTHFLOW, MUDFLOW); and landslides, which are perceptible and involve relatively dry masses of earth debris (BLOCK SLUMPING, DEBRIS FALL, DEBRIS SLIDE, ROCK FALL, ROTATIONAL SLIP, SLIP-2, SLUMPING, SUBSIDENCE-2). COLLUVIUM, SCREE.

mass production the making of one article, or type of goods, in large numbers by a standardized process.

matched samples in statistics, a pair or set of SAMPLES-3 in which each member of one sample is matched by a corresponding member in every other sample by reference to qualities other than those being immediately investigated. The object of matching is to obtain better estimates of differences by eliminating the possible effects of other variables. Assessments of significance may prove difficult if the members of the second sample have to be selected purposively in order to match those of the first, instead of being chosen at random. PAIRING, RANDOM SAMPLE.

material index a measure of materials used in manufacturing industry, calculated by the total weight of localized materials used per product divided by the weight of the finished product. Most manufacturing industries have an index greater than 1.0 and are described as 'weight-losing'. A material index exceeding 1 suggests MATERIAL ORIENTATION; less than 1 indicates MARKET ORIENTATION. HEAVY INDUSTRY, LIGHT INDUSTRY.

materialism **1.** in ONTOLOGY, the theory that matter is the basic reality of the universe, that nothing exists except matter, that consciousness and will can be shown to be the products of material agencies, and thus that they too are derived ultimately from matter. The possibility of disembodied minds and mental states is therefore excluded unless they are identified with states of the brain and the nervous system. The DIALECTICAL MATERIALISM of Marx and Engels introduces an element of evolution to that standard theory of materialism: it maintains that mind, while originating in matter, is distinct in nature from it. HISTORICAL MATERIALISM, IDEALISM, REALISM **2.** the belief that material things are more valuable than spiritual things.

material orientation the tendency of a manufacturing industry to be located close to sources of the materials used in its processes. MARKET ORIENTATION.

mathematical model a MODEL-2 which expresses relationships between variables in terms of numerical values (e.g. REGRESSION MODEL). SYMBOLIC MODEL.

matrix (late Latin, womb; pl. matrices) **1.** a place or medium within which something is bred, formed, developed, produced; a place or point of origin and growth **2.** the material on which a LICHEN grows **3.** the fine-grained material of a rock within which something (e.g. coarser particles, FOSSILS, METALS etc.) is embedded (GANGUE) or to which something adheres. In SEDIMENTARY ROCKS the matrix is usually the cementing (CEMENT-2) material, in IGNEOUS ROCKS it is the GROUND-MASS of CRYSTALLINE or glassy minerals **4.** the impression left in a rock after a FOSSIL etc. has been removed **5.** in mathematics, an ordered ARRAY of mathematical elements conveniently arranged for the carrying out of various operations on it, e.g. in a square or rectangular arrangement of rows and columns of quantities or symbols.

mature *adj.* having reached a stage of full natural development, applied to **1.** a landscape exhibiting the features resulting from a long CYCLE OF EROSION when dissection of the original surface is complete, little trace of it remaining; such a landscape may be maturely dissected by RIVERS which are still young **2.** a river characteristic of such a landscape, in which erosion is at a minimum because the stream has acquired a normal fall in its bed **3.** a shoreline in a condition of approximate equilibrium between erosion, weathering and transportation **4.** a soil with well-developed characteristics produced by the processes of soil formation, and in equilibrium with its environment, or a soil with well-developed horizons. OLD AGE, SENILE RIVER, SOIL HORIZON, YOUTH.

maximum-minimum thermometer a THERMOMETER used to register the maximum and the minimum ambient air TEMPERATURES reached over a selected period of time. A common type consists of a column of MERCURY housed in and partly filling a graduated, U-shaped tube, the rest of each side of the tube being topped-up with a transparent liquid. One side of the U is completely filled; but a small space is left at the top of the other side. Small metal needles rest on top of each end of the mercury, inside the tube; they are mobile but stick to the tube sides when not being actively pushed by the column of mercury. When the temperature rises, the liquid on the side of the U which is completely full expands, exerting pressure on the column of mercury, which accordingly rises on the other side of the U (i.e. the side with the space in it), pushing up the needle, which sticks to the tube to register the maximum temperature reached. When the temperature drops, the liquid in the full side contracts, the mercury rises under the weight of the liquid in the side with the space, the needle in the full side rises and sticks to the tube to register the minimum temperature reached. As the mercury swings up and down, the needles stay in the ultimate positions to which they were pushed. They have to be reunited with the mercury by a magnet operated by hand or by a push-button device.

maximum sustainable yield the highest yield obtainable within a specified period from a renewable NATURAL RESOURCE without jeopardizing the future productivity of that resource. SUSTAINABLE DEVELOPMENT.

maximum thermometer a THERMOMETER used to register the highest ambient air TEMPERATURE reached over a selected period of time. There are many types in use. One consists of a sealed glass, graduated tube containing MERCURY; on the top of the column of mercury lies a metal needle. As the temperature rises the mercury expands and pushes the needle upwards; but the needle is arranged in such a way that it cannot fall with the mercury when the temperature falls and the mercury contracts. It is left standing, stuck to the sides of the tube, and the lower end of it (i.e. the end which touched the mercury) registers the highest temperature reached. The needle remains set in this position until reunited (usually manually, with a magnet) with the column of mercury.

MAXIMUM-MINIMUM THERMOMETER, MINIMUM THERMOMETER.

mayen (Switzerland: German) an intermediate shelf, between ALP-1 and valley floor, where cattle stay for a while on their upward journey in May and their downward trek in September. TRANSHUMANCE, VORALP.

MDC **1.** moderately developed country **2.** more developed country, as defined by United Nations, 1980, with reference to the USSR, Japan, Australia, New Zealand and all the countries in North America and Europe. HDC, LDC, UNDERDEVELOPMENT.

meadow **1.** a piece of land, of any size, permanently covered with grass which is mown for hay **2.** any piece of enclosed grassland but especially one low-lying or close to a river. GRAZING, PASTURE, WATER MEADOW.

mean **1.** something occupying (or, as *adj.*, applied to something occupying) a position midway between two extremes in number, quantity, degree, kind, value etc. **2.** an average. ARITHMETIC MEAN.

meander a loop-like bend, a pronounced bend or loop in the course of a sluggish river or of a valley (MEANDERING VALLEY). The river itself develops the curve of a meander by LATERAL EROSION, the bank on the concave side (the outside) of the curve being eroded by the current (RIVER-CLIFF), while deposition is taking place on the convex side (the inside) (SLIP-OFF SLOPE). Eventually the meander becomes so well-developed that it is nearly circular. At this stage the river may break across the strip of land (the NECK) separating the stream, a cut-off (OXBOW) may be formed, and the river flows on a straight course. DIVAGATING MEANDER, INCISED MEANDER, POOL AND RIFFLE; and the terms below relating to meander.

meander bar POINT BAR.

meander belt the part of the flat floor of a valley across which a stream and its channel wind, i.e. the area between the outer banks of successive MEANDERS.

meander core the piece of land in the centre of an INCISED MEANDER nearly encircled by the river, or completely so if the river has broken through the MEANDER NECK. It is usually the remnant of the spur that helped to cause the meander. Fig 29.

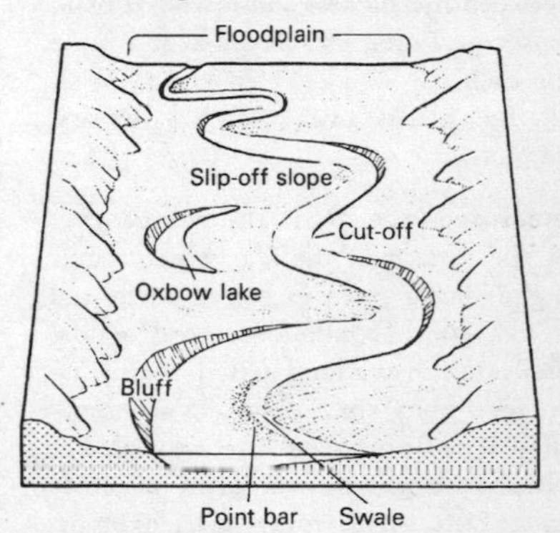

Fig 29 Meanders

meandering valley a winding valley with large, sweeping curves that cut into the solid rocks that contain it. It has extensive deposits of alluvium on the floor over which existing streams have created meanders smaller than those of the valley as a whole. MEANDER, MISFIT RIVER.

meander neck the strip of land separating the stream on each side of a well-developed MEANDER. OXBOW.

meander scar a discernible depression, infilled with deposits and vegetation, marking the channel of a cut-off meander (OXBOW).

meander scroll POINT BAR.

meander terrace a terrace formed on one bank of a river as it meanders in its valley and at the same time (due to REJUVENATION) erodes the valley floor. In flowing across the valley the current on the outside of the curve of the meander cuts down into the former level of the river's FLOODPLAIN, so that part of that higher level is left as a terrace.

mean diurnal range the mean difference between the highest and lowest rainfall, or temperatures, etc. recorded at a place for each day of a selected period (usually four weeks) over a number of years. RANGE-4.

mean sea-level MSL, the average level of the surface of the sea (which varies slightly from place to place), calculated from a series of continuous records of tidal oscillation over a long period, the standard level from which all heights are calculated. The British standard is calculated at Newlyn, Cornwall, and forms the Ordnance Datum (OD) from which heights on British maps are measured.

means of production in Marxism, materials and machinery. The materials are seen as the object of labour (either natural raw materials such as minerals, to which labour is to be applied; or objects such as manufactured components, or harvested crops, on which some labour has already been used). The machinery (i.e. a scythe as well as a complicated machine) is seen as the means or tools of labour, used to transform the materials. CONSTANT CAPITAL, MODE OF PRODUCTION.

mean solar day SOLAR DAY.

mean solar time an average or MEAN SOLAR DAY of 24 hours, a useful measure because the period of time between two successive daily returns of the real sun to the meridian (LOCAL TIME) is not always the same. APPARENT TIME, ROTATION OF THE EARTH.

measurement 1. the magnitude, length, degree etc. of something in terms of a selected unit **2.** in statistics, at its simplest, the classification of individuals, groups or other units and the placing of them in previously defined CATEGORIES-4 (termed categorical or qualitative measurement; if a number can be applied meaningfully to each category it is termed a quantitative measurement). Four scales or levels of measurement are usually employed, each with its particular properties, a particular series of statistical techniques being applied at each level. The scales are the NOMINAL SCALE, ORDINAL SCALE, INTERVAL SCALE, RATIO SCALE (from the simplest to the most precise). CLASS-5, CLASS INTERVAL, VARIABLE-3.

mechanical weathering the disintegration of rock by the agents of weathering (e.g. EXFOLIATION, FREEZE-THAW, GRANULAR DISINTEGRATION, SPALLING etc.) which cause internal and external stresses without chemical alteration of the rock, as opposed to chemical weathering (CORROSION). HYDRATION, ORGANIC WEATHERING, WEATHERING.

mechanization the introduction and use of machines in an industrial process in order to enhance, lighten, or replace human power, an integral part of INDUSTRIALIZATION. INDUSTRIAL REVOLUTION.

medial moraine, median moraine the debris lying centrally in a line along the surface of a GLACIER, occurring when the LATERAL MORAINES of two confluent glaciers meet.

median 1. something situated in the middle 2. in statistics, the centrally occurring value in a data set which is arranged in rank order, i.e. the value above and below which lie 50 per cent of the observations in a distribution. If there is an even number of observations, the median lies midway between the two centrally occurring values. The upper quartile covers the highest 25 per cent of the values, the lower quartile the lowest 25 per cent; and the INTERQUARTILE RANGE is the difference between the lowest number in the upper quartile and the highest number in the lower. ARITHMETIC MEAN, CENTRAL TENDENCY, MODE.

medical geography the study of the geographical aspects of health and the provision of health care. It covers the study of the spatial distribution of human disease and causes of death, together with the factors of the environment conducive to human health and sickness. It includes deaths from disease (MORTALITY), illness not necessarily fatal (MORBIDITY), the diseases permanently located in certain areas (ENDEMIC disease), and the wider spread (PANDEMIC) or sudden outbreak (EPIDEMIC) of disease. It is thus concerned with aetiology (ETIOLOGY) and EPIDEMIOLOGY.

Mediterranean climate the western margin warm temperate climate (one of KOPPEN'S C climates), occurring on the coastal lands round the western Mediterranean and on the narrow western coastal margins of continents in latitudes 30° to 40° (California, central Chile, South Africa, Australia). It is a climate with wide variations, but is generally characterized by mild wet winters (average temperature for coldest month usually over 6°C: 43°F) and hot dry summers (warmest month usually over 21°C: 70°F) with a high amount of sunshine. Some of the most marked variations, especially in precipitation and temperature, occur in the Mediterranean area itself. The regime is due to the dominance of subtropical high pressure systems (ANTICYCLONE) in summer, and the passage of DEPRESSIONS-3 associated with moist winds from the oceans in winter.

megalith a large stone used as a monument or in construction. Some cultures have been marked by the use of such stones (e.g. Stonehenge in England).

megalithic *adj.* applied to the large monumental constructions termed MEGALITHS, to the people who put them up, and to the period in which they were constructed (Megalithic Age), i.e. the cultures of NEOLITHIC to BRONZE AGE times.

megalopolis a very large and spreading urban complex, with some open land, formed when built-up areas, widespread over an extensive area, enlarge to such an extent that they become linked together, as in the northeastern seaboard of the USA, or along the northern shore of the Inland Sea of Japan. ECUMENOPOLIS.

melting point the temperature at which a SOLID becomes LIQUID under normal pressure.

melt-water, meltwater water derived from the melting (MELTING POINT) of snow and ice, e.g. from the SNOUT of a GLACIER.

melt-water channel a channel in the solid rock or in drift deposits in a once-glaciated area, unrelated to the trend of the present drainage pattern, and in some cases cutting across it. DRAINAGE-2.

mental map COGNITIVE MAP.

Mercalli scale, modified a scale formerly used in measuring EARTHQUAKE intensity, based on the observed effects on buildings, etc., ranging from I (detectable only by seismograph reaction) to XII (catastrophic, i.e. the total destruction of buildings). It superseded the ROSSI-FOREL SCALE, and has now in turn been superseded by the RICHTER SCALE.

mercantile *adj.* pertaining to, concerned with, engaged in, trade and commerce.

mercantile model, mercantilist model an approach to the study of urban systems suggested in 1970 by J. E. Vance, American geographer, as an alternative to CENTRAL PLACE THEORY. To summarize, it views wholesaling as the key urban function, and explains the development of an URBAN SYSTEM-1 in terms of trading. His study was particularly related to the evolution of the MERCANTILE cities of the USA Atlantic seaboard.

mercantilism **1.** trade and commerce **2.** a theory popular among European nations during the sixteenth and seventeenth centuries that the economic and political strength of a country lay in its acquiring large quantities of gold and silver, to be achieved by restricting imports, developing production for exports, and prohibiting the export of gold and silver.

mercury Hg, a silver-coloured, poisonous, metallic ELEMENT-6, liquid at temperatures above −38.8°C (−37.8°F), boiling at 356.9°C (674.4°F) under normal pressure. It is used in industrial processes and in BAROMETERS and THERMOMETERS.

mere a large pond or shallow lake, occupying a hollow in glacial drift, especially in TILL OR BOULDER CLAY. The term is used especially in place-names in Cheshire and Shropshire in England, relating not only to water-filled KETTLE HOLES but also to subsidence hollows resulting from the removal of subterranean salt.

meridian terrestrial, one of the lines of LONGITUDE which link the North Pole to the South Pole and cut the equator at right angles, i.e. half of one of the GREAT CIRCLES, the other half being termed the antimeridian. The prime, standard, initial or zero meridian, 0°, is usually considered to be the meridian passing through Greenwich, and meridians are numbered from it to 180° east or west of it. MERIDIONAL-2.

meridian *adj.* pertaining to noon, especially to the position of the sun at noon.

meridional *adj.* **1.** of, pertaining to, characteristic of the south, especially of the inhabitants of the south, particularly of southern Europe **2.** of or pertaining to a MERIDIAN.

meridional flow an atmospheric circulation in which the dominant flow of air is from north to south, or from south to north, across the PARALLELS OF LATITUDE, in contrast to ZONAL FLOW.

merino a Spanish breed of SHEEP with very fine white wool (botany wool), extensively crossbred with other breeds of sheep to improve fleeces.

mesa (Spanish, table) a high, extensive TABLELAND or an isolated flat-topped hill (broader than a BUTTE), the remnant of a plateau that has been subjected to DENUDATION in a semi-arid region. It consists of horizontal strata capped by a more resistant stratum, and one or all sides slope steeply or form cliffs. MESETA.

meseta (Spanish) **1.** the high TABLELAND

of the heart of Spain **2.** an alternative to MESA.

mesic *adj.* applied to conditions in which MESOPHYTES flourish. HYDRIC, XERIC.

mesoclimate the CLIMATE of a local area marked by its deviation from the norms of the MACROCLIMATE in which it lies. In area it is larger than a LOCAL CLIMATE or a MICROCLIMATE.

Mesolithic *adj.* of or pertaining to the culture period midway in the STONE AGE between the PALAEOLITHIC and the NEOLITHIC, from about 8000 to 6000 BC in Europe, when the bow and arrow came into use, the dog was domesticated, pottery came to be made, England was cut off from continental Europe by the rising sea, and there was marked climatic change. ATLANTIC STAGE.

mesopause the layer in the ATMOSPHERE-I some 80 km (50 mi) above the earth's surface, the boundary between the MESOSPHERE and the THERMOSPHERE, and the layer with the lowest temperature. Fig 4.

mesophyte a plant growing under medium conditions of moisture, where there is neither an excess nor a deficiency of water. Mesophytes are typical of the CONIFEROUS FOREST growing in high latitudes (TAIGA) and high altitudes, and include pine species, fir, spruce, larch. HYDROPHYTE, HYGROPHYTE, TROPOPHYTE, XEROPHYTE.

mesosphere the zone in the ATMOSPHERE-I extending from the STRATOPAUSE to the MESOPAUSE, between some 50 to 80 km (30 to 50 mi) above the earth's surface, in which the temperature, having reached its maximum in the stratopause, decreases with height to the mesopause, where it is at its lowest. Fig 4.

metal **I.** an ELEMENT-6 of high specific gravity, with atoms structured in such a way that they easily lose electrons to form positively charged IONS, an element that is usually DUCTILE, MALLEABLE, and a good conductor of heat and electricity **2.** a compound or alloy of such an element. PRECIOUS METAL, ROAD METAL.

metamorphic *adj.* pertaining to, characterized by, formed by METAMORPHISM, applied especially to rocks and rock formations.

metamorphic aureole the zone of COUNTRY ROCK surrounding an IGNEOUS INTRUSION in which the country rock is metamorphosed by heat and migrating fluids (THERMAL or CONTACT METAMORPHISM) from the intrusion. BATHOLITH, MAGMA. Fig 7.

metamorphic rock a well-defined, new type of rock derived from pre-existing rock by mineralogical, chemical and structural changes in the pre-existing rock brought about by its contact with a great heated mass of MAGMA (THERMAL or CONTACT METAMORPHISM) or by folding and pressure (DYNAMIC METAMORPHISM). METASOMATISM.

metamorphism a change of form, applied in geology to the process of the transformation of pre-existing rock into a new, well-defined type of rock by ENDOGENETIC processes, i.e. by the action of heat (THERMAL OR CONTACT METAMORPHISM, METAMORPHIC AUREOLE) or by severe compressional earth movements (DYNAMIC METAMORPHISM, REGIONAL METAMORPHISM) or by both, which modifies or changes the texture, composition, physical or chemical structure of the pre-existing rock. AUTOMETAMORPHISM, FOLIATION, HYDRO-

THERMAL, MAGMA, METAMORPHIC, METASOMATISM, PNEUMATOLYSIS.

metaphysics (Greek, the words of Aristotle that come after the *Physics* in the sequence of his work) **1.** the branch of philosophy concerned with the first principles of things, with what really exists in the world (e.g. the concepts of time, space, being, substance). The investigation is usually conducted by rational argument, not by mystical intuition, and may be described as transcendent (regarding reality as being above and beyond normal experience, e.g. as in the view of the universe supplied by supernatural religion) or immanent (regarding reality as consisting only of objects of experience). COSMOLOGY, EPISTEMOLOGY, IDEALISM-2, ONTOLOGY **2.** with 'of', the theoretical principles or philosophical explanation of a particular branch of knowledge.

metasomatism the change in a pre-existing rock brought about by a solution from external sources which percolates through the rock and (introducing material from other series of rocks) causes chemical reactions by which one mineral in the pre-existing rock is partly or wholly altered into other minerals, or is replaced by another mineral of different composition, without a change in texture. PNEUMATOLYSIS is not involved. HYDROTHERMAL, PETRIFICATION.

meteor a solid body moving rapidly in space glowing when it enters the earth's ATMOSPHERE-1 on account of the heat generated by friction with the atmosphere, and appearing as a 'shooting star' or 'falling star'. The friction causes most meteors to disintegrate to dust (meteoric dust), but some solid remnants may reach the earth's surface as METEORITES.

meteoric dust the dust in the ATMOSPHERE-1 derived from disintegrating METEORS and trapped in the earth's gravitational field, a component of the dust present everywhere in the universe.

meteoric water water on the earth's surface which is derived from PRECIPITATION, as distinct from CONNATE WATER, JUVENILE WATER.

meteorite a solid extra-terrestrial body which reaches the earth's surface, typically formed of METALS (iron-nickel), or SILICATES or a combination of both described as stony, stony-iron or iron according to composition, a fragment of a broken ASTEROID. Meteorites vary in size, from as small as a dust particle to the exceptionally large 10 km (6 mi) diameter example that fell at the Gulf of Mexico some 65 million years ago. Most meteorites reaching the earth's surface weigh about one kg (2.2 lb).

meteorological screen STEVENSON SCREEN, a white painted wooden box on legs, about 1 m (3 ft) above the ground, designed to shelter meteorological instruments from strong wind and solar and terrestrial radiation. The roof is usually insulated, the sides are louvred (one usually hinged to act as a door), so that air can flow through freely; and the THERMOMETERS within (usually dry and wet bulb, maximum and minimum thermometers) are supported on a frame, standing free from the roof, floor and sides, so that they give shade readings as accurately as possible. Some types also house a hygrograph, recording changes in RELATIVE HUMIDITY and a thermograph, a self-recording thermometer.

meteorology the scientific study of the processes and physical phenomena operating in the earth's ATMOSPHERE-1 in the short term, as distinct from CLIMATOLOGY, which is concerned with the long term. Weather forecasting depends on meteorological studies, and on the whole meteorologists confine their attention to the layers of the earth's atmosphere where the weather that affects the earth's surface is generated, i.e. to the TROPOSPHERE and STRATOSPHERE. MICROMETEOROLOGY.

metre m, the basic SI unit of length, defined in 1960 from a wavelength in the spectrum of krypton, equivalent to about 39.37 in. A metre is divided into 10 decimetres (rarely used) or 100 centimetres (cm) or 1000 millimetres (mm). In measures of area 1 square metre (10.8 sq ft) equals 10 000 square centimetres or 1 million square millimetres. In measures of volume 1 cubic metre (35.315 cu ft) equals 1 million cubic centimetres.

metropolis 1. the mother city of a colony or settlement, hence the capital or chief city of a country, serving as the seat of government or of ecclesiastical authority, or as the main commercial centre **2.** the largest town or agglomeration of people in an extensive area. ECUMENOPOLIS.

metropolitan *adj.* of, pertaining to, characteristic of, or constituting, a METROPOLIS in the limited senses defined under (**1**), but also as applied to those large urban centres etc. specified under (**2**).

mica a group of transparent SILICATE minerals, common in ACID IGNEOUS ROCKS, in METAMORPHIC ROCKS, and in derived sediments, which, having perfect CLEAVAGE-1, can easily be split into very thin, tough, pliable, lustrous plates, used as an electrical insulator and as a heat-resistant substitute for glass. The main micas are the dark BIOTITE; the light-brown, green or red muscovite (both occurring in GNEISS and SCHIST); and the yellow phlogopite (rich in magnesium and present in MARBLE and PERIDOTITE).

micro- (Greek) small, as opposed to MACRO-, large, e.g. MICROCLIMATE.

microclimate the CLIMATE of a very small area, the modification of the general climate produced by conditions in the immediate environment of a subject, e.g. of a cereal crop, of a building. The area involved is smaller than that of LOCAL CLIMATE and is confined to the very shallow layer of the atmosphere close to the ground. MACROCLIMATE, MICROCLIMATOLOGY, MICROMETEOROLOGY.

microclimatology the scientific study of MICROCLIMATES. MICROMETEOROLOGY.

microfinance, microcredit a financial service inaugurated by Bangladeshi economist, Professor Muhammad Yunus, through the Grameen Bank, 1976, designed to help impoverished individuals and small groups in Third World countries. Very small personal loans, with minimal rate of interest, are advanced to encourage income-generating activities, to give support in times of stress (e.g. drought, house repairs), to protect borrowers from extortionate interest on loans from other providers, to finance weddings and funerals. Advances range from as little as fifteen to hundreds of dollars, depending on need and local rates of exchange. Savings are protected, accumulated capital used for further loans. Capital is initially provided by various small banks, NGOS (e.g. Oxfam, Action Aid), care institutions, international agencies (e.g. European Union, British Overseas Development Administration, UN Conference on Trade and Industry etc.) and the WORLD BANK.

micrometeorology the detailed scientific study of the layer of the ATMOSPHERE-1 nearest to the earth's surface, i.e. from ground level up to about 1.2 m (4 ft), important in the study of MICROCLIMATES. MICROCLIMATOLOGY.

micronutrient an ELEMENT-6 or combination of elements needed by an organism in order to maintain health, but only in very small quantities. The micronutrients needed by plants in varying amount are boron, cobalt, copper, iron, manganese, molybdenum, zinc. MACRONUTRIENT, TRACE ELEMENT.

microorganism any organism too small to be seen by the unaided eye, being of microscopic or even smaller size (e.g. a BACTERIUM).

Middle East part of the earth's surface EAST of Europe, now usually taken to include parts of southwest Asia and northeast Africa, stretching from Turkey through Iran, Iraq and the countries of Arabia to Sudan and Egypt and including the countries bordering the eastern shores of the Mediterranean. FAR EAST, NEAR EAST.

midlatitude *adj.* applied to the latitudinal zone lying between 23°30' and 66°30' in the northern and southern hemispheres. TEMPERATE.

midlatitudes LATITUDE.

midnight sun the SUN-2 still to be seen shining above the horizon at midnight in latitudes greater than 63°30' north or south, in the northern hemisphere between mid-May and the end of July, and in the southern between mid-November and the end of January. SOLSTICE.

mid-ocean ridge OCEANIC RIDGE.

migrant one who migrates. EMIGRANT, EXILE, EXPATRIATE, IMMIGRANT, MIGRATION, REFUGEE.

migration 1. the act or process of moving from one place to another with the intent of staying at the destination permanently or for a relatively long period of time **2.** of humans, such a movement from one area (usually the home area) to work or settle in another. EMIGRANT, EXILE, EXPATRIATE, IMMIGRANT, PUSH-PULL THEORY, REFUGEE **3.** of animals, e.g. birds, the seasonal movement from one region to another **4.** of plants, the movement to extend habitat.

mile (Latin mille passus or passuum, a thousand paces) mi, a British unit of linear measurement **1.** statute mile, 1760 yards or 5280 feet (1609.35 metres) **2.** geographical mile, one minute of arc measured along the equator, 6087.2 ft, rounded to 6080 ft (1852 m) **3.** nautical mile, one minute of arc or $\frac{1}{21\,600}$ of a GREAT CIRCLE, standardized in Britain as one minute of arc at 48°N, 6080 ft, equivalent to 1.1516 statute mile (1853.25 m) **4.** international nautical mile (used by the USA and other countries) 6076.1033 ft, approximately 1.15 statute mile (1852 m). KNOT **5.** in measures of area, 1 square mile (2.59 square kilometres) equals 640 ACRES each of 4840 square yards or 43 560 square feet.

millet any of several small-grained cultivated cereals, the most widely grown food grains in tropical regions, being drought-resistant, tolerant of poor soils, storing well, some having a higher mineral content than that of other cereals, e.g. BULRUSH MILLET, FINGER MILLET. Locally important in Africa are Job's tears, hungry rice and TEFF. SORGHUM is sometimes classified as a millet. The most important millets of temperate regions are common millet, little millet, foxtail millet, Japanese millet.

milli- m, prefix, one thousandth, attached

to SI units to denote the unit × 10^{-3}, e.g. millimetre (mm), one thousandth part of a METRE. CENTI-, HECTO-, KILO-.

millibar mb, a unit of atmospheric pressure indicated by a BAROMETER, equal to 1000 DYNES per square centimetre of MERCURY; 1000 mb equals 1 BAR-3. The formula for conversion is not universally applicable, but at 0°C (32°F) at latitude 45°, 29 in mercury = 982 mb; 30 in = 1016 mb; 31 in = 1049 mb; and 1000 mb = 750.1 mm (29.531 in). Millibars are commonly used in drawing weather charts, isobars (ISO-) being drawn at 2 or 4 millibar intervals.

million city a city with a resident population of one million people or more.

Millstone Grit a hard, coarse-grained, SANDSTONE occurring in Britain in the central Pennines and in Northumberland, under the Coal Measures at the base of the Upper CARBONIFEROUS. It was probably a delta-deposit laid down in a shallow sea.

Minamata disease a disease of the central nervous system caused by MERCURY poisoning. The name is derived from a bay and town in Japan where many people were poisoned in 1959 by eating fish and shellfish which had ingested dimethyl mercury, present in the sediments of the bay as a result of effluent discharged by a nearby factory. POLLUTION.

mine an excavation deep in the ground, made for the purpose of extracting MINERALS, together with its shaft and associated buildings. MINING, OIL WELL, OPENCAST MINING, PIT, QUARRY.

mineral **1.** a naturally occurring inorganic HOMOGENEOUS-2 substance, usually CRYSTALLINE with a definite chemical composition (capable of being expressed by a chemical formula, or varying only within certain limits); hence the distinction from a rock, which is commonly (apart from rock salt) a mixture of minerals. On this definition water (ice) is a mineral, but (being of organic origin) petroleum and natural gas are not, although they are commonly described as such, as are some organically-derived limestones and siliceous rocks **2.** loosely applied to any mineral deposit won by mining, e.g. metallic ore, coal.

mineral horizons one of the two major classes of SOIL HORIZON, the other class being the O HORIZONS. The mineral horizons consist mainly of inorganic matter. MINERAL SOIL.

mineralization **1.** the process whereby gases and water from the heated lower layers of the earth's crust by passing through fissures, cracks etc. cause changes in the rocks, and the deposition of minerals of economic importance. METASOMATISM **2.** the replacement of the organic parts of a plant or animal by MINERALS-1 during decomposition or fossilization. FOSSIL.

mineral oil any oil of so-called MINERAL-1 origin, notably PETROLEUM.

mineral soil a soil low in HUMUS, consisting mainly of material of MINERAL-1 origin. MINERAL HORIZONS, ORGANIC SOIL.

mineral spring a SPRING-2 containing a high proportion of mineral salts in solution.

minifundio (Spanish) **1.** a very small farm in Latin America **2.** a very small plot of land cultivated by a Spanish settler for personal use in return for military service to the Spanish crown. LATIFUNDIA.

minimax location the site for a firm

selected by an entrepreneur looking for the minimum cost and the likelihood of maximum profit for the enterprise. LOCATION THEORY.

minimum thermometer a THERMOMETER used to register the lowest ambient air TEMPERATURE reached at a particular place over a selected period of time. One type commonly used consists of a horizontal glass tube filled with alcohol with a dumb-bell shaped metal marker held at one end of the alcohol by surface tension. As the temperature falls the alcohol contracts and the marker is drawn to the level of the lowest temperature registered, where, even if the temperature rises, it sticks until reunited with the alcohol in its original position (by the use of a magnet). MAXIMUM THERMOMETER, MAXIMUM-MINIMUM THERMOMETER.

mining **1.** deep excavating in the earth in order, by underground workings, to extract MINERALS-1, METALS, METALLIC ORES. A distinction is sometimes made between mining (also termed deep mining) by underground workings and QUARRYING (also termed surface mining), conducted at the surface (PIT, QUARRY). OPENCAST MINING is associated with extensive surface workings, whereas quarrying is more limited in extent **2.** the extraction of non-renewable NATURAL RESOURCES.

minor intrusion in geology, an IGNEOUS INTRUSION that is small when compared with a large, deep-seated plutonic intrusion. Examples include DYKES, SILLS, small LACCOLITHS and VEINS.

minute a unit of measurement applied to **1.** time, one-sixtieth of an hour **2.** an angle, one-sixtieth of an (angular) degree; one-sixtieth of a degree of latitude or longitude.

mirage an optical phenomenon in which distant objects may be seen inverted, as if mirrored in water, or suspended in mid-air. This is due to the unusual distribution of density in the ATMOSPHERE-1. For example, when the air near the ground is greatly heated by conduction and becomes less dense, rays of light from above (approaching at a slight angle) are refracted towards the observer, i.e. they are bent upwards so that the sky appears as a glistening sheet of water on, for example, a road surface. In high latitudes, when a warm layer of air rests on a cold layer, the light rays may be bent down from a warm layer, so that an inverted or even double image of a distant object appears.

misfit river, misfit stream, underfit river a river that appears now to be too small for its valley, due to BEHEADING by another stream, or a change of climate, or the valley's being enlarged and broadened by glaciation, or because the volume of water has been reduced by seepage through the thick alluvium of the flood-plain. MEANDERING VALLEY.

mist obscurity in the lower layers of the ATMOSPHERE-1 caused by particles of condensed moisture held in suspension, limiting vision (officially) to between one and two kilometres (about 1000 to 2000 yards). FOG, HAZE.

mist forest, cloud forest hygrophilous forest (HYGROPHYTE) occurring on mountain slopes in tropical regions, where mist or cloud is constant or frequent.

mistral (French) a powerful, cold, dry northwesterly or northerly wind, blowing from the high Massif Central towards the relatively warm Golfe du Lion, particularly affecting the Rhône delta and the north coast of the Mediterranean. It is particularly common in winter when the cold air from the winter high pressure system lying over central Europe is channelled through the lower Rhône valley to the low pressure area lying over the western Mediterranean. BORA.

mixed cultivation the growing of two or more crops intermingled on the same field or plot, especially a mixture of tree and ground crops.

mixed economy an economic system in which some parts operate according to the forces of the free market (production, distribution and exchange being in private hands), while other parts are in the hands of the government (i.e. there is more STATE INTERVENTION in a mixed economy than there is in a true MARKET ECONOMY). CENTRALLY PLANNED ECONOMY, PLANNED ECONOMY, STATE CAPITALISM.

mixed farming AGRICULTURE in which both crops and livestock are produced on an individual farm (not to be confused with MIXED CULTIVATION). ARABLE LAND, CROP FARMING, FARMING, LIVESTOCK FARMING.

MNC, MNE MULTINATIONAL COMPANY, CORPORATION OR ENTERPRISE.

mobile *adj.* **1.** capable of moving or being moved from place to place **2.** moving or moved with ease, fluid **3.** of organisms, having the power of moving from one place to another.

mobile dune a coastal DUNE partially fixed by vegetation but still liable to DEFLATION-1 and BLOW-OUTS-1, a type transitional between a FORE-DUNE and a STABILIZED DUNE.

mobile industry FOOTLOOSE INDUSTRY.

mobility the state or quality of being MOBILE, e.g. of people moving readily from one place to another and from job to job, in search of employment or higher incomes; or of individuals in relation to the degree to which each one has access to travel facilities (personal mobility); or of individuals and households in being able to move between social classes and income groups (social mobility).

mode in statistics, the value of the variable occurring most frequently in a set of observations. ARITHMETIC MEAN, CENTRAL TENDENCY, MEDIAN.

model **1.** a three-dimensional representation, more or less to scale, of something that exists or is to be constructed, e.g. of a building, which may be to scale in every dimension and detail; or of a landscape, with ground measurements to scale but the vertical scale exaggerated to show ALTITUDE-3; or a working scale model, such as one representing tides, which includes processes **2.** a representation of some aspects of reality, selected and brought together to show certain of its properties, and providing a working HYPOTHESIS against which reality can be tested. On a range of abstraction these are, at the first level, the iconic model, the most realistic, a scaled-down representation of reality (e.g. with real phenomena shown in their characteristic form, but scaled-down). At the second level is the analogue model in which real phenomena are represented by different but analogous phenomena (e.g. clusters of people shown by clusters of points). The final level of abstraction, the furthest from reality, is reached by the symbolic model, in which real phenomena are represented by mathematical expressions.

mode of production (one of the central concepts in Marxism) the economic base of a society, the way in which the productive activities in the society are organized, and thereby affect the social as well as the economic relations in that society. Marx identified the historical succession of modes as primitive communal, slave, feudal, capitalist, state capitalist, socialist, communist, the changes of mode being brought about by class conflict, e.g. between landlord and peasant, capitalist (BOURGEOISIE-2) and workers (PROLETARIAT-2) as each class tries to gain control of the MEANS OF PRODUCTION, asserting that the other class is inept at providing society with an acceptable level of subsistence. Marx maintained that the proletariat would eventually overcome the capitalist ruling class (DIALECTICAL MATERIALISM) and that a classless communist society would emerge. CAPITALISM, COMMUNISM, FEUDALISM, HISTORICAL MATERIALISM, POST-INDUSTRIAL SOCIETY, SOCIALISM.

moder in soil science, a HUMUS layer intermediate in composition between MOR and MULL-3, occurring where decomposition is greater than in mor, but has not advanced so much as in mull, but where there is some mixing with particles from the underlying MINERAL SOIL, due to the presence of soil fauna.

moderately developed country MDC.

modern *adj.* of the present day, up-to-date, contemporary, differentiated in thought from the past, particularly the recent past. MODERNISM, MODERNITY, POSTMODERNISM.

modernism, Modernism 1. the mental acceptance of MODERNITY-2, of being MODERN **2.** Modernism, the culture of innovation and technological change that emerged in Europe and spread internationally in the latter part of the nineteenth century, beginning to fade in the late 1930s, and gradually merging into POSTMODERNISM. To select some elements from the socio-economic-political background that appear in geographical studies, this was a period influenced by POSITIVISM and LOGICAL POSITIVISM, a time of mass production (FORDISM); the consolidation of the power of European countries in their African colonies; the destruction and aftermath of the First World War; high unemployment in Europe; the rise of COMMUNISM-2 and FASCISM; the emergence of the USA as a leading world power; KEYNESIANISM; rapid progress in scientific, technological, psychological and sociological research etc. To generalize, the promoters and followers of Modernism rejected traditional, representational and romantic art in favour, for example, of impressionism, cubism, surrealism and abstract art in the visual arts; jazz and atonality in music; FUNCTIONALISM and unornamented, pure geometrical forms (in concrete) in architecture that came to be institutionalized as the International Style, and high-tech glass and steel structures; among other styles in writing, the stream-of-consciousness style is associated with Modernism. HUMANISTIC GEOGRAPHY.

modernity 1. the state or quality of being MODERN **2.** the unique forms of social system, activities, technology, attitudes and values that characterize a MODERN society, expressed particularly in the visual arts, architecture, literature and music of the time: e.g. in eighteenth-century Europe, expressed by the promoters and followers of the Enlightenment (similar, in part, to the humanists of the RENAISSANCE), with its rejection of religious beliefs and prejudices and emphasis on democratic, rational, scientific and materialistic approaches, at a time when great advances in science and technology gave rise to INDUSTRIALIZATION, rapid population increase and urban growth, changes in land use patterns and land management, and an ANTHROPOCENTRIC attitude that viewed the NATURAL ENVIRONMENT as a source of COMMODITIES. These elements combined to promote international trade and the spread of CAPITALISM with its cumulative economic power. The writings of Newton, Locke and Hume were influential in England; Voltaire, Descartes, Diderot (and the Encyclopedists) in France. The Classical music of the time gradually evolved from the Baroque; in architecture there was a revival of classical forms; in the visual arts David expressed the high moral tone of the movement, Hogarth portrayed the decadence of the affluent urban society of the time. Twentieth-century examples of MODERNITY-2 are MODERNISM and POSTMODERNISM. CLASSICAL ECONOMIC THEORY, HUMANISTIC GEOGRAPHY, UTILITARIANISM.

modernization **1.** in general, the process or act of making up-to-date, of changing something from the past so that it becomes in harmony with current taste, thinking, technology **2.** in society, a process of social change which commonly accompanies or follows INDUSTRIALIZATION and may include an increase in social mobility, the blurring of boundaries between social classes, the advancement of education, the development of social services, and the adoption of procedures in government more effective than those previously prevailing.

mogote (Cuba: Spanish) a karst INSELBERG, a steep-sided limestone residual hill in KARST, rising from a nearly flat alluviated plain, formerly termed a haystack hill in some areas, e.g. in Puerto Rico. The term was originally local to Cuba, where it referred to residual hills of folded limestone, but it is now applied internationally to karst residual hills in tropical regions.

Mohorovičić Discontinuity, Moho the boundary surface between the MANTLE of the earth and the rocks of the earth's surface, lying at a depth of some 40 km (25 mi) under the continents but only some 6 to 10 km (4 to 6 mi) under the ocean. In the Discontinuity, owing to the different densities of the crust and the mantle, there is a very sharp change in the rate of travel of EARTHQUAKE waves (PUSH WAVE, SHAKE WAVE): they accelerate.

Moh's scale HARDNESS OF MINERALS.

moisture index THORNTHWAITE'S CLIMATIC CLASSIFICATION.

molecular *adj.* pertaining to, involving, or consisting of, MOLECULES.

molecular attraction the attraction of MOLECULES for each other, especially, in HYDROLOGY, the attraction of the molecules of the surfaces of solid rocks for the molecules of water, and of the water molecules for each other, the means by which some GROUND WATER is held in fine-grained rocks. CAPILLARITY.

molecule two or more ATOMS linked by chemical bonding and constituting the smallest group of combined atoms of an ELEMENT-6 or of a compound which can exist freely (FREE-2) while retaining the characteristic properties of the substance.

mollisols in SOIL CLASSIFICATION, USA, an order of soils characteristic of grassland, with a thick, organically-rich surface layer. It includes soils with widely varied profiles, all structurally well developed, e.g. BRUNIZEM, CHERNOZEM, RENDZINA.

monadnock a residual hill, a remnant of erosion, left standing above the general level of a denuded plain (PENEPLAIN). The term is derived from the name of a mountain of this character in New Hampshire, USA.

monetarism the economic theory which asserts that the level of activity of an economy can be controlled by controlling the money supply. KEYNESIANISM.

monocline 1. (British) a FOLD-2 in which the bend is in only one direction (i.e. a fold with one limb, a single bend in horizontal beds), the rock stratum, through TENSION in the earth's crust, changing its dip by increasing the steepness of inclination, and then levelling out again or resuming its original dip. It is termed a monocline because only one fold, or one half of a fold, is presented instead of the two occurring in an arch or trough **2.** (American) a synonym for HOMOCLINE (British usage), i.e. a structure of several beds dipping evenly in one direction. Fig 24.

monoculture cultivation in which a single crop predominates and is planted successively on the same land, in contrast to a ROTATION OF CROPS.

monopoly 1. the exclusive control of the supply of a product or service in a particular market by a single supplier, who thus dominates the market **2.** an exclusive right to conduct a particular business or provide a particular service, granted by a ruler, government, etc. **3.** a COMMODITY-1 under exclusive, single control **4.** a single supplier who has exclusive control. DUOPOLY, OLIGOPOLY; PERFECT COMPETITION.

monsoon (Arabic mausim, season) originally applied to the regular winds of the Arabian sea, blowing for six months from the northeast, six months from the southwest. Now generally applied to those and some other winds that blow with considerable regularity at definite seasons of the year due to the seasonal reversal of pressure over land masses and their neighbouring oceans, combined with the influence of JET STREAMS. In the typical area of the Indian subcontinent and southeast Asia it is the seasonal inflowing moist winds that bring rain, hence the monsoon season is termed the RAINS, and the term monsoon is applied to the rains without reference to the winds.

monsoon forest the FOREST-1 of the tropical monsoon lands where the annual rainfall is between 1000 and 2000 mm (40 and 80 in) and there is a marked dry season. It consists of BROADLEAVED TREES that lose their leaves in the hot dry season (February to May in the Indian subcontinent and Burma). The trees, mainly HARDWOOD, do not grow so closely together as those in EQUATORIAL FOREST, and the number of species is low.

montane *adj.* of or pertaining to mountain regions, applied particularly to the vegetation growing on high land below the TREE LINE.

montane forest the FOREST-1 of the cool uplands in the zones of tropical and equatorial climates. MOSSY FOREST.

moon 1. a natural satellite of any planet **2.** the earth's only natural satellite, appearing to move in the CELESTIAL SPHERE, in relation to the stars, from west to east, responsible with the sun for tidal action on the earth. It revolves round the earth in a slightly elliptical orbit, the distance from the earth varying from 348 285 km (216 420 mi) surface to surface PERIGEE to 398 587 km (247 67mi) APOGEE. The diameter is about one-quarter of that of the earth, its mass 1/81 of the earth. It has no atmosphere, no water, and shines by reflecting light emitted by the sun. It revolves round the earth every 27 days 7 hours 43 minutes 15 seconds (sidereal month), and one revolution related to the sun on average 29 days 12 hours 44 minutes (lunar or synodic month, i.e. from one 'new' moon to the next); the LUNAR DAY is about 24 hours 50 minutes. Because it rotates on its own axis once in each revolution of its orbit (i.e. the period of rotation is the same as the period of revolution), the same face is always seen from the earth. The changes in aspect seen from the earth are due to changes in the relative position of the earth, moon and sun, and are termed phases, the sequence being: new moon, invisible to faintly visible, the moon lying between the earth and the sun (CONJUNCTION) so that viewed from earth it is not illuminated by the sun's light, although it may reflect a faint glow of light from the earth; the first quarter (QUADRATURE), when the moon has moved through about one-quarter of its orbit round the earth, and appears as a semicircle, having 'grown' from a crescent, bow facing west; gibbous moon, the phase reached when the moon has passed through another eighth of its orbit, so that three-quarters of its face as seen from earth is illuminated by the sun; full moon, when the earth lies between the moon and the sun (OPPOSITION) so that viewed from earth the whole face of the moon is illuminated by light from the sun. From new moon to full moon the moon is said to be waxing; from full moon to new moon it is waning. SYZYGY, TIDE.

moor, moorland open, unenclosed land, generally elevated, with acid peaty soil, which is not good pasture though used to some extent as ROUGH GRAZING. Different types are distinguished by dominant plants, hence cotton grass moor, heather moor, etc.

mor in soil science, raw HUMUS, low in animal life, acidic and crumbly, unmixed with and sharply demarcated from the underlying MINERAL SOIL. MODER, MULL-2,3.

moraine (French) an accumulation of unstratified (STRATIFIED) debris, especially boulders and coarse material, carried down and deposited by a GLACIER or ICE SHEET. Some of the debris falls from above, from the mountain slopes, on to a glacier (SUPERGLACIAL), some is plucked (PLUCKING, SUBGLACIAL) from the floor beneath a glacier or ice sheet by ice action. On the surface of a VALLEY GLACIER there are usually LATERAL MORAINES on each side, and a MEDIAL MORAINE if two valley glaciers have joined. ENGLACIAL moraines are those enclosed in the ice. The debris may be deposited as GROUND MORAINE when the ice melts, or as a TERMINAL or END MORAINE. The term applies both to the material and to the feature produced. PUSH MORAINE, RECESSIONAL MORAINE, SUBGLACIAL MORAINE. Fig 20.

morbidity the incidence of disease in a population.

more developed country MDC. UNDERDEVELOPMENT.

morphology the scientific study of the form, structure, origin and development of organisms; or of the external structure of rocks in relation to form; or of landforms or topographic features resulting from erosion. GEOMORPHOLOGY.

mortality **1.** the number of deaths in a particular period or place **2.** the death-rate (BIRTH-RATE). MORBIDITY.

mosaic **1.** in aerial photography, a composite photographic representation of an area obtained by joining together individual photographic prints **2.** in ecology, a vegetation pattern in which community types are interspersed.

moss **1.** a class of primitive land plants distributed throughout the world, growing on rocks, walls, trees, heaths, damp ground **2.** a term applied to a BOG or SWAMP, especially in northern England and parts of Scotland, dominated by bog moss (SPHAGNUM); and also to summit plateaus of the southern Pennines where the moorland is dominated by cotton grass (*Eriophorum vaginatum*) and some *Sphagnum*; and to coastal marshes with FEN PEAT, e.g. Solway Moss.

mossy forest FOREST-1 in which the living and dead trees and ground surface support a strong growth of MOSSES-1, occurring particularly on uplands supporting MONTANE FOREST, e.g. in the humid cloudy conditions of the windward side of a high OCEANIC ISLAND in the TROPICS.

mother-of-pearl cloud nacreous cloud, a rare, iridescent, usually LENTICULAR CLOUD, occurring in the upper layers of the STRATOSPHERE when atmospheric pressure and temperature are low.

mottled *adj.* marked with blotches, with patches of differing colours, e.g. in a GLEY SOIL.

moulin (French, a mill) glacier mill, a steep shaft or vertical circular hole carved through the ice of a GLACIER or ICE SHEET by a stream of melt-water as it swirls, laden with rock debris, down a fissure in the ice.

mountain any natural elevation of the earth's surface with a summit small in proportion to its base, rising more or less abruptly from the surrounding level. In Britain the term is commonly restricted to an elevation exceeding 600 m (2000 ft), the term HILL being applied to a lower elevation; but mountain may be applied to an elevation even under 300 m (1000 ft) if it rises sufficiently abruptly from the surrounding level.

mountain building OROGENESIS.

mountain chain a complex series of roughly parallel mountain ranges (RANGE-2) forming a connected system, i.e. consisting of more than one mountain range. CHAIN.

mountain wind ANABATIC, FOHN, KATABATIC.

mud **1.** an unconsolidated rock of clay and/or silt grades (GRADED SEDIMENTS) with much water, e.g. as commonly deposited in estuaries, lakes, lagoons and at depths under the ocean. It may be partly consolidated in certain geological formations to form MUDSTONE, resembling a soft shale, but non-plastic. MUD FLAT **2.** a manufactured SLURRY-1 used in the process of sinking BORES-2.

mud flat an expanse of fine clay or silt (GRADED SEDIMENTS) deposited by FLOCCULATION-2 in estuaries and sheltered coastal areas (e.g. behind a sand-spit), covered by water at high tide and sometimes colonized by hygrophilous plants, such as MANGROVE in tropical areas, various grasses in cooler regions. HYGROPHYTE.

mudflow a moving mass of soil made fluid by rain or melting snow, i.e. a mud avalanche. AVALANCHE, EARTHFLOW, MASS MOVEMENT.

mudstone a general term applied to an unlaminated, non-plastic, indurated SEDIMENTARY ROCK, consisting of CLAY MIN-

ERALS and other clay-grade constituents. GRADED SEDIMENTS, INDURATION.

mud-volcano **1.** a cone of mud associated with escaping gases in the earth's surface, formed when gases, trying to escape through a stratum of wet clay, whisk the clay to a soft slurry which is ejected at the surface with hisses and bubbles, though usually quite cold, to build a cone **2.** an ejection of hot mud from the volcanic vent, building a small ephemeral cone. VOLCANO.

mulga (Australia: aboriginal term) a scrubby tree or spiny shrub, *Acacia aneura*, dominant over large areas of arid Australia, hence the term mulga scrub. BRIGALOW, MALLEE, MALLEE SCRUB.

mull **1.** (Scottish, from Gaelic) a promontory or headland in Scotland **2.** (Swedish) in soil science, mild HUMUS derived mainly from leaf-mould, occurring as a surface layer in DECIDUOUS forests **3.** in soil science, a loose, crumbly humus layer mixed with the underlying MINERAL SOIL, occurring where decay is rapid in the presence of a plentiful soil fauna, including especially earthworms. MODER, MOR.

multiband spectral photography photography which makes use of a combination of selected narrow spectral bands in the VISIBLE LIGHT and the NEAR-INFRARED wavelengths to give simultaneously two or more images of the same area. ELECTROMAGNETIC SPECTRUM, REMOTE SENSING.

multifinality the end state, one of dissimilarity, achieved by systems founded on similar initial conditions and undergoing change. EQUIFINALITY.

multinational, multinational company/Corporation (MNC), multinational enterprise (MNE) a very large business enterprise which has subsidiary companies, branches, offices, factories etc. in very many countries, and commonly ranges world-wide, that is controlled from headquarters in the country of origin, as distinct from a transnational corporation (TNC) which operates in just two or more countries, and locates any of its functions (including its headquarters) in the most advantageous location.

multinational *adj.* of, pertaining to, or consisting of, many nationalities or ETHNIC-2 groups. MULTINATIONAL.

multiple deprivation a state in which people (especially a family) who are disadvantaged in one respect are also likely to be disadvantaged in others, e.g. those disadvantaged by poverty and hence a low income, are likely to have poor housing, poor health etc. DEPRIVATION, POVERTY CYCLE.

multiple land use the use in common of a tract of land for two or more purposes, one of which is in many cases recreational (RECREATION), e.g. as in NATIONAL PARKS in Britain. PARALLEL LAND USE.

multiplier effect the way in which an increase or decrease in activity acts as a stimulus to the initial effect of that activity, and thereby multiples its effect, e.g. the opening of a new factory will give rise to employment additional to that in the factory itself by stimulating employment in firms providing goods and services to the factory, and in the local services and shops which meet the needs of those working in the factory. All this additional employment increases purchasing power, which gives rise to further employment in a wide range of other firms, and so on. Reverse effects are generated by the closure of a factory, resulting in an increase in unemployment.

multiracial *adj.* of, pertaining to, or consisting of, people of several ETHNIC-2 or cultural groups, coexisting amicably and cooperatively together, each group having equal rights and opportunities.

multispectral *adj.* applied in REMOTE SENSING to a device that makes use of several wavebands in recording images. ELECTROMAGNETIC SPECTRUM.

multispectral scanner in REMOTE SENSING, a scanning device that operates simultaneously in various wavebands in recording images. ELECTROMAGNETIC SPECTRUM, LANDSAT, PIXEL.

multispectral sensing in REMOTE SENSING, the recording of images by one or more SENSORS operating in several wavebands. ELECTROMAGNETIC SPECTRUM.

multivariate analysis **1.** a statistical analysis of data in which more than one type of measurement or observation is involved, the number of VARIABLES being greater than two **2.** in REGRESSION ANALYSIS, an explanation of VARIANCE in terms of several variables, taking into account not only the relationship of INDEPENDENT to DEPENDENT variables, but also the interrelationship of independent variables. FACTOR ANALYSIS.

municipal *adj.* of, relating to, or carried out by, local self-government, especially of a town or city.

muskeg (Canada: Algonquian origin) an undrained basin in subarctic or transition forest region in Canada, filled in with PEAT and bog moss (SPHAGNUM). There are dense stands of tamarack and black spruce around the margins, the trees declining in height towards the centre of the bog.

mutation a changing or being changed, especially a sudden change in the CHROMOSOMES of a cell, the changes in the DNA of individual genes (gene-mutation) being the most common. Mutations occurring in the gametes (reproductive cells) or their precursors can produce an inherited change in the characteristics of the organisms that develop from them; thus mutation is a potent force in evolution.

mutualism **1.** broadly, any association of one organism with another of a different species **2.** SYMBIOSIS, pure, but also modified to the extent that neither partner is vitally important or totally beneficial to the life of the other **3.** COMMENSALISM **4.** the concept that mutual dependence is an essential, basic factor if social well-being is to be achieved.

mutually exclusive categories in statistics, two CATEGORIES characterized by the fact that an observation may fall in either one of them, but not in both.

mycorrhiza the association of a FUNGUS with the root of a higher plant. There are two main types: endotrophic, in which the vegetative part of the fungus (i.e. the mycelium) is within the cortex cells of the root; and ectotrophic, in which the mycelium is external, covering the smaller roots completely. The association is not always mutually beneficial; but in some cases it seems to be helpful, or even vital, to the host, e.g. mycorrhizal fungus is essential to the growth of seedling pine trees. MUTUALISM-1, MYCOTROPHIC.

mycotrophic *adj.* applied to a plant involved in MYCORRHIZA.

N

nadir **1.** the lowest point **2.** the point on the CELESTIAL SPHERE directly opposite to the ZENITH **3.** in REMOTE SENSING, the point on the ground vertically underneath the centre of the sensor.

NAFTA North American Free Trade Area, a trading union comprising USA, Canada and Mexico.

nālā (Indian subcontinent: Urdu-Hindi) a dry river bed, or one with an intermittent stream; commonly anglicized as nullah. ARROYO, NULLAH, WADI.

nanoplankton, nannoplankton the extremely small microscopic organisms, both PHYTOPLANKTON and ZOOPLANKTON, occurring in bodies of fresh or salt water. Fossil nanoplankton are common in PELAGIC sediments. DIATOM, DIATOM OOZE, PLANKTON.

nappe (French, a tablecloth) **1.** a very large overfold in the earth's crust, an OVERTHRUST mass of rock in a near horizontal fold that has moved forward for many kilometres from its 'roots', covering the formations beneath (as a cloth over a table). It may be either the hanging wall of a low-angled THRUST FAULT or a RECUMBENT FOLD in which the reversed middle limb has been sheared out by great pressure. As a result of later DENUDATION a piece of nappe may be left isolated as a nappe outlier (German klippe) **2.** applied less precisely, especially in France, to any overlying sheet of rock, e.g. a LAVA flow, equivalent to the German decke.

narrow gauge of railway, a gauge less than 143 cm (4 ft 7 in). RAILWAY GAUGE.

narrows a constricted passageway in a STRAIT, in part of a river, in a valley, or in a PASS.

NASA National Aeronautics and Space Administration, the civilian agency concerned with the space exploration programme of the USA, established by act of Congress, 1958. LANDSAT.

nation the largest SOCIETY-1,2,3 of people, generally linked by common descent, historical, ethnic and possibly linguistic ties, having common interests of place and land, and usually recognized as a separate, political entity. It is sometimes defined as an independent political unit, but a nation does not necessarily enjoy statehood (STATE-2) or political autonomy: it may exist as an historical community or be identified by its cultural ties. NATION STATE.

national a member of a NATION.

national *adj.* of or pertaining to a NATION.

National Aeronautics and Space Administration NASA.

national grid **1.** on Ordnance Survey maps in Britain at present, the metric GRID-1 based on the transverse Mercator projection. The axes are 2°W and 49°N, and from their intersection at TRUE ORIGIN the FALSE ORIGIN is transferred 400 km W and 100 km N. Drawn on the metric system, with 500, 100 and 1 km squares, one reference system covers the whole of Britain, the grid lines corresponding with sheet lines **2.** a network of transmission lines linking the main generating stations to distribution centres in a country or region in order to maintain a constant supply (e.g. of electricity, of water) to the consumer. Fig 30.

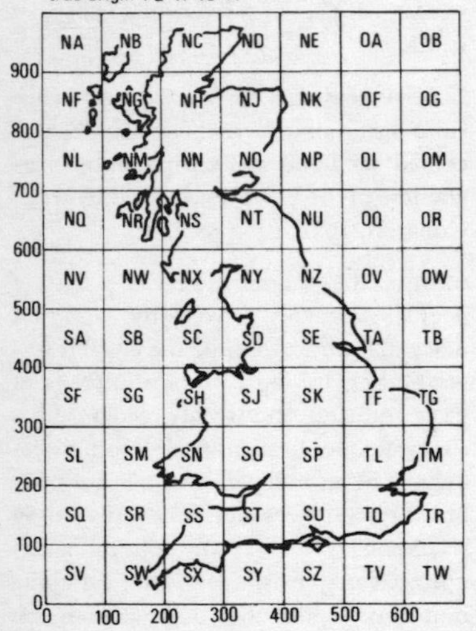

Fig 30 The national grid of the British Ordnance Survey

nationalism **1.** the devotion to one's NATION **2.** national aspiration **3.** the advocacy of national unity or independence **4.** the process whereby a nation has been established as an independent political unit.

nationalization the act of putting privately controlled or owned property (e.g. an activity, industry, land) under public control or ownership, i.e. under the control or ownership of the state, with or without compensation. The term expropriation is applied to nationalization without compensation. PRIVATIZATION.

national park an extensive area of countryside officially designated by government in order to protect and conserve its special natural features (scenic beauty, native flora and fauna, display of geological phenomena, etc.) and in some cases its historical associations, for public enjoyment and for scientific purposes. The concept originated with the Yellowstone National Park, USA, in 1871; and national parks, varying in type, are now to be found in most countries. COUNTRY PARK, NATIONAL PARKS.

National Parks in Britain (England and Wales). Ten National Parks (Lake District, Yorkshire Dales, North York Moors, Peak District, Snowdonia, Brecon Beacons, Exmoor, Dartmoor, Cheviots, Pembrokeshire Coast) were designated in 1965 under the National Parks and Access to the Countryside Act 1949. They are areas of natural beauty but for the most part the land is farmed or otherwise used, villages and other settlements lie within their boundaries, and normal country life and their use for recreation continues (MULTIPLE LAND USE). Building development is controlled. The same Act set up the Areas of Outstanding Natural Beauty (AONB). HONEYPOT.

national separatism the aspiration of a NATION to form its own independent SOVEREIGN STATE. Some writers attribute the impetus to UNEVEN DEVELOPMENT.

nation state a SOVEREIGN STATE most of the members of which constitute a NATION.

native **1.** a person born in a given place or country **2.** one of the original inhabitants of a country **3.** a plant or animal originating in an area.

native *adj.* **1.** applied to metals occurring in nature or in a pure state **2.** inherent, belonging to a person or thing by nature **3.** belonging to someone by birth, e.g. native land **4.** of or relating to a NATIVE-1 of a particular place **5.** of or belonging to a plant or animal originating in an area.

NATO North Atlantic Treaty Organization, an alliance for mutual defence, the treaty being signed on 4 April 1949 by Belgium, Canada, Denmark, France, Iceland, Italy, Luxembourg, the Netherlands, Norway, Portugal, the UK, the USA. Greece and Turkey were effectively admitted in February 1952, the Federal Republic of Germany in May 1965, Spain 1982. Headquarters in Brussels. WARSAW PACT.

natural *adj.* **1.** pertaining to, existing in, or formed by NATURE-1, i.e. not ARTIFICIAL or supernatural (inexplicable in terms of the known laws governing the material universe) **2.** in chemistry, found in the earth's crust.

natural change the net change in the total population of an area arising from the balance of births and deaths. BIRTH-RATE.

natural gas a mixture of combustible gaseous HYDROCARBONS and non-hydrocarbons occurring, frequently with PETROLEUM, in the rocks of the earth's crust, a source of ENERGY, used as a fuel or as a raw material in the PETROCHEMICAL industry. It is transported by pipeline; or in liquid form (LNG, liquefied natural gas) in carriers at a temperature of –160°C (–260°F). GAS.

natural increase of population the rate of population growth shown by subtracting the number of deaths from the number of births.

naturalization **1.** the admittance of a foreigner to the citizenship of a country **2.** the introduction of an animal or plant to a habitat in which it is not native, but where it can flourish and reproduce.

natural landscape the LANDSCAPE as unaffected by human activities, i.e. the physical landscape (including relief and NATURAL VEGETATION) as opposed to the CULTURAL LANDSCAPE. But human activities have been so widespread that little 'natural landscape' thus defined still exists, and it can be said that nearly all landscape is now cultural; thus it is perhaps preferable to refer to the natural and cultural elements in the landscape.

natural region **1.** a part of the earth's surface characterized by a comparatively high degree of uniformity of structure, surface form and climate within it **2.** a part of the earth's surface possessing a unity based on any significant geographical characteristics, whether physical, biological or human, or any combination of these, in contrast to an area demarcated by a boundary imposed for political or administrative purposes, regardless of any geographical unity. REGION-1.

natural resources the wealth supplied by nature, available for human use, and deemed to be useful, including energy, mineral deposits, soil fertility, timber, water power, fish, wildlife and natural scenery etc., the list being indeterminate because the assessment of what constitutes a RESOURCE-1,2 is constantly changing. Natural resources are now commonly classified as: flow (those that are renewable, being always available but open to human modification in that they may be depleted, sustained or increased by human activity, e.g. amenity landscape, soils, forests); stock (non-renewable, e.g. minerals); and continuous (always available and independent of human action, e.g. solar and tidal energy). ALTERNATIVE TECHNOLOGY, CONSERVATION, RESOURCE MANAGEMENT.

natural selection the mechanism of evolutionary change suggested by Charles Darwin in his theory of evolution (EVOLUTION-2) and also by the English naturalist Alfred Russell Wallace, 1823–1913 (WALLACE'S LINE), i.e. that of the many variant offspring produced by a generation of organisms only those most fitted to their environmental conditions will survive and breed, transmitting their advantageous characteristics to their offspring, the weaker, less well adapted variants failing in competition with them, and thus not perpetuating their disadvantageous characteristics.

natural vegetation the plant-association which is primarily due to nature rather than to human activity. But little of the world's vegetation is entirely unmodified by human activities, which include the burning of plant cover, the introduction of alien species (e.g. rabbits), or the grazing of LIVESTOCK. Thus a large part of the so-called natural vegetation is at best only SEMI-NATURAL. The term is therefore now usually applied to all vegetation not deliberately managed or controlled in farming activities, i.e. it includes the 'natural' as well as the 'semi-natural'. SUCCESSION-2.

nature 1. the physical universe, including the laws and forces ruling changes within it, excluding objects made by human beings **2.** the essential, fundamental character of something or someone.

nature reserve an area of land or water managed for the protection and conservation of its animal and plant life and its physical features.

nautical mile KNOT, MILE.

naze, nose, ness, nore a headland or promontory. NESS.

neap tide a TIDE in which the difference between HIGH WATER and LOW WATER is small, the high tide being lower and the low tide being higher than usual (TIDAL RANGE). It occurs twice a month, about the time of the first and last quarters of the moon, when the earth, sun and moon lie at right angles to each other (QUADRATURE), with the effect that the gravitational pull of the sun opposes that of the moon instead of reinforcing it. MOON, SPRING TIDE, SUN.

Near East EAST-2 of western Europe, a term formerly applied to the territory of the Ottoman Empire in the eastern Mediterranean region and to the Balkan states, or to Palestine and the adjacent lands facing the Mediterranean, a term now usually superseded by MIDDLE EAST.

nearest-neighbour analysis a statistical technique used to describe a point pattern, involving measuring the distance between each point and one or more of its neighbours. The observed point pattern is then compared with a theoretical random pattern, which allows the non-randomness of the observed point pattern to be judged. The average of the distances between each observed point and its nearest neighbour is divided by the expected random spacing, to give the statistic *Rn*, with values varying (not in linear progression) from 0, indicating maximum clustering, through 1 (random distribution) to 2.15, indicating that all points are uniformly distributed throughout the area. Problems arise over the spacing, shape and size of areas. It is always important to be consistent in defining the area to be analysed in comparative studies, and in some cases it may be necessary to measure not only the proximity of the nearest neighbour to each observation but also the distance to the second, third, or *n*th nearest neighbour. For example, if observations occur in widely scattered pairs over the specified area, measurement of the distance to the second nearest neighbour will be needed. The technique was originally used by botanists, and later applied to the study of geographical problems.

nebula a cloud of gas and/or dust, luminous in many cases, occurring in the interstellar medium of a GALAXY.

neck 1. of land, isthmus or promontory, a narrow stretch of land with water on each side **2.** a narrow stretch of woodland or of ice **3.** a high level pass, especially the narrowest part. VOLCANIC NECK.

negative area a term sometimes used to suggest a range of environmental factors which render an area unfit for human habitation, i.e. ANECUMENE as opposed to ECUMENE.

negative movement of sea level, the lowering of the sea level in relation to the land caused by **1.** a global lowering of ocean level (EUSTATISM); or **2.** a local vertical movement such as WARPING, TILTING or isostatic recovery (ISOSTASY) of the land. KNICKPOINT, POSITIVE MOVEMENT of sea level, REJUVENATION.

negentropy a measure of order or organization in a SYSTEM-1,2,3. ENTROPY.

nehrung (German) a long sandspit separating a HAFF OR LAGOON from the sea, common along the south coast of the Baltic sea. LIDO.

neighbourhood a small district inhabited by people, in which there are close, everyday social contacts and within which the individual feels secure, 'at home', i.e. the home territory. The boundaries are indeterminate and more readily discernible by those outside than by those inside the district.

neighbourhood effect 1. the influence of the local residential area on the decisions and behaviour of a person living in it. **2.** the effect of proximity in the DIFFUSION-1 of INNOVATION, new adopters being most likely to be near to existing users, the likelihood of adoption of innovation decreasing with increased distance.

neighbourhood unit in town planning, a physical and social unit within a large town, the unit being self-contained in the sense of having its local shops (selling particularly CONVENIENCE GOODS), banking, postal and other service facilities and social amenities, but depending on the main centre of the town for other than daily needs. A neighbourhood unit may arise naturally from the absorption of a village within an expanding town.

nekton, necton a collective term for animals that live at various depths in the ocean or in lakes, i.e. in the PELAGIC ZONE, and swim actively, in contrast to PLANKTON which float, and BENTHOS which live on the bottom.

nematoda a phylum (CLASSIFICATION OF ORGANISMS) of unsegmented worms, some free-living in soil, water and other liquids, many being PARASITES of animals and plants (e.g. hookworm in human beings, eelworm in root crops), and important as decomposing agents in HUMUS. DECOMPOSITION-1.

neoclassical economic theory a body of

economic theory introduced in the latter part of the nineteenth century (e.g. by W. S. Jevons) and forming the dominant economic analysis used (usually with modifications) in capitalist societies today (MARKET ECONOMY). Outlined briefly, in general it accepts CLASSICAL ECONOMIC THEORY (apart from Ricardo's LABOUR THEORY OF VALUE), but refines and extends it. For example, neoclassical economic theory, related to a free-enterprise, capitalist system (CAPITALISM-1), is founded on the idea that the maximization of PROFIT-2 and UTILITY by a large number of small producers and consumers who do not have power to influence to any great extent the operation of the market in which they act (perfect competition) benefits the entire community. It assumes that the whole economic system is regulated by the interaction of supply and demand in the market place. Business decisions are made primarily on the basis of consideration of production processes to be used and the scale of output, not of the location of plant. Business buys or hires land, labour and capital (FACTORS OF PRODUCTION) and uses them in production processes in a way designed to maximize profits. The prices of production factors and of the finished goods sold are beyond the control of business. The public offers for sale (to business) labour, land, capital goods; and the interaction of supply and demand for these determines prices paid as wages, rent, interest (i.e. the distribution of income). The public uses income to buy goods and services chosen to maximize personal satisfaction or UTILITY; and this all-powerful consumer demand, interacting with the costs at which business can supply goods, determines the prices of those goods and services. The theory also assumes that markets are self-regulating, that they automatically adjust to changes and always tend to move towards equilibrium at a price which balances supply and demand. Thus neoclassical economics focuses on individual decisions and the aggregates of those decisions, generally ignoring social costs and benefits, and concentrating on the analysis of cost, profit, revenue and utility.

neoclassical theory of regional development the theory that imbalance in growth and well-being between regions or between cities is temporary and will be resolved eventually by the effect of market forces alone. It assumes perfect price flexibility, and perfect mobility of labour and capital.

neo-colonialism, neocolonialism 1. the situation in which a foreign power intervenes in the economic, and sometimes the political, affairs of another country, in some cases to the resentment and annoyance of some nationals of that country. The intervention does not necessarily stem from a former colonial relationship **2.** the transfer of power from external colonial control to internal control accompanied by the preservation of the trade and investment (sometimes also of military, fiscal and political) relations existing before independence was gained from the dominant, external colonial power. COLONIALISM.

neolith (Greek, new stone) a polished stone tool of the last period of the STONE AGE.

Neolithic *adj.* of or pertaining to the last period of the STONE AGE (succeeding the PALAEOLITHIC and the MESOLITHIC) from about 6000 to 3000 BC in Europe and western Asia, when NEOLITHS came into use as well as implements of polished bone and horn, animals came to be domesticated, crops cultivated, weaving undertaken, and the wheel used. The long BARROWS and MEGALITHS of Britain are associated with this cultural period. BRONZE AGE.

neritic *adj*. associated with shallow water, especially with shallow coastal water. OCEANIC.

neritic province, neritic zone one of the zones of the aquatic environment based on depth of water, variously defined, but commonly applied to the LITTORAL and SUBLITTORAL marine zones between LOW WATER mark and depths of 180 to 365 m (100 to 200 fathoms: 590 to 1200 ft), or the edge of the CONTINENTAL SHELF. PELAGIC.

ness (Scotland and eastern England, also naze, nore, nose) a headland or cape, a SPUR of a mountain ridge; used especially in place-names, apparently where Scandinavian influence was strong.

nesting the enclosing of objects by a succession of similar, ever-larger ones, e.g. in CENTRAL PLACE THEORY, the pattern displayed when low order trade areas lie within the boundaries of high order trade areas.

net NETWORK.

net primary production in ecology, PRIMARY PRODUCTION.

net radiation a measure of the difference between incoming and outgoing RADIATION.

network **1.** the actual structure that forms a net, i.e. the knotted yarn of a fishing net, or the veins of a leaf etc. **2.** any set of interlinking lines (links or routes) that cross or meet one another (at NODES, junctions, terminals) in the manner of those in a net, e.g. a railway network, a GRID-3 **3.** a system with its unit members interlinked in some way, e.g. an information network **4.** a chain of radio or television stations etc. GRAPH-2, INTERNET, NETWORK CONNECTIVITY, NETWORK DENSITY, NETWORK SOCIETY, TOPOLOGICAL DIAGRAM.

network connectivity the extent to which movement is possible between different parts of a NETWORK-2,3 and, if movement occurs, the extent to which it is direct. Directness is expressed as the ratio between route distance and GEODETIC DISTANCE (a ratio termed the route factor, or index of circuity). Fig 31. ALPHA INDEX, BETA INDEX.

network density the length of links, routes, edges (GRAPH-2) of a NETWORK-2,3,4 per unit area.

network society a phenomenon of the POST-INDUSTRIAL SOCIETY, in which individuals anywhere in the world who have access to the appropriate technology can communicate with each other, with institutions and with information services, transcending time, space, national, economic and social boundaries. GLOBALIZATION, INTERNET.

neutrality **1.** the quality or state of not taking the side of or assisting either of two opposing sides in a dispute, controversy, war etc. **2.** the quality or state of having no distinctive colour or other quality **3.** in chemistry, the state of being neither ALKALINE nor ACID (pH) **4.** in international law, the status of a state (or a NATION) which has a declared policy of nonparticipation. The state may have adopted this policy of neutrality from choice (e.g. Sweden, Switzerland in the Second World War); or it may be imposed on the state by others (e.g. Austria, following the Second World War).

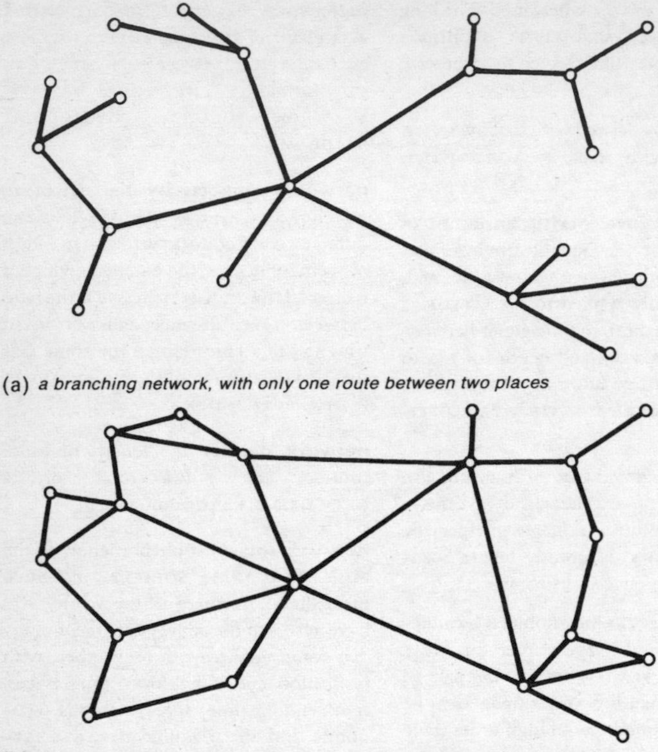

Fig 31 Examples of networks

neutral stability, indifferent equilibrium the state of a parcel of air in balance with its surroundings, i.e. when a saturated parcel has an ENVIRONMENTAL LAPSE RATE equal to the SATURATED ADIABATIC LAPSE RATE or when an unsaturated parcel has an environmental lapse rate equal to the DRY ADIABATIC LAPSE RATE.

nevados (Ecuador: Spanish) a cold wind blowing regularly down the valleys of the high Andes in Ecuador, caused partly by radiation at night, partly by the cooling of air on contact with ice and snow. KATABATIC.

nevé, névé (French) glaciologists prefer the term FIRN. The correct spelling is nevé.

newly industrializing country/economy NIC/NIE.

New Red Sandstone the red SANDSTONE deposited in the Permian and Triassic periods (GEOLOGICAL TIMESCALE) in Europe. OLD RED SANDSTONE.

newton n, the unit of force in the SI system, defined as the force needed to accelerate a mass of 1 kg by 1 m per second per second. 1 newton = 10^5 dynes, 10 n = 1 bar; 1 n per square m = 1.4504 × 10^4 lb per sq in; 1 lb per sq in = 6894.8 n per square m.

New Town, new town a town designated in Britain under the New Towns Act 1946, planned as a well-balanced, self-contained unit to include housing, employment, educational facilities, social amenities etc. primarily to relieve population pressure in OVERCROWDED cities and CONURBATIONS. In practice these New Towns have tended to become major urban centres, GROWTH CENTRES, providing employment, shopping facilities and other services for people living outside the designated area of the town. The term new town (without initial capital letters) is sometimes more generally applied to planned, self-contained towns in Britain and elsewhere, which (like the New Towns) have not evolved slowly but start life in a mature state, being built specifically for a particular purpose, e.g. to bring to life a hitherto sparsely-settled, stagnant region by acting as a development centre; for political considerations, to form a new administrative centre (e.g. Brasilia, Canberra); for research and scientific purposes to act as a centre for those engaged in research and development in a particular field of study (the 'academic' towns of the former USSR); or to cater for tourists. Such planned towns are not recent phenomena: many small examples were built in western Europe between the twelfth and fourteenth centuries (e.g. the bastides in southern France); most were defended by a wall, ditch, occasionally by a castle; most had a rectilinear street pattern centred on a market square (unlike the haphazard layout of small medieval settlements).

New World a term commonly applied to the continents of North and South America and the Caribbean islands, the western HEMISPHERE.

NGO non-governmental organization, a voluntary body, usually with international membership, recognized officially by United Nations, so that it may give evidence, act as consultant, and attend meetings of UN committees.

NIC new industrializing country. AIC, INDUSTRIALIZATION, LDC, NIE, UNDERDEVELOPMENT.

niche 1. in ecology, the specific part of a HABITAT occupied by an organism, where it can exist and develop. RANGE-7 **2.** the role played by an organism in the ECOSYSTEM **3.** in geology, a small recess or shelf in a rock face.

niche glacier a small CIRQUE GLACIER, lying in a funnel-shaped hollow high on a steep mountain slope. Alternative terms are cascade glacier, cliff glacier, wall-sided glacier.

nickpoint KNICKPOINT.

NIE newly industrializing economy. NIC.

night the time during which the SUN is below the horizon, the opposite of DAY.

nimbostratus a low, thick, dark grey mass of CLOUD from which continuous rain or snow usually falls. It is commonly associated with the WARM FRONT of a DEPRESSION-3, having thickened from ALTOSTRATUS.

nimbus CLOUD, CLOUD FORMS.

nitrification the process by which AEROBIC soil bacteria convert organic compounds of NITROGEN (which cannot be absorbed by green plants) into nitrates (which can be absorbed by green plants). DENITRIFICATION, NITROGEN CYCLE.

nitrogen a colourless, odourless gaseous ELEMENT-6, the main constituent (some 78 per cent by volume) of the ATMOSPHERE-1, an essential constituent of living organisms, an essential MACRONUTRIENT for plants. Atmospheric nitrogen is the main raw material used in the manufacture of fertilizers, nitric acid and ammonia; and naturally occurring potassium nitrate (saltpetre) is used as FERTILIZER. ACID RAIN, NITROGEN CYCLE, NITROGEN FIXATION.

nitrogen cycle the circulation of NITROGEN atoms through ECOSYSTEMS brought about by natural processes in which living organisms play the major part. Inorganic nitrogen compounds (nitrates) are converted into organic nitrogen compounds by AUTOTROPHIC plants which either die, decay or are eaten by animals. These organic nitrogen compounds then pass, through the excreta or by the death and decay of the animals or plants, to the soil or water, where they are converted back to inorganic nitrogen compounds by nitrifying bacteria (NITRIFICATION), again becoming a necessary nutrient for green plants. But some of the atmospheric nitrogen is processed by nitrogen-fixing bacteria and blue-green algae (NITROGEN FIXATION) to form organic nitrogen compounds; and some nitrogen passes to the atmosphere by means of denitrifying bacteria (DENITRIFICATION) which convert some of the nitrates to atmospheric nitrogen. LEGUMINOSAE. Fig 32.

nitrogen fixation 1. the process by which atmospheric NITROGEN is converted into organic nitrogen compounds by some blue-green ALGAE and by nitrogen-fixing BACTERIA living in the soil. Some of these bacteria live independently in the soil, others symbiotically (SYMBIOSIS) with leguminous plants (LEGUMINOSAE) in NODULES-2 on their roots, absorbing atmospheric nitrogen and from it forming organic nitrogenous compounds which enrich the soil. NITROGEN CYCLE **2.** the industrial conversion of atmospheric nitrogen into useful compounds (e.g. ammonia and ammonium compounds) used in fertilizer, explosives.

nivation 1. a general term for the effects produced by snow and nevé (FIRN) in the weathering and sculpture of rocks in contrast to those produced by glacier ice **2.** the rotting and disintegration of rocks underlying and round the edges of a patch of snow lying in a hollow (nivation hollow or nivation niche) brought about by FREEZE-THAW and CHEMICAL WEATHERING (sometimes termed snow patch erosion). This enlarging of the hollow may lead to the formation of a CIRQUE (nivation cirque).

noble metal a METAL resistant to OXIDATION or CORROSION, i.e. GOLD and PLATINUM.

noctilucent cloud a luminous blue-silver cloud occurring mainly in midlatitudes and high latitudes in the STRATOSPHERE, thought to consist of ice crystals or METEORIC DUST.

nocturnal *adj.* of or relating to night; in zoology, active mainly at night.

nodal *adj.* of, like, or situated at a NODE or nodes.

node 1. a central point in any system or complex. **2.** in a NETWORK-2,3, the point at which routes (edges, links) meet.

nodule 1. a small rounded mass, e.g. a

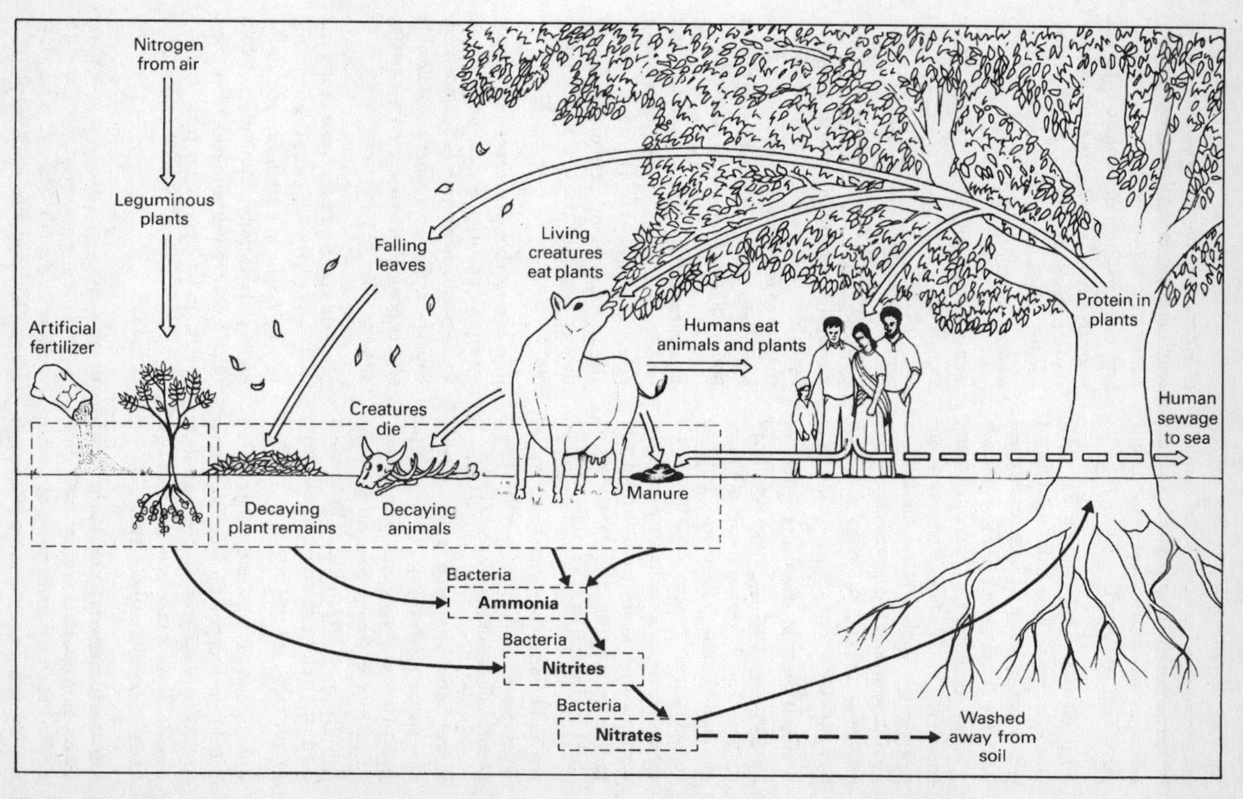

Fig 32 The nitrogen cycle

discrete rounded concretion of minerals in a sedimentary rock **2.** a root nodule, such as a mass on the root of a leguminous plant (LEGUMINOSAE).

nomadism a type of human life style, now rare, based essentially on constant movement in search of sustenance, especially for grazing animals. Most nomads wander within defined areas, often using regular routes. The term semi-nomadism is sometimes applied to the life style of nomads who use fixed quarters in the wet season but migrate in the dry season to find pasture.

nominal scale in statistics, a system of categorization, and the simplest form of measurement. It has a low level of differentiation, the observations being put into convenient, MUTUALLY EXCLUSIVE and exhaustive classes according to a particular attribute, without any particular order or preference. BAR GRAPH, CATEGORICAL DATA ANALYSIS, MEASUREMENT-2.

nominal variable a VARIABLE, sometimes termed an attribute, which may be placed only on a NOMINAL SCALE.

nomothetic approach a law-seeking approach, concerned with the search for general laws or theories, not with particular cases or situations (in contrast to the IDIOGRAPHIC APPROACH). The method used can be either deductive or inductive. DEDUCTION, INDUCTION.

nonaligned *adj.* applied to a nation state that does not support the policies of either of two opposed groups of powers (e.g. historically, of NATO or of the WARSAW PACT countries). NEUTRALITY.

non-basic activity, non-basic function, non-basic industry in urban development, a manufacturing or service activity within a city or urban area which meets needs within that city or urban area, resulting in an internal exchange of revenue. It is sometimes termed 'secondary' basic industry in contrast with the 'primary' basic industry (BASIC ACTIVITY). BASIC–NON-BASIC RATIO, ECONOMIC BASE THEORY, URBAN ECONOMIC BASE.

non-governmental organization NGO.

non-hydraulic civilization HYDRAULIC CIVILIZATION.

nonnull hypothesis 1. in general, an hypothesis alternative to the one under test, i.e. to the NULL HYPOTHESIS-1 **2.** an hypothesis under test where the effect is not equal to zero (NULL HYPOTHESIS-2).

nonparametric test a statistical test which does not need an estimate of a POPULATION PARAMETER or, generally, any assumptions as to the distribution of the scores. DISTRIBUTION-4, PARAMETRIC TEST.

non-renewable resources any of the NATURAL RESOURCES classified as stock, assessed to be finite and which, once used, could be replaced only after a considerable span of geological time, e.g. FOSSIL FUELS, MINERALS.

Norfolk rotation a four-year crop rotation (ROTATION OF CROPS) originating in East Anglia in eastern England in the eighteenth century, comprising wheat (which needs much nitrogen), a root crop, barley, and a leguminous crop (LEGUMINOSAE) which replenishes the NITROGEN in the soil.

norm 1. that which is usual, expected, average, conforming to an accepted standard, measure or pattern. The *adj.* is NORMAL **2.** a standard or ideal to which people think behaviour ought to conform or which is laid down by a legislating authority, *adj.* NORMATIVE.

normal **1.** the usual or average state, level etc. **2.** in mathematics, a line drawn at right angles to the tangent line at a point on a curve.

normal *adj.* **1.** standard, regular, conforming to a NORM-1 **2.** in mathematics, forming a right angle **3.** approximating to an average or statistical NORM-1.

normal distribution normal probability distribution, in statistics, a continuous FREQUENCY DISTRIBUTION of infinite range, represented on a graph by a symmetrical curve (FREQUENCY CURVE) shaped like a bell, and termed a normal curve, most of the values being grouped around the mean, with a generally regular fall in the numbers of values away from the mean as distance increases in both directions. The normal distribution is important in PROBABILITY theory, and is used to estimate the closeness of the results of a sample to the true figure of the POPULATION-4.

normal fault a FAULT caused by TENSION in which the INCLINATION of the FAULT PLANE is at an angle of between 45° and the vertical and the direction of the DOWNTHROW is the same, i.e. the beds abutting the fault of its upper face (HANGING WALL) are displaced downwards relative to those against the lower face (FOOTWALL). This is not, as the use of 'normal' might imply, the most usual, the most common type of fault. Fig 22.

normal watershed a WATERSHED in a mountainous region which runs along the crest of the highest range of a mountain chain. ANOMALOUS WATERSHED.

normative *adj.* **1.** of or relating to a NORM-2, thus concerned with rules, regulations, proposals **2.** establishing a NORM-2.

normative approach an approach which concentrates on what ought to occur in certain circumstances rather than on what actually occurs.

normative explanation explanation based on the assumption that the NORM-2 rules BEHAVIOUR, and from that deducing spatial consequences.

norte (Mexico and Central America: Spanish) **1.** a cold northerly winter wind, sometimes strong, blowing in Mexico and Central America especially over the coast lands (a continuation of the NORTHER), causing a sudden, often great, drop in temperature **2.** applied by some authors to a similar wind blowing in eastern Spain during winter. PAPAGAYO.

north **1.** one of the four CARDINAL POINTS of the COMPASS, directly opposite the south, lying on the left side of a person facing due EAST **2.** towards or facing the north, the northern part, especially of a country. BRANDT REPORT.

north *adj.* of, pertaining to, belonging to, situated towards, coming from, the NORTH, e.g. of winds blowing from the north.

North American Free Trade Area NAFTA.

North Atlantic Treaty Organization NATO.

northeast trades TRADE WIND.

norther a cold, northerly winter wind, blowing in the rear of a DEPRESSION-3 over the southern USA, especially Texas and the Gulf of Mexico, sometimes violent, in some parts dry and dusty, in others accompanied by thunderstorms and hail, bringing a sudden great fall in temperature that devastates fruit crops. NORTE, PAPAGAYO.

northern hemisphere the half of the earth north of the equator.

Northern Lights AURORA.

northing **1.** the second part of a grid reference (GRID-1), i.e. the distance north on a map as measured from a point fixed in its southwest corner. EASTING **2.** nautical, a sailing northwards, or the distance so travelled since the last reckoning point.

north magnetic pole MAGNETIC POLE.

North Pole the geographical North Pole, the northern extremity of the earth's axis. MAGNETIC NORTH, MAGNETIC POLE, POLE, TRUE NORTH.

North-South BRANDT REPORT.

northwester, nor'wester a strong wind or gale blowing from the northwest, e.g. in South Island, New Zealand, where it is a FOHN-type wind (hot and dry).

nosology (Greek nosos, disease) a scientific or systematic classification of diseases.

nuclear energy the ENERGY released by the fission or fusion of atomic nuclei, a very important power source. ELECTRICITY, NUCLEAR FISSION, NUCLEAR FUSION.

nuclear family a FAMILY-1 unit comprising husband, wife and children. EXTENDED FAMILY.

nuclear fission the act or process of splitting an atomic nucleus into smaller nuclei. It is usually achieved by the bombardment of the large nucleus with high energy particles, which releases a great amount of energy, available for industrial or scientific purposes, or for a war weapon. NUCLEAR ENERGY, NUCLEAR FUSION, NUCLEAR REACTOR.

nuclear fusion the combination of light atomic nuclei to form heavier ones under very high temperatures and pressure, the loss of mass occurring in the process releasing a great amount of energy as RADIATION, e.g. as in a hydrogen bomb. NUCLEAR ENERGY, NUCLEAR FISSION.

nuclear reactor the equipment in which NUCLEAR FISSION is initiated, carried out and controlled.

nuclear winter a climatic anomaly which it is thought might occur if sufficient nuclear explosions on earth created enough dust in the ATMOSPHERE-1 to cut off the sun's radiation (SOLAR RADIATION) from the earth's surface. Severe wintry conditions would eventually prevail, with catastrophic effects on ECOSYSTEMS.

nucleated settlement a rural settlement comprising **1.** a cluster of dwellings, in contrast with a DISPERSED SETTLEMENT **2.** a cluster of dwellings with an organizational centre.

nuée ardente (French) a dense mass of exceedingly hot, gas-charged and gas-emitting fragmental lava, with particles separated by compressed gas, usually incandescent but sometimes dark, emitted from some volcanic eruptions, and rolling downwards at high speed in explosive, devastating blasts, e.g. the town of St Pierre, Martinique, was destroyed by such phenomena in 1902 when Mount Pelée erupted (PELEAN).

nugget a small lump of a native precious metal, especially of GOLD.

nullah (Anglo-Indian corruption of nālā) **1.** the dry bed of an intermittent stream **2.** in Hong Kong, an artificial water supply channel or a drainage channel, constructed of concrete (presumably so-named because either channel may be periodically dry).

null hypothesis **1.** in general, an HYPOTHESIS under test, as distinct from alternative hypotheses that are under consideration. NONNULL HYPOTHESIS **2.** an HYPOTHESIS about to be tested statistically which states that no difference between the groups being tested or that no relationships between the variables will be discovered. All hypotheses to be tested statistically are stated in this form. A null hypothesis rejected on the grounds that a difference or correlation does exist provides evidence of an ALTERNATIVE HYPOTHESIS. STATISTICAL TEST.

nunatak, nunataq (Inuit; Swedish pl. nunatakker) an island of rock or a mountain peak rising above a glacier or land ice.

nutrient a substance that serves as a food, a term applied especially to such a substance used by a PLANT.

O

OAS Organization of American States, a group of American nation states, the members of which on 30 April 1948 signed the Charter of the Organization of American States, aiming to achieve an order of peace and justice, promote American solidarity, strengthen collaboration among the member states and defend their sovereignty, territorial integrity and independence. The aims were extended on 14 April 1967: to promote Latin American economic integration and foreign trade, to raise agricultural productivity and living standards, and to expand other social development programmes. The member countries are Argentina, Barbados, Bolivia, Brazil, Chile, Colombia, Costa Rica, Dominican Republic, Ecuador, El Salvador, Grenada, Guatemala, Haiti, Honduras, Jamaica, Mexico, Nicaragua, Panama, Paraguay, Peru, Surinam, Trinidad and Tobago, USA, Uruguay, Venezuela.

oasis **1.** an area in a hot desert where the presence of water at a suitable level permits sustained plant growth. Such oases vary in extent from a very small area supporting a few palm trees (typical of the Sahara) to a tract of hundreds of square kilometres supporting a large settled agricultural population (e.g. those near the rivers Nile and Euphrates) **2.** a patch of ice-free land in an icebound landscape.

oat, oats (pl. is common use) *Avena sativa*, a cultivated grass, and one of the main CEREAL-1 crops of temperate regions, absent from Mediterranean and tropical lands. Oats existed in Europe in the BRONZE AGE. They grow under conditions similar to those needed by BARLEY and WHEAT, but need less heat and more moisture. Oats are fed to livestock as grain, processed pellets, hay, or in green stage; the grain is used for human food (oatmeal, rolled oats). Oat flour is an antioxidant, used as a preservative in the preparation of some foods (e.g. some margarines).

OAU Organization of African Unity, an organization founded by 30 African countries in May 1963 to which all independent African nation states now belong, aiming to advance the unity and solidarity of African countries, to coordinate political, economic, cultural, health, scientific and defence policies, and to eliminate colonialism.

objective *adj.* **1.** not influenced by personal feelings **2.** having a real existence external to an observer, or pertaining to an object or event independent of the feelings or imagination of an observer. HUMANISTIC GEOGRAPHY, SUBJECTIVE.

oblate *adj.* of a SPHEROID, flattened at the poles. EARTH-1, GEOID.

oblique fault a FAULT in which the STRIKE lies at an oblique angle to the strike of the bed it traverses.

obsequent fault-line scarp a FAULT-LINE SCARP which has 'turned around' so that it is now facing in the reverse direction to that produced by the initial earth movement. It occurs when prolonged denudation takes away the resistant strata from the higher side of the fault-line scarp to expose the underlying less resistant strata; in comparison with this the strata on the DOWNTHROW side may be more resistant, so that it persists (as the surface is generally lowered) along the line of the original fault but facing in the opposite direction to that of the original fault-line scarp. RESEQUENT FAULT-LINE SCARP.

obsequent stream, obsequent river, obsequent drainage a natural water flow, a stream, or a river that flows in the direction opposite to that of the DIP of the rock strata. An obsequent stream flows in an obsequent valley. CONSEQUENT STREAM, SUBSEQUENT STREAM. Fig 42.

Occident, the originally Europe as opposed to Asia (the ORIENT). Now extended to those parts of the world peopled by Europeans or by those of European descent. ORIENTALISM.

occlusion, occluded front an atmospheric phenomenon occurring when an advancing COLD FRONT overtakes a WARM FRONT, thereby raising the WARM SECTOR of the depression and cutting it off (occluding it) from the earth's surface, to form an occluded front, reduced at the earth's surface to a line, termed the line of the occlusion. The cold air continues to advance and meets the cold air ahead of the warm front. If the overtaking cold air is colder than this cold air ahead, it forms a cold occlusion; if it is not so cold, a warm occlusion; if there is very little difference, a neutral occlusion. Fig 33.

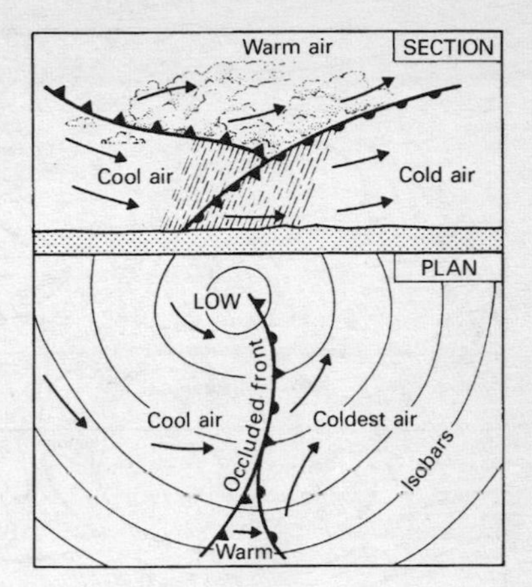

Fig 33 An occlusion: section and plan

ocean 1. the body of salt water which covers 70.78 per cent of the earth's surface **2.** one of the main areas of the whole, i.e. the Atlantic, Pacific, Indian, Arctic, to which some would add the Southern or Antarctic south of 50°S. Sometimes the Atlantic and the Pacific are divided into North and South, the equator forming the conventional limit between the North and South Atlantic and the North and South Pacific. In the South Pacific the ocean extends to within 490 km (305 mi) of the South Pole at the snout of the Scott Glacier.

ocean current CURRENT-3. Fig 34.

ocean floor all the bed of the ocean below LOW WATER mark.

ocean floor spreading, sea floor spreading OCEANIC RIDGE, PLATE TECTONICS.

oceanic *adj.* **1.** of, pertaining to, or occurring in, the ocean **2.** associated with the deep sea as opposed to NERITIC.

oceanic basin the great depression occupied by an OCEAN-2 but excluding the CONTINENTAL SHELF.

oceanic climate a climate that is equable

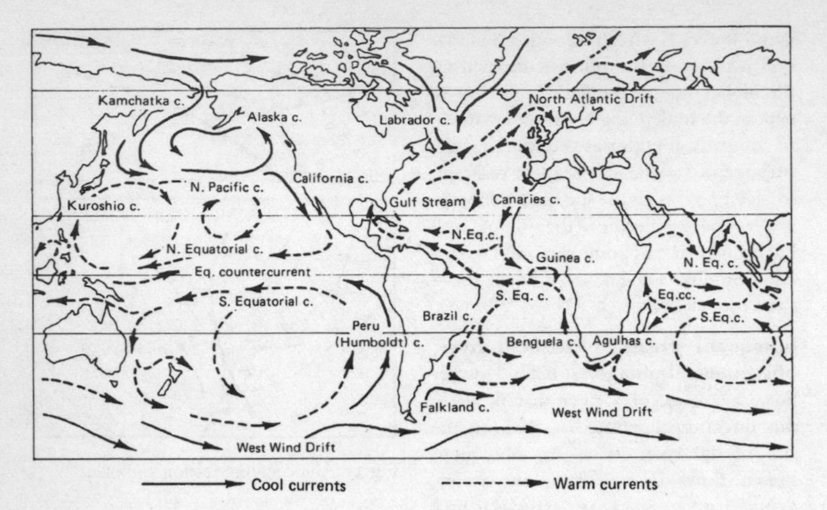

Fig 34 The ocean currents of the world in January

owing to the moderating influence of surrounding waters, i.e. with a moderate range of temperature in summer and winter. CONTINENTAL CLIMATE, INSULAR CLIMATE, MARITIME CLIMATE.

oceanic crust that part of the crust of the earth which lies under the floor of the OCEAN BASINS. PLATE TECTONICS.

oceanic island an island in the ocean, far away from any continent, as opposed to a CONTINENTAL ISLAND.

oceanic mud the BATHYAL deposits on the CONTINENTAL SLOPE consisting of Blue, Coral, Green and Red Muds derived from clay particles eroded from the land, which are larger than those of the OOZE of the ocean floor.

oceanic province the part of the PELAGIC DIVISION, PELAGIC ZONE constituting the deep sea, the edge of the CONTINENTAL SHELF forming the boundary between it and the NERITIC PROVINCE.

oceanic ridge a mid-ocean ridge, broadly the symmetrical ridge formed on the OCEAN FLOOR where, according to the theory of PLATE TECTONICS, lithospheric plates drift apart (ocean floor spreading, sea floor spreading), the edges of the plates lift to form the ridge and hot MAGMA surges through the weakened crust, cooling quickly to make new OCEANIC CRUST, each plate being enlarged by an identical amount, the boundary between them being termed constructive. Not all oceanic ridges are now active. Studies in the PALAEOMAGNETISM of rocks on each side of the ridge show that the pattern of the direction of magnetism fossilized in the rocks is exactly the same on each side of the ridge, i.e. on the margin of each plate, the youngest rocks (the outcome of the most recent outpouring of magma) being nearest to the crest of the ridge. Reversals in the direction of the earth's magnetic field have been recorded, and by checking the magnetic pattern of the rocks against this timetable the spreading of the ocean floor, and thus of plate movements, can be calculated. The evidence provided by mid-ocean ridges thus strongly supports the theory of plate tectonics. MAGNETIC ANOMALY.

oceanic trench, ocean trench in the theory of PLATE TECTONICS, a subduction zone, a long narrow depression in the ocean floor (TRENCH-1) where, as lithospheric plates converge under the ocean, one dives under the other, the edge of the descending plate going steeply and deeply into the MANTLE where it is absorbed (BENIOFF ZONE). Friction arising from the grinding together of the two plates leads to EARTHQUAKES. The SEDIMENTARY ROCKS of the diving plate are scraped off, the rocks of the OCEANIC CRUST descend into the mantle, but not so deeply as does the plate: they are less dense than the rocks of the mantle and, deformed by heat, they rise, molten, to the ocean floor, where they erupt as LAVA and combine with the scraped-off sediments to form an ISLAND ARC, an arc of volcanic islands near the weakened margin of the overriding plate, e.g. the Aleutian and the Japanese islands. ANDESITE, COLLISION ZONE.

OD, Ordnance Datum the mean sea-level at Newlyn, Cornwall, calculated from hourly observation of the tide in the period 1915 to 1921, from which heights shown on British maps are measured. BENCH MARK.

ODECA Organización de los Estados Centroamericanos (Organization of Central American States), Central American Common Market (CACM), an organization set up in December 1960 with the aims of furthering economic, cultural and social cooperation among the members, i.e. El Salvador, Guatemala, Honduras and Nicaragua, joined by Costa Rica in 1962. Of the Central American countries only Panama has not joined.

OECD Organization for Economic Cooperation and Development. In 1961 the OEEC (Organization for European Economic Cooperation) which had existed since 1947 gave place to OECD, headquarters in Paris. The members, Australia, Austria, Belgium, Canada, Denmark, the then Federal Republic of Germany, Finland, France, Greece, Iceland, Irish Republic, Italy, Japan, Luxembourg, the Netherlands, New Zealand, Norway, Portugal, Spain, Sweden, Switzerland, Turkey, UK, USA (Yugoslavia, with special status, took part in activities), aim to develop each other's economic growth and social welfare, coordinate policies, and work together in helping developing countries.

OEEC Organization for European Economic Cooperation, an organization established in 1947–8 by sixteen European countries with the aim of assisting in the distribution and administration of resources sent to Europe under the MARSHALL PLAN, and of coordinating their economic activities. It was superseded in 1961 by the OECD.

office park, business park a tract of land, in some cases park-like in character (PARK-2), with good transport facilities and communications, planned and in some cases officially designated for groups of modern office buildings. INDUSTRIAL PARK, PARK-6, RESEARCH AND SCIENCE PARK.

offshore *adj.* applied to **1.** movement away from the shore towards the sea **2.**

located at a point or in an area relatively near to the shore, e.g. offshore island, offshore fishing zone **3.** the zone to seaward of the FORESHORE and the BACKSHORE.

offshore bar BAR.

O horizons organic horizons, one of the two major classes of soil horizon used by some soil scientists in their SOIL CLASSIFICATION, the other class being the MINERAL HORIZONS. O horizons correspond, broadly, to the A_{00} and A_0 horizons. They are formed of accumulated material derived from plants and animals and lie over the predominantly mineral or inorganic horizons. The organic material in the uppermost layer, O_1, is easily recognizable to the unaided eye; but in the layer below, O_2, it has decomposed, is not so easily identifiable, and is termed HUMUS. Some soil scientists identify the O layer as: L (the LITTER); F (partly decomposed litter); H (decomposed litter). Below the O horizons lie the A_p and A_h horizons (SOIL HORIZON), soil layers roughly corresponding to A_1, and profile grading to A_2.

oil crude oil, mineral oil, PETROLEUM.

oil cake a general term for the crushed and pressed seeds remaining after the oil has been expressed from any of the OILSEEDS (VEGETABLE OILS), the 'cake' being used as cattle food and occasionally as fertilizer.

oil crops plants that yield oils used either for food products (e.g. margarine) or for industrial purposes (e.g. soap-making). DRYING OIL, OIL CAKE, VEGETABLE OILS.

oil dome an underground, roughly hemispherical structure in gently flexed sedimentary strata containing an accumulation of PETROLEUM.

oilfield an area containing one or more subterranean pools of PETROLEUM (OIL-POOL) associated with one geological phenomenon, a term applied especially when the petroleum is being exploited.

oil palm a tree native to western tropical Africa, now cultivated there and elsewhere in tropical Africa and in similar climatic conditions in Asia (e.g. Malaysia, Indonesia), the largest producer of vegetable oil measured in yield per ha (per acre), hence valuable commercially. The oil is obtained from the fibrous pulp of the fruit and from the kernel (palm kernel oil, the more valuable), and is used in the making of industrial products (e.g. soap) and food products (e.g. margarine, for which palm kernel oil is preferred).

oil-pool a separate reservoir of PETROLEUM or NATURAL GAS occurring, with water, in the pores and fissures of SEDIMENTARY ROCKS.

oil-sand, tar-sand porous sandstone or sand at or near the surface of the land, impregnated with viscous HYDROCARBONS (BITUMEN-2), occurring particularly in Canada (Athabasca valley). The Athabasca tar-sands are estimated to constitute one of the largest oil reserves in the world, but they are expensive to exploit commercially.

oilseed any of the plant seeds that yield useful oil. DRYING OIL, VEGETABLE OILS.

oil-shale bituminous shale, i.e. black, brown or green SHALE containing HYDROCARBONS from which oil and gas can be distilled at very high temperatures.

oil well a shaft sunk into the ground, in many cases to a great depth, for the purpose of extracting PETROLEUM.

old age senility, the final and declining

stage of development in theories of growth, e.g. of a RIVER, YOUTH.

Old Red Sandstone a series of DEVONIAN (GEOLOGICAL TIMESCALE) SANDSTONES, red, brown, white, with CONGLOMERATES, LIMESTONE, MARLS-1, SHALE, thought to have been deposited in large lakes. NEW RED SANDSTONE.

Old World the eastern HEMISPHERE, especially Africa (with Madagascar), Europe, and (eastwards) Asia as far as the Malay archipelago.

oligopoly in economics, the control of a market by a few producers, none being dominant. DUOPOLY, MONOPOLY, PERFECT COMPETITION.

oligotrophic *adj.* applied to a freshwater body in which the plant nutrients are low and the oxygen content is high, especially at lower levels in the summer, and which can therefore support only a few organisms. Most oligotrophic lakes are deep, with steep sides, the water being clear and poor in dissolved minerals. EUTROPHIC.

olive *Olea europaea*, a small, slow-growing, long-lived, EVERGREEN tree with oval leaves thickly covered with silky hairs to prevent loss of moisture. Native to the Mediterranean region, it is cultivated there and in other warm temperate and subtropical areas for the sake of its edible nutritious fruit, yielding oil, used for culinary, medicinal and cosmetic purposes, the inferior grades being used as lubricants and in soap-making. The olive is sometimes used as an INDICATOR PLANT for the true Mediterranean climate.

olivine a group of rock-forming silicate minerals (SILICA-2), usually dark green, consisting of silicates of MAGNESIUM and IRON, occurring in BASIC and ULTRABASIC ROCKS, and probably a major component of the SIMA. FERROMAGNESIAN MINERAL, PERIDOTITE.

omnivore an animal, sometimes termed a DIVERSIVORE, that eats plants and animals, as opposed to a CARNIVORE or HERBIVORE. FOOD CHAIN.

onion weathering EXFOLIATION, SPHEROIDAL WEATHERING.

onset and lee in glaciology, the result of the action of a GLACIER on a rock mass over which it has passed, the rock being smoothed by ABRASION on the part facing upstream (the onset part) and marked by the effect of PLUCKING on the downstream (the lee) side. ROCHE MOUTONNEE.

ontology the branch of METAPHYSICS concerned with the essence of things, the science or study of being, the theory of existence, or, more precisely, the theory concerned with what really does exist as distinct from that which seems to (but does not) exist or from that which can be correctly said to exist but only if it can be thought of as a whole composed entirely of ELEMENTS-1 that really do exist. COSMOLOGY, IDEALISM, MATERIALISM.

oolite (Greek oos, egg; lithos, stone) a sedimentary rock, usually CALCAREOUS, e.g. limestone, with very small concretions (OOLITH) resembling the roe of fish, hence the name. Oolite (with capital initial letter) is the formation name applied to the upper part of the JURASSIC (GEOLOGICAL TIMESCALE) in England, on account of the oolitic texture of its limestones.

oolith a small rounded grain of rock (smaller than a PISOLITH) ranging in diameter from some 0.25 to 2 mm, consisting of a particle of shell or mineral (e.g. quartz) enclosed by concentric layers of other material, and resembling a fish-egg. OOLITE.

oolitic limestone OOLITE.

ooze a deposit on the floor of the deep ocean far from land, classified as biogenic or non-biogenic according to its origin. The biogenic oozes consist mainly of minute organic remains (e.g. DIATOMS, globigerina, pteropods, radiolara, that give the ooze a specific name); at an even lower level lie the non-biogenic oozes (RED CLAY), consisting of wind-blown VOLCANIC ASH that has settled on the ocean surface and sunk, meteoric dust, and material carried by ICEBERGS etc.

OPEC Organization of Petroleum Exporting Countries, an organization established in 1960 by what were then the chief oil-producing countries, with the aim of formulating a common policy in respect of dues paid to them by the oil companies, and of their share in the capitalization of the oil companies. CARTEL.

opencast mining (American, open-cut mining or strip mining) the form of excavation used when extensive mineral deposits, e.g. of coal, brown coal or iron ore, lie near the earth's surface, the mineral strata being exposed by the removal of the overlying strata (the overburden) and worked mechanically on a large scale (a QUARRY being more limited in extent). MINE, MINING, PIT.

open field common field, a large field which, before the INCLOSURE of the village lands of England (mainly in the seventeenth and eighteenth centuries), was worked by the villagers in common. Each village usually had two or three open fields. FIELD SYSTEM.

open hearth process a process for STEEL production. It incorporates a reverberatory furnace with a cup-shaped hearth for melting and refining PIG IRON, iron ore and scrap iron, the air employed to remove the carbon being played over the molten metal (instead of being blown through it as in the Bessemer process). Neither the open hearth nor the Bessemer process removes any phosphorus present in the pig iron (phosphorus makes steel brittle).

open system a SYSTEM-1,2,3 characterized by the supply and escape of energy and material across its boundaries (unlike a CLOSED SYSTEM). Such a system regulates itself and may eventually reach a STEADY STATE (e.g. a CENTRAL PLACE). ECOSYSTEM.

opisometer a small instrument used in measuring distances on a map. A calibrated dial is attached to and records the revolutions of a toothed wheel which is small enough to be pushed along and negotiate any bends on the specified course (of a river, road, etc.).

opportunity cost in economics, cost expressed in terms of the best of the alternative opportunities foregone, a measure of the loss or sacrifice involved in using a resource or a location for one particular purpose rather than for another. ECONOMIC RENT.

opposition the position of a heavenly body when it is directly opposite another one, e.g. when three celestial bodies share a common line, the two outer bodies are in opposition to each other and to the central body. Thus, when viewed from the earth, the position of a heavenly body when its direction is opposite to that of the sun. Full MOON and SPRING TIDES occur when the sun and moon are in opposition to the earth. CONJUNCTION, QUADRATURE, SYZYGY.

optical stimulated luminescence OSL.

optimization model any MODEL-2 concerned with making the best possible choice from a set of alternatives. OPTIMIZER CONCEPT.

optimizer concept the theory that people arrange themselves in space in order to make the most favourable use of available resources and demands.

optimum, optimal *adj.* the most favourable or the best possible in the given circumstances (e.g. in constraining conditions) or for a particular purpose.

optimum city size a city of a size above which the benefits (the advantages of further growth) would be outweighed by the costs (the disadvantage of such growth). The criteria used may be economic, social, psychological or political, and may refer to the city itself or to the city in relation to a region or to the country in which it lies.

optimum location the site for a firm which is central, minimizes the costs of the necessary spatial relations and (if selling prices for the product vary) maximizes the differences between costs and revenues (the classical theory of firm location formulated by A. Weber in 1909). BID PRICE CURVE.

optimum population 1. the number of individuals that can be accommodated in an area to the maximum advantage of each individual **2.** the number of human inhabitants considered to be the most favourable for the full use of all the resources available in an area, so that the standard of living achieved is adequate or as high as possible **3.** the number of human inhabitants who would produce the highest total production or real income per head in an area. OVERPOPULATION, UNDERPOPULATION.

orbit the closed course of a heavenly body, especially the closed path of a planet around the sun or of a satellite (natural or artificial) around another heavenly body.

orbit of the earth the counterclockwise closed elliptical path of the EARTH-1 around the sun, taking 365.26 days to complete. The velocity of the earth is not constant, it varies in some sections of the orbit, the average speed being 106 000 km (66 000 mi) per hour. The APHELION occurs about 4 July when the earth is some 152 mn km (94.5 mn mi) distant from the sun, the PERIHELION about 3 January, when the distance is 147 mn km (91.5 mn mi). Fig 18.

order 1. sequence, systematic ranking, arrangement, the way in which one thing follows another **2.** in biology, one of the groups in the CLASSIFICATION OF ORGANISMS. STREAM ORDER.

ordered attribute synonym for ORDINAL VARIABLE. ATTRIBUTE.

order of goods in CENTRAL PLACE THEORY, Christaller's ranking (according to their scarcity in a CENTRAL PLACE SYSTEM) of the retail and service facilities (to which he applied the term goods) carried by a CENTRAL PLACE. High order goods are specialized and scarce, middle order goods less specialized, low order goods generally available everywhere. Thus the order of goods is linked to CENTRAL PLACE HIERARCHY, higher order goods being available in higher order centres with larger COMPLEMENTARY REGIONS, lower order goods (e.g. a CONVENIENCE GOOD) being available in lower order centres with a smaller complementary region. RANGE-9.

ordinal *adj.* of a number, showing position or order in a series, arrangement according to order or rank. INTERVAL, NOMINAL.

ordinal scale in statistics, a MEASUREMENT scale which allows individuals in a data set to be placed in correct relative order even if absolute values are unknown; thus the distance between categories or ranks is unspecified. A VARIABLE measured on this scale is termed an ordinal variable or ordered attribute. BAR GRAPH.

ordinal variable in statistics, an ordered attribute, a VARIABLE measured on the ORDINAL SCALE. ATTRIBUTE.

ordinate in mathematics, the vertical or *x*-coordinate in a plane coordinate system. ABSCISSA.

Ordnance Datum OD.

Ordnance Survey OS.

Ordovician *adj.* of the second oldest period of the PALAEOZOIC era, and of the system of rocks formed at that time. GEOLOGICAL TIMESCALE.

ore an indefinite term applied to any solid naturally-occurring mineral aggregate of economic interest from which one or more valuable constituents (especially metallic minerals) may be obtained by treatment.

organ in biology, a unit which is both structural and functional, is adapted for some specific and essential function, and forms part of an animal or plant, e.g. a heart, a leaf.

organic *adj.* **1.** of or pertaining to an ORGAN of an animal or plant **2.** having the characteristics of, being related to, belonging to, or derived from living animals and plants and their remains **3.** in chemistry, applied to compounds that contain, or are derived from, CARBON (excluding the simpler compounds such as carbon dioxide), the naturally occurring constituents of animals and plants; hence organic chemistry, the branch of chemistry concerned with the scientific study of organic substances, i.e. with the compounds of carbon (excluding the oxides of carbon, carbonic acid and the carbonates) **4.** arising from the natural processes of or the substances produced by plants and animals. INORGANIC.

organic acids acids produced during the first stage of the decomposition of organisms, before HUMIFICATION. They include acetic, lactic and oxalic acids.

organic carbon CARBON present in soils derived from organic material rather than from minerals.

organic deposit a sediment formed by and consisting of the remains of organisms.

organic farming agriculture practised without the use of 'artificial' (i.e. chemical) fertilizers or pesticides, only ORGANIC-4 manure being added to the soil (and, in some cases, only pesticides such as pyrethrum being used).

organic horizons O HORIZONS.

organic soil a soil consisting mainly of ORGANIC-4 material, as distinct from a MINERAL SOIL. MINERAL HORIZONS, O HORIZONS.

organic weathering the disintegration of rocks brought about mechanically by the penetration of plant roots, or chemically by the ORGANIC ACIDS and HUMIC ACID created by plants such as MOSSES which, growing in crevices and hollows, steadily enlarge their niches as the acids they produce rot the rock. Burrowing and grazing animals also contribute to these processes. CHEMICAL WEATHERING, MECHANICAL WEATHERING, WEATHERING.

Organization (Organisation) for Economic Cooperation and Development OECD.

Organization for European Economic Cooperation OEEC. OECD.

Organization of African Unity OAU.

Organization of American States OAS.

Organization of Central American States (Spanish) Organización de los Estados Centroamericanos. ODECA.

Organization of Petroleum Exporting Countries OPEC.

Orient, the (to orient, to cause to face eastwards, especially to build a church with its altar at the eastern end) the countries lying to the east of Europe, especially the FAR EAST, the opposite of OCCIDENT.

orientalism a fanciful concept held in the imagination of some people of the OCCIDENT that romanticizes and fears the peoples and cultures of the ORIENT.

orientation (French orienter, to place facing east; medieval maps have east at the top) the positioning of, or the position of, someone or something in relation to the points of the compass, especially the positioning of a map or of a surveying instrument in the field so that a north-south line on the map lies parallel to the north-south line on the ground.

origin **1.** the point in time or in space at which the existence of something begins. FALSE ORIGIN, SEISMIC FOCUS, TRUE ORIGIN **2.** the source, the cause, or the beginning of something.

orogenesis, orogeny tectonic activity and mountain building. Orogenesis is usually applied to the process; orogeny to the great periods of mountain building, e.g. the ALPINE, ALTAIDES, ARMORICAN, CALEDONIAN, HERCYNIAN, VARISCAN periods. EPEIROGENETIC.

orographic precipitation, orographic rain relief rainfall, precipitation caused by the cooling of moisture-laden air as it rises over a high relief barrier, occurring particularly on the high ground and windward slopes that face a wind blowing steadily from a warm ocean. It leads to RAINSHADOW on the lee side of the barrier; but the orographic factor may be just additional to other rain-inducing factors, e.g. a high relief barrier may slow down the progress of a DEPRESSION-3, lengthening the duration of the CYCLONIC (FRONTAL) RAIN. THUNDERSTORM.

orography **1.** a branch of physical geography concerned with the knowledge and description of the surface relief of the earth **2.** more specifically, the scientific study of the relief of mountains and mountain systems. A term now rarely used. BATHYOROGRAPHICAL.

OS Ordnance Survey, the official British map-making authority. It publishes the largest range of maps in Britain, from detailed maps on the scale of 1:1250 for urban areas to a world map at 1:40 million scale, with 1:2500; 1:10 000; 1:25 000; 1:50 000 and 1:250 000 maps in between in the main series. Survey information is held on computer. It seems likely that maps-on-demand to meet the needs of the individual user should soon be possible.

oscillatory wave theory of tides oscillation wave theory of tides, the hypothesis that the surface of the ocean can be divided into tidal units (amphidromic systems) each with a centre (a node), termed an AMPHIDROMIC POINT. Within each unit the water oscillates in response to its depth, the relative positions of the earth, moon, sun (CONJUNCTION, OPPOSITION, SYZYGY) and the gyratory movement resulting from the earth's rotation (CORIOLIS FORCE). At the nodes the water stays almost level, rotating in an anticlockwise direction in the northern hemisphere, clockwise in the southern; but from the nodes CO-TIDAL LINES radiate outwards and the height of the tidal rise grows (STANDING WAVE). In the English channel a KELVIN WAVE results, the theoretical amphidromic point, the degenerate amphidromic point, being in Wiltshire.

OSL optical stimulated luminescence, a dating technique using instruments that reveal when a buried surface (e.g. of a soil, of an artifact) was last exposed to sunlight.

osmosis the process whereby the solvent (e.g. water) from a weak (dilute) SOLUTION (termed the hypotonic solution) separated by a semi-permeable membrane from a stronger (i.e. more concentrated) solution (termed the hypertonic solution) passes through the membrane so that the two solutions reach a state of equilibrium in concentration (solutions of equal concentration being termed isotonic). Similarly a pure fluid (e.g. water) will pass through the membrane to a solution (e.g. a sugar solution). If pressure is applied to the stronger solution the process stops. This pressure is termed osmotic pressure and the more concentrated the solution the greater is the pressure needed to stop the process. The movement of water in living organisms occurs, in the main, through osmosis, most cell membranes being semi-permeable to some extent.

outback in Australia, back country, i.e. the areas remote from the main settlements.

outcrop in geology, the part of a rock body, STRATUM or VEIN which reaches the earth's surface and is exposed, or covered (e.g. by superficial soil, vegetation or buildings). EXPOSURE-2.

outflow cave an effluent cave, a CAVE-1 from out of which a stream flows, or formerly flowed.

outlier 1. a mass of rock (not necessarily elevated above the surrounding country) normally of stratified sediments, completely surrounded by older rocks, usually as a result of circumdenudation. By analogy applied to other units separated from the main mass, e.g. to a hill or hills separated from a main highland even if the feature is not an outlier in the geological sense **2.** an outlying area in which a particular CULTURE-1 is dominant but which lies as an island in an extensive area of another culture/other cultures. The outlier is linked to its distant CORE AREA by long-distance routes. DOMAIN-2, SPHERE-5.

outport a port situated nearer the sea and consequently more accessible to vessels (especially those of greater size and draught) than the main port to which it is subordinate. It may eventually supplant the older main port if the latter cannot be modified to meet changing needs in size and port facilities.

output a PRODUCT or PRODUCTION which can be measured.

outwash apron, fan, plain, frontal apron, valley train material (clay, sand, gravel) washed out by melt-water flowing from GLACIERS and ICE SHEETS and deposited, beyond the ice, sometimes over wide areas (FLUVIO-GLACIAL). Sorting and re-sorting takes place in this process, the coarser material being deposited near to the ice, the finer being carried farther before settling. Fig 20.

overbank stage in river flow, the stage when a river overflows the banks of its normal channel and spreads in flood on to the FLOODPLAIN, usually carrying material which is deposited there and termed overbank deposit. BANKFULL STAGE, FLOOD STAGE, STREAM STAGE.

overburden the overlying soil and rock which has to be removed in OPENCAST MINING or STRIP MINING before the seam of coal or bed of ore is exposed.

overcast *adj.* cloudy, sometimes applied as a noun to total cloud cover.

overcropping the overplanting and taking of too many harvests from the soil without restoring its fertility, thus leading to the creation of poor soils which lack sufficient plant nutrients. OVERGRAZING, OVERSTOCKING.

overcrowding the cramming into a given space of more people, more organisms or more things than there is room for, or than is allowed or desirable. The term is applied especially to an excessive number of people living in a specific dwelling, usually measured by the number of persons per room, a matter of judgement varying in one social context to another (e.g. a dwelling in London, a dwelling in Hong Kong). OVERPOPULATION.

overfold an overturned asymmetrical ANTICLINE or an anticline of which one limb is inverted and lies beneath the other. RECUMBENT FOLD.

overgrazing, overstocking the putting of so many animals on land that the pasture or other vegetation cover is damaged, sometimes beyond recovery. OVERCROPPING.

overland flow surface runoff, the unhampered, unchannelled, downslope movement of a broad expanse of shallow water produced by sudden, heavy precipitation, often leading to SOIL EROSION. If the flow is due to the fact that the rate and amount of rainfall exceeds the rate at which the soil can absorb it (INFILTRATION CAPACITY, INFILTRATION RATE) it is termed infiltration-excess overland flow; but if it is due to the SATURATION of the soil over which it flows, it is termed saturation overland flow. Fig 41.

overlay a transparent sheet laid over a map, photograph, diagram etc. It may bear information to be used in conjunction with that shown on the map, photograph, or diagram, or on another overlay; or it may be used for rough work or for correcting the material beneath. SYNOPTIC MAP.

overpopulation too many people, usually applied to the population in an area where the available RESOURCES are inadequate for the support of the great number of people living there, i.e. when the number exceeds that of the OPTIMUM POPULATION, with the result that the standard of living declines and economic and social aspirations cannot be realized. Some authors relate the term to production as well as to available resources, distinguishing absolute overpopulation (where the absolute limit of production has been reached, but living standards stay low) from relative overpopulation (where present production does not support the population but greater production is feasible). BIRTH CONTROL, UNDERPOPULATION.

overspill of population, people in excess

of the number that can be properly housed and served in an area, and who accordingly have to be accommodated elsewhere, e.g. in NEW TOWNS, or, in Britain, in EXPANDED TOWNS.

overthrust in geology, the THRUST of the rocks of the upper limb of a FOLD along a horizontal or near horizontal plane over the lower limb of the fold, caused by folding so intense that the rocks may fracture. NAPPE.

oxbow, oxbow lake a cut-off or mortlake, a crescent-shaped lake formed when a river breaks across the neck of a well-developed MEANDER. Fig 29.

oxidation the process of combining with, or being combined with, OXYGEN. In the chemical WEATHERING of rocks (CORROSION) oxidation occurs particularly in rocks containing IRON because iron takes up oxygen very readily, thus the FERROUS state changes to the oxidized FERRIC state with a yellowish-brown, crumbly crust.

oxide a compound of OXYGEN with another element. OXIDATION.

oxisols in SOIL CLASSIFICATION, USA, an order of soils occurring in tropical and subtropical areas, well-weathered and including most bauxite and lateritic soils. LATERITE.

oxygen a colourless, odourless, invisible gaseous ELEMENT-6, the most abundant in the earth's crust, forming about eight-ninths by weight of water, nearly one-half by weight of all the rocks of the earth's crust, and about 20 per cent of the earth's ATMOSPHERE-1. It is essential to plant and animal life, combines easily with other elements, forming OXIDES. In industry it is used in cutting, welding, in producing flames of high temperature in steel making, and in the manufacture of nitric acid from ammonia, etc. FREE OXYGEN.

ozone an allotropic form of oxygen, O_3, i.e. oxygen with MOLECULES consisting of three atoms instead of the usual two. It has a strong smell and is present in very small quantity in the ATMOSPHERE-1 of the earth, the main concentration occurring between heights of 30 and 80 km (20 and 50 mi) above the earth's surface. Its presence there is vital to atmospheric processes and to the existence of life on earth, because it absorbs harmful short-wave ULTRAVIOLET RADIATION while allowing the beneficial longer ultraviolet radiation from the sun to pass to the earth's surface. Its own existence results from the absorption by OXYGEN of the short-wave ultraviolet radiation from the sun, a process leading to the rise in temperature that occurs at the STRATOPAUSE. Fig 4.

P

Pacific suite a PETROGRAPHIC PROVINCE distinguished by the PACIFIC TYPE OF COAST, intense folding, and rocks rich in calcium. ANDESITE LINE, ATLANTIC SUITE, SPILITIC SUITE.

Pacific type of coast CONCORDANT COAST.

pack animal an animal (usually an ass or mule) that carries goods, fodder etc., usually loaded on a pack saddle (pannier).

pack ice any area of sea ice, whatever the form, however distributed, other than FAST ICE. A collection of pack ice with visible limits is termed a patch. A long strip, from a few kilometres to more than about 100 km (60 mi), is termed a belt.

paddy PADI.

padi (Malay; commonly anglicized as paddy) **1.** RICE as a plant **2.** the grain in the seed-head of the plant **3.** the unhusked grain (hence paddy field, paddy cultivation, paddy harvest etc.). The term 'rough rice' was formerly used internationally for unhusked rice. FAO now applies padi to unhusked rice to distinguish it from milled rice (rice grain from which the husk has been removed).

pahoehoe (Hawaiian) a newly consolidated LAVA flow with a smooth, glassy or ropy surface. Its chemical composition is in many cases identical to that of AA. PILLOW LAVA, ROPY LAVA.

pairing in statistics, a method of control whereby people or things are selected from the whole population and matched in pairs, each pair being characterized by the same quality or qualities. This gives rise to paired, matched or linked data (linked data is the term applied to data from two batches of data where each value in one batch is naturally linked with a unique value in the other). MATCHED SAMPLES.

palaeo- (Greek palaios, ancient) as prefix, ancient.

Palaeocene, Paleocene *adj.* of the earliest of the epochs or of the series of rocks of the TERTIARY period. GEOLOGICAL TIMESCALE.

palaeoecology the scientific reconstruction of past environmental conditions as revealed by fossil records. PALAEONTOLOGY.

palaeolith, paleolith an unpolished, chipped stone implement used in the second period of the STONE AGE.

Palaeolithic, Paleolithic *adj.* of or pertaining to the second period of the STONE AGE, preceded by the EOLITHIC age and followed by the MESOLITHIC age, ending some 8000 years ago, coinciding in the main with the PLEISTOCENE glacial epoch (GEOLOGICAL TIMESCALE). It is the period when, it can be confidently stated, human beings began to shape tools.

palaeomagnetism, paleomagnetism sometimes termed archaeomagnetism, the fossil magnetism evident today in IGNEOUS ROCKS. When cooled, igneous rocks (containing iron oxides) retain the magnetism present in them at the time of their cooling. This indicates the direction of the earth's MAGNETIC FIELD at that time and provides an historical record of the shifting position of the MAGNETIC POLES. MAGNETIC ANOMALY, MAGNETIC REVERSAL, OCEANIC RIDGE, PLATE TECTONICS, TERRESTRIAL MAGNETISM.

palaeontology, paleontology the scientific study of FOSSIL remains of past geological periods. The study of fossil plants is now usually designated palaeobotany, of fossil animals, palaeozoology.

Palaeozoic, Paleozoic *adj.* of or pertaining to the first geological era following the PRECAMBRIAN (GEOLOGICAL TIMESCALE), when seed-bearing plants, fish, reptiles, amphibians appeared, and invertebrates began to develop (some becoming extinct towards the end of the era).

palsa (Swedish pals, from Finnish palsa, elliptical) a dome-like peat hillock, containing several lenses of ice, varying in form and size, up to 3 m (10 ft) or more in height, often surrounded by open water, occurring in northern Sweden and in the TUNDRA elsewhere. The term should not be equated with PINGO.

palynology the study and analysis of pollen grains preserved in soil or peat and of FOSSIL POLLEN and other microfossils highly resistant to acids found in sedimentary rocks of all ages, the findings providing valuable environmental indicators and aids to stratigraphic correlation. STRATIGRAPHY.

pampa, pampas (South America: Spanish) the midlatitude or temperate grassy plains stretching from southern Brazil through Uruguay into the heart of Argentina, the western part (dry pampa) being largely arid and covered with semi-desert grassy vegetation, the eastern part (humid pampa) having a higher rainfall and covered with tall, coarse grass (pampa). Some of the dry pampa and much of the humid pampa has now been ploughed and planted with European grasses or wheat and other crops, and extensive areas are used for cattle rearing. GRASSLAND-I.

pampero (South America: Spanish) a strong southwest wind associated with the cold front in the rear of a LOW PRESSURE SYSTEM moving eastwards, blowing in the PAMPAS area, particularly in Argentina and Uruguay, and on the adjoining coasts, sometimes accompanied by rain, thunder, lightning, taking the form of a LINE SQUALL, sweeping up dust from the pampas and causing a great fall in temperature as the storm passes. It occurs most frequently in summer, and results from the northward advance of a polar air mass.

pan 1. in soil science, HARD PAN **2.** (Afrikaans, pl. panne) any shallow, generally rounded hollow occurring in arid and semi-arid regions which may in some cases hold water only in the rainy season, but in others throughout the year **3.** more specifically, the flat central part of such a depression which may be briefly or seasonally flooded. PLAYA **4.** a container in which gold-bearing sand or gravel is washed in order to separate the valuable GOLD from the other alluvial deposits.

pandemic *adj.* applied to a disease so widespread that it affects the whole world or a whole continent or a whole country. ENDEMIC, EPIDEMIC.

Pangaea the name given by A. Wegener in his theory of CONTINENTAL DRIFT to a great land mass, the supercontinent of PRECAMBRIAN times, probably split in two parts, GONDWANALAND in the south being separated by a vast ocean, TETHYS, from LAURASIA in the north. It is thought that Pangaea started to break up some 190 mn years ago, perhaps first in what is now the Gulf of Mexico. PLATE TECTONICS.

panhandle a narrow protruding strip of land, especially a narrow strip of the territory of a state protruding into that of another state, or between the territory of another state and the sea, e.g. Alaska Panhandle (US territory) lying between Canada and the Pacific ocean.

pantograph an instrument used in the mechanical copying of a drawing or map on any scale selected, consisting of hinged rods arranged to form a parallelogram and rotating about a fixed point.

papagayo (Mexico: Spanish) a dry, strong and cold northerly to northeasterly wind blowing over the plateau of Mexico in winter, causing the temperature to fall suddenly to a low level. A continuation of the NORTHER of the USA, it is due to the movement of air from the high Mexican plateau to the low pressure area lying over the Gulf of Mexico and the Caribbean sea. NORTE.

parabolic dune a crescent-shaped SAND DUNE that faces the wind, i.e. the curve is convex, with a steep face downwind, the 'horns' pointing into the wind, i.e. in the opposite direction to those of a BARCHAN. Parabolic dunes are found particularly on sandy shores and plateaus inland where sudden wind eddies and BLOWOUTS whisk away the central part of the dune and carry it downwind.

paradigm **1.** in general, an accepted example or pattern **2.** specifically, an approach, a school of thought with its associated methodology, accepted by a group of leading scholars as being of particular importance, and used by them and others for a while in their field of study. It points to the kinds of phenomena which should be investigated and to the best means of carrying out the investigations, and it owes its pre-eminence to its perceived ability to be more successful than its competitors in solving some problems regarded as acute at a particular time. PARADIGM SHIFT.

paradigm shift a significant change of approach in a field of study occurring when a prevailing PARADIGM-2 fails to meet the needs of changing thought or new theories, falls from favour, and a new paradigm emerges which is, by consensus, accepted.

parallel drainage a DRAINAGE-2 pattern in which the channels of streams and their tributaries are almost parallel to one another, due to the control of the trend of the underlying structure, or to a uniform slope gradient. Fig 17.

parallel land use a type of MULTIPLE LAND USE in which the tract of land is not used in common for two or more purposes, but in which the uses are kept spatially separate, e.g. a limited recreational strip bordering the highway through a commercial FOREST-1.

parallel of latitude a line drawn on a map to link all points on the earth's surface with the same angular distance north or south of the equator (LATITUDE), termed a parallel because each is a line encircling the earth parallel to the equator. Because the earth is a SPHEROID the circles become smaller from the equator towards the poles. Thus each is a SMALL CIRCLE, but the equator (0° latitude) is a GREAT CIRCLE. Parallels of latitude are marked off in ninety divisions, i.e. ninety degrees, from the equator to each of the poles, each degree being subdivided into 60 minutes, each minute into 60 seconds; and the parallels of latitude meet MERIDIANS (lines of LONGITUDE), which are not parallel to one another, at right angles. Fig 27(b).

parallel retreat of slope a progressive backward movement of a slope with very little change of GRADIENT, the result of weathering and erosion. The theory that the form of a slope may alter little despite a long period of erosion is favoured by many slope authorities. REPOSE SLOPE.

parameter in mathematics **1.** a quantity which is constant in the case under consideration but which may vary from case to case **2.** an unknown quantity which may vary over a certain set of values **3.** a VARIABLE of which other variables are taken to be FUNCTIONS-3 **4.** a descriptive summary of some characteristic of a POPULATION-4, e.g. MEAN, MEDIAN, STANDARD DEVIATION **5.** any of the factors in a whole that helps to define main characteristics or limits. In statistics the term parameter commonly occurs in expressions defining FREQUENCY DISTRIBUTION (e.g. POPULATION PARAMETER) or in models describing a STOCHASTIC situation (e.g. REGRESSION parameter).

parametric test a statistical test which involves the estimation of a POPULATION PARAMETER and also some assumptions about the data, i.e. that the data are measured on an INTERVAL SCALE and that the POPULATIONS-4 from which the samples are drawn have a NORMAL DISTRIBUTION and equal variances. NON-PARAMETRIC TEST.

parasite an organism which lives on a living organism of a different species (the host) from which it draws most or all of its food sometimes, but not always, to the detriment of the host. ECTOPARASITE, ENDOPARASITE, EPIPHYTE.

parasitic cone ADVENTIVE CONE.

parent material in soil science, the little altered but weathered bedrock, or the transported glacial or alluvial material, or an earlier soil, on which soil-forming processes work to make the soil layers. SOIL, SOIL HORIZON.

park 1. in medieval England, a piece of land held by the lord of the manor by royal grant for keeping animals (e.g. deer), for a source of food, for hunting, differing from a CHASE or FOREST-2 in being enclosed (and also differing from a forest in not being subject to Forest Laws); later applied to **2.** a large, enclosed piece of ground, usually landscaped, comprising grassland and widely planted trees, used for private recreation and the grazing of deer, cattle etc. and attached to a large house and later to **3.** a public park, an enclosed recreational area in or near a town, laid out with flowerbeds, paths, lawns, perhaps a small ornamental lake etc. **4.** in the concept of a NATIONAL PARK, an extensive area of countryside to be protected and conserved for public benefit, originating in the Yellowstone National Park, USA, in 1871 **5.** in the mountains of Colorado, Wyoming, USA, a high, enclosed valley with a flat, grassy floor, frequently with a RANCH **6.** a tract of land, in some cases park-like (PARK-2) in character, set aside and officially designated (e.g. by planning authorities) for a particular purpose, e.g. a COUNTRY PARK, INDUSTRIAL PARK, OFFICE (BUSINESS) PARK, RESEARCH AND SCIENCE PARK **7.** in Scotland and Ireland, a paddock or field, a small enclosed area of land usually used for pasture, sometimes for crops.

parkland 1. see applications PARK-1,2 **2.**

by analogy with a typical English park (consisting of grassland with scattered trees, PARK-2), the tropical SAVANNA lands of grassland with scattered trees, especially in Africa (also termed park savanna).

paroxysmal volcanic eruption a sudden violent eruption of a VOLCANO, usually of one that has been dormant for a long period. PLINIAN ERUPTION.

partial drought in Britain, a period of 29 consecutive days on some of which slight rain may fall, but during which the daily average does not exceed 0.25 mm (0.01 in). DROUGHT.

particle 1. a very small piece of matter **2.** in physics, a piece of matter assumed to have mass but to be so small that it does not have dimensions, i.e. it has finite mass but is of infinitesimal size. GRADED SEDIMENTS.

particulate *adj.* **1.** existing as a very small, separate particle **2.** of or relating to such a very small, separate particle. Particulate matter in smoke emissions, fumes, dust etc. is an agent of POLLUTION.

part-time farmer a FARMER who has a regular occupation, other than that of FARMING, to provide an income. The two main types are the HOBBY FARMER, who farms primarily for pleasure, not for a living; and the worker-farmer, whose main occupation is in the town but who relies on the output of a small holding, worked in any spare time, for a food supply and an important auxiliary source of income.

pascal pa, the unit of pressure in SI, equal to one NEWTON per sq METRE.

pass a narrow gap or COL in a mountain range providing a passageway through the barrier.

passive glacier a GLACIER with a low rate of accumulation (ALIMENTATION) and ABLATION because it receives only light snowfall and undergoes little melting in summer. It flows very slowly and transports little ice and debris. ACTIVE GLACIER.

pastoral *adj.* of, pertaining to, or characterized by, the care of grazing animals. GRAZING, LIVESTOCK FARMING.

pastoralism LIVESTOCK FARMING.

pasture land covered with growing grass and/or other HERBS-1 on which LIVESTOCK can feed, as distinct from a MEADOW-1 where the vegetation is mown for hay or silage. GRAZING, LIVESTOCK FARMING, ROUGH GRAZING.

patch cutting the FORESTRY practice of felling trees in a defined area within a FOREST-1 so that the tree cover may regenerate naturally, inward towards the centre of the cleared area from the surrounding forest. The aim is to conserve the forest as a FLOW RESOURCE and to maintain a sustained yield (SUSTAINED YIELD FORESTRY). CONSERVATION, RESOURCE CONSERVATION, RESOURCE MANAGEMENT.

paternoster lakes a chain of lakes in a glaciated valley caused by the damming action of morainic ridges or rock bars, so-named because the chain resembles the string of beads that constitute a rosary. Such lakes occur especially where the downward slope of the valley floor takes the form of a series of steps.

pathetic fallacy the attribution of human characteristics or emotions to natural phenomena, e.g. an angry sea.

patterned ground ground that is embellished with circles, polygons, nets, stripes etc., varying in size and occurring in (but not wholly confined to) regions of present or former PERMAFROST. The rock fragments may or may not be sorted, e.g. in a polygonal pattern the smaller fragments may appear near the centre and increase in size outwards, or the pattern may be outlined by vegetation, no sorting of fragments having taken place; and on a slope there may be step-like forms, or stripes consisting of lines of stones, graded as the slope flattens. It has been suggested that the processes of freeze-thaw and SOLIFLUCTION, moisture conditions and the extent (or non-existence) of vegetation cover contribute to the development of the patterns. PIPRAKE.

pays (French) a small NATURAL REGION in France, of no administrative significance, demarcated by the unity of some physical feature or features (e.g. of geology, relief, land use) and/or of social or cultural features. A term used by geographers writing in English who wish to avoid the problem of defining 'natural'.

pea a climbing or bushy leguminous ANNUAL plant (LEGUMINOSAE), probably (but not certainly) native to western Asia, now widely grown for the sake of its pods, containing seeds also termed peas, used as human food since ancient times, the whole plant also being used for animal feed and for GREEN MANURE.

peak **1.** a pointed top or projection **2.** the highest point, the maximum **3.** the more or less pointed prominent summit of a mountain, often used in place-names; but this does not apply to The Peak of Derbyshire, England, which is a high, flat-topped plateau of MILLSTONE GRIT, an area scheduled as a NATIONAL PARK **4.** a high isolated mountain.

peasant very loosely applied to an agricultural worker, usually (but not always) farming at or near subsistence level and either holding a proprietary right over the land farmed or being an employed labourer. Some specialists in LAND TENURE restrict the term to the farmer holding a proprietary right. PEASANT FARMING, SUBSISTENCE AGRICULTURE.

peasant farming a farming activity carried on by a small family unit at or near subsistence level on a small area of land, e.g. as distinct from that on a large farm with employed labour, organized primarily for commerce. In peasant farming the family provides the labour. Production is primarily for the benefit of the family but in most cases (particularly if the land is leased from a landlord) surplus products are sold in the market (in order to pay the rent). If the land is leased from a landlord the tenancy is usually held by the family group, rather than by an individual. If the land is owned by the peasant the right of ownership is usually vested in the family group, not in an individual. FARMING, LAND TENURE, PEASANT, PLANTATION-2, SUBSISTENCE AGRICULTURE.

peat a dense deposit of dead vegetable matter only partially decomposed (DECOMPOSITION) mainly because it has accumulated in water or in very damp conditions where oxygen is deficient (ANAEROBIC). Dark brown or black, partially converted to carbon, it forms the first stage in the development of LIGNITE, BROWN COAL and COAL; and may be neutral or alkaline (FEN PEAT) or acid (BOG PEAT). It is burnt as fuel, or applied to the soil to improve the texture or raise its water-retaining property.

pebble a small stone, naturally rounded by the action of water or wind, diameter between that of GRAVEL and COBBLE.

There is some confusion over precise size, but a pebble is commonly defined as having a diameter lying approximately between 10 and 50 mm (0.4 to 2 in); on the WENTWORTH SCALE it lies between 2 to 64 mm (0.08 to 2.5 in). The term is sometimes wrongly applied (as 'angular pebbles') to the fragments in a BRECCIA.

ped a naturally-formed aggregate of soil particles. CRUMB.

pedalfer in soil science, a soil from which the base compounds CALCIUM CARBONATE and MAGNESIUM carbonate have been leached, and which contains accumulations of ALUMINIUM, and IRON compounds. Such soils occur especially in a humid climate with annual precipitation over 610 mm (24 in), and include BROWN EARTHS and PODZOLS. Pedalfer is applied to one of the two types of ZONAL SOIL, the other being PEDOCAL. SOIL CLASSIFICATION.

pedestal, pedestal rock a residual, columnar mass of weak rock capped with a harder rock. Opinions differ as to whether it is formed by differential WEATHERING helped by rainwash, or by wind ABRASION.

pediment an eroded rock platform, cut into the BEDROCK, usually slightly concave and triangular in shape, bare or carrying a very small quantity of rock debris, extending over a considerable area at the foot of an abrupt mountain slope or face, the lower edge sloping gently away. Pediments form basal slopes of transport over which occasional rainstorms carry weathered material derived from the steeper slope above, and are characteristic of ARID and semi-arid lands. A pediment should not be confused with a BAJADA, formed by deposition. Fig 35.

pediplain, pediplane a gently undulating or level surface, either exposed or lightly covered with a thin layer of alluvium, possibly formed by the merging

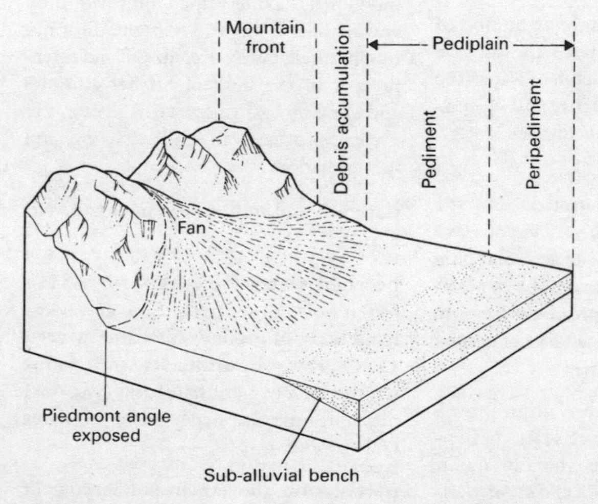

Fig 35 A pediment

of several PEDIMENTS in the course of erosion. Fig 35.

pedocal in soil science, a soil, only slightly leached (LEACHING), containing an accumulation of CALCIUM CARBONATE and MAGNESIUM carbonate, and occurring mainly in climates with an annual precipitation below 610 mm (24 in), insufficient to remove the soluble constituents. Pedocal is applied to one of the two types of ZONAL SOIL, the other being PEDALFER. SOIL CLASSIFICATION.

pedogenic *adj.* **1.** of or pertaining to soil formation **2.** of or pertaining to the effect caused by soil factors. EDAPHIC.

pedology the scientific study of the morphology, composition and spatial distribution of soils, including their classification and, in a general way, their uses.

pelagic *adj.* of, pertaining to, occurring in, living in, the mass of open water of the ocean or a lake, i.e. away from the land and the floor of the ocean or lake.

pelagic division, pelagic zone one of the two chief divisions of the aquatic environment based on depth of water (the other being the BENTHIC DIVISION), consisting of the whole mass of open water.

Pelean *adj.* of or pertaining to a type of volcanic eruption characterized by the extrusion of a very acid, very viscous lava which tends to consolidate as a solid plug and by the violent emission of NUEE ARDENTE, causing widespread devastation. HAWAIIAN, STROMBOLIAN, VULCANIAN OF VESUVIAN ERUPTION.

pelosols one of the seven major groups in the 1973 SOIL CLASSIFICATION of England and Wales, clay soils which are argillic (ARGILLACEOUS), CALCAREOUS or noncalcareous, with a brown, grey or red and commonly MOTTLED layer below the surface.

peneplain, peneplane almost a plain, an almost featureless, gently undulating land surface, the penultimate stage before the stage of the plain without relief in a CYCLE OF EROSION or denudation.

peneplanation the process of forming a PENEPLAIN. The spelling peneplaination is, by general agreement, incorrect. CYCLE OF EROSION.

peninsula almost an island, a tract of land, large or small, projecting into a body of water, nearly surrounded by water or having water on at least three sides, so that the greater part of the boundary is coastline. ISTHMUS.

peninsular *adj.* of, pertaining to, belonging to, a peninsula.

perception in general, sensory experience which has acquired meaning or significance, i.e. a process in which the individual, having gained information by any or all of the senses (sight, hearing, touch, smell, taste), organizes it and interprets it in the light of his/her attitudes and experiences. COGNITION stresses the mental processes of understanding and interpretation.

perched block ERRATIC, ERRATIC BLOCK.

perched water table GROUND WATER (often isolated) separated from an underlying body of ground water (the WATER TABLE proper) by unsaturated rock, forming part of a zone of saturation (VADOSE) different from that occupied by the water table proper.

percolation the downward seeping of water through pores, joints and other interstices in soil and rock, sometimes accompanied by LEACHING.

percoline a SOIL HORIZON along which water seeps laterally. THROUGHFLOW.

perennial *adj.* applied particularly to plants that continue in growth from year to year. The aerial parts of perennial herbaceous plants (HERB-1) die down after seed production and are replaced by new shoots in the following year. The new shoots of woody perennials start from the permanent woody stems above ground, hence the great size of some woody perennials, e.g. TREES, SHRUBS. ANNUAL, BIENNIAL, EPHEMERAL.

perennial irrigation a system of IRRIGATION in which water is artificially made available to plants, not seasonally, as in BASIN IRRIGATION, but throughout the year. The methods used to maintain a regular water supply range from the simple lifting of stream water by ARCHIMEDES' SCREW, SAQIA, SHADUF or water pump, by sinking of wells and by the creation of ponds (TANK) to the building of dams and barrages on rivers to hold back enough water to last from one flood period to the next, thus providing a constant supply for the canals and channels by which the water is led on to the dry land.

perennial stream a stream that flows continuously throughout the year, in contrast to an INTERMITTENT STREAM.

perfect competition a theoretical market condition in which there are many independent buyers and sellers, each of the sellers holding only part of the goods involved, all resources being perfectly divisible and mobile. The sellers are not allowed to collude. Buyers and sellers are fully informed of prices throughout the market, and none of them individually can influence prices. DUOPOLY, MARGINAL UTILITY, MARKET ECONOMY, MONOPOLY, OLIGOPOLY.

pericline a small ANTICLINE in which the strata are arched up into a dome so that the beds dip away on all sides from a central point. CENTROCLINE.

peridotite a group of coarse-grained ULTRABASIC IGNEOUS ROCKS, usually dark green, consisting mainly of OLIVINE with or without FERROMAGNESIAN MINERALS, and without FELDSPAR.

perigean tide the tidal condition characteristic of a period when the moon is at its PERIGEE, so that its gravitational pull is great, resulting in a high tide that is higher, a low tide that is lower, and a TIDAL RANGE greater, than usual. APOGEAN TIDE.

perigee the point in the orbit of any of the earth's planets or satellites when it is nearest to the earth, applied particularly to the orbit of the MOON around the earth. APOGEE, APSIS.

periglacial *adj.* applied to areas adjacent to an ICE SHEET or GLACIER of the past or of the present, and to all phenomena associated with such a situation.

perihelion the point nearest to the SUN in the orbit of a comet or planet around it. The earth arrives at its perihelion about 3 January, when it is some 147.3 mn km (91.5 mn mi) from the sun. APHELION, APSIS.

period in geology, a division of geological time, part of an ERA. The rocks laid down in a period constitute a system. GEOLOGICAL TIMESCALE.

permafrost permanently (i.e. for a continuous period of at least two years) frozen soil, subsoil and, some authors add, bedrock (ACTIVE LAYER). PATTERNED GROUND, PIPRAKE.

permafrost table the surface between the upper limit of the PERMAFROST and the lower limit of the overlying ACTIVE LAYER.

permeability the quality or state of being PERMEABLE.

permeable *adj.* capable of being wholly penetrated by a fluid, of allowing the passage of a fluid, of being SATURATED. The opposite condition is termed IMPERMEABLE.

permeable rock a rock that allows the free passage of water through it owing to its POROSITY, e.g. sandstone, oolitic limestone. Some writers include also rock with joints, bedding planes, cracks, fissures etc. that allow the free passage of water, defining the porous rock as being of primary permeability, the rock with joints etc. of secondary permeability. Other authors distinguish the secondary group as being pervious (PERVIOUS ROCK). IMPERMEABLE ROCK, IMPERVIOUS ROCK.

Permian *adj.* of the latest period (of time) or system (of rocks) of the PALAEOZOIC era, when amphibians declined, reptiles developed, sandstone strata were deposited. GEOLOGICAL TIMESCALE.

Persian wheel SAQIA.

persistent *adj.* lasting, enduring despite opposition or difficulties, repeatedly occurring, applied to **1.** leaves which, though withered, stay on a plant **2.** an organism which continues to occupy an area in which conditions are now generally hostile to the species to which it belongs **3.** chemicals which, because of their stability, do not readily decompose **4.** meteorological conditions which continue for more than the normal or expected length of time, e.g. a persistent high pressure system (HIGH).

pervious rock some authors use this term as synonymous with PERMEABLE ROCK, but others restrict it to a rock that allows the free passage of water through it owing to the presence in it of joints, cracks, fissures etc., e.g. chalk, carboniferous limestone. IMPERMEABLE ROCK, IMPERVIOUS ROCK, POROSITY.

pest an insect or other animal considered by human beings to be harmful to them and their activities, e.g. to the growing of crops, food storage etc.

pesticide a chemical substance used in killing insects or other animals regarded by human beings as being harmful to them and their activities.

petrification the turning of organic material into stone or into a substance nearly as hard as stone, the original organic tissue being very slowly replaced by deposits of minerals such as silica, agate or calcium carbonate carried in solution. FOSSIL.

petrochemical a chemical derived from PETROLEUM or NATURAL GAS or from a derivative of either, extensively used industrially in the production of plastics, synthetic fibres, drugs, detergents, fertilizers, fungicides, insecticides, pesticides etc.

petrographic province a region in which the various IGNEOUS ROCKS are so related by marked specific peculiarities as to differentiate them from other assemblages of igneous rocks of other regions or cycles. The term 'suite' has been applied to rocks showing such affinities with each other, and three suites are usually identified: ATLANTIC, PACIFIC, SPILITIC.

petroleum crude oil, mineral oil, a mixture of HYDROCARBONS in a solid (BITUMEN), liquid or gaseous state (NATURAL GAS), occurring in porous SEDIMENTARY ROCKS interbedded with SHALES and other rocks. It originates from the alteration of vegetable and animal remains entombed in the sediments when these were deposited, especially under brackish water conditions. The term is usually applied specifically to the liquid form, commonly trapped, with natural gas, at the top of the arch (dome) of an ANTICLINE between impervious beds. This petroleum, obtained by drilling into the OIL DOME, is subsequently distilled in a refinery, yielding fractions making PETROL, paraffin oil, diesel fuel, fuel oil, lubricating oils and heavy fuel oils; ASPHALT and paraffin wax are obtained from the residue. The hydrocarbons in the heavier fractions are further 'cracked' to make them suitable for, and to increase the amount of fuel for, internal combustion engines. OIL FIELD, OIL POOL, OIL-SAND, OIL-SHALE, OIL WELL.

pH potential hydrogen, the standard measure of ACIDITY or alkalinity (BASE-2) of a substance, based on the activity of HYDROGEN IONS in a litre of a solution (or of a pure liquid), expressed in gram equivalent per litre. The pH values range from 0 to 14.0, NEUTRALITY-3 being 7.0 (the pH of pure water at 25°C: 77°F). Numbers lower than neutrality signify increasing acidity, the higher numbers increasing basicity (alkalinity). Knowledge of the pH of the soil under cultivation is most important in agriculture and horticulture. ACID RAIN, ACID SOIL, ALKALI SOIL, ALKALINE SOIL.

phacolith, phacolite a lens-shaped intrusion of IGNEOUS ROCK occupying the saddle (the crest) of an ANTICLINE or the keel (the trough) of a SYNCLINE, differing from a LACCOLITH in that it is shallower and the laccolith has a flat base.

phenomenal environment the ENVIRONMENT seen as including not only natural phenomena (PHENOMENON-1) but also the effects of human activity, i.e. the BIOTIC and physical environment combined with the environments altered by, and in some cases almost entirely created by, human beings: the 'real' world as distinct from the 'perceived' world represented by the BEHAVIOURAL ENVIRONMENT.

phenomenalism in philosophy, the theory that phenomena (PHENOMENON-3) are the only things that can be known, that everything else is either non-existent or inaccessible to the human mind. METAPHYSICS.

phenomenon pl. phenomena **1.** a fact or event that can be described and explained scientifically **2.** a person, fact, thing, or event that is extraordinary **3.** in philosophy, something known by (sensed) perception, i.e. not by thought or intuition.

phosphate any of the substances obtained naturally from the weathering of rock rich in PHOSPHORUS (phosphate rock) or from GUANO, or industrially by processing phosphate rock or basic slag from blast furnaces. Phosphates, an important source of PHOSPHORUS, are used as a soil FERTILIZER and in industrial processes. SUPERPHOSPHATE.

phosphorus a nonmetallic ELEMENT-6 existing in several structural forms, in nature only in combined state, chiefly in phosphate rock. All living cells contain organic compounds of phosphorus (it is an essential MACRONUTRIENT), the inorganic compounds being important constituents of minerals, soil, bones, teeth. It is used in making FERTILIZER, detergents and matches. PHOSPHATE.

phosphorus cycle the circulation of atoms of PHOSPHORUS brought about by natural processes, the major part being played by living organisms which contain organic compounds of phosphorus. On their death their tissues decompose, the phosphorus returns to the soil in a form suitable to sustain plants, and the plants provide food rich in phosphorus for animals.

photic region, photic zone (Greek phōs, light) the layer of water of a lake or of the ocean indicated by the penetration of light and the distribution of plants. It includes the DISPHOTIC ZONE and the EUPHOTIC ZONE. PELAGIC.

photochemical *adj.* of or relating to the action of light on chemical properties or reactions. PHOTOSYNTHESIS, SMOG-2.

photogrammetry **1.** the technique of using photographs to obtain measurements of the subject of the photograph **2.** the science or art of making a topographical map from air photographs.

photoperiod the duration of daylight (when direct SOLAR RADIATION reaches the earth's surface). The relative length of dark and light periods affects the behaviour and growth of animals and plants. DAYLENGTH.

photoperiodism the responses of organisms to DAYLENGTH.

photosynthesis the SYNTHESIS-1 by living cells of complex organic compounds from simple inorganic compounds, with the aid of light energy, the process by which the living cells of green ALGAE and higher green plants manufacture CARBOHYDRATES from CARBON DIOXIDE and water, using energy absorbed from sunlight by CHLOROPHYLL, and releasing OXYGEN (AUTOTROPHIC, PHOTOTROPHIC). The products of photosynthesis directly or indirectly supply nearly all plants and animals with the energy they need for metabolism; thus, with a few minor exceptions, all other forms of life depend for their existence on the photosynthesis carried out by green plants (FOOD CHAIN, PRIMARY PRODUCTION). The notable exceptions are bacteria (BACTERIUM) in the OCEANIC CRUST which seem to draw energy from methane and minerals in the crust.

phototrophic *adj.* applied to an organism that obtains energy from sunlight. AUTOTROPHIC, CHEMOTROPHIC, HETEROTROPHIC, PHOTOSYNTHESIS.

phreatic surface GROUND WATER table. PHREATIC WATER.

phreatic water GROUND WATER below the PHREATIC SURFACE, i.e. water in the zone of saturation. If several wells were sunk and the water levels in them linked by a flat or undulating surface, the ground water table, i.e. the phreatic surface, would be the result.

physical geography one of the two main divisions of GEOGRAPHY (the other being HUMAN GEOGRAPHY), concerned with the study over time of the characters, processes and distribution of the 'natural' phenomena in the space accessible to human beings and their instruments, i.e. in the ATMOSPHERE-1, BIOSPHERE, HYDROSPHERE, LITHOSPHERE.

physical weathering mechanical weathering, the disintegration of rocks caused by the weather (e.g. frost and temperature change) without any chemical change taking place. EROSION, WEATHERING.

phytoplankton minute plants, many microscopic, mostly algae (e.g. DIATOMS), which float in bodies of fresh or salt water, which are basic in the FOOD CHAIN, and on which nearly all animal life in the open sea depends. PLANKTON, ZOOPLANKTON.

pidgin a spoken language that incorporates the vocabulary of two or more languages and an extremely modified grammar of one.

pie diagram, pie graph DIVIDED CIRCLE DIAGRAM.

piedmont at the foot of a mountain, term used as a noun and applied to the gentle sloping ground lying between the steep slope of a mountain and the plain below, including the PEDIMENT and the BAJADA; or, very widely, as an *adj.*, e.g. piedmont alluvial plain, piedmont depressions etc. Fig 35.

piedmont glacier an extensive sheet of ice covering low ground at the foot of a mountain range, formed by the merging of a number of parallel VALLEY GLACIERS. EXPANDED-FOOT GLACIER.

piezometric level, piezometric surface the level to which water from a confined aquifer will rise under its own pressure in a borehole, the underground level to which water will rise under its full head in an aquifer.

piggyback transport a road-rail transport system in which goods are carried by road vehicle from the factory to a railway station, where the road vehicle is hoisted on to a goods train, transported for the major leg of the journey, off-loaded at a convenient station, to continue its journey to its ultimate destination by road. If an articulated vehicle is used only the trailer may be carried by train, to be reconnected to a tractor at the station at the end of the rail leg.

pig iron cast iron, impure iron with high combined carbon content, sometimes also with sulphur and phosphorus, the product of a BLAST FURNACE. The molten metal from the blast furnace is run into channels or lateral moulds known as pigs, hence the name (in the large modern integrated steel works the molten metal is fed directly to the steel converter). The high content of carbon and phosphorus make pig iron (cast iron) very brittle. These impurities are removed by puddling, a process in which the cast iron is re-melted and stirred when molten, so that the carbon escapes, to be burnt in the heated air of the furnace, the product being termed wrought or malleable iron. Wrought iron is tolerably hard, but not flexible, elastic or hard enough to be used in making machinery, construction equipment, cutlery and so on: for those purposes it must be converted into STEEL.

pillow lava LAVA which has consolidated so as to resemble a jumbled mass of pillows due to the fact that MAGMA (especially BASALTIC magma) was extruded under water on a sea floor, or flowed into the water before consolidation, so that the water cooled the outer skin quickly and the lava formed a rounded mass which partly collapsed. AA, PAHOEHOE.

pingo (Inuit, a conical hill) an isolated, more or less conical mound of unconsolidated gravel or earth, varying greatly in height between 6 and 90 m (20 to 300 ft), generally with a core of clear ice, occurring in TUNDRA lands in Alaska, Arctic Canada, Greenland, Siberia when water under the surface freezes in autumn. The ensuing ice core expands and raises the surface to form a dome. The top of the pingo may later collapse, forming a crater which may become partially filled with water. Some authors distinguish a closed-system pingo (formed from the freezing of an isolated body of water) from an open system pingo (formed from the freezing of ground water at its pressure head). PALSA.

pinnate *adj.* resembling a feather.

pinnate drainage a distinctive DENDRITIC pattern of DRAINAGE-2 that resembles a feather in that the tributaries to the main stream are closely spaced and meet the main stream at acute angles. It is likely to occur if the slopes on which the tributaries develop are unusually steep. Fig 17.

pioneer community in ecology, SUCCESSION-2.

pipe **1.** in a VOLCANO, the vent opening into the CRATER **2.** in chalk country, a vertical joint made wider by CARBONATION-SOLUTION and filled with sand and gravel **3.** a mass of mineral ore or diamond-rich rock, e.g. KIMBERLITE, shaped like a column, in some cases formed by FLUIDIZATION **4.** a tunnel in the soil, commonly in the interface between horizons (SOIL HORIZON), resulting from THROUGHFLOW.

pipeline a long stretch of linked pipes with pumps at intervals, used to carry gases, liquids, or solid material in the form of SLURRY-2.

piprake (Swedish) the layer or layers of perpendicular ice needles (very small thin, sharp-pointed pieces of ice), varying in length from 1 or 2 mm to nearly 0.5 m, formed in PERIGLACIAL conditions on the surface of the ground, i.e. the limit surface between the soil and free atmosphere, or just below the surface, where layers of needles may be separated by a thin MINERAL SOIL layer. They contribute to FROST HEAVING and the making of PATTERNED GROUND.

pisciculture the breeding and rearing of fish for commercial purposes. AQUACULTURE, FARM, FARMER, FISH FARM.

pisolite, pisolith pea stone, a type of LIMESTONE made up of rounded bodies the shape and size of a pea, diameter 2 to 10 mm (0.08 to 0.40 in), i.e. larger than the OOLITHS of an OOLITE. Jurassic pea grit is typical; and other rocks may show pisolitic structure, e.g. some LATERITES and IRON ores.

pit **1.** a surface working, a deep hole in the ground from which minerals or other materials are extracted, e.g. clay, gravel **2.** a coal mine, including the associated pithead buildings and the shaft. MINE, MINING.

pitch **1.** any of the several dark, sticky, resinous substances, solid at low temperatures, plastic at high, which are a residue formed in the distillation of TARS, turpentine etc., but especially the naturally occurring nearly black substance termed ASPHALT. When an oil-sand has been exposed by natural erosion the oil gradually evaporates, leaving such natural pitch as a residue (as in the pitch lake of Trinidad). BITUMEN-2 **2.** a resin derived from certain CONIFEROUS trees, e.g. pitch pine **3.** pitch of a fold, the direction in which the AXIS OF A FOLD dips.

pitchblende a mineral, a complex oxide of uranium with small quantities of other elements from which URANIUM is extracted.

pixel picture element, in REMOTE SENSING, using MULTISPECTRAL SENSING, the very small basic unit from which a satellite image may be constructed. The area it can represent ranges from 5 m^2 to 10 km^2 (6 sq yd to 4 sq mi), the smaller the area the better the RESOLUTION-1,2 at ground level.

place **1.** a particular part of space, an area or volume of SPACE-2, unoccupied or occupied, e.g. by a person, object or organism **2.** a particular area in space, e.g. town, village, district etc., real or as perceived (MENTAL MAP), where people and environment interact over time to give it characteristics distinct from those of surrounding places **3.** POSITION in a hierarchy, scale, orderly arrangement, or in space. LOCATION, SITUATION.

placer an alluvial deposit of sand or gravel with particles of valuable minerals, especially of GOLD, weathered from rocks or veins and washed down by a stream.

place utility **1.** the measure of approval accorded by an individual to a location in his/her ACTION SPACE. Dissatisfaction with place utility, expressed by a low measure, may result in MIGRATION-2 **2.** the VALUE-1,3 of a PLACE-1,2 brought about by efficient TRANSPORT facilities.

plagioclase feldspar a FELDSPAR with an oblique cleavage, containing SODIUM and CALCIUM.

plagioclimax (Greek plagios, oblique) the stage reached by a plant community that is stable and in equilibrium with the existing conditions of its environment but which has been deflected or drawn back by the influence of BIOTIC FACTORS from what would have been its natural CLIMAX.

plagiosere a plant SUCCESSION-2 which has been deflected by BIOTIC FACTORS from its normal course of development, resulting in the formation of a PLAGIOCLIMAX rather than a CLIMAX.

plain a term with wide variations in use, especially in compound terms, but generally applied to an unbroken flat or gently rolling land surface, lacking prominent elevations and depressions, especially one of low elevation. Some authors restrict the application to such a feature with a horizontal structure. HIGH PLAINS.

plan **1.** a map of a small area on a large scale on which everything is drawn precisely to scale **2.** a large scale detailed chart **3.** a drawing to scale representing a horizontal section of a solid object **4.** a formulated scheme of action, or the way in which it is proposed or intended to carry out some proceeding or course of action to achieve some goal, e.g. in connexion with economic and social development. LAND USE PLANNING, PLANNING, PLANNING REGION, REGIONAL PLANNING.

planalto (Brazil: Portuguese) a high plain or plateau, especially the great Brazilian plateau.

planet **1.** a heavenly body that is not a meteor, comet, or artificial satellite, which revolves around a star by which it is illuminated **2.** specifically, such a body revolving about and illuminated by the sun in the earth's SOLAR SYSTEM. ASTEROID, EARTH.

plane table, plane-table a small drawing board mounted on a tripod and used in the field with an ALIDADE in topographical survey.

planetary *adj.* of or pertaining to a planet or to the planets, specifically to the earth as a planet.

planetary winds the major latitudinal winds (TRADE WINDS, WESTERLIES etc.)

in the ATMOSPHERIC CIRCULATION in contrast to local winds (LAND BREEZES, SEA BREEZES etc.).

plankton the collective term for all forms of minute plant and animal life (especially of the many microscopic organisms freely floating in the ocean or in fresh water). It does not include the larger plants, e.g. floating SEAWEED. All plankton are of great ecological and economic importance, in that they are the essential food for fish, marine mammals, sea birds etc. (FOOD CHAIN). AEROPLANKTON, DIATOM, NANOPLANKTON, PHYTOPLANKTON, POTAMOPLANKTON, ZOOPLANKTON.

planned economy an economic system in which the central government controls capital, labour, production, distribution of goods etc. to meet the needs of a comprehensive economic plan, instead of allowing free play of market forces (supply and demand). MIXED ECONOMY, PRIVATE SECTOR.

planning **1.** a method for outlining or defining goals and ways of achieving them **2.** the drawing-up and implementation of a PLAN-4. BLIGHT, LAND USE PLANNING, PLANNING REGION, REGIONAL PLANNING.

planning region a specific unit area for which there is a PLAN-4 for economic and social development. LAND USE PLANNING, REGIONAL PLANNING.

plant **1.** any member of the kingdom Plantae (CLASSIFICATION OF ORGANISMS), the typical plant being a living organism lacking means of independent locomotion, having no central nervous system, the cell walls consisting of cellulose, nutrition depending on simple gaseous or liquid substances (principally carbon dioxide and water) which, with the aid of CHLOROPHYLL functioning in light, the organism builds up into sugars and other complex materials (FOOD CHAIN, PHOTOSYNTHESIS); but many plants, e.g. fungus, do not exhibit those characteristics and some microscopic organisms, e.g. some BACTERIA, seem to be as much ANIMAL-1 as plant. ECOLOGY, PARASITE, VEGETATION **2.** in industry, the buildings, machinery and other equipment used in the manufacture of goods or the production of power.

plantation **1.** something that is planted, hence applied particularly to a group of growing trees which has been planted to provide timber and wood pulp, and/or to function as part of a land reclamation scheme **2.** a farm or estate, especially one in the tropics, with large scale planting of CASH CROPS, especially on a monocultural basis (MONOCULTURE) under scientific management, often with a large labour force, sometimes with facilities and plant to process the crop, e.g. as on tea or rubber plantations. The term plantation agriculture or plantation system implies rather more: the use of a large labour force (in many cases formerly of slave labour) to produce one or two commercial crops on a large scale, in contrast to PEASANT FARMING **3.** historically, a colony or settlement in a new or conquered country.

planter **1.** one who manages a PLANTATION-2 **2.** historically, a colonist. PLANTATION-3.

plant indicator INDICATOR PLANT.

plant succession SUCCESSION-2.

plastic *adj.* of or pertaining to any substance which changes shape under pressure and keeps the new shape when the pressure is eased. ELASTIC.

plateau (French) a markedly elevated tract of comparatively flat or level land, a tableland, usually bounded on one or more sides by steep slopes which drop to lower land, or by steep slopes rising to a mountain ridge. Some large plateaus consist of LAVAS, especially of BASALTS, poured out through fissures, hence the term plateau lava (PLATEAU BASALT). Originally French, with plural plateaux, the term is regarded as having been absorbed into the English language, the plural becoming plateaus. INTERMONTANE, PLATEAU GRAVEL, WATERFALL.

plateau basalt BASIC LAVA, very fluid at high temperatures, that reaches the surface through fissures in the earth's crust (FISSURE ERUPTION) and spreads evenly over an extensive area. It builds up a lava PLATEAU, especially if there are successive eruptions. FLOOD BASALT.

plateau gravel gravel (small stones, sand, grit) occurring either as an extensive sheet on the surface of a PLATEAU or as a capping on hills which represent the dissected remains of former plateaus. Such gravel deposits provide important evidence in the study of the chronology of DENUDATION and of geomorphological history.

plateau lava PLATEAU BASALT.

plate tectonics the theory that the EARTH'S CRUST (the LITHOSPHERE) consists of several large and some small, rigid, irregularly-shaped plates (lithospheric geotectonic plates) which carry the continents (continental crust) and the ocean floor (oceanic crust) and float on the ASTHENOSPHERE, moving very slowly, the movement probably resulting from currents in the asthenosphere. The theory is supported by evidence from studies in PALAEOMAGNETISM. Boundaries between plates are termed constructive (OCEANIC RIDGE), destructive (COLLISION ZONE, OCEANIC TRENCH, SUBDUCTION ZONE), conservative (TRANSFORM FAULT), EARTHQUAKES being associated with the destructive and conservative. BENIOFF ZONE, PLUME. Figs 36, 37.

platinum a silvery-white metallic ELEMENT-6, heavy, ductile, malleable, highly resistant to acids, corrosion, and heat, a good conductor of electricity, used in the manufacture of chemical and electrical apparatus, jewellery and, in the form of black powder (platinum black), as a CATALYST especially in hydrogenation or oxygenation, e.g. in PETROLEUM refining.

playa (Spanish, shore) **1.** a flat basin in an arid land which may become covered with water periodically, so that a shallow, usually salt, lake forms amidst flats of saline and alkaline mud **2.** the fluctuating lake so formed.

Pleistocene *adj.* with several applications **1.** the most common and now generally accepted, of or pertaining to the earlier epoch (time) or series (rocks) of the QUATERNARY period or system. GEOLOGICAL TIMESCALE **2.** of the last 1 to 2 mn years (or of the last 600 000 years), including the HOLOCENE, i.e. coinciding with the appearance of human beings on the earth **3.** of the last 1 to 2 mn years (or the last 600 000 years) excluding the Holocene, i.e. characterized by the formation of glaciers, coinciding with the most recent GREAT ICE AGE.

Plinian eruption a PAROXYSMAL VOLCANIC ERUPTION, the name derived from Pliny the Younger (c. 62 to c. 114 AD), a Roman author and administrator, who saw and recorded such an eruption of Vesuvius in 79 AD. VULCANIAN ERUPTION.

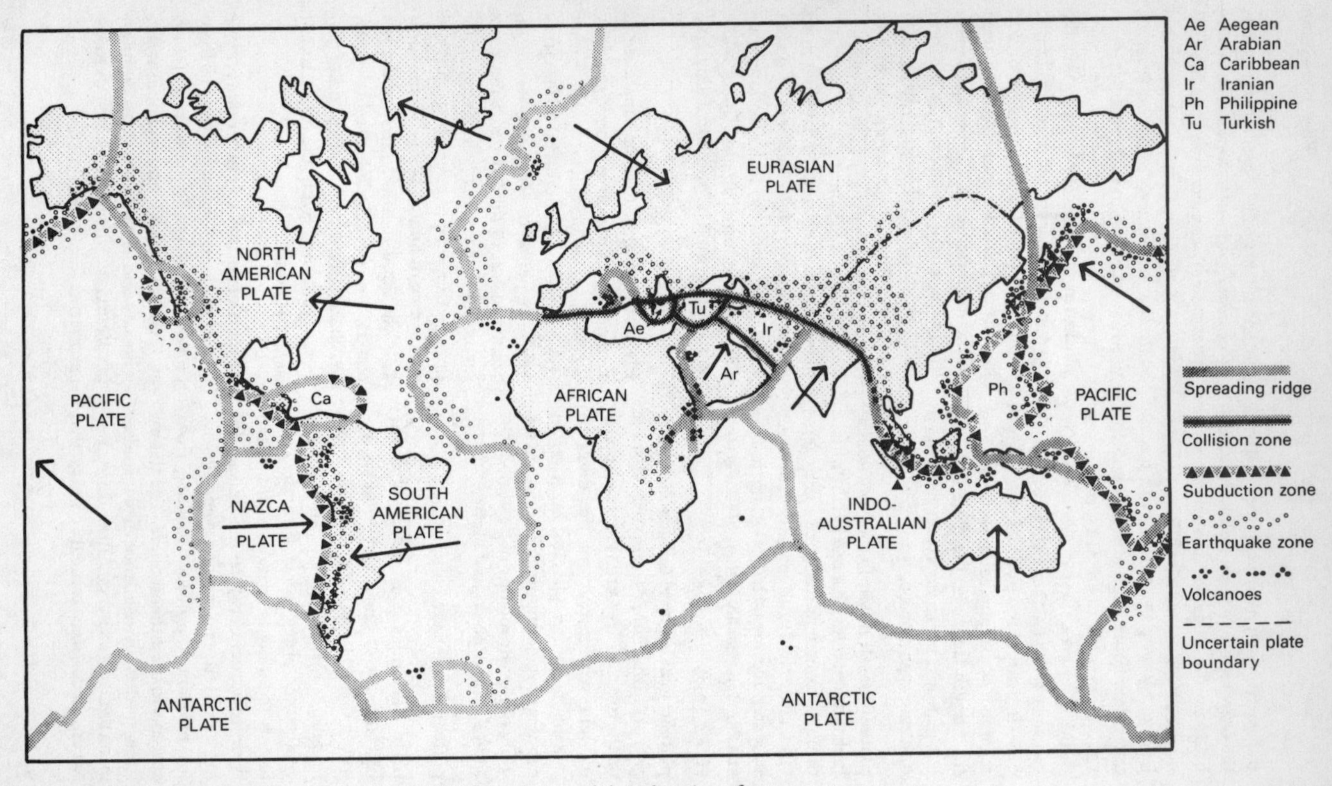

Fig 36 Plate tectonics: tectonic plates, their probable boundaries and their direction of movement

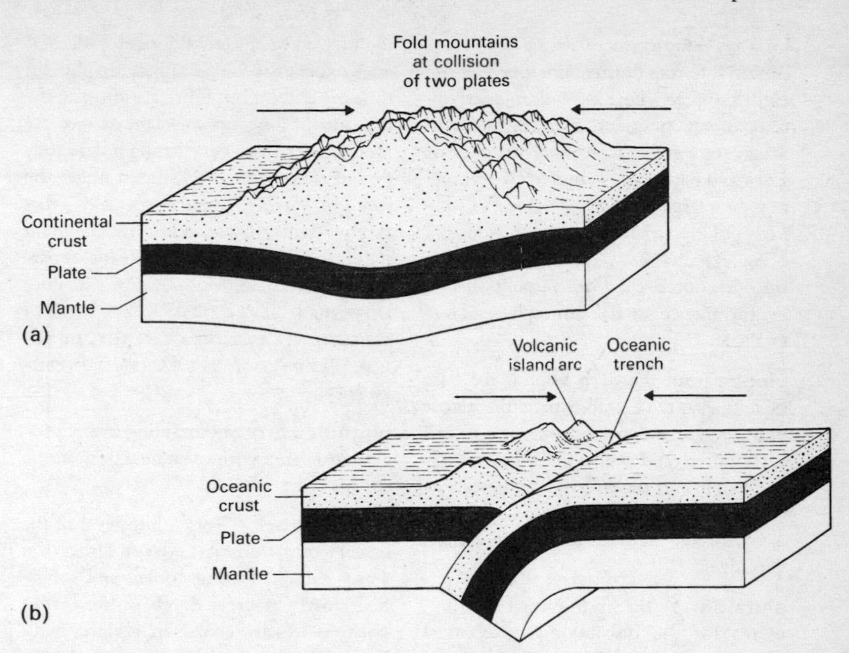

Fig 37 Plate tectonics: converging plates
(a) plates carrying continental crust, colliding and making fold mountains
(b) plates carrying oceanic crust, converging to form an oceanic trench and a volcanic island arc as one plate dives under the other, the crust being consumed in the mantle

plinth the lower and outer portion of a SAND DUNE, beyond the SLIP-FACE boundaries, which has never been subjected to sand AVALANCHES.

Pliocene *adj.* of or pertaining to the most recent epoch (time) or series (rocks) of the TERTIARY period or system. GEOLOGICAL TIMESCALE.

plucking 1. one of the main erosion processes carried out by a GLACIER, effecting the removal of rocks from its valley floor. Water enters cracks in the rocks of the floor, freezes and detaches rock fragments which become frozen to, and carried away by, the under surface of the glacier as it moves along **2.** applied sometimes to QUARRYING-2, one of the erosion processes carried out by a stream.

plug a roughly cylindrical mass of volcanic rock marking the neck (VOLCANIC NECK) of an ancient volcano, sometimes left isolated, exposed by denudation of the rest of the CONE.

plume 1. in PLATE TECTONICS, the upward movement of magma in a convection current caused by a locally very hot area lying between the CORE and the MANTLE of the earth which melts the rocks of the mantle and results in a localized swelling on the earth's surface, followed by a cracking open of rocks. There are certain localities (termed hot spots) on the earth's surface where this volcanic activity is relatively frequent, e.g. Hawaii **2.** a stream of effluent consisting of gases or gases and particulates emitted by a chimney, or a larger ribbon-like cloud of similarly polluted air produced by an industrial complex, city etc., drifting downwind from its source, the form varying, shaped by turbulence in the atmosphere. POLLUTION.

plunge pool the deep pool at the base of a WATERFALL into which the water plunges, cut out by the whirling round of boulders and stones to form a large POTHOLE-2. If the stream later deserts its course the plunge pool remains as a nearly circular lake, cut in the bedrock and often very deep.

pluralism **1.** the institutional arrangements for the distribution of political power **2.** an approach to the study of political systems which argues that power should be diffused in a political system, that the system should be open and accessible to all its members, and that it is possible by democratic processes to reach compromise agreements when disagreement and conflicts arise in society **3.** the existence or the toleration of different attitudes and beliefs within a group or an institution etc., or of these as well as different ethnic or cultural groups within a society. DUALISM.

pluralistic integration INTEGRATION in society in which distinct groups coexist within a common framework of political and legal rights.

plural society a society within which there are two or more elements, social orders or cultural groups which, in many areas of social behaviour, do not mix because each group wishes to retain its identity, its beliefs, traditions etc. All societies except the very simplest are pluralist to some extent, but if in a nation state the demand of a group to retain its separate identity becomes very strong a delicately poised unity may break down under the strain of warring factions, especially if that group demands separate, independent statehood. Such a demand for separate territorial and political sovereignty is termed SEPARATISM. See ACCULTURATION, CULTURE CONTACT, DUALISM, PLURALISM, PLURALISTIC INTEGRATION.

plutonic *adj.* of or pertaining to any process associated with great heat deep in the earth's crust.

plutonic rock (Greek Pluto, god of the underworld) an INTRUSIVE IGNEOUS ROCK which, having cooled and solidified slowly at great depth in the earth's crust, is usually coarse in texture, with large CRYSTALS. EXTRUSION, EXTRUSIVE ROCK, GABBRO, GRANITE, INTRUSION, VOLCANIC ROCK.

pluvial, pluviose *adj.* of, pertaining to, due to, or characterized by, rain. Frequently used in connexion with postglacial climatic fluctuations (a pluvial period being a long period when rainfall was distinctly heavier than that in those relatively dry stages that preceded and followed it), the term came to be used as a noun, i.e. a pluvial period became a pluvial.

pneumatolysis the process in which chemical changes are brought about in rocks by the action of heated gases and vapours from the earth's interior, usually in the later stages of a cycle of IGNEOUS activity, affecting both the IGNEOUS ROCK itself and the surrounding COUNTRY ROCK, and usually resulting in the formation of new minerals, including metalliferous ores. AUTOMETAMORPHISM, HYDROTHERMAL, METASOMATISM.

podzol, podsol (Russian podzol, ash) podzol is now the preferred spelling, a ZONAL SOIL developed under cool moist coniferous forest climatic conditions (in which precipitation is fairly low but the evaporation rate is also low) especially from sandy parent rock, e.g. in the heathlands of western Europe and the CONIFEROUS FOREST region of northern Europe and Canada. Downward leaching produces an acid A HORIZON with three layers, a top thin surface layer of black organic matter, below which lies a partly leached layer and under that a heavily leached, ash-coloured layer (hence the name) devoid of iron compounds and lime. The two layers of the B HORIZON consist of an upper layer of partly cemented leached material and a lower layer rich in iron oxide, clay and humus. The C HORIZON below is of little altered parent rock. Podzols are not generally good agricultural soils, but they can be upgraded by the application of lime and fertilizers. GREY-BROWN PODZOL, LEACHING, PODZOLIC SOILS, RED-YELLOW PODZOLIC SOIL, SOIL, SOIL ASSOCIATION, SOIL HORIZON, SPODOSOLS.

podzolic soils soils characterized by a dark B HORIZON rich in aluminium and/or iron oxide and humus. The overlying leached A HORIZON and the organic surface layer typical of a podzol may be absent. They range from well drained to badly drained and cover true PODZOLS as well as brown podzolic soils (well drained loamy or sandy soils, lacking an organic or bleached A horizon, pH 5 to 6, developed in humid temperate climates, transitional between BROWN EARTHS and PODZOLS), GLEY podzols (poorly drained soils developed on land with intermittent waterlogging, typically MOTTLED in the gleyed grey horizon, with an overlying dark brown to black A horizon which may or may not have the bleached horizon and/or organic layer of the true podzol), and stagnopodzols (soils with the typical thin peaty surface layer of the true podzol, overlying an intermittently waterlogged gleyed bleached horizon, below which lies a thin iron pan layer, rich in iron oxide and/or a more crumbly layer). HARD PAN, LEACHING, WATERLOGGED.

podzolization 1. a general term applied to the process or processes by which soils are depleted of BASES-2, become acid, and come to have developed eluvial A HORIZONS and illuvial B HORIZONS. ELUVIATION, ILLUVIATION **2.** specifically, the process by which a PODZOL is developed, including the more rapid removal of IRON and ALUMINA than of SILICA from the surface horizons; but also applied to similar processes at work in the formation of certain other soils of humid regions.

point bar the deposit of sand or gravel developing on the inside of a MEANDER bend, growing by the ACCRETIONS which accompany the migration of the meander. It is sometimes termed a meander bar, meander scroll or scroll meander, but point bar is now usually preferred. Fig 29.

Poisson distribution a theoretical discontinuous FREQUENCY DISTRIBUTION which is positively skewed and truncated (NORMAL DISTRIBUTION) and in which, if perfect, the MEAN and the VARIANCE are equal, i.e. the square root of the STANDARD DEVIATION is equal to the square root of the variance. The Poisson distribution is most likely to be produced by a random process in which there is only a small probability of an event occurring, i.e. the probability of the event occurring is very small indeed in comparison with the probability that it will not occur. BINOMIAL DISTRIBUTION.

polar *adj.* of or pertaining to those parts of the earth, or of the celestial sphere, close to the POLES-4 **2.** of or pertaining to the POLE-3 of a MAGNET.

polar air mass an AIR MASS with cool temperatures, symbol P, originating in midlatitudes (40° to 60°) over the ocean (polar maritime air mass, Pm) or over the continental interior (polar continental, Pc). The term was introduced originally to stress the difference in temperature with a TROPICAL AIR MASS. It should not be confused with the ARCTIC AIR MASS, symbol A, or the ANTARCTIC AIR MASS, symbol AA, air masses which originate near the poles.

polar front the FRONT or frontal zone where the polar maritime and tropical maritime (POLAR AIR MASS, TROPICAL AIR MASS) meet, over the North Pacific and North Atlantic oceans. Disturbances along this front, which shifts over a broad zone, play a major part in determining north European weather. ARCTIC FRONT, ATLANTIC POLAR FRONT, POLAR FRONT JET STREAM.

polar front jet stream a JET STREAM formed along the POLAR FRONT. Fig 5.

polar high a persistent area of high atmospheric pressure (ANTICYCLONE, HIGH) located over the POLAR region of ANTARCTICA. Fig 5.

polar low a persistent area of low atmospheric pressure (LOW) located in the upper atmosphere over high LATITUDES. Fig 5.

polar wind a very cold wind blowing from the north or south POLAR-1 region.

polder (Dutch) low-lying land, usually below sea-level, reclaimed from the sea, lake, or floodplain of a river by embanking (DIKES-1) and draining, often supported by continued pumping. The work of creating polders is termed empoldering, and occasionally the tract enclosed is termed an empolder.

pole 1. one of two extremes **2.** a point of attraction **3.** one of the two points at the ends of a MAGNET where its power of pulling iron towards itself is greatest **4.** either end of the axis of a sphere, e.g. the northern or southern extremity of the axis of rotation of the earth, i.e. the NORTH POLE, SOUTH POLE, the earth's geographical poles. MAGNETIC POLE, TRUE NORTH, TRUE SOUTH.

Pole Star the bright star in the constellation Ursa Minor, seen in the ZENITH at the NORTH POLE. It is used to find TRUE NORTH from any point in the NORTHERN HEMISPHERE, the height of the Pole Star above the horizon seen from that given point being equal to the LATITUDE. SOUTHERN CROSS.

polje (Slavic, field or cultivated area) **1.** in KARST, a large or very large (e.g. with an area of some 400 sq km: 155 sq mi) closed depression in limestone, usually elliptical, with a flat floor either of bare limestone or covered by alluvium, sometimes marshy, sometimes with an intermittent lake, generally surrounded by steep limestone walls **2.** in the former Yugoslavia, any enclosed or nearly-enclosed valley.

pollarded *adj.* applied to **1.** a living tree cut back to the main trunk so that the new growth forms a thick head of many branches, not to be confused with coppiced (COPPICE) **2.** an animal from which the horns have been removed.

pollen the very fine powder (microspores) carrying the male element of seed-producing plants, each grain of pollen (i.e. each microspore) containing a (much reduced) male gametophyte. The pollen grains are carried by wind, insects, birds or water to the female parts of the plant, where pollen tubes containing the male nuclei develop and eventually penetrate the embryo sac.

pollen analysis PALYNOLOGY.

pollen count a measure of the POLLEN in the atmosphere in a given volume of air during 24 hours.

pollution **1.** the direct or indirect process by which any part of the ENVIRONMENT is affected in such a way that it is made potentially or actually unhealthy, unsafe, impure or hazardous to the welfare of the organisms which live in it, i.e. the results are harmful (TERATOGENIC POLLUTION). The term is sometimes also applied, loosely, to such processes if they give rise to results which are merely objectionable **2.** the state of being so harmfully affected. Pollution usually occurs as a result of the presence of too much of some substance, or the excessive occurrence of a process or action, in an inappropriate place at an unsuitable time, such as oil spillage, sewage outfall, or industrial effluent in a river, lake or sea, e.g. mercury in the sea (MINAMATA DISEASE); sulphates etc. in the atmosphere (ACID RAIN); soil nutrients causing EUTROPHICATION; industrial heat causing THERMAL POLLUTION; noise of flying aircraft in residential areas near airports at night; obnoxious fumes from an industrial process in a residential area (PLUME-2).

pollution haven a location that lacks stringent anti-pollution legislation. POLLUTION.

polygon **1.** a plane figure enclosed by five or more straight lines **2.** in soil, a polygonal pattern varying in diameter from a few millimetres to tens of metres, usually with a slightly concave or convex surface, visible on the soil surface in areas of PERMAFROST or of very cold winters (PATTERNED GROUND), but also in other areas where contraction has taken place, e.g. in PLAYAS, in arid areas, in deserts.

polynya (Russian poluinya) a large area of open water surrounded by sea ice, sometimes bounded on one side by the coast, and occurring annually in the same region, notably off the mouths of big rivers.

pond an area of still water, smaller than a lake, lying in a natural hollow (POND *verb*) or in a depression formed by digging or by embanking a natural hollow (and specifically named according to use, e.g. fish-pond).

pond *verb* to dam a stream, to check a flow, to form a pond, e.g. when the normal flow of a stream is interrupted (e.g. by an uplift of part of the stream bed or by an obstruction which may include a strong flow of water from a side valley), the water of the main stream is said to be ponded back, forming a lake or large pond.

ponente a westerly wind, usually cool, usually bringing dry weather, blowing on the coasts of Corsica and Mediterranean France.

ponor (Slavic, pl. ponore) an AVEN, a vertical or steeply inclined shaft in limestone country, leading from a SWALLOW-HOLE or from the ground surface to an underground cave, and through which water may pass.

pool **1.** a small body of standing water, artificially impounded or occurring naturally, specifically named to accord with use or formation, e.g. swimming-pool, WHIRLPOOL **2.** a still, usually deep, area in a stream or river.

pool and riffle a pattern developed on the bed of a stream by a sequence of alternating scoured pools and shallow gravel bars termed RIFFLES, even if the channel is straight and the bed uniform. The distance between the pools depends on the width of the stream (commonly the distance is 5 to 7 times the width of the stream). If the bed is easily eroded the channel begins to wind, pools develop on the outer (concave) bank, riffles grow on the inner side to become POINT BARS, the beginning of MEANDERS.

population **1.** the total number or a specified group of people or of animals or of plants living in an area at a particular time **2.** the process of providing an area with inhabitants. DEPOPULATION **3.** a group of individuals regarded without consideration of the interrelationships within the group **4.** in statistics, a term synonymous with AGGREGATE-1, any finite or infinite collection of individuals under consideration (not necessarily a collection of living organisms) of which the qualities may be estimated by SAMPLING.

population density the number of individuals occupying a particular unit area. DENSITY-1.

population explosion a great, rapid, increase in POPULATION-1,3 in a specific area, or in the world, e.g. as experienced in the human population in the twentieth century.

population potential **1.** the number of people who might be able to live in a certain area at a standard of living that is reasonable in relation to the resources available in that area. CARRYING CAPACITY-3,5 **2.** a measure of the accessibility of a particular mass of people to a point.

population pyramid a BAR GRAPH in the form of a pyramid drawn to express the age and sex composition of a human POPULATION-1. The age groups are shown on the vertical scale, commonly graduated into ten-year intervals, youngest at the base, and the number or percentage of males and females within each of the age groups on the horizontal scale, the males lying traditionally to the left, females to the right of a line drawn perpendicular to the horizontal axis, and expressing zero (percentages or numbers increasing to the left for males, to the right for females). The result is shaped more or less like a pyramid, but is rarely symmetrical. If the pyramid has a wide base and tapers to a pointed top, it is termed expansive (denoting an expanding population, with many children and a declining death rate); if its shape resembles a tall dome rather than a pyramid, it is termed stationary (denoting a stable, slowly growing population, with a decline in mortality and a low birth rate); if the shape is broadly oval with a pointed top, the base cutting the oval below its widest part, it is said to be constrictive (denoting a declining population, with a birth rate lower than the death rate). In addition to birth and death rates, migration and the tendency of females to outlive males affect the shape of population pyramids. BIRTH RATE. Fig 38.

pore water pressure the pressure on rock and soil particles produced by water con-

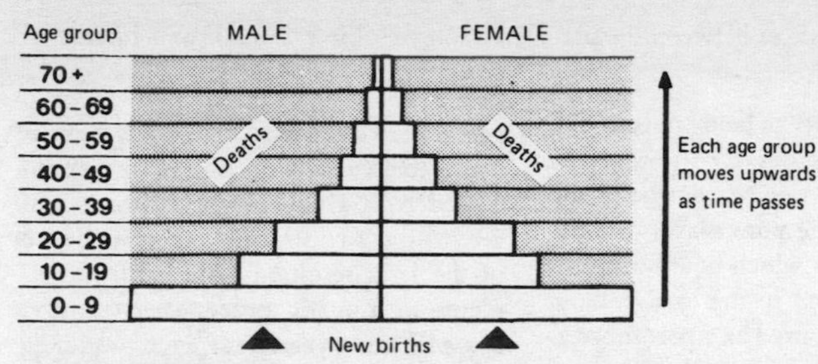

(a) High death rate

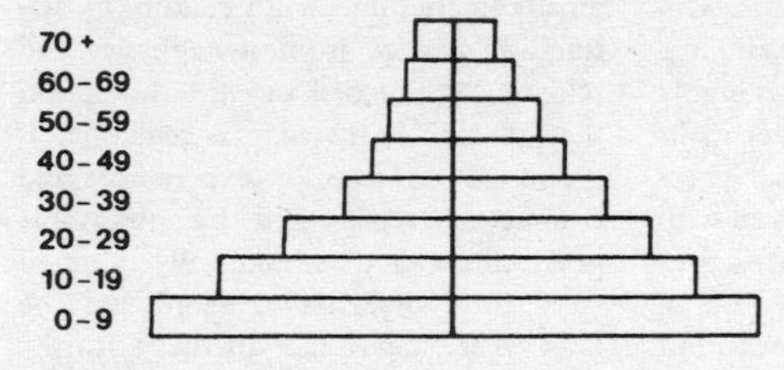

(b) Effect of reduction in death rate

(c) Effect of reduction in birth rate (after 10 years)

(d) Effect of reduction in birth rate (after 40 years)

Fig 38 Population pyramids, showing the effects of reductions in birth and death rates

tained in the interstices between the particles.

porosity the quality of being porous, full or abounding in pores. The porosity of rocks, i.e. the ratio or percentage of the total volume of the pore spaces (minute interstices through which liquids or gases can pass) in relation to the total volume of the rock, is measured by a porosimeter. Sand, gravel, sandstones, with open textures and coarse grains, are typical porous rocks. Porosity is quite different from perviousness (PERVIOUS ROCK, PERMEABLE ROCK). Dry CLAY, for example, is highly porous and will hold much water in its pores, but when saturated the small spaces between the grains become blocked with water held by SURFACE TENSION, preventing the passage of water. To be an AQUIFER or source of water a rock must be both porous and pervious. Porosity may be increased by LEACHING or decreased by COMPACTION. IMPERMEABLE.

porphyry a HYPABYSSAL rock with large crystals in a crystalline or glassy groundmass (MATRIX-3) of finer grain. The hard Egyptian porphyry composed of red or white FELDSPAR crystals embedded in a red or purple glassy base was much used by the Romans as a decorative material for buildings, vessels, ornaments. ACIDIC ROCK.

port a term loosely applied to a place where ships may anchor to load or unload cargo, or to the HARBOUR itself, or to the harbour and adjoining settlement, quays, handling facilities for cargo, warehouses, transport termini etc. ABANDONED DOORSTEP, ENTREPOT, FISHING PORT, FREE PORT, OUTPORT, RIVER PORT, TREATY PORT.

position 1. the PLACE-1 occupied by a person or an object in relation to another person or object **2.** rank in a hierarchy. LOCATION, SITUATION.

positive movement of sea level, the raising of the sea level in relation to the land caused by a global raising of the ocean level (EUSTATISM) or by the subsidence of the land resulting from earth movements or isostatic processes. ISOSTASY, NEGATIVE MOVEMENT.

positivism a philosophy developed originally in the nineteenth century to distinguish science from metaphysics and religion. It is based on the assumption that all true knowledge is scientific, is grounded on facts or experience, and completely represented by observable phenomena and scientifically verified facts, their objective relations and the laws that determine them. LOGICAL POSITIVISM, RELEVANCE, SCIENTIFIC LAW.

possibilism a philosophical concept which, while accepting that every ENVIRONMENT has its limits which restrict human activity, argues that within those limits there is a set of opportunities which offer human beings freedom of choice of action in contrast to DETERMINISM or ENVIRONMENTALISM. PROBABILISM.

post-Fordism a term connoting the system of manufacturing processes and practices that developed in western industrialized countries in the mid-1970s (FORDISM). Its keynote is its flexibility, in machinery and work practices. Highly skilled, versatile workers operate advanced, automated machinery, making individually tailored products for niche markets; component parts may be provided by specialist companies able to deliver just in time (at the time the parts are required). Management (incorporating the financial structure), research and development, production and marketing interact with labour.

postglacial *adj*. occurring after a glacial period, generally applied to all time succeeding the glacial epoch (PLEISTOCENE) in the QUATERNARY period. GEOLOGICAL TIMESCALE.

post-industrial city a city (a large town) which it is thought will become typical of the POST-INDUSTRIAL SOCIETY. It is presumed that employment will be mainly in the TERTIARY and especially in the QUATERNARY INDUSTRIES, population density will be low, differences in living standards and SOCIAL STATUS low, recreational facilities abundant, personal (rather than public) transport the norm. It is suggested that Canberra, Australia, is a prototype. INDUSTRIAL CITY, PRE-INDUSTRIAL CITY.

post-industrial society a form of society which some authors see as evolving in the industrialized societies of western Europe, Japan and North America in the latter part of the twentieth century as a result of changes in shape, pattern, and framework of their societies. This new form is variously seen to involve a pronounced shift in employment from manufacturing to service industries (SECONDARY INDUSTRY, TERTIARY INDUSTRY, QUATERNARY INDUSTRY); the rise to pre-eminence of professional and technically qualified people; or the rejection by some young people of materialism; or the decline of the role of the wage-earners as the agent of change, resulting from technological progress. In this concept, the stage preceding the post-industrial is termed the industrial (employment mainly in SECONDARY INDUSTRY) and, preceding that, the pre-industrial (employment mainly in PRIMARY INDUSTRY). NETWORK SOCIETY, POST-INDUSTRIAL CITY, POSTMODERNISM.

postmodernism the swiftly changing MODERNITY-2 that emerged from MODERNISM about the 1960s and 1970s. To select some elements from the socio-economic-political background that appear in geographical studies, this was the time of the decline of COMMUNISM-2 and the dissolution of the USSR; the independence of former European colonies; rising nationalism; the spread of market economies, MULTINATIONALS and the growing strength of FAR EASTERN economies; re-structuring of industry (POST-FORDISM); the NEW INTERNATIONAL DIVISION OF LABOUR; rapid progress in HIGH TECHNOLOGY and information technology (INTERNET); in space research (REMOTE SENSING) and in BIOTECHNOLOGY, including GENETIC ENGINEERING; the institution of the WELFARE STATE; the spread of the feminist movement (FEMINIST GEOGRAPHY); GLOBALIZATION; the growing power of special interest groups; mass consumerism influenced by the visual images of the MASS MEDIA; and concern for the future of the global NATURAL ENVIRONMENT evidenced by interest in BIODIVERSITY, CONSERVATION, POLLUTION, SUSTAINABLE DEVELOPMENT, respect for ECOSYSTEMS and ENDANGERED SPECIES (CITES). Postmodernism expresses these swift changes by experimental, individualistic and ephemeral visual art; minimalism in music; and architecture that tends to renounce that of MODERNISM in favour of broken or gentler outlines, some ornamentation and, in some cases, a vernacular style. EARTH SUMMIT, RIO, ECOTOURISM, HABITAT I and II, HUMANISTIC GEOGRAPHY, POST-INDUSTRIAL SOCIETY.

potamobenthos organisms living on the bed of a river. BENTHOS.

potamoplankton minute living organisms floating in sluggish rivers and streams. PLANKTON, POTAMOBENTHOS.

potash 1. loosely applied to any POTASSIUM salt or compound **2.** specifically a potassium carbonate, K_2CO_3, especially an impure form. FERTILIZER.

potassium a silvery-white, soft metallic ELEMENT-6, rapidly oxidizing in air, occurring in plants and animals, and in combined form in minerals. An essential plant NUTRIENT, it is used in the compound POTASH in mineral fertilizers; its salts are extensively used in medicine and industry. The rate of radioactive decay (RADIOACTIVITY) in potassium-argon is sometimes measured in DATING, especially in dating rocks. FELDSPAR.

potato *Solanum tuberosum*, a descendant of the PERENNIAL herb native to South America, now cultivated in most areas except the arctic and equatorial regions, one of the most extensively grown and important food crops of the world, yielding higher food value per area cultivated than any CEREAL crop. The potato has fibrous roots and many underground stems culminating in swollen tips that make the edible tubers rich in carbohydrates with some protein and potash. The shoots that develop from the 'eyes' of these tubers generate new growth.

potential evapotranspiration the highest amount of water that could be evaporated or transpired from plants from a given area if the plants had an unlimited water supply. EVAPOTRANSPIRATION, THORNTHWAITE'S CLIMATIC CLASSIFICATION.

potential instability the original condition of an AIR MASS before it becomes conditionally unstable (CONDITIONAL INSTABILITY) as a result of being lifted over a relief barrier or over a mass of cooler air at a FRONT.

pothole 1. applied loosely, to any deep hole, vertical cave system, or underground CAVE-1, in LIMESTONE country, hence the term potholing for caving or exploring underground caverns etc. **2.** in studies of erosion, a hole, more or less circular, worn in rocks by whirling stones, as in the BEDROCK of the channel of an eddying swift stream.

poultry chickens, ducks, geese, turkeys and other domesticated birds raised for food, i.e. for the eggs they produce as well as for their flesh. BATTERY SYSTEM, LIVESTOCK.

poverty 1. a deficiency in, a lack of, or an inadequate supply of, something **2.** of soil, unproductive **3.** the state or quality of being poor, a relative term without precise meaning. At subsistence level, only those with insufficient food and shelter for survival may be said to be in a state of poverty; but poverty is usually defined in sociology as a concept of relative deprivation, i.e. the absence or inadequacy in the lives of the poverty-stricken of the diets, amenities, standards, services and activities which are common or customary in the society in which they are living. Official definitions of poverty are customarily given in social welfare schemes: such definitions are often governed by political considerations.

poverty cycle the recurrent transmission of POVERTY-3 and DEPRIVATION from one generation to the next. This has been explained in terms of the assumed passing-on from parents to children of negative attitudes towards work, education and the law, as a consequence of which the children lack adequate equipment to compete for jobs. The assumption is that children of poor parents start school at a disadvantage, may not be encouraged and supported by their parents while at school, achieve little, leave school with few or no qualifications, can find only low-paid work, if at all, and stay poor, their children in turn inheriting their disadvantages. MULTIPLE DEPRIVATION.

power in physics, the rate at which work is done (as distinct from ENERGY, the capacity for doing work), the transfer of energy. GEOTHERMAL ENERGY, HYDROCARBON, HYDROELECTRIC POWER, NUCLEAR FISSION, NUCLEAR FUSION, SOLAR BATTERY, TIDAL POWER STATION, WATER POWER, WIND POWER.

pozzolana, pozzuolana (Italian Pozzuoli, a town near Naples) fine VOLCANIC ASH used in HYDRAULIC CEMENT and some mortars.

prairie an extensive area of unbroken grassland, generally without trees, occurring in midlatitudes in North America and considered by some authors to be equivalent to the STEPPE of Europe, the PAMPA of South America, the VELD of southern Africa. Prairie occurs where summer rains are light (total annual rainfall 250 to 500 mm: 10 to 20 in, with some local drought) and summer temperatures are high, conditions well suited to grain crops. Thus little remains of the original prairie, most having been ploughed and sown to cereals or grasses finer than the indigenous. Some authors distinguish long grass prairie from short grass prairie, equating the latter with steppe.

prairie soil 1. brunizem, a ZONAL SOIL developed in midlatitude subhumid temperate areas that were formerly under PRAIRIE grasses but which are now sown and cultivated. It is very similar to CHERNOZEM, but is dark brown and slightly acidic on the surface; some leaching has occurred but there is no great accumulation of calcium carbonate in the B HORIZON **2.** a general term for all dark soils of treeless plains. MOLLISOLS.

Pre-Boreal, Preboreal a climatic phase, with cold, dry conditions favourable to the growth of birch-pine forest in England, pine forest in Scotland, following the QUATERNARY glacial epoch (GEOLOGICAL TIMESCALE), lasting in eastern England until about 7500 BC. The sea level was more than 60 m (195 ft) below the present level, thus the British Isles were part of continental Europe. BOREAL, MESOLITHIC, SUB-ATLANTIC, SUB-BOREAL.

Precambrian, Pre-Cambrian *adj.* of or pertaining to all geological time and rocks before the Cambrian period (time) or system (rocks) (GEOLOGICAL TIMESCALE), a period spanning probably more than 4000 million years. The rocks have been subjected to much upheaval and change, but a few FOSSILS-1 have been found in some of the unmetamorphosed rocks; and some Precambrian rocks are visible in northwest Scotland and in Wales.

precession of the equinoxes the gradual change in the annual occurrence of the EQUINOXES, due to the relative change of position of the ECLIPTIC and the EQUATOR as the axis of the earth's rotation describes a cone-shaped rotation (similar to the movement of a gyroscope). The conical rotation is caused by gravitational forces between the earth and the sun, the earth and the moon. It is calculated that the equinoxes swing round the ecliptic once in 25 800 terrestrial years.

precious metal a METAL-1, especially

GOLD, PLATINUM, SILVER, prized for its high value.

precipitation 1. in meteorology, the deposition of moisture on the surface of the earth from atmospheric sources (METEORIC WATER), including DEW, HAIL, RAIN, SLEET, SNOW **2.** in chemistry, the formation, the settling out, of solid particles in a SOLUTION.

precipitation-day a period of 24 hours, commencing normally at 0900 hours, on which rainfall exceeds 0.25 mm (0.01 in) or snowfall 0.25 cm (0.1 in). It is a more appropriate and precise term than rain-day, especially when applied to areas with much snowfall, and is generally now the preferred term. WET-DAY.

pre-industrial city a CITY-1 of the past or present serving a population of which most are engaged in agriculture. It has characteristics markedly different from those of an INDUSTRIAL CITY typical of the highly developed industrialized countries of today. For example, the pre-industrial city is usually identified by the non-existence of a CENTRAL BUSINESS DISTRICT; in some cases one may exist but if it does it is not dominant. Usually there is no specialization of land use, the urban layout is relatively un-ordered, the street markets, shops, workshops and homes being mixed together (although a particular craft may be concentrated in a particular district). The poorest live on the periphery, not in the INNER CITY zone. There is a marketing economy based on crafts, little specialization of labour in the large population, and in most cases family ties and relationships that ensure rigid social, ethnic, and tribal segregation, often on occupational lines. POST-INDUSTRIAL CITY.

preservation the act of keeping in existence, preventing ruin and decay. The term usually implies the maintenance of something in its present form (or as close to that form as is possible), without change; whereas the work of CONSERVATION, aware of present and future needs, is more positive, forward-looking and flexible, and does not rule out necessary change.

pressure 1. the action of exerting a steady force on something, of trying to force or persuade. PRESSURE GROUP **2.** in climatology, meteorology, ATMOSPHERIC PRESSURE.

pressure gradient BAROMETRIC GRADIENT.

pressure group a group of people united by shared attitudes and goals who try to obtain decisions favourable to their interests by gaining access to and influencing the governmental process.

pressure release the outward force of pressure set free from inside a rock mass when the overlying strata is removed by DENUDATION. This commonly occurs in massive, unjointed rock, e.g. in granite.

pressure system a distinct atmospheric circulation system of high or low pressure. ANTICYCLONE, COL, DEPRESSION, RIDGE OF HIGH PRESSURE, SECONDARY DEPRESSION, WEDGE OF HIGH PRESSURE.

pressure wave PUSH WAVE.

prevailing wind a WIND blowing most frequently from a specific direction in a particular area. DOMINANT WIND.

primary *adj.* of the first in order of time of origin or precedence.

primary industry, primary activity, primary sector an activity concerned with the collecting or making available of material provided by nature, i.e. agriculture, fishing, forestry, hunting, mining and quarrying. INDUSTRY, SECONDARY INDUSTRY, TERTIARY INDUSTRY, QUATERNARY INDUSTRY.

primary labour market employment in well-paid, secure and pensionable work, as distinct from the secondary labour market, with poorly-paid, insecure and, in some cases, part-time work.

primary production in ecology, the total quantity of organic matter as it is newly formed by PHOTOSYNTHESIS. It is estimated that only some one-tenth of one per cent of the SOLAR energy reaching the earth is fixed in plants by photosynthesis. The matter remaining after the energy needs of the plants have been met is termed the net primary production. FOOD CHAIN, PHOTOSYNTHESIS, PRODUCTION, SECONDARY PRODUCTION.

primary wave of earthquakes, PUSH WAVE.

primate city a CITY-1 which is first in the rank of city sizes in a country, being larger than any other, its great size giving it an impetus to become even larger, so that the gap between its population total and that of the city or cities of the second rank is exceptionally wide. According to the RANK-SIZE RULE, the primate city will be more than twice the size of the second largest city; according to CENTRAL PLACE THEORY it will be three times the size of the second rank city. Some authors regard the primate city as being equivalent to a METROPOLIS. GATEWAY CITY.

prime meridian, initial meridian the MERIDIAN from which LONGITUDE is measured, i.e. 0°, the longitude of Greenwich.

primeur (French, anything new or early) a fruit or vegetable cultivated in such a way that it is ready for market before, or very early in, its normal season.

prisere primary SERE, a natural plant SUCCESSION-2, originating on a bare area and progressing to a CLIMAX.

prismatic compass a magnetic COMPASS used in surveying, in which a prism is incorporated so that an image of the angle of the bearing can be read by the operator while using sights through a telescope.

private sector one of the two major divisions of a national economy, the other being the PUBLIC SECTOR. The private sector covers consumer expenditure for goods and services and business expenditure for plant and equipment, and thus includes the economic activities of private persons, industrial and commercial companies, institutions (insurance companies, building societies, finance houses), trade unions, charities and churches etc. MIXED ECONOMY, PLANNED ECONOMY, PUBLIC SECTOR.

privatization denationalization, the reverse of NATIONALIZATION, i.e. the removal of state control or ownership of an activity or property in order to allow it to pass into private hands.

probabilism **1.** in philosophy, the doctrine that no knowledge is absolutely certain and that PROBABILITY provides sufficient grounds for action **2.** in geography, a modification of POSSIBILISM, that everywhere there are possibilities but some are more probable than others.

probability **1.** generally, loosely applied to the degree of belief that an event may occur **2.** in statistics, the likelihood of an event occurring, expressed as a ratio between the number of the actual occurrences of the event and the average number of cases that would favour its occurrence, taken over an infinite series of these cases. The probability of an event occurring lies within the range 0 to 1, zero indicating absolute impossibility, 1 indicating absolute certainty, the probability of the occurrence of most events lying between the two.

produce agricultural or horticultural products.

producer goods, producers' goods goods such as tools and raw materials needed by a manufacturer to make other goods. Thus producer goods do not satisfy the needs or desires of the individual person directly (CONSUMER GOODS), but only indirectly, in the production of other, final products. CAPITAL GOODS, CONSUMER DURABLES, CONSUMER GOODS.

product something grown, manufactured or created, the end result of a process. OUTPUT.

production **1.** that which results from a process, effort, action. PRODUCT **2.** the total OUTPUT, especially of something grown or manufactured, measured in absolute terms. FOOD CHAIN, PRIMARY PRODUCTION, SECONDARY PRODUCTION.

production rate in ecology, the number of organisms formed in an area in a given period of time. Gross production rate is the rate of assimilation shown by organisms of a given TROPHIC LEVEL-1. Net production rate is the gross production rate less loss of matter brought about by respiration, decomposition and predation. FOOD CHAIN, PRIMARY PRODUCTION, PRODUCTION-2, SECONDARY PRODUCTION.

productivity of land **1.** actual productivity, the equivalent of YIELD-1 **2.** potential productivity, the hypothetical yield under stated conditions, especially under the best conditions possible (land potential is the term sometimes used in this sense, LAND CLASSIFICATION).

profile **1.** the shape of something, viewed from the side **2.** a short, concise descriptive written sketch **3.** a side elevation or section, e.g. the shape shown in outline where the plane of a SECTION cuts vertically the surface of the ground, i.e. producing the surface outline alone, as in a RIVER PROFILE. SOIL PROFILE.

profit **1.** an excess of income over expenditure, especially in a particular transaction over a period of time **2.** the surplus money produced by industry after deductions have been made for the cost of wages, raw materials, rents, charges.

proglacial *adj.* before, in advance of, to the front of, a GLACIER, applied to time or position. PROGLACIAL LAKE.

proglacial lake a stretch of water ponded during a glacial period between an ice front and rising ground, e.g. a morainic ridge, the sediments in the lake consisting of FLUVIOGLACIAL deposits. FLUVIOGLACIAL DEPOSITION.

progradation a process in which the shore is extended seaward by the action of waves on a seashore, continuing so long as the current builds up the sea bottom offshore. AGGRADATION, BEACH RIDGES.

progressive wave applied to ocean waves, a wave propagated in a channel of (theoretically) infinite length, its wave length being the distance between two successive crests, and its period the time taken to move one wave length.

proletariat **1.** the lowest CLASS-3 in

ancient Rome **2.** the lowest SOCIAL CLASS in a modern society, particularly (in Marxist theory) the wage earners possessing neither property nor capital and living by the sale of their labour. BOURGEOISIE, MODE OF PRODUCTION, POST-INDUSTRIAL SOCIETY.

promontory a headland, a cliff, usually rocky, projecting into the sea, often associated with offshore rocks and STACKS.

property **1.** possessions, a thing or things owned **2.** real estate (i.e. immovable property, e.g. land or buildings as distinct from personal possessions) or a piece of real estate (especially a building such as a house) **3.** an attribute, a characteristic, a quality.

protalus rampart an accumulation of coarse angular rock debris, resembling a MORAINE, consisting of material that has slipped down from perennial banks of snow, and lying parallel to the slope that produced it. TALUS.

proton a positively charged, fundamental particle present in all ATOMS.

proven reserves RESERVES of which the existence and the location are known and the quantity can be estimated, but which may be exploited only if the demand is great enough to make it economically feasible to do so. RESOURCES.

psychic income factors which, reflecting personal psychological and mental attitudes, give the decision-maker the greatest comfort, happiness, pleasure, satisfaction in living and/or working in a particular location.

psychosphere the human mind, human thought and culture (as studied by psychologists, anthropologists and other social scientists) as an environmental factor, to be taken into account together with the lithosphere, atmosphere, hydrosphere and biosphere in environmental studies. TECHNOSPHERE.

public goods social goods, goods that cannot be subdivided for distribution and sale to individuals, e.g. national defence, fire services, hospital services, street lighting etc. In Britain (apart from national defence) they are usually subject to public control via local government; and what is included/excluded is often a political decision.

public housing social housing, housing units financed directly by government agencies or local authorities and rented to tenants.

public participation participation by the public in political decision-making, especially in relation to policies which have a direct effect, e.g. land use planning policy.

public sector, government sector one of the two major divisions of a national economy, the other being the PRIVATE SECTOR. The public sector covers the official economic activity of central and local government, the nationalized industries and any corporations partly financed by the government, but excludes the consumer expenditure of government employees. In a MIXED ECONOMY it usually represents some 25 per cent of all economic activity; in countries with a CENTRALLY PLANNED ECONOMY it usually represents 100 per cent.

public utility a business concern which provides and administers a public service, e.g. by providing gas, electricity, water, telephones, railways etc., which requires special wayleaves or other rights or compulsory powers over the use of land.

pulse a leguminous food plant (LEGUMINOSAE), e.g. butter bean, chick pea, black and green grams, lablab, lentil, pea, or its edible seeds.

pumice solidified froth on the surface of an ACID LAVA flow, formed by the escape of gases and vapours from the cooling lava, to give a very light, fine-grained, cellular acid rock, capable of floating in water and used as an abrasive in smoothing, cleaning and polishing. Pumice dust forms one of the components of RED CLAY-1 on the deep ocean floor.

punctuated equilibrium the term popularly applied to the theory of allopatric speciation which suggests that the evolution of species occurs not as a steady, continuous process as proposed by Darwin but by a process of sudden leaps forward, followed by a calm period with little change; that new species develop rapidly from a small subpopulation of ancestors, often in an isolated area at the limit of the ancestral range, thereby splitting the lineage. CLADISTICS.

push moraine, push-moraine (less frequently push-ridge moraine, or shoved moraine) mounds of sand and gravel pushed into broad, smooth, massive, parallel ridges, frequently arc-shaped, on the margin of ice as it advanced over glacial drift from an earlier glaciation. It may be composed of SUPERGLACIAL and ENGLACIAL material dumped in front of the ice (dump moraine); or of local glacial and non-glacial debris which the advancing ice piled up in its path. MORAINE.

push-pull theory of MIGRATION-1,2, a theory which suggests that people are pushed by adverse conditions (e.g. overpopulation, poverty, political repression, war, dislike of a development scheme) to leave an area, and are at the same time attracted to another area by what are perceived as favourable conditions (e.g. the likelihood of a better job, higher wages, freedom of movement) in another area. STRENGTH THEORY, WAGE DIFFERENTIAL THEORY.

push wave P-wave, a primary wave, a pressure or compressional wave, a shock wave produced by an EARTHQUAKE, a compressional vibration resembling a sound wave, which passes through solids, liquids and gases. It is termed push wave because each particle is displaced by the wave along the direction of its movement through the earth's crust, mantle and core. GUTENBERG DISCONTINUITY, LONGITUDINAL WAVE, MOHOROVICIC DISCONTINUITY, SHAKE WAVE, TRANSVERSE WAVE, WAVE.

puszta (Hungarian, waste) Hungarian STEPPE region, the open treeless plains of temperate grassland in the heart of Hungary, also termed alföld.

puy (French) a small hill, the cone of an extinct volcano, rising from a plateau in the Auvergne, France. The term does not have a precise definition: the puy may be composed of ash or cinder, of acid lava in the shape of a dome, or it may have a double cone; and there may be evidence of PELEAN or STROMBOLIAN volcanic activity. The term is applied elsewhere to similar hills.

P-wave PUSH WAVE.

pyramid peak, pyramidal peak a HORN.

pyroclast fragmental material, derived from MAGMA as well as from the wall of the VENT of a VOLCANO, ejected by the explosion of rapidly freed gases in an explosive volcanic eruption. It consists of ash, bombs, cinders, dust, lapilli, fragments of older rocks etc. CLAST, CLASTIC, PUMICE, SCORIA, TEPHRA, and entries under VOLCANIC.

pyroclastic *adj.* applied to a rock formed from the debris of an explosive volcanic eruption. CLAST, CLASTIC, PYROCLAST.

Q

qanat FOGGARA, KAREZ.

quadrat one of the equal-sized sample areas (usually a square) used in QUADRAT ANALYSIS. In plant ecology a quadrat of one metre square is commonly used for sampling in order to gain an accurate statistical record of the composition of low plant cover, but for tree cover the quadrat has to be much larger.

quadrat analysis a statistical technique derived from one used in plant ecology (QUADRAT). The area covered by a point pattern is divided into equal-sized sample areas, and the number of points occurring within each is counted. The observed distribution can then be tested against a theoretical pattern of distribution.

quadrature 1. in astronomy, the position when a celestial body lies at 90° or 270° to another, e.g. when the sun, earth and moon form a right angle, the earth being the apex. This occurs twice each LUNAR MONTH with the result that the tide-producing effects of the moon and sun are in OPPOSITION, producing tides of low range. NEAP TIDES **2.** one of the two points on the orbit of a celestial body midway between the syzygies (SYZYGY). CONJUNCTION, MOON.

quagmire in general applied to any soft, wet ground, but specifically a quaking bog, an area of soft, wet ground so soft that it quakes or trembles when trodden on. BOG.

quake an EARTHQUAKE.

quaking bog QUAGMIRE.

qualitative *adj.* relating to, concerned with, or involving, QUALITY.

qualitative variable in statistics, usually, a VARIABLE-2 which may be measured on nominal or ordinal scales, commonly termed an ATTRIBUTE-2. Some statisticians however apply the term qualitative variable to a variable measured on the nominal scale only, maintaining that the use of ordinal, interval or ratio measurement (MEASUREMENT-2) to display increasing quantities of an attribute involves QUANTIFICATION. Thus a quantitative variable may refer to a variable measured on ordinal, interval or ratio scales, or a variable measured on interval or ratio scale only. This difference in approach arises from the fact that all variables can be represented by numbers.

quality 1. grade, degree of goodness, worth **2.** an attribute, trait, characteristic.

quality of life the degree of goodness of the conditions of life and of the life-style of a person. Objectively it may be possible to assess the degree of goodness by the extent to which SOCIAL WELL-BEING is achieved. But 'one person's meat is another's poison', and judgement as to what should be included among the factors contributing to social well-being (apart from the basic sufficiencies of shelter, food, water, clothing etc.), let alone the evaluation of the importance of each factor, can only be subjective, varying from one individual to another, from one society to another, from one time to another.

Perhaps the term quality of life may therefore best be interpreted as people's subjective feelings of satisfaction with their living conditions and life-style.

quango a quasi-autonomous non-government organization, a government appointed body or agency in the UK which is financed by, but is not part of, a government department.

quantifiable *adj.* measurable in terms of quantity, conceivable or treatable as a QUANTITY.

quantification the action of measuring QUANTITY, the expression of a property or QUALITY in numerical terms.

quantitative *adj.* relating to, concerned with, quantity or with the measurement of QUANTITY.

quantitative revolution a movement in approach to geographical studies in the 1950s and 1960s concerned with the use of statistical and mathematical methods and techniques to analyse associations in the attempt to produce objective systems of classification and theories of spatial organization.

quantitative variable QUALITATIVE VARIABLE.

quantity 1. an amount, sum or number **2.** a great deal, very many **3.** the property of things that can be measured **4.** in mathematics, anything which is measurable, or a figure or symbol used to represent this.

quarry an open excavation on the surface of the earth, worked usually for the extraction of rocks and certain non-metallic minerals. MINE, MINING, OPENCAST MINING, PIT.

quarrying 1. the extracting activity conducted in a QUARRY, limited in extent when compared with OPENCAST MINING **2.** the eroding of its channel by a young stream by the lifting effect of water as it penetrates cracks in rocks, termed PLUCKING-2 by some authors.

quartile one of the four equal parts of a data distribution. INTERQUARTILE RANGE, MEDIAN.

quartz silicon dioxide, one of the commonest minerals in nature. It is characteristic of ACID IGNEOUS ROCKS, is resistant to chemical WEATHERING, is abundant in SEDIMENTARY and METAMORPHIC ROCKS, often fills JOINTS, VEINS and cavities, sometimes mixed with other MINERALS, as well as ORES, so stable that it usually constitutes the majority of sand grains in a sand or SANDSTONE. When pure (rock crystal) it is like clear glass, but much harder (no 7 on the HARDNESS scale), and occurs in hexagonal crystals and massive form, usually colourless and transparent, but also in coloured (amethyst), translucent and opaque forms. Rigid, it does not easily expand and is therefore used in making heat-resistant apparatus. It is also used in radio-transmitters and astronomical clocks because the opposing faces of the crystals are able, under pressure, to take up opposite electrical charges; conversely there is a change of volume if an electromotive force is applied.

Quaternary *adj.* **1.** of the most recent period (time) or system (rocks) of the Cainozoic era, i.e. following the PLIOCENE of the TERTIARY period or system (GEOLOGICAL TIMESCALE), and including the PLEISTOCENE and HOLOCENE, the time when human beings appeared on earth **2.** applied by some geologists to the fourth era in the sequence of geological time, i.e. post-Cainozoic, post-PLIOCENE, commencing about the time of the onset of the most recent ICE AGE.

quaternary industry, quaternary activity, quaternary sector the activities in the TERTIARY INDUSTRY which are concerned with research, with the assembly, processing and transmission of information, and with administration, including the control of other industrial sectors. INDUSTRY, PRIMARY INDUSTRY, SECONDARY INDUSTRY.

quicksand a thick mass of unstable, fine loose sand, sometimes mixed with mud, supersaturated with water, occurring on some coasts and near river mouths, liable to suck down any heavy object that comes to rest on it. SUPERSATURATION.

R

race 1. classificatory term, in broad terms equivalent to a SUB-SPECIES, but sometimes specifically and unscientifically applied to a distinct group of the single species *Homo sapiens* (from which all persons living today are descended). Such a group is identified as possessing well-developed and primarily heritable physical characteristics which differ from those of other groups (e.g. head shape, hair character, skin colour, facial features etc.), a biological concept of little value when applied to the world population of today, there having been much migration and interbreeding since the time of the appearance of *Homo sapiens* on earth. RACISM **2.** swiftly flowing water in a narrow channel, natural (as in a river) or artificial, controlled to provide power, as in the channel leading river water to the wheel of a watermill (head-race) or from the mill (tail-race) **3.** tidal race, a rush of seawater through a restricted channel occurring where there are considerable differences in the tides at each end of it **4.** an offshore current flowing strongly round a headland.

racism 1. the assumption that the abilities and characteristics of a person are determined by RACE-1, and that biologically one race is inherently superior to another. 'Race' in this context is often arbitrarily extended to include religious sects, as well as national, linguistic and cultural groups. ETHNOCENTRISM **2.** a political programme or social system based on such assumptions.

radar acronym of Radio Detection and Ranging, a device for determining the presence, distance or speed of movement of an object by means of transmitting microwaves at it and measuring by electronic devices the speed of the microwaves' return after reflection from the object. It is used especially in air and sea navigation, in tracking satellites and missiles and in automatic guidance; in land use and geological studies, and in measuring and locating atmospheric phenomena (thus of great benefit to weather forecasters). REMOTE SENSING, SONAR.

Radburn layout in town planning and urban studies, a planned urban layout, developed by Clarence Stein, applied in Radburn, New Jersey, USA, in 1928, which separates pedestrians from vehicles by arranging that 'superblocks' of housing, shops, offices, schools etc. enclose a central green or pedestrian space. Each superblock has its peripheral ring roads, off which come service cul-de-sacs. The central green or pedestrian space has pedestrian access only, by underground passages or surface walks.

radial drainage a pattern of DRAINAGE produced when streams flow down and outward from a central dome or cone-shaped upland, occurring especially in mountainous areas or on volcanic cones. Fig 17.

radiance in REMOTE SENSING, the spatial distribution of RADIANT power density.

radiant in physics, a point from which radiant energy is emitted. RADIATION.

radiant *adj.* emitting energy in the form of ELECTROMAGNETIC WAVES. ENERGY, RADIATION.

radiant energy the ENERGY originating from the sun.

radiant exitance in REMOTE SENSING, the measure of RADIANT ENERGY (ENERGY) per unit area leaving the surface or object which is undergoing sensing.

radiant flux, radiant power in REMOTE SENSING, the time-rate of the flow of RADIANT ENERGY (ENERGY) measured in watts. IRRADIANCE, RADIANCE, RADIANT EXITANCE.

radiation the act or process of radiating, i.e. of propagating or transmitting energy in the form of particles or electromagnetic waves, e.g. an electric light bulb emits radiation in the visible and infra-red regions of the ELECTROMAGNETIC SPECTRUM; radiant ENERGY is emitted by the SUN (INSOLATION, SOLAR CONSTANT, SOLAR RADIATION); and there is loss of heat from the surface of the earth by ground radiation, i.e. long-wave radiation emitted by land or water surfaces and passing up into the ATMOSPHERE (TERRESTRIAL RADIATION), especially on a clear night. ATMOSPHERIC WINDOW, CONVECTION, GREENHOUSE EFFECT, THERMAL CONDUCTION.

radiation fog, ground fog a layer of white FOG occurring particularly in low-lying areas when the weather is settled, the sky is clear, the air near the earth's surface is calm and moist and there is little wind, e.g. in Britain, in spring and autumn. In such conditions the ground surface cools quickly at night by RADIATION, cooling the overlying moist air, which flows downward by gravity to the hollows where, cooled to DEW POINT, it is condensed. Such fog usually disappears when the sun rises, but it may persist if the layer is thick and it lies under a layer of temperature inversion (INVERSION OF TEMPERATURE). It may also lead to SMOG.

radioactivity, radioactive decay the property possessed by some natural elements, and many synthetic elements, of spontaneously decaying and in the process emitting charged particles from the nuclei of their atoms accompanied by electromagnetic radiation, usually at a constant known rate (HALF-LIFE) to form more stable ISOTOPES. DATING, POTASSIUM, RADIOCARBON DATING, RADIOMETRIC AGE.

radiocarbon dating a method of radiometric DATING used to determine the age of organic remains, such as wood, bone, shells etc. (effective for material up to 3000 years in age, some would say fairly accurate up to 20 000 years, the accuracy diminishing beyond 30 000 years). Very broadly, the method is based on the assumption that carbon 14, a rare radioactive isotope of CARBON incorporated in organic matter, diminishes at a known rate after the death of the organism. Carbon 14, with a HALF-LIFE of about 5600 years, is formed in the upper atmosphere by the bombardment of nitrogen by sub-atomic particles, oxidizes to carbon dioxide, enters the CARBON CYCLE on earth and is absorbed by all living organisms. The proportion of carbon 14 to carbon 12 is constant in the atmosphere and in living organisms. But when an organism dies it ceases to absorb carbon dioxide and the proportion of carbon 14 to carbon 12 decreases with the decay of the unstable radioactive isotope. With the knowledge of the half-life of carbon 14 and the ratio of carbon 14 to carbon 12 in a sample, it is thus theoretically possible to calculate the age of the sample at the time of death. But recent research suggests that the production of carbon 14 is not constant through time, and that to reach a more satisfactory measure radiocarbon years should be compared with known tree ring dates, thus giving 'real years'. DENDROCHRONOLOGY.

radiometer in REMOTE SENSING, any device that measures/records RADIANCE from, or IRRADIANCE on to, a surface.

radiometric age the age of a substance, e.g. rocks, measured in years before the present as revealed by the time taken for a particular ratio of daughter atoms to parent atoms to be formed by natural radioactive decay of the parent atom (RADIOACTIVITY). The HALF-LIFE of a particular natural radioactive element (commonly uranium 238 or 235, or POTASSIUM) is used in the calculation, which is based on the assumption that the system is a closed one (CLOSED SYSTEM) and that none of the daughter atoms existed in the original material. RADIOCARBON DATING.

radium a strongly radioactive white metallic element which disintegrates into radon (a radioactive gaseous element). First isolated by Pierre and Marie Curie, it is used in radiography and radiotherapy, and in luminous paints. RADIOACTIVITY.

radon RADIUM.

rail (railway) gauge the width in the clear between the top flanges of the rails, the parallel metal lines on which trains run. The standard gauge used through Europe (except Spain, Portugal, the former USSR and certain European countries linked to the former USSR) as well as throughout North America and parts of Australia is 1435 mm (4 ft 8½ in). Broad gauges of 1600 mm (5 ft 3 in) and 1650 mm (5 ft 6 in) are used in Spain, Portugal, the former USSR, parts of the Indian subcontinent, Australia and South America. Narrower gauges, especially 1066 mm (3 ft 6 in) or less, are used in South Africa and parts of Australia; and the metre gauge (3 ft 3⅜ in) in many parts of the world, including the Indian subcontinent.

rain drops of water, large enough to fall under the influence of gravity from CLOUDS to the earth's surface, formed by the coalescence of water droplets produced by condensation of WATER VAPOUR in the atmosphere. CONVECTIONAL RAIN, CYCLONIC RAIN, DRIZZLE, MIST, OROGRAPHIC PRECIPITATION, PRECIPITATION-1, RAINS.

rainbow an arc of concentric bands of light in the colours of the spectrum seen in the sky when the sun is behind and RAIN is in front of an observer, caused by the reflection and refraction of sunlight in the water drops. The larger the drops, the brighter the colours.

rain-day, rainy day, day of rain in UK, a period of 24 hours, commencing normally at 0900 hours, on which 0.25 mm (0.01 in) or more of RAIN is recorded. PRECIPITATION-DAY.

raindrop RAIN.

raindrop erosion soil erosion caused by large raindrops falling on bare earth, occurring especially in tropical and semi-arid areas. The raindrops dislodge soil particles, or cause soil COMPACTION-2, which leads to increased surface RUNOFF.

rainfall the quantity of RAIN falling in a certain time within a given area, usually expressed in millimetres (mm) or inches (in). Unless otherwise stated in statistics, snow and hail (converted to water equivalent) are included. For calculation of averages a period of 35 years is commonly used, though many records are based on a shorter period. DROUGHT, PRECIPITATION-1, RAIN-DAY, RAIN SPELL.

rain forest, rain-forest FOREST-1 composed mainly of EVERGREEN hygrophilous trees (HYGROPHYTE). If the term is strictly applied it should be restricted to such a forest growing in moist TROPICAL lowlands only in non-seasonal tropical climates, i.e. with evenly distributed rainfall (EQUATORIAL FOREST), e.g. of the Amazon and the Congo (Zaire) basins, the islands of Borneo and New Guinea. But if more generally applied, it may include the somewhat less luxuriant EVERGREEN forest with some DECIDUOUS species growing at low altitudes on tropical mountains, or the evergreen forests of OCEANIC subtropical climates in south-western China, southern Africa, eastern extratropical Australia, and New Zealand. CANOPY, CLOUD FOREST, TROPICAL FOREST.

rain gauge an instrument used in measuring rainfall, consisting of a funnel with the diameter of the mouth 12.5 cm (5 in) or 20 cm (8 in), the rim ideally 30 cm (12 in) from the ground, fitting closely into a collecting container, the water collected being periodically measured in a vessel graduated to the area of the mouth of the funnel. It should be positioned at a distance from the nearest building, ideally at a distance equal to twice the height of that building.

rains, the the rainy season or wet season, also termed the MONSOON season in India and other monsoon countries. In India the rains 'break' between early June and early July, especially about 15 June, and last until September–October.

rainshadow an area with a relatively small average rainfall occurring on the LEE side of a high land barrier, e.g. a mountain barrier. The high land gives rise to OROGRAPHIC PRECIPITATION on the windward slopes, thereby reducing the moisture content of the air-stream on the leeward, which is warmed and dried further as it descends. CHINOOK, FOHN.

rain spell in Britain, a period of at least fifteen consecutive RAIN-DAYS on each of which at least 0.25 mm (0.01 in) of rain falls.

rain-wash, rainwash hill wash, hill-wash **1.** the surface creep of soil and weathered rock down a slope under the influence of gravity aided by rainwater. MASS MOVEMENT **2.** material which originates in this way.

raised beach 1. a former beach of a sea (or lake), sometimes with a cliff at the rear and a wave-cut platform in front, covered with ancient beach deposits, now situated above the present sea (or lake) level as a result of NEGATIVE MOVEMENT of sea level or of ISOSTASY. If it lies 40 to 45 m (130 to 150 ft) above the present sea level, it is usually termed a marine erosion surface, marine platform, marine terrace. The term raised beach does not apply to a beach abandoned by a lake which is drying up or being drained **2.** the ancient beach deposits themselves. Fig 39.

raised bog a deep, lens-shaped BOG occurring in a shallow basin having grown above the WATER TABLE and therefore dependent on the precipitation-evaporation rate. The growth thickens towards the centre, e.g. as in central Ireland. It may provide PEAT for fuel, used domestically and formerly in power stations. BLANKET BOG.

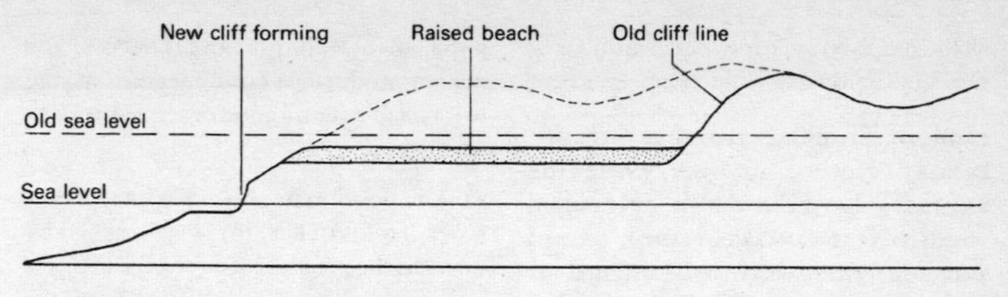

Fig 39 Section through a raised beach

rake (Scottish) **1.** a well-defined group of patches of different types of vegetation over which hill sheep move in a fairly regular daily pattern **2.** a sloping terrace on a mountain side or rock face.

ranch a FARM where cattle, horses or sheep are bred and reared, especially on a large scale in areas once covered with grassland in southern and western USA, southwest central Canada, Argentina, Uruguay. The animals used to roam to find food on the open RANGE-3, but now they are mainly confined to enclosures and fed on fodder crops grown on the farm. DUDE RANCH.

ranching the activity of breeding and rearing animals on a large scale on a RANCH.

R and D research and development, activities that together are essential to the continuing progress and success of an industrial enterprise. Research is concerned with the investigation of the commercial use of new materials and processes, development with the feasibility of using the findings of that research, or with new ways of exploiting existing materials and/or processes (INNOVATION). Most large industrial firms have R and D departments as part of their structure, but some use the expertise emanating from other sources, e.g. from researchers in RESEARCH AND SCIENCE PARKS or universities. HIGH TECHNOLOGY.

randkluft BERGSCHRUND.

random *adj.* applied in mathematics to numbers as likely to come up as any others in a set.

random error in statistics, a deviation from observed true value, occurring as if it had been chosen at random from a PROBABILITY-2 distribution of such errors. Over a large number of cases random errors balance out because, unlike SYSTEMATIC ERRORS, they do not occur predominantly in one direction.

random numbers sets of numbers drawn purely by chance, e.g. by rolling a dice. RANDOM SAMPLING.

random sample a probability sample, a SAMPLE-3 which has been selected by a method of random selection (RANDOM SAMPLING).

random sampling, random selection probability sampling, probability selection, a method of selecting sample units based on the theory of PROBABILITY, so that each sample unit has a fixed and known chance of selection. Random selection is usually carried out by reference to tables of random numbers which are generated by a computer in such a way that all the numbers appear an equal number of times in the overall table. *See* p. 456.

random sampling error the SAMPLING ERROR occurring in cases where the SAMPLE-3 has been drawn by random selection (RANDOM SAMPLING), termed sampling error where it is assumed or understood that random selection has been used.

randstad (Dutch) in the Netherlands, a ring-city, a circular CONURBATION consisting of Amsterdam, Haarlem, Leiden, 's-Gravenhage, Rotterdam and Utrecht, interspersed with agricultural land devoted mainly to market gardening and glasshouse cultivation and containing within the central area agricultural land cultivated to meet the food needs of the urban ring.

range 1. a row or line of things **2.** a single line of mountains forming a connected system. CHAIN, MOUNTAIN CHAIN **3.** a natural or semi-natural grazing area, usually unenclosed, over which animals may roam in search of food. RANCH **4.** the difference between the least and the greatest of a series of numerical values, e.g. of pressure, elevation, temperature, the difference made clear by the context, e.g. the difference between average day temperatures and average night temperatures (the daily range), or the difference between average winter and summer temperatures (annual range). ABSOLUTE RANGE, MEAN DIURNAL RANGE **5.** the maximum attainable distance, e.g. of a missile **6.** the maximum distance a vehicle can travel without refuelling **7.** in ecology, the limited area within which an organism is distributed, i.e. the limit of its habitat which affords it an appropriate niche; or the period within which it occurred, e.g. in PALAEONTOLOGY **8.** in mining, a mineral belt, e.g. the Mesabi range **9.** of a good, in CENTRAL PLACE THEORY, the maximum distance consumers are willing to travel to obtain a good or service. ORDER OF GOODS.

rank 1. position in a hierarchy. RANKING, RANK-SIZE RULE **2.** the stage reached by COAL in the progress of its change in physical composition, from LIGNITE (with the lowest CARBON content) to ANTHRACITE (with the highest).

rank correlation CORRELATION, SPEARMANN'S RANK CORRELATION COEFFIENT.

ranker in soil science, soils with an AC horizon developed on silicate (SILICA) rocks or sediments, the content of the organic matter in the A HORIZON being much higher than in the A horizon of non-calcareous soils. SOIL, SOIL ASSOCIATION, SOIL HORIZON.

ranking the action of arranging, assigning RANK, thus the arrangement of numerical data in sequence, in rank order, according to a specific quality, without making any assumptions about the intervals between the ranks.

rank-size rule a rule describing the distribution of town or city sizes in an area. It states that if a set of towns in an area is ranked in descending order of size of population, the population of any given town will be inversely proportional to its rank in the list, i.e. the population of any given town tends to be equal to the population of the largest town in the set divided by the rank of the given town; e.g. if the population of the largest town numbers 100 000, the population of the fifth largest town will be 20 000 (100 000 divided by 5). PRIMATE CITY.

rape *Brassica napus*, an ANNUAL herb, origin uncertain, grown widely in Europe,

North America and Asia for cattle feed, for spring 'greens' for cooking, in seedling form for salad; or dug in as a GREEN MANURE. The seeds are pressed for their oil, used in cooking and as a lubricant, the residue being pressed to form nutritious cattle cake (OIL CAKE).

rapids (rarely rapid) part of a stream where the water is relatively shallow and the rate of flow is accelerated by continuous and unbroken increased slope of the stream bed or by a gently dipping outcrop of hard rocks, causing the flow of water to become swift, turbulent and broken. CASCADE, WATERFALL.

ratio the relation between two quantities of the same kind, indicated by a colon (e.g. 3:2), and expressed by dividing the magnitude of one by that of the other. Thus 3:2 where the division is implied but not carried out, or as $3/2 = 1.5$, indicating that for each unit in the denominator there are 1.5 in the numerator.

ratio scale in statistics, a MEASUREMENT scale which is the same as, but more precise than, the INTERVAL SCALE, because it has a true, fixed zero point. Thus equal proportional variations in the data correspond to equal absolute variations on the scale. A VARIABLE measured on this scale is termed a ratio variable.

ravine a small, narrow valley with steep sides, larger than a GULLY, smaller than a CANYON.

raw data DATA.

raw material(s) the basic commodity or commodities, natural (e.g. a plant product such as cotton, or a mineral) or partially processed, i.e. the product of another activity (e.g. wood pulp, wheat flour), which are to be transformed by some industrial or manufacturing process into some further product before being used. The term is sometimes loosely applied to include the source of energy employed, e.g. coal, petroleum.

reach an uninterrupted stretch of water, e.g. in a straight section of a river, especially if navigable, between two bends, or between locks on a canal.

reafforestation REFORESTATION.

realism **1.** an attitude based on facts and reality, as opposed to one founded in imagining and emotions etc. **2.** in philosophy, as opposed to IDEALISM, the doctrine which maintains that material things, the object of sense perception, have a real existence, i.e. that they exist outside the mind; or the doctrine which holds that ideas, or universals (a general proposition, concept or idea), have an absolute existence outside the mind. MATERIALISM-1.

Réaumur scale a temperature scale, now obsolete, introduced by René-Antoine Ferchault de Réaumur, French physicist, 1683–1757, in which the ice point, the freezing point, of water at one atmosphere is 0° and the steam point, the boiling point, is 80°; thus $1°C = 33.8°F = 0.8°R$.

recessional moraine stadial moraine, one of a succession of TERMINAL MORAINES which mark the temporary limit of an ice sheet, formed as it pauses in retreat, or sometimes as it slightly readvances and again retreats. MORAINE.

recharge a process in which water in an AQUIFER, e.g. in an ARTESIAN BASIN, is replenished by the sinking of precipitation into the land surface. ARTIFICIAL RECHARGE.

reclamation the act of reclaiming, winning back. LAND RECLAMATION.

recreation a leisure-time activity undertaken for the sake of refreshment or entertainment (e.g. in TOURISM), in many cases away from home, and in the countryside bringing about MULTIPLE LAND USE.

rectangular drainage a DRAINAGE-2 pattern formed usually under the influence of a rectilinear joint pattern, the tributaries meeting larger streams mainly at right angles, and all streams having sections of approximately the same length between junctions. Fig 17.

recumbent fold an OVERFOLD. NAPPE. Fig 24.

recycling the long-established practice of collecting and purifying waste materials and converting them to new and useful products, e.g. as in the making and use of SHODDY, the reconstitution of glass bottles, the recovery of metals from scrap, the re-use of waste paper and rags in paper making.

red clay 1. a fine-grained, soft deposit, mainly of hydrated silicate of ALUMINA, rich in IRON oxides, occurring on the floor of the deepest parts of the ocean (ABYSSAL ZONE, OOZE) where the water is so deep that the calcareous shells of microscopic organisms and even siliceous shells are dissolved before they reach the bottom. Thus the red clay, in which MANGANESE NODULES occur, is derived mainly from volcanic (PUMICE) and METEORIC DUST, material carried by icebergs, and insoluble remains of marine life (e.g. sharks' teeth) **2.** applied loosely to any red-coloured clay.

red earth a term loosely and unscientifically applied to a red-coloured ZONAL SOIL with clay, quartz and iron compounds, occurring in tropical areas (e.g. in Brazil, Guyana, eastern Africa, southern Deccan in India, Sri Lanka), resulting from chemical WEATHERING in conditions of high humidity and high temperatures in a region with a markedly seasonal rainfall. It can be 15 m (50 ft) in thickness. The term should not be used as a translation of TERRA ROSSA. LATERITE.

redevelopment in urban areas, URBAN RENEWAL.

red-line district in an urban area, originally an area delimited on a map by a red boundary line and regarded by mortgage controllers as being in decline, unstable in social and economic terms, a poor security risk, in which property is therefore considered to be unacceptable as a loan security. The practice of such delimiting is termed redlining. GATEKEEPERS.

red rain rain coloured by red dust carried by high-level winds from an arid to a more humid area, e.g. from the Sahara to southern Europe.

red snow 1. snow coloured red by the presence of various organisms, found in snowfields throughout the world **2.** specifically, snow coloured by the presence of the red algae which occurs in acidic alpine and arctic environments.

red-yellow podzolic soil a soil with an ABC horizon (SOIL HORIZON) sequence. The profile has an A_2 horizon unless it has been removed by erosion. The clayey B HORIZON is 'blocky' in structure. PODZOL, SOIL, SOIL ASSOCIATION, SOIL CLASSIFICATION.

reef 1. a mass of rock or coral, sometimes of shingle or sand, occurring in the sea, usually covered at high tide, but often partly exposed at low tide. ATOLL, BARRIER REEF, CORAL REEF (includes fringing reef), STACK **2.** a bed or vein of metal ore or metal, particularly of gold-bearing QUARTZ.

reflection the process by which a beam of particles or a wave (e.g. visible light), in collision with an opaque surface, may be deviated or reversed in direction. Reflection is regular from a smooth (especially a polished) surface, diffuse (not coherent) from a rough surface. The proportion of RADIANT ENERGY incident upon a surface which is reflected or scattered by it is termed reflectance.

reforestation, reafforestation the planting of trees on land previously forested but from which the trees have been removed by natural causes or by cutting, burning or other means. AFFORESTATION.

refraction the change in direction of the path of a ray occurring when ELECTROMAGNETIC WAVES (e.g. visible light) or other energy-bearing waves, pass obliquely from a less dense to a denser medium. REFLECTION.

refugee a person who, owing to religious persecution or political troubles, seeks shelter or protection from danger in a foreign country. EMIGRANT, EXILE, EXPATRIATE, IMMIGRANT, MIGRANT, MIGRATION.

reg a stony desert, a desert plain covered with tightly packed, wind-scoured gravel, e.g. as in Algeria. The gravel may be cemented by salts drawn to the surface in solution by CAPILLARITY and precipitated by evaporation, thereby forming a DESERT PAVEMENT. ERG, HAMADA.

regelation the re-freezing of ice which has melted under pressure, the re-freezing taking place as pressure is released (the melting point of ice is lowered by pressure, but it rises when pressure is released). The process is a contributory factor to the flow of a glacier: the melt-water flows down to a place where pressure is less, where it again freezes.

regime **1.** the seasonal fluctuation in respect of precipitation, or of the volume of a glacier, or of the volume of water in a river **2.** recurring seasonal pattern of climatic changes.

region **1.** an area of the earth's surface with one or more features or characteristics (natural or the result of human activity) which give it a measure of unity and make it differ from the areas surrounding it. According to the criteria used in the differentiation, a region may be termed cultural, economic, morphological, natural, physiographic, political etc.; and a region may be identified by single, multiple or 'total' attributes. COMPLEMENTARY REGION, NATURAL REGION **2.** the area or space surrounding a specific place, e.g. the London region **3.** an area which is a unit of administration, e.g. a planning region. CHOROGRAPHY, CHOROLOGY.

regional development the economic and cultural growth in a REGION-3, large or small, especially in one suffering serious economic problems, usually stimulated and organized, sometimes directed, sometimes financed, by direct or indirect government action. REGIONAL PLANNING.

regionalism **1.** the local feeling of group consciousness associated with a particular geographical area, e.g. the South, the West, the Middle West **2.** the French movement of the late nineteenth century directed to the revival of REGIONAL identities and feelings, sometimes associated with political overtones **3.** the movement to decentralize central government, placing it at a level intermediate between that of the state and the small local government units **4.** in economic and social planning

(PLAN-3), the selection of a REGION-3 to serve as the basic area for future development. REGIONAL PLANNING.

regional metamorphism the alteration of pre-existing rocks by a combination of pressure (DYNAMIC METAMORPHISM) and heat (THERMAL METAMORPHISM) over a very extensive area, associated with an OROGENY, and leading to the formation of a wide range of new rocks and minerals.

regional planning 1. comprehensive PLANNING (i.e. concerned with socio-economic and political affairs) on a spatial basis, the area concerned ranging from a CITY-1 and its surrounding rural HINTERLAND-2 or several cities and their overlapping hinterlands, as distinct from town planning, which is localized, concerned with small areas **2.** such comprehensive planning within such an area by central government with the aim of reducing inequalities between REGIONS-3 in a nation state **3.** sometimes applied as equivalent to economic planning in REGIONAL DEVELOPMENT, although the economic region may differ in design and size from the region identified or specified by the city regional planners. LAND USE PLANNING.

regolith a general term applied to the mantle of loose material (soil, sediments, broken rock, volcanic ash, wind-blown material etc., i.e. the soil and weathered rock) overlying the solid BEDROCK. MANTLE-2, WEATHERING.

regression 1. in general, the act of moving back, the tendency to move back **2.** in astronomy, the movement of a heavenly body, e.g. a planet, in a direction opposite to the normal, e.g. as in the PRECESSION OF THE EQUINOXES **3.** in biology, the return to an earlier or less complex form **4.** in statistics, the relationship between the mean value of a random variable and the corresponding values of one or more other variables. REGRESSION ANALYSIS.

regression analysis a statistical technique which aims to explain the variation in an observed quantity in terms of the dependence of one VARIABLE (the DEPENDENT VARIABLE) on one or more other variables (the INDEPENDENT VARIABLES).

In LINEAR-4 regression analysis the data are plotted on a graph, the dependent variable on the vertical axis, the independent variable(s) on the horizontal axis, the trend of the points on the graph indicating the relationship between dependent and independent variables. A straight line is fitted as close to the data points as possible to give the best description of the trend (termed the regression line, or the line of best fit, the distances between the points and the line being termed regression residuals). The line is placed so that the sum of the squares of the deviations of the points from the line is the smallest possible (termed the least squares method). The line can then be used to predict expected values of one variable given the value of the second variable. SLOPE-3. Fig 40.

Reilly's law of retail gravitation a law postulated by W. J. Reilly, New York, in 1931, that two cities attract retail trade from an intermediate city or town in the vicinity of the breaking point, approximately in direct proportion to the populations of the two cities and in inverse proportion to the square of the distances (distance via most direct improved automobile highway) from these two cities to the intermediate town. GRAVITY MODEL.

rejuvenation becoming again youthful; in geology and geomorphology, the development of younger surface forms, appropriate to the earlier stages of the

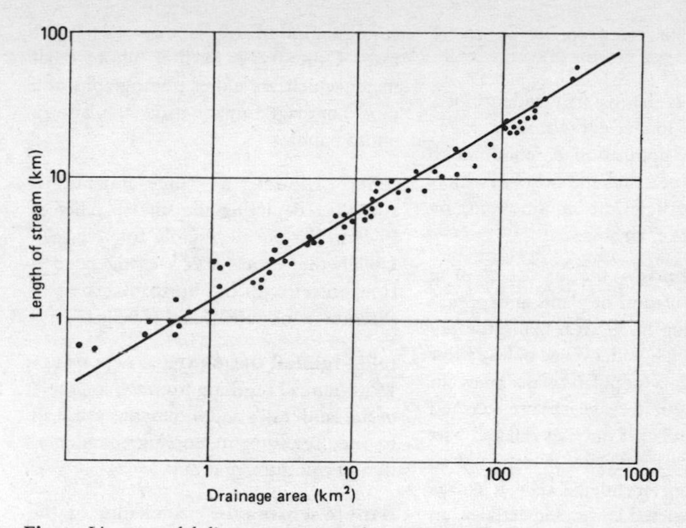

Fig 40 Linear model: linear regression relating the length of streams to drainage areas

CYCLE OF EROSION, occurring when a comparatively well-advanced cycle is interrupted by an increase in the rate of erosion, especially erosion by a RIVER, due to causes defined as dynamic, eustatic (DIASTROPHIC EUSTATISM, GLACIO-EUSTATISM) or STATIC, resulting in such river features as ALLUVIAL TERRACES, INCISED MEANDERS, KNICKPOINTS, TERRACES-2.

relative humidity RH, the ratio between the actual amount of moisture in the air and that which would be present if the air were saturated at the same temperature, expressed as a percentage. Air with an RH of 60 may be considered as approximately separating DRY from 'moist' atmosphere. ABSOLUTE HUMIDITY, DRY-BULB THERMOMETER, HUMIDITY, HYGROMETER, SPECIFIC HUMIDITY, WATER VAPOUR, WET-BULB THERMOMETER.

relative relief relative altitude, the relation of the highest and lowest points in a land area, the difference between the two extremes being termed the amplitude of relative relief. AVAILABLE RELIEF, RELIEF.

relaxation time the lapse of time during which a physical SYSTEM-1,2,3, having been upset by changes in the factors controlling or influencing it, readjusts to those changes and reaches a new state of EQUILIBRIUM.

relevance 1. the quality of being connected with, pertinent to, the subject **2.** specifically, the degree to which something known or being discussed or studied has a bearing, particularly a practical bearing, on some current issue in society, e.g. the contribution geographers may make in helping to find practical solutions to current environmental or social problems. The consideration of relevance gained favour in 1970s study as a reaction to the scientific approach of POSITIVISM, deemed to be inappropriate in social studies.

relic, relict 1. the material evidence of a thing that no longer exists **2.** in ecology, an organism, population or community the remnants of which still exist but which was at an earlier time common in, or characteristic of, an area.

relic distribution the RANGE-7 of a RELIC-2 population of plants and/or animals surviving in an area, being the remains of a much wider range of an earlier time. If such a population occupies an area throughout its range it may be termed an absolute relic; if only an isolated part of the area, a local relic; if restricted to a single region, an endemic relic; if its area has been restricted by human activity, an anthropogenic relic; and if it establishes a secondary distribution by occupying a suitable habitat it is termed a migrant relic.

relict boundary a BOUNDARY that can still be discerned in the CULTURAL LANDSCAPE despite the fact that it has been abandoned and no longer serves a political purpose.

relict sediments sediments deposited under conditions different from those of the present.

relief 1. the physical shape of the surface of the earth, its mountains and valleys, plains and plateaus, the physical landscape **2.** often applied loosely to indicate inequalities or variations in shapes and forms of the earth's surface. The use of the terms 'high relief' and 'low relief' is best restricted to areas which respectively show a great or little variation in altitude. TOPOGRAPHY should not be confused with relief. AVAILABLE RELIEF, RELATIVE RELIEF.

relief map a map depicting the surface configuration of an area, e.g. a CONTOUR map. Other types include photo-relief maps, which are either photographs of a model or diagrammatic maps simulating a photograph.

relief model a three-dimensional MODEL-1 depicting the surface relief of an area, but not necessarily true to scale. Usually the vertical scale is exaggerated in comparison with the horizontal so as to accentuate mountains and plateaus.

relief rainfall OROGRAPHIC PRECIPITATION, rainfall resulting from the RELIEF-1 of the land. Hills and mountains cause air to rise, resulting in cooling, condensation of moisture, and rain.

remote sensing the examination of, the obtaining of information about, an object or phenomenon at a distance from it, without physical contact with it, particularly by devices based on the ground, by SENSORS carried aboard ships or aircraft, or by spacecraft or satellites orbiting the earth, which gather data in digital form at a distance from their source. The data collected by such equipment may be based on measurements of variations in electromagnetic radiation, in acoustic energy, or in gravitational or magnetic force fields. Computers are commonly used in the retrieval and storage processes. The information obtained from remote sensors orbiting the earth in this way has greatly advanced the understanding and knowledge of the earth's surface, its geological structure and mineral resources, the circulation of the ocean, and atmospheric phenomena. AERIAL PHOTOGRAPH, ELECTROMAGNETIC SPECTRUM, GROUND INFORMATION, ENHANCEMENT, FALSE COLOUR, IMAGE, INFRA-RED RADIATION, MULTISPECTRAL SENSING, PIXEL, RADAR, RADIOMETER, REFLECTION (reflectance), RES-

OLUTION-1,2, SIGNATURE, SURVEY-4, SONAR, SYNOPTIC IMAGE, THERMAL INFRA-RED SENSING.

Renaissance, Renascence the artistic and literary revival which began in Italy in the fourteenth century and profoundly affected thought and the arts in Europe for some two centuries thereafter. Broadly, it was characterized by the spread of HUMANISM, a return to classical values (which weakened the grip of the medieval church), realism in the visual arts, an architecture radically different from that typical of the preceding centuries, a spirit of objective scientific enquiry, the birth of printing, and the founding of centres of learning. It opened the way for the development of the modern world. HUMANIST, HUMANISTIC, MODERNITY-2.

rendzina, rendsina in soil science, a group of INTRAZONAL SOILS, shallow and calcareous, with an AC horizon, usually with a brown or black friable surface, humus content MULL-2,3, underlain by light grey or yellow calcareous material, CALCIUM CARBONATE being distributed throughout the profile. It develops from relatively soft parent rock, under grass and forest vegetation in humid and semi-arid regions. It is well developed in England on Chalk, but it also occurs on harder limestones. MOLLISOLS.

renewal in urban areas, URBAN RENEWAL, REDEVELOPMENT.

rent 1. an item of income or payment to any factor of production which is limited in supply **2.** any surplus earned on account of superior quality or ability **3.** popularly, the price paid for the use of any durable GOOD (usually of land or buildings) at certain specified or customary times, or for a certain period of time. ECONOMIC RENT, LAND RENT, LAND TENURE.

repose slope a slope, usually steep, with its steepness governed by the ANGLE OF REPOSE of the superficial layer of debris, which maintains its angle in its retreat. PARALLEL RETREAT OF SLOPE.

representative fraction RF, the fraction expressing the ratio between the distance measured between two points on a map and the corresponding distance measured on the ground, the usual way of expressing the scale of a map. The numerator is one (indicating one unit, such as an inch or centimetre), the denominator is the number of units on the ground which this represents. Thus an RF of 1/1 000 000 (often written 1:1 000 000, or 1:1 mn) means that 1 in or 1 cm on the map represents one million in or cm on the ground. Since a mile is 63 360 in, this is approximately 16 miles, and a map of 1:1 000 000 is roughly 16 miles to one inch.

reproduction rate any of the calculations used to indicate trends in numbers of human population. They vary from simple and crude to complex and refined. The simplest is the gross reproduction rate, calculated by the number of all females of reproductive age (15 to 50 years) expressed as a ratio of the actual number of females born; or by the number of daughters a woman would produce if throughout her life she gave birth in accord with current (age-specific) fertility rates (FERTILITY-3). Net reproduction rate involves a more refined calculation, the factors modifying the calculated gross reproduction rate being taken into account, e.g. deaths of females before and during reproductive age, infertility, un-mated state, numbers of reproductive males. If continued over a period of time a net reproduction rate of less than one leads to population decline; of one, to a stationary population (STABLE POPULATION); and if more than one, to an increasing population. BIRTH-RATE, NATURAL INCREASE.

research and science park a tract of land, commonly park-like (PARK-2,6), planned and officially designated for the accommodation of individuals and organizations engaged in research, development and scientific production. In some cases these activities (especially the last) overlap with those found in an INDUSTRIAL PARK, so the difference between these two types of park is not everywhere distinct. R AND D.

resequent, re-consequent drainage a DRAINAGE-2 pattern in which streams lie approximately along the line of former CONSEQUENT streams, after long-term denudation of a folded area. They thus appear to be consequent but, due to some agency such as RIVER CAPTURE, they have in fact developed (with synclinal valleys and anticlinal ridges) from a SUBSEQUENT drainage pattern.

resequent fault-line scarp a FAULT-LINE SCARP facing in the original direction of the DOWNTHROW, produced by erosion along the OBSEQUENT FAULT-LINE SCARP.

reserve **1.** something held back, kept in store for future use. RESERVES, RESOURCES **2.** reservation, an area of land demarcated for a special use, e.g. to protect the habitat of rare species, a nature reserve.

reserves part of a RESOURCE considered to be unstable in current conditions of technical skill and economic and social needs. The existence and location of some reserves is known (PROVEN RESERVES), but of others it may be hypothetical (UNDISCOVERED RESERVES).

reservoir **1.** a container or receptacle, natural or artificial, in which a liquid or gas collects and is stored **2.** a natural or artificial lake, the water of which is used in irrigation, in producing hydroelectricity, in manufacturing processes and for domestic purposes **3.** a highly porous (POROSITY) PERMEABLE ROCK mass which is able to transmit as well as to hold a fluid **4.** a reserve supply of something, e.g. a relict species. RELIC.

residual deposit an accumulation of rock waste, from clays to boulders in grain-size, caused by the WEATHERING, the disintegration, of rocks in situ. RESIDUAL SOIL.

residual soil a SEDENTARY SOIL, a soil resting on the material from which it was formed. SECONDARY SOIL.

resistant rock a rock with physical and chemical properties which make it able to withstand the processes of EROSION and WEATHERING.

resolution in REMOTE SENSING **1.** the ability of a SENSOR to make a distinction between objects which have similar temperatures, or between objects which have similar spectral characters **2.** the minimum distance (linear or angular) between two objects at which they continue to appear distinct and separate on an image or photograph.

resource, resources **1.** a source of supply or support, the means (or collective means) of meeting a need or deficiency, especially an economic or social need or deficiency **2.** a stock (or stocks) which can be used if necessary, the total stock consisting of the total amount of a substance, i.e. the accessible, the inaccessible, and the unusable, at the current stage of technology. NATURAL RESOURCES, RESERVE, RESERVES, RESOURCE CONSERVATION, RESOURCE MANAGEMENT, SUSTAINABLE DEVELOPMENT.

resource conservation the careful

management and maintenance of NATURAL RESOURCES. CONSERVATION, RESOURCE, RESOURCE MANAGEMENT.

resource development the bringing of a RESOURCE-2 into production, the realization of its potential.

resource management the skilful control of a RESOURCE-1,2 by those who ensure that it is used economically and with forethought, who determine the present need for and the value of such a use, balance benefits and costs, take into account environmental constraints, social, economic and political implications, technological inventiveness, national policy; and possible future needs, technology and uses. A decision not to develop a resource at a particular time is also part of management. CONSERVATION, NATURAL RESOURCES, RESOURCE CONSERVATION, RESOURCE DEVELOPMENT, SUSTAINABLE DEVELOPMENT.

resource orientation the tendency of a firm or a particular industry to be located close to the source of its raw material(s) when the source of the RAW MATERIALS is localized, not widespread like that of UBIQUITOUS MATERIALS. MARKET ORIENTATION.

respiration any of the processes by which an organism takes in air or dissolved gases, uses one or more of them in chemical reactions which produce energy, and expels the unused parts of the air or gases together with the by-products of the chemical changes. Plants and animals take in OXYGEN and expel CARBON DIOXIDE produced by the oxidation of CARBON compounds (e.g. glucose) in the system. In daylight green plants use carbon dioxide from the air to form starch and expel oxygen as a by-product. AEROBIC RESPIRATION, ANAEROBIC RESPIRATION, PHOTOSYNTHESIS.

resurgence the emergence of an underground stream (DISAPPEARING STREAM) from a cave, commonly occurring near the point where an IMPERMEABLE layer underlying the PERMEABLE layer through which the stream has passed intersects the surface.

retrogressive method a method of approaching an understanding of the past that begins by a careful, detailed analysis of the identifiable factors which have created the present (particularly applied to past landscapes). RETROSPECTIVE METHOD.

retrospective method a method of approaching an understanding of the present based on a study of the past which, it is maintained, illuminates present conditions. HISTORICAL GEOGRAPHY has been described as retrospective geography. RETROGRESSIVE METHOD.

reverse(d) fault a thrust fault, a FAULT in which the FAULT PLANE dips to the upthrow side (THROW OF FAULT), the result of older beds on one side of the fault plane being compressed and THRUST over younger beds on the other side. Fig 22.

RF REPRESENTATIVE FRACTION.

RH RELATIVE HUMIDITY.

rhyolite a fine-grained to glassy, ACIDIC extruded IGNEOUS ROCK of QUARTZ and alkali FELDSPARS, with FERROMAGNESIAN MINERALS, corresponding in chemical composition to PLUTONIC granite. The banding in some cases indicates the flow of the rhyolite as molten MAGMA which, even at very high temperatures, is much less fluid than basic lavas (BASALT), tending to consolidate in masses.

ria a funnel-shaped indentation on a coast,

narrowing inland, increasing in depth seaward, a drowned river valley, occurring particularly along coasts of the ATLANTIC TYPE as a result of a rise in sea level (EUSTATISM). A ria differs from a FJORD in being shorter and lacking the irregularities of depth characteristic of the fjord. The stream that made the original valley, and which flows into the head of the ria, is clearly too narrow in relation to the present size of the inlet.

ribbon development the building of houses etc. along each side of the main roads extending outwards from a built-up area, typical in Britain between 1920 and 1939.

rice *Oryza sativa*, a CEREAL-1, native to Asia, cultivated in wet tropical areas (especially in monsoon Asia) and in some subtropical areas for its seed (PADI), used for human food, the staple cereal in the growing countries. There are many different varieties in cultivation, the high-yielding strains produced by hybridization playing an important part in the GREEN REVOLUTION. Upland or hill rice (dry rice) is grown without irrigation on tropical hillslopes; but most rice is grown in flat fields which can be flooded, either through heavy rainfall or by irrigation water. In Asian countries the seed is usually germinated in a nursery field from which the young plants are transplanted to larger fields when they are some 15 to 25 cm (6 to 10 in) in height (other crops may be growing in the larger fields whilst the seeds are germinating in the nurseries). Transplanting, by hand, is usually done into the mud of the flooded fields. As the plants grow, rapidly, the water is absorbed and by harvest time the fields are almost dry. Reaping is still mainly by hand. In some parts of the tropics three, or even more crops a year are obtained, two are common, and in many parts of the monsoon lands rice as the summer crop can be grown on the same land as wheat or barley as winter crops. With intensive cultivation under these methods yields may be very high; but very high yields are also obtained from rice grown under mechanized conditions in California, Spain and northern Italy. Poorer qualities of rice are used for making starch; the hulls are fed to cattle, as are the stalks, which are also used in making paper, thatching, footwear; fermented rice kernels provide alcoholic beverages.

Richter scale a scale, in Arabic numerals from 0 to over 8, used in measuring the magnitude of EARTHQUAKES, based on instrumental recordings of a standard SEISMOGRAPH 100 km (62 mi) from the EPICENTRE, the larger numbers being applied to the larger disturbances, 7 being a major earthquake. The Richter scale superseded the MERCALLI SCALE and the ROSSI-FOREL SCALE.

ridge a term loosely applied to any long, narrow, steep-sided rise in the land, sometimes to a small feature in a mountain RANGE-2, but infrequently to the range itself and never to a mountain chain. OCEANIC RIDGE.

ridge and furrow a method of working heavy soils, attributed to the Anglo-Saxons in England, but continuing after INCLOSURE, to promote surface drainage. The land was ploughed so as to form broad ridges separated by furrows which served as drains to draw off water. Where such land is now in grass, the ridge and furrow can, in many cases, still be seen.

ridge and valley a form of relief in which ridges and valleys lie close together and almost parallel, e.g. the ridge and valley region of the Appalachians, where the

RESISTANT ROCKS (sandstones, quartzites and conglomerates) form the ridges and the valleys have been carved through the more easily eroded shales and limestones.

ridge of high pressure a long, narrow region of HIGH atmospheric pressure, broader than a WEDGE, lying between two LOW pressure areas, and responsible for a brief period of fine weather in a generally rainy period. ATMOSPHERIC PRESSURE.

riffle 1. a rocky obstruction in a river bed, or the riffled water resulting from such an obstruction **2.** a shallow gravel bar over which water flows rapidly with a ruffled surface, formed in a stream with a gravel bed which, with intervening pools, gives rise to a POOL AND RIFFLE pattern.

rift valley a long, narrow section of the earth's crust let down between two parallel series of FAULTS with THROWS in opposite directions, thus appearing as a long, flat-floored valley with steep sides, as in the rift valleys of east Africa, or in the Midland Valley of Scotland.

rill 1. a small natural STREAM-1 of water **2.** in soil erosion, a small erosion channel.

rill erosion the removal of soil by RILLS which, if persistent, may enlarge the rill channels so much that they unite to form GULLIES, leading to gully erosion. SOIL EROSION.

rime an accumulation of white, opaque, granular ice tufts formed on the windward side of objects with a temperature below that of freezing point when supercooled droplets (e.g. in cloud or freezing fog) are blown against them. It resembles HOAR FROST. SUPERCOOLING.

ring-city RANDSTAD.

ring-dyke a dyke (DIKE) in the area surrounding a circular or dome-shaped igneous INTRUSION. The vertical and outward pressure of the MAGMA forming the intrusion causes fractures in the COUNTRY ROCK-1, into which the magma flows and solidifies, forming ring-dykes if the fractures are vertical.

ring road a road encircling a built-up area, used as a bypass or as a service road for the area.

rip a disturbance and turbulence, rough water, sometimes in a river, more commonly in the sea, occurring when two tidal streams flowing from different directions meet; or a strong outflowing surface or near-surface current meets the incoming surf; or when a tidal stream suddenly flows into shallow water; or when a tidal or river current flows over an irregular floor, especially one with abrupt changes in depth.

riparian *adj.* of, pertaining to, situated on, or associated with, a river bank, applied especially in legal terms, e.g. riparian rights, riparian land, riparian states. RIVERINE, RIVERSIDE.

ripple, ripple-mark, sand ripple a series of small more or less parallel ridges, produced especially on sand by the wind, the current of a stream, or by waves on a shore.

rise, swell on the deep sea floor, a gently sloping, long, broad elevation rising from the deep sea floor, its summit far below the surface of the water.

risk 1. the chance of danger, injury, loss etc. **2.** a situation in which there is the possibility of several possible outcomes occurring as a consequence of a decision or action taken when the probabilities of the possible outcomes are known and can be calculated for each individual outcome (e.g. as in insurance risk). UNCERTAINTY.

river a general term applied to a natural STREAM-1 of water flowing regularly or intermittently over a bed, usually in a definite channel, towards the sea, a lake, or an inland depression in a desert basin, or a marsh or another river. Stages of development may be recognized, in youth with swift streams actively eroding steep-sided V-shaped valleys; maturity, middle age, with broad open valleys with gentler slopes and the beginning of MEANDERS; and old age (SENILE RIVER) with broad valleys and a sluggish flow of water. Not all rivers fit this ideal model, e.g. the flow may be sluggish but the stream not old (CYCLE OF EROSION). The point or points of origin are termed the SOURCE of a river, the path it follows is its course, the part where it enters a sea or lake its mouth. Most rivers have an upper mountain course, a middle or plains course, and a lower or estuarine-delta course. Rivers are especially active as eroding agents in REJUVENATION, particularly in STATIC REJUVENATION. See BED LOAD, DELTA, DISSOLVED LOAD, DRAINAGE-2, STREAM-1, STREAM ORDER, SUSPENDED LOAD and entries qualified by river.

river bank the rising land bordering a river.

river basin all the area of land drained by a river and its tributaries.

river bed, riverbed the channel in which a river flows or once flowed.

river capture, river piracy the action by which a river, by rapid headward erosion, captures and diverts to itself the headwaters of another stream, thereby enlarging its own drainage area and diminishing that of the other. The stream of which the headwaters have been captured is said to be beheaded; the point at which the capture occurs is the elbow of capture; self-shortening of the course of a river is autopiracy. River capture may give rise to MISFIT STREAMS and WIND GAPS.

river cliff in a MEANDER of a river in a young river valley, the steep, concave (outer) curve cut by the stream current, facing the gentle SLIP-OFF SLOPE of the spur opposite.

riverine, riverain *adj.* **1.** of, pertaining to, situated on, living on, the banks of a river **2.** of or relating to a river or its vicinity. The application is less restricted than that of RIPARIAN.

river port a PORT situated on a river, usually at the point farthest from the mouth where the water is deep enough for navigation by trading vessels.

river profile a section of a river valley, either longitudinal (along the course of the river, showing the slope from source to mouth, RIVER) or transverse (CROSS PROFILE OF A VALLEY, i.e. across the valley at right angles to the stream). The long PROFILE-3 ('long' is preferred to 'longitudinal') usually takes in the actual length of the centre of the stream, and the height of the surface at mean level, measurements being adjusted for minor variations of level. The height of the floodplain (especially useful in comparing the present profile with reconstructed profiles) is sometimes preferred; in this case minor windings are ignored because such features cannot be reconstructed for past phases of the river at higher levels. TALWEG.

riverside a general, imprecise term, the land alongside a river. RIPARIAN, RIVERINE.

river terrace a part of the former floodplain of a river, left on the side of a river valley as the stream cut down its bed and now appearing as a generally flat, step-like strip on the side of the valley, at a level higher than that of the present channel. Such a terrace is usually built up of gravel, coarse sand and alluvium deposited by the river when it was flowing at the level of the terrace. Thus the terrace represents a part of the valley floor at that time, and may not be perfectly flat. The term is applied to a rock BENCH as well as to a gravel-covered terrace. ALLUVIAL TERRACE.

Riviera 1. the coastal strip with its numerous resorts, bordering the Mediterranean from Marseille in southern France to Genova in Italy **2.** by analogy, applied to resort coasts elsewhere, e.g. Cornish Riviera, the southern coast of Cornwall, England.

road metal not metal, but broken, tough, hard stone which breaks into angular pieces without splintering or making much dust, used for surfacing roads. MACADAM.

Roaring Forties the latitudes between 40° and 50°S where the westerly winds are not obstructed by land as they blow strongly and regularly over the ocean, bringing gales, rough seas and rain associated with the regular procession of DEPRESSIONS-3 moving from west to east. The term is occasionally applied to the winds. WESTERLIES.

robber economy the working of RESOURCES-2, especially of stock resources which, once used up, are not renewed, in contrast to the development of renewable resources (NATURAL RESOURCES). Sometimes also applied to the needless destruction of resources for the sake of quick profits, especially if it is unlikely that the resource will recover in the future.

roche moutonnée (French) a rock mass forming a hillock, resulting from ice action, seen in most glaciated valleys. The upstream side is smoothed and rounded owing to the effect of ABRASION and commonly shows glacial striation (STRIAE); the downstream side is steeper and rougher, owing to PLUCKING. ONSET AND LEE.

rock 1. in general use, a large, hard, consolidated, compact part of the earth's crust, also a large piece of this material protruding from the land or sea **2.** in geology, any naturally formed aggregate of mineral particles, whether it is hard, relatively soft, unconsolidated or incompact, which constitutes an integral part of the lithosphere (thus includes mud, clay, sand, coral etc.). Rocks are classified by the manner of formation (IGNEOUS, METAMORPHIC, SEDIMENTARY) or by age (GEOLOGICAL TIMESCALE, STRATIGRAPHY). COUNTRY ROCK.

rock creep the slow movement of rock blocks down a slope under the influence of gravity. MASS MOVEMENT.

rock fall a free fall of individual boulders or blocks of bedrock down any steep slope under the force of gravity. MASS MOVEMENT, ROCK SLIDE.

rock flour, rock-flour very finely powdered rock material produced by the grinding action of a glacier when, with rocks frozen into its mass, it abrades its bed. The process is mechanical, there is little or no chemical action, hence rock flour has the same mineralogical composition as that of the rocks from which it is formed. It is a major constituent of BOULDER CLAY. ABRASION.

rock salt sodium chloride, a mineral occurring in a clear or white to brownish crystalline mass and forming, with BRINE, the source of SALT-2 used in commerce and industry.

rock slide, rock-slide the sliding of an individual rock mass down a gentle gradient, e.g. down a bedding joint or fault surface, under the force of gravity. MASS MOVEMENT, ROCK FALL.

roller a term popularly applied to the ocean swell which even in calm weather gives rise to very large BREAKERS along coasts, due to the length of the FETCH.

root crop a term applied particularly to plants with a root system incorporating swollen fleshy parts which serve as a store of food for animals, grown especially for the feeding of animals and, because they can be stored, much used for winter feed for cattle and sheep, e.g. in midlatitudes, turnips, Swedish turnips or swedes and mangolds (formerly called mangelwurzels). In midlatitudes root crops feature in the ROTATION OF CROPS (NORFOLK ROTATION) and those grown mainly for human consumption include sugar beet, potatoes, carrots, parsnips, turnips. Tropical root crops include arrowroot, CASSAVA, sweet potatoes, yams.

ropy lava, corded lava LAVA which has solidified so that the surface of the flow is glassy and smooth with surface shapes resembling ropes or cords. PAHOEHOE.

Ro-Ro roll-on/roll-off technique, CONTAINER.

Rossby waves long waves with a wavelength of some 2000 km (1250 mi), very large-scale westerly movements in the airflow of the upper ATMOSPHERE-1 (in the middle and upper TROPOSPHERE), stretching from sub-polar to tropical latitudes. There are generally about four such waves in the northern hemisphere, four in the southern.

Rossi-Forel scale a scale formerly used in the measuring of the intensity of EARTHQUAKE shocks, superseded by the modified MERCALLI SCALE c. 1931, now superseded by the RICHTER SCALE.

Rostow model a LINEAR MODEL devised by W. W. Rostow, American economic historian, to identify stages of growth in a capitalist economy, based on growth in Europe, North America and Japan. His stages are (1) traditional society, mainly agricultural, little science or technology; (2) extractive industries develop, agriculture and transport improve; (3) manufacturing develops; (4) economic growth spreads; (5) high production and mass consumption of goods.

rotational slip the downward movement of a mass of rock or ice on a slip-plane, in which the solid mass appears to rotate on a pivot as it descends (an action which may, in the case of ice, contribute to the basin shape of a CIRQUE), and which usually results in the solid mass presenting a well-defined, uphill-facing, back-slope. MASS MOVEMENT.

rotation grass, rotation grassland, temporary grassland grass which is sown and grown for one or more years in rotation with other crops. LEY, ROTATION OF CROPS.

rotation of crops a farming system in which a systematic succession of different crops is grown on the same piece of land so that the maximum use is made of soil nutrients but soil fertility is not exhausted. LAND ROTATION, MONOCULTURE, NORFOLK ROTATION.

rotation of the earth the revolving movement of the EARTH on its axis from west to east, from which it appears (from the earth) that the sun, moon and stars move round the earth from east to west. The average period of rotation of the earth in relation to the sun, i.e. the time interval between two successive crossings of the sun over a meridian, is 24 hours (MEAN SOLAR TIME); but the rotation period measured by two successive crossings of a meridian by a selected star is less, being 23 hours 56 minutes 4.09 seconds (sidereal day). The velocity of the earth's rotation is some 1690 km (1050 mi) per hour at the equator, some 845 km (525 mi) per hour at 60°N and S, and zero at the poles. AXIS OF THE EARTH.

rough grazing unimproved, usually unenclosed grazing, including many different types of natural and semi-natural vegetation, e.g. moorland, scrubland, salt marsh, mountain pasture. In Britain the distinction between rough grazing (usually unenclosed) and improved grazing (usually enclosed in fields) is clear, the two categories being shown separately in official statistics. In other countries enclosed grassland for grazing does not exist, and statistics merely separate pasture from arable: thus statistics for grassland, pasture, grazing are not internationally comparable.

rubber **1.** an elastic substance made from the sap (LATEX) of a number of different tropical trees and climbers. Formerly, 'wild rubber' was collected from such plants in the equatorial forests of South America and tropical Africa; it was superseded by rubber from plantations of the Brazilian rubber tree, *Hevea brasiliensis*, in Malaysia (with some in Indonesia, Thailand, Sri Lanka, Nigeria). The trees are 'tapped' by cutting a thin slice of bark, the white sap is collected, curdled by the addition of a little acid, and stickiness countered by the addition of sulphur. It was this treatment of rubber with sulphur (vulcanizing) which opened up the possibility of its extensive use in manufacturing **2.** a synthetic substitute for the natural product.

runnel **1.** a small stream **2.** the small channel in which a small stream flows.

runoff **1.** all the water flowing from a drainage area **2.** that part of the PRECIPITATION which runs off the land surface into streams, in contrast to those parts which either soak into the ground or evaporate (although some authors include water which returns to the surface by seepage and from springs). OVERLAND FLOW, THROUGHFLOW. Fig 41.

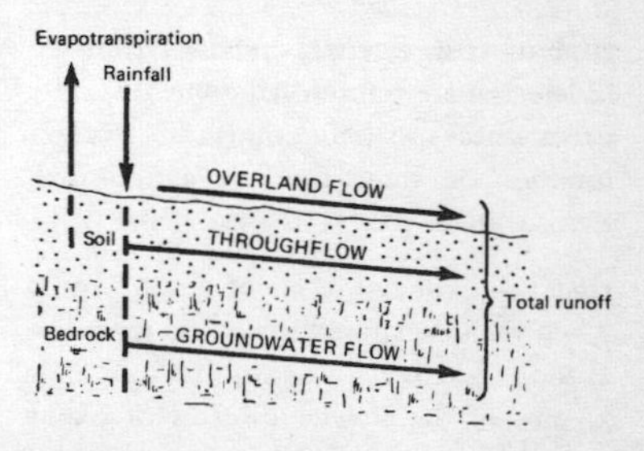

Fig 41 Runoff

rural *adj.* of, belonging to, relating to, characteristic of, the country or country life, in contrast to the town or urban life. It is now recognized that in industrialized countries the distinction between rural and URBAN is blurred. Some authors write of a rural-urban continuum, or URBAN-RURAL CONTINUUM, and the creation of a rural-urban complex as 'rurbanization'. RURBAN, RURBAN FRINGE.

rural population **1.** broadly, the population living in the countryside, not in a town **2.** more specifically, that part of the population defined on the basis of various criteria, e.g. functional, those maintained by the exploitation of the intrinsic resource of the land (agrarian, or agrarian and mining); landscape-sociological, those living in a 'non-built-up' area; statistical, the size of the agglomeration, or the density of population; socio-psychological, many 'primary' social contacts **3.** in Britain, before the reform of local administration in 1974, the population living in the administrative units known as rural districts, many of which included fairly large towns or suburbs of towns.

rural-urban continuum RURAL, URBAN-RURAL CONTINUUM.

rurban *adj.* having the characteristics of country and town combined.

rurban fringe, rural-urban fringe an indeterminate transitional zone around a town where urban functions and activities impinge on those that are agricultural, rural. URBAN-RURAL CONTINUUM.

rust 1. a hydrated oxide of IRON formed when iron is exposed to air and moisture **2.** the red-brown colour of that substance **3.** any of the several diseases of plants caused by an order of parasitic fungi, or the parasitic fungus itself.

rutin a glycoside, present in plants, especially in buckwheat and tobacco leaves, used in the treatment of radiation injuries and of hypertension.

ruware a rock pavement, a low rounded, in some cases elongated, exposure of rock, rising from a plain in a tropical area. Opinions differ as to whether it is an early stage in the formation of a dome-shaped INSELBERG, or the result of the wearing down of an inselberg combined with the merging of surrounding PEDIMENTS.

rye *Secale cereale*, a tall-growing CEREAL with dark grain, cultivated in the colder parts of Europe as far north as the Arctic Circle, and in mountainous areas up to 4270 m (14 000 ft). It tolerates the poor soils unsuited to most of the other midlatitude grain crops. The grain yields a dark-coloured flour, used in making bread (which is heavy, with a rather sour flavour) and crispbread which keeps well and is favoured in low-calorie diets. The grain is also used in making whisky, gin, beer. Young plants are fed to livestock, the straw is used in thatching, bedding, paper-making.

S

sagebrush a semi-desert type of vegetation in western North America dominated by sagebrush, *Artemisia tridentata*, in association with other small-leaved shrubs.

Sahara (Arabic sahra) a desert, a plain, a term applied to the greatest of all deserts, the Sahara, northern Africa (which should therefore not have 'desert' added to its name).

Sahel (Arabic shore) the semi-arid vegetation zone lying between the SAHARA and the savanna lands to the south in north Africa. Prone to long periods of drought, it is characterized by a short uncertain rainy season and a long dry season. The original CLIMAX was thorn woodland, but the vegetation now consists of patches of poor wiry and tussocky grass, acacia and thornbushes. DESERTIFICATION.

St Elmo's fire small flickering flames around the tops of tall objects, such as mastheads, visible at night, occurring in stormy weather, associated with the passing of a FRONT. The phenomenon constitutes a brush discharge, occurring when a stream of molecules of air, electrically charged, is repelled by a sharp-pointed, charged conductor.

salina (Spanish) a PLAYA with a high concentration of salts, enclosed from the sea.

saline *adj.* salt, or of, containing, or tasting SALT-2; or relating to or being characteristic of chemical SALTS-1.

saline soil an INTRAZONAL SOIL with a high concentration of soluble salts, occurring in areas with high evaporation, e.g. in hot deserts or in dry areas with high summer temperatures (as in cool temperate continental interiors), and especially resulting from the drying of salt lakes that once lay in inland drainage basins. Irrigation, unless carefully managed in such areas, may increase salinity. The salt solution is drawn upwards in the soil by CAPILLARITY, it dries out and forms a surface crust over a salt-impregnated more granular layer. SOIL, SOIL CLASSIFICATION.

salinity the degree of concentration of common salt in a solution, determined by measuring the density of the solution, and usually expressed in parts per thousand by mass.

salinization the PRECIPITATION-2 of soluble salts within the soil.

salt 1. in chemistry, a chemical compound derived from acids formed when all or part of the replaceable HYDROGEN ATOMS in a MOLECULE of the ACID are replaced by a METAL, directly or indirectly. Classified as normal salt when all the replaceable hydrogen atoms have been replaced by a metal, acid salt when only part of all replaceable hydrogen atoms have been replaced **2.** common salt, sodium chloride, a white crystalline compound, widespread in nature as a solid (ROCK SALT) around margins of salt lakes, in SALT DOMES, or in solution in seawater, and present in all animal fluids. It is obtained commercially by the evaporation of BRINE from seawater, from brine wells and salt lakes; or from the solid deposits in rock salt mines. It is used as a food seasoning and preservative, and widely as a raw material in industrial processes and manufacturing industry, e.g. the chemical industry, in glass and soap making etc.

saltation **1.** leaping, jumping **2.** the mode of transportation of sediments bouncing along a surface, e.g. of material by a river, whereby particles such as small pebbles make intermittent leaps from the bed of the stream. BED LOAD **3.** the similar movement of grains of sand propelled by the wind in a hot desert.

salt dome, salt plug an almost circular mass of ROCK SALT or of other salt forced upwards from a great depth in the earth's crust and, being plastic under pressure, squeezed towards the weak part of the sedimentary cover, thus often topped by limestone cap rock. Oil and gas fields are in many cases associated with salt domes, which extend to great depths in the earth's crust.

salt flat a stretch of unbroken, salt-encrusted horizontal land, the bed of a former salt lake, now dried out, usually permanently but sometimes only temporarily.

saltings a slightly elevated natural area of salt marsh, with muddy channels, supporting HALOPHYTES, and covered by the sea at high water. The distinction is not always made between saltings and SALT MARSH.

salt marsh a natural coastal marsh, supporting HALOPHYTES, regularly covered by the sea at high water. Some authors use the term to include the elevated area (SALTING), others exclude it. Salt marshes may be enclosed for grazing, for LAND RECLAMATION, for the recovery of salt, and when protected in this way the term sea-marsh is sometimes applied.

salt mine a MINE where a natural deposit of rock salt is worked. SALT-2.

salt pan **1.** a small undrained natural basin in which water evaporates, leaving a deposit of salt **2.** a shallow vessel in which salt water accumulates and from which SALT-2 is obtained by evaporation.

samiel KHAMSIN.

sample **1.** a small part taken from the whole by which the characteristics of the whole can be deduced **2.** an individual portion, a specimen, by which the quality of more of the same sort can be judged **3.** in statistics, a part of a POPULATION-4 or a subset from a set of units, deliberately selected with the object of investigating the properties of the parent POPULATION-4 or set. RANDOM SAMPLE, SAMPLING, STRATIFICATION.

sampling **1.** in general, the judging of the quality etc. of the whole by examining a part **2.** in statistics, the process of selecting a part or a subset in order to judge the quality, characteristics etc. of the whole by investigating the properties of the part. QUADRAT, RANDOM SAMPLE; and SAMPLE-2.

sampling error in statistics, the difference between a POPULATION-4 value and an estimate of it derived from a SAMPLE-3. No sample, however carefully selected, can be a perfect representation of the population from which it is drawn; but if the sample units are selected at random from the population (RANDOM SAMPLE) the sampling error can be calculated. The greater the precision of the sample, the less the sampling error. Sampling error does not include errors due to imperfect selection, BIAS-2 in response or estimation, mistakes in observation and recording etc. Errors arising after the sampling has been done are termed non-sampling errors. RANDOM SAMPLING, RANDOM SAMPLING ERROR.

samun, samoon (Iran) a warm, dry FOHN-like wind in Iran, descending from the mountains of Kurdistan; not to be confused with SIMOOM, SIMOON.

sand 1. comminuted rock or mineral fragments of small size commonly, but not necessarily, SILICEOUS-1, ranging from 2 to 0.2 mm diameter, classification internationally agreed as coarse (2 to 0.2 mm), fine (0.02 to 0.002 mm) **2.** a soil of which 90 per cent or more is sand. ARENACEOUS-1, GRADED SEDIMENTS.

sandbank an accumulation of sand, piled up by the action of waves or currents, occurring in a river (POINT BAR) or by the sea and exposed at low water.

sand drift a formation of blown sand, occurring in the lee of a gap between two obstacles, caused by funnelling of the wind or by the concentration of the sand stream on the windward side from a broad to a narrower front.

sand-dune a general term for a mound or ridge of loose, well sorted sand, piled up by the action of wind on sea coasts or in hot deserts. DUNE.

sandplain (Australia) a large, sand-covered plain of uncertain origin, a term used especially in western Australia.

sandstone a porous, ARENACEOUS SEDIMENTARY ROCK, widespread and laid down throughout geological time, formed mainly from rounded grains of QUARTZ and various minerals, varying considerably in colour, laid down in shallow seas, estuaries and deltas, along shallow coasts, in hot deserts, to be consolidated, compacted, and cemented by such substances as CLAY or SILICA which affect the colour of the sandstone and can be used as criteria in classification, e.g. calcareous, siliceous, ferruginous, or dolomitic sandstone. CAMBRIAN, NEW RED SANDSTONE, OLD RED SANDSTONE.

sandstorm a phenomenon in arid or semi-arid regions caused by a very turbulent wind which, passing over sandy soil, lifts and carries clouds of sand for a relatively short distance and rarely above a height of 15 to 30 m (50 to 100 ft). DUST-STORM.

sandur (Icelandic) **1.** a general term for sandy ground, sand flat, sand bank **2.** an alluvial outwash sand plain (OUTWASH APRON) formed by glacier streams flowing from a glacier edge to the sea.

Santa Ana, Santa Anna a hot, dry wind, commonly dust-laden, blowing from the north and northeast down the Sierra Nevada and over the south Californian deserts. It blows mainly in winter, but also occurs in spring, when it may damage the blossom or young fruit of fruit trees.

sapropel sludge or mud which collects in swamps or shallow marine basins, rich in organic matter, formed by slow ANAEROBIC decomposition of remains of small organisms, e.g. of DIATOMS and PLANKTON, which, if compressed by accumulated sediments, may form PETROLEUM compounds.

saprophyte an organism, usually a plant, obtaining organic matter in solution from decaying or decayed organic matter, e.g. a fungus living on dead wood; or yeasts, which include among their activities the production of alcohol. EPIPHYTE, PARASITE.

saqia (Sudan: Arabic) a simple animal-powered mechanical device, also known as the Persian wheel, used to raise water from a river to irrigate the land above. Types vary, but a common, simple one, resembling the hand-powered SHADUF, incorporates a bucket suspended from the end of a pole, the pole pivoting on the top of a support mounted vertically on the ground and attached to a system of cog-wheels operated by a circling animal. The bucket is dipped in the water and, when full, swung round by the cog-wheel system to be tipped and emptied, usually into a channel carrying water to the cultivated land. In the Persian wheel a series of buckets is fixed round the circumference of a vertically mounted wheel, similarly operated by a cog-wheel, animal-powered system.

sargasso gulfweed, a tropical brown seaweed which gives its name to the calm Sargasso sea, characterized by the mass of sargasso which floats on it, supporting a variety of marine organisms, some of which are unique and peculiar to it.

sarsen a mass of hard SANDSTONE, varying in constituents, especially one formed in the EOCENE, left as a residual mass on the surface of southern England. It is widely used as a building stone, and in prehistoric times in megalithic monuments. MEGALITH.

sastrugi, zastrugi (Russian pl.) preferred spelling zastrugi, wavelike ridges of hard snow formed by the action of wind carrying ice particles, occurring on a level surface (e.g. of a snow field or ice field), the axes of the ridges lying at right angles to the wind.

satellite **1.** a natural or artificial celestial body constrained by GRAVITATION and moving in ORBIT around another more massive heavenly body, e.g. one of the artificial satellites in orbit around the earth, particularly one used in REMOTE SENSING **2.** a state which depends economically and politically on another, more powerful, state.

satellite town a self-contained town, i.e. with its own industry, etc., especially a NEW TOWN, in some cases in the style of a GARDEN CITY, associated with a major city with which it has good communications.

saturated *adj.* applied to **1.** a SOLUTION having the highest possible amount of a SOLUTE in a specified amount of the SOLVENT at a given temperature **2.** the ATMOSPHERE-1 when it cannot hold any more WATER VAPOUR, i.e. when the number of molecules of water going in coincides with the number going out. If cooling occurs at that stage CONDENSATION results, giving MIST, CLOUD or RAIN **3.** a rock holding in its interstices the maximum possible amount of water. WATER TABLE **4.** a rock having the maximum possible amount of combined SILICA (if the rock is oversaturated the excess silica occurs as free QUARTZ). SATURATION LEVEL.

saturated (or wet or moist) adiabatic lapse rate the rate of loss of TEMPERATURE-2 with increasing height occurring in a moist or SATURATED-2 body of air as it ascends ADIABATICALLY. The rate is governed by the WATER VAPOUR content (itself governed by temperature) because latent heat is given up as CONDENSATION takes place (the higher the temperature the greater the content of water vapour, the greater the release of latent heat). Thus the rate can vary between 0.4°C and 0.9°C per 100 m (20°F and 5.3°F per 1000 ft). CONDITIONAL INSTABILITY,

DRY ADIABATIC LAPSE RATE, ENVIRONMENTAL LAPSE RATE, LAPSE RATE, NEUTRAL STABILITY.

saturation the act of completely filling, or of completely satisfying, or the result of those acts. SATURATED.

saturation level, saturation point, saturation stage the level or point or stage at which a SOLUTION, the ATMOSPHERE-1, or a substance, or an area etc. reaches a stage of SATURATION, e.g. when the POPULATION-1,3 of an area equals its CARRYING CAPACITY-1,2,3,5, or when the DIFFUSION-1 of INNOVATION is complete.

savanna a term with a wide variety of applications, but best restricted to the natural, open, tropical grassland with scattered trees and bushes (mainly XEROPHYTES) covering vast areas in Africa, South America and northern Australia between the EQUATORIAL FORESTS and the hot deserts, a region with its particular soil conditions and a regular climatic regime. Rain falls in the hot summer and leads to a sudden luxuriant growth, but this withers in the drying winds of the winter with its low rainfall, or is scorched in frequent extensive fires. CAMPO, LLANO.

savanna woodland park-like woodland with xerophytic undergrowth. PARKLAND-2, XEROPHYTE.

scabland a landscape with flat-topped hills or plateaus, bare or covered with a thin soil consisting of angular debris formed in situ and supporting a sparse vegetation, formed by the glacial erosion of a hard basalt surface, occurring particularly in northwestern USA. Compare BADLANDS, associated with soft sediments.

scalded flat 1. in Australia, a low plain with soil impregnated with salt **2.** in soil science, specifically a red-brown soil which has lost part of its A HORIZON.

scale 1. a level of representation of reality, the proportion of a representation to the object it represents **2.** linear scale, the indication on a MAP or PLAN-1,2,3, of the ratio between a given distance on the map or plan and the corresponding distance on the earth's surface. This is shown by a graduated line, or by the REPRESENTATIVE FRACTION; or it is expressed in words **3.** an arrangement of marks spaced at intervals to represent a series of numerical values and used in measuring temperature (THERMOMETER), length, angles etc.

scarp, scarp-face, scarp-slope the abrupt, sometimes cliff-like face or slope terminating an elevated surface of low relief, i.e. an ESCARPMENT-2, the steep slope of a CUESTA. Escarpment is the preferred term.

scarp-foot spring a SPRING-2 near the foot of an ESCARPMENT, occurring particularly where chalk, limestone or sandstone overlies clay.

scarpland, scarplands a region characterized by a number of parallel or subparallel scarped ridges separated by VALES, as in the scarplands of England, also termed scarp-and-vale terrain. Fig 42.

scattergram a graph showing the way in which a DEPENDENT VARIABLE, plotted on the ordinate or vertical axis (the *y*-axis), relates to an INDEPENDENT VARIABLE plotted on the abscissa or horizontal axis (the *x*-axis). The scattered points may lie in such a way that they form, for example, a LINEAR-4 or a CURVILINEAR RELATIONSHIP; or they may not form a pattern at all, indicating an absence of any relationship. Scattergrams are therefore

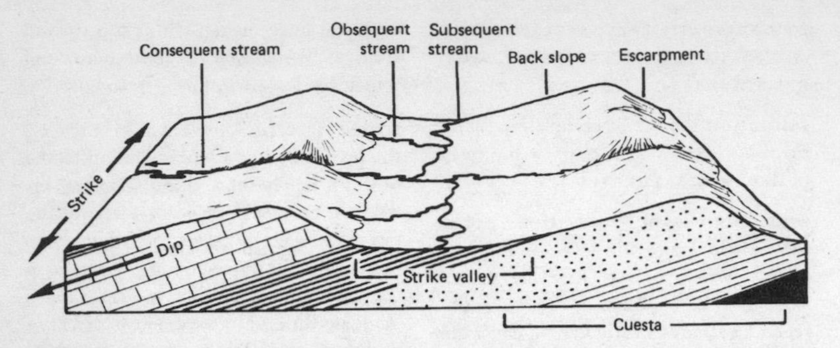

Fig 42 Scarp and vale terrain and characteristic streams

often used to discover if there are relationships which would not be revealed in an ordinary CORRELATION COEFFICIENT.

schist a foliated METAMORPHIC ROCK in which the various minerals have crystallized or been recrystallized into thin layers, lying parallel to each other, a rock which will split into more or less irregular flakes, owing to the occurrence in it of such minerals as MICA. The texture is independent of the BEDDING PLANES of the original rock, the grains are of medium size, not so coarse as those of GNEISS; and the type of schist is distinguished by the dominant mineral, e.g. mica-schist. Nearly all types of SEDIMENTARY and IGNEOUS rocks will become schists if subjected to sufficient heat and pressure.

science (Latin scientia, knowledge) **1.** the condition or fact of knowing **2.** knowledge gained by detailed observation, by DEDUCTION of the laws governing changes and conditions and by testing these deductions by experiment **3.** a branch of study, especially one concerned with facts, principles and methods.

science park RESEARCH AND SCIENCE PARK.

scientific *adj.* **1.** of, pertaining to, used in, SCIENCE-1,2 **2.** of or using methods based on well-established facts and conforming to well-established laws **3.** using knowledge made available by scientists.

scientific law a general statement of fact methodically established (according to the orthodox view) by INDUCTION, on the basis of observation and experiment (EMPIRICISM). Scientific laws are usually rooted in DETERMINISM-1, and are universal in their application in that they usually make statements which cover all members of a particular class of things. But they stray into PROBABILITY in statistics in making statements about a methodically estimated proportion of the class of things under observation. Some philosophers of science do not agree that the method of establishing a scientific law is wholly inductive. They see two phases in the scientific method, the first comprising inspired guesswork, leading to the formation of an HYPOTHESIS; and the second, the confirmation of the hypothesis by induction.

scoria pl. scoriae **1.** a mass of volcanic rock, fine-grained and resembling clinker from a furnace, the holes being caused by the expansion of gases and steam imprisoned in the LAVA and the rapid cooling of its surface. AA **2.** an accumulation of similar clinkery material which has been ejected from a volcano as PYROCLASTS.

Scotch mist mizzle, precipitation resembling mist and drizzle, occurring when CLOUD lies near the ground, especially common in hilly or mountainous areas, e.g. in Scotland, hence the name.

scour the strong, erosive action of a current or flow of water, e.g. of the tide, of a river, in clearing away deposits such as mud and sand; or of boulders frozen into the base of an ice sheet or glacier. ABRASION.

scree **1.** a slope consisting of an accumulation of loose angular rock debris of any size and commonly formed by frost action from the parent rock, lying at a uniform angle (commonly of some 35°) at or near the foot of a steep cliff, rock-buttress, mountain etc. **2.** the angular rock debris itself **3.** a synonym for TALUS. ANGLE OF REPOSE, MASS MOVEMENT, REPOSE SLOPE.

scroll **1.** on a FLOODPLAIN, a narrow stretch of floodplain added to the outer end and downstream side of spurs between enclosed meanders **2.** a type of POINT BAR, sometimes termed a meander bar or meander scroll, a low, narrow ridge running in line with the curve of a MEANDER, formed when the river overflows its banks. OVERBANK STAGE.

scrub **1.** vegetation consisting of dwarf or stunted trees and shrubs, often very thick, XEROPHILOUS and growing on poor soil, or in a semi-arid area, or in an exposed position. It may be a natural CLIMAX or a transitional stage in a PLANT SUCCESSION, leading to woodland **2.** the land covered with this type of vegetation **3.** sometimes applied to RAIN FOREST in Queensland, Australia.

scrub forest FOREST-1 consisting of malformed, small or stunted trees and shrubs.

scrub woodland an open cover of the trees and shrubs characteristic of SCRUB FOREST.

sea **1.** in general, applied to the great body of salt water on the earth's surface (the OCEAN-1), i.e. as opposed to land **2.** one of the smaller bodies of salt water of the ocean with a proper name, e.g. Mediterranean sea, China sea **3.** a large body of inland salt water, e.g. Sea of Aral, Dead Sea, Salton sea.

sea breeze, lake breeze a local breeze (BEAUFORT SCALE) which blows usually during the afternoon from the sea to the land, owing to the differential heating and cooling of land and water. The heating of the land by day causes the ascent of warmed air in a small low pressure area, and cooler air from the sea flows inland for a short distance to take its place. A sea breeze occurs in calm, settled weather where temperature changes are regular, and especially in equatorial latitudes. A similar breeze is associated with large lakes. LAND BREEZE.

sea floor spreading OCEANIC RIDGE, PLATE TECTONICS.

sea-fret (southwestern England) a salt mist moving inland from the sea, often very destructive of vegetation.

sea-level, sea level MEAN SEA-LEVEL.

seamount a topographical feature rising from the ocean floor, an isolated peak, usually a volcano, with a pointed summit (as opposed to a flat-topped GUYOT), the summit usually lying well below the ocean surface, e.g. sometimes 3000 m (nearly 10 000 ft) below.

search space the locations within an area where an individual (or an organization) searches in order to meet a specific need or specific needs (e.g. for housing), based on information from that individual's current AWARENESS SPACE.

SEASAT 1 the first of the NASA SATELLITES-1 to be launched (1978), equipped with RADAR sensors, to obtain information for oceanographic research. LANDSAT, REMOTE SENSING.

seashore 1. a general term applied to land immediately adjoining the sea **2.** the land between the lowest water line of the SPRING TIDE and the highest limit of storm waves.

season 1. a division of the year associated with the duration of daylight and/or characteristic climatic conditions related to changes in the intensity of SOLAR RADIATION brought about by the inclination of the earth's axis to the plane of the ECLIPTIC and the elliptical nature of the ORBIT OF THE EARTH around the sun. Defined astronomically there are four divisions in the year, lasting from EQUINOX to SOLSTICE, from solstice to equinox (sequence repeated). In midlatitudes the seasons are associated with the life cycle of plants, winter (dormant), spring (sowing), summer (growing and ripening), autumn (harvesting), conventionally and arbitrarily defined as WINTER (December to February), SPRING (March to May), SUMMER (June to August), AUTUMN (September to November) in the northern hemisphere, the reverse in the southern. In tropical regions the seasons are linked to rainfall, the year being commonly divided into two, the wet (rainy) season and the dry season; in monsoon regions there are commonly three seasons, termed cold, hot, rainy; in polar regions the periods of change between winter and summer are so brief as to be scarcely noticeable, so there are in effect only two seasons; and in equatorial regions there is little differentiation of season **2.** a period of time most favourable to something, e.g. strawberry season, when that fruit is at its best; or to some activity, e.g. the football season.

seaway 1. in general, a way across the open ocean used regularly by shipping **2.** a SHIP CANAL large enough for ocean-going vessels, notably the St Lawrence Seaway in Canada.

seaweed any marine plant, but especially the red, brown, or green ALGAE living in or by the sea, many edible and used directly as human food, particularly in Japan and China; or as a fertilizer; or as a source of useful substances (e.g. iodine or the gelatinous agar-agar, used in making jellies and as the base of media for bacteriological culture).

sebka (north Africa: Arabic) a salt-encrusted mud flat, a closed depression in an arid area which becomes temporarily marshy after the (rare) rain. PLAYA.

secondary consequent stream a tributary to a SUBSEQUENT STREAM which, although formed after the subsequent stream, flows parallel to the original CONSEQUENT STREAM.

secondary depression in meteorology, a relatively small area of low atmospheric pressure associated with a main primary DEPRESSION-3, encircling the latter in an anti-clockwise direction in the northern hemisphere (clockwise in the southern) as it moves along its course. It may be linked to the primary depression, appearing in the isobars (ISO-) as a protuberance, or it may be self-contained, with closed isobars; in either case its pressure may be lower than that of the primary, which it may eventually absorb.

secondary industry, secondary activity, secondary sector MANUFACTURING INDUSTRY, i.e. INDUSTRY concerned with transforming material provided by PRIMARY INDUSTRY into something more directly useful to people, e.g. manufactured goods, construction work, electric power production (but some countries include the last in primary industry). QUATERNARY INDUSTRY, TERTIARY INDUSTRY.

secondary labour market PRIMARY LABOUR MARKET.

secondary production in ecology, the energy lost in the RESPIRATION of the consumer levels in a FOOD CHAIN. PRIMARY PRODUCTION, PRODUCTION.

secondary soil a transported soil, soil formed on transported material, as opposed to a RESIDUAL SOIL. SEDENTARY SOIL.

secondary vegetation a general term applied to the natural plant-cover growing on land that was once cleared of vegetation, used for a while, and then abandoned. NATURAL VEGETATION, SEMI-NATURAL VEGETATION.

secondary wave of EARTHQUAKE. SHAKE WAVE.

second home a dwelling of a freeholder or leaseholder (LAND TENURE) whose usual, commonly larger, residence is elsewhere, and who occupies the second home at weekends and/or holiday periods.

Second World the countries which adopted a centrally planned communist system after the Second World War, including the countries of eastern Europe linked economically with the USSR. FIRST WORLD, THIRD WORLD.

section **1.** in geometry, the plane figure resulting from the cutting of a solid by a plane; hence **2.** the formation revealed by a cut, or representation of a cut, made vertically through a landform, rock or soil so as to show the surface and subsurface layers, e.g. in a geological section, the surface layer and the underlying strata. In that example the representation of such a cut may be small scale and generalized (diagrammatic section), or have an accurate surface profile with the strata in diagrammatic form (semi-diagrammatic). A totally accurate representation, without exaggeration of the vertical scale, would give a misleading impression of the dip of the strata (VERTICAL EXAGGERATION). CROSS SECTION, SOIL PROFILE.

sector model sectoral model, of land use, a MODEL-2 developed on the assumptions that the arrangement of routes radiating from a city centre conditions the structure of the city and its growth; and that differences in accessibility between the radial routes lead to variations in land value and consequently of land use in the sectors created by these routes, the outer arc of each sector tending to repeat the pattern of its earlier growth.

secular *adj.* **1.** of a change or event occurring very rarely, e.g. once in a century or other very long period of time **2.** continuing, lasting over such a long period **3.** concerned with temporal, worldly matters rather than with religion.

sedentary *adj.* being established in, staying in, one place, i.e. not migratory, not transported.

sedentary agriculture farming as practised by a settled farmer in one place, as distinct from SHIFTING CULTIVATION.

sedentary soil a soil formed from the decay and decomposition of the solid rocks on which it lies. SECONDARY SOIL.

sedge any member of Cyperaceae, a very large family of coarse grass-like herbs, usually with solid, three-sided stems, commonly growing in wet places, although some are XEROPHILOUS. BOG PEAT.

sediment 1. matter which, owing to its greater density, naturally sinks by gravitation to the bottom of any undisturbed liquid with which it was formerly mixed, e.g. as in a SOLUTION (COLLOID) **2.** in geology, unconsolidated particles or grains of rocks deposited by river, ocean, ice, wind. GRADED SEDIMENTS, SEDIMENTARY ROCKS, SEDIMENTATION.

sedimentary rock a rock consisting of material derived from pre-existing rocks (i.e. from SEDIMENTS-2) or from organic debris, laid down in layers, in some cases with FOSSILS, some being consolidated (e.g. by COMPACTION-1), others unconsolidated. They may be mechanically formed (CLASTIC rock) and ARENACEOUS (e.g. SANDSTONE), ARGILLACEOUS (e.g. SHALE), or of larger-sized grains (e.g. CONGLOMERATE). They may be organically formed and CALCAREOUS, FERRUGINOUS, SILICEOUS-1 or CARBONACEOUS (e.g. the COALS). They may be formed by chemical processes (e.g. FLINT). Those formed by drying-up include ROCK SALT. The layers vary greatly in thickness and may lie horizontally or at an angle; considerable tilting reveals that they were disturbed after the period of deposition. In all sequences of sedimentary rocks the oldest bed lies at the bottom, the youngest at the top, unless the order is upset by folding, faulting or other disturbance. LAW OF SUPERPOSITION.

sedimentation 1. the downward movement of finely divided solid particles through a fluid under the influence of GRAVITATION-1 **2.** the act or process of deposition, of settling as a SEDIMENT-2.

seed bed an industrial area near the CENTRAL BUSINESS DISTRICT where rents are low and premises cheap and easily converted to small factories, conditions which favour the setting-up of new small manufacturing enterprises. Those that flourish transplant themselves to larger premises farther from the central area. They and those that fail are soon replaced by new, hopeful, industrialists. There is thus a rapid turnover of manufacturing establishments in the seed bed area; but such areas are squeezed out by policies of URBAN RENEWAL and CENTRALIZATION-1.

seeding of clouds the dropping of chemical particles (e.g. solid carbon dioxide, silver iodine) from aircraft on to clouds in order to stimulate CONDENSATION and lead to rainfall.

seepage the very slow percolation or oozing out of a fluid through a porous body (POROSITY), or along a fault or joint-plane, e.g. the slight oozing out of PETROLEUM on the ground surface, an important indication of the presence of oil in rocks below; or the percolation of surface water into the soil; or the slow oozing out of ground water at the surface when the flow of water is insufficient and the pressure not high enough to form a SPRING-2.

segregation 1. separation of sub-groups from the major group in a SOCIETY-2,3, especially involving the establishment by law or custom of separate facilities for different social or ethnic groups, as for whites and blacks in the former Republic of South Africa. APARTHEID **2.** the process by which individuals and groups settle in areas already housing people with tastes, preferences, or social characteristics similar to their own. GHETTO.

seiche a periodic or occasional, brief, undulation of the water in a restricted area, e.g. in a lake, estuary or bay, apparently caused by abrupt changes in ATMOSPHERIC PRESSURE, or by wind, or EARTHQUAKE.

seif-dune (Arabic seif, sword) a SAND DUNE piled up longitudinally as a steep-sided ridge, sometimes stretching over many kilometres, and lying parallel to the direction of the prevailing wind.

seine a large fishing net held vertically in the water by floats fixed to one edge and weights to the other. The fish (especially DEMERSAL fish) are trapped as the ends are drawn together.

seismic *adj.* relating to, characteristic of, produced by, movement within the earth, e.g. an EARTHQUAKE.

seismic focus, seismic origin the place in the earth's crust, under the surface, from which an EARTHQUAKE shock originates. DEEP FOCUS, EPICENTRE.

seismic wave **1.** a shock wave generated by an underground explosion or EARTHQUAKE **2.** a TSUNAMI.

seismograph a scientific instrument used in recording the duration, magnitude and direction (horizontal and vertical) of earth tremors, natural (EARTHQUAKE) or ARTIFICIAL.

seismology the scientific study of EARTHQUAKES and of other movements of the solid earth, including earth tremors produced artificially. SEISMOGRAPH.

selva (South America: Portuguese and Spanish) **1.** dense EQUATORIAL FOREST of the Amazon region **2.** such equatorial forest growing elsewhere **3.** the Amazon region in which it grows.

semi-arid climate, semi-desert the transitional zone lying between SAVANNA and the true hot desert (SAHEL), or between the hot desert and a Mediterranean climatic region, characteristically supporting patchy XEROPHILOUS vegetation.

semi-natural vegetation vegetation not actually planted by human hand, but resulting directly or indirectly from the activities of human beings or their livestock, e.g. SECONDARY VEGETATION. NATURAL VEGETATION.

semiotics the theory of symbols, the systems of signs by which an individual or a culture expresses meaning.

senile river a RIVER system in old age in the CYCLE OF EROSION, i.e. when all slopes have been worn down and the products of decomposition accumulated, covering any irregularities in the surface, so that the FLOODPLAIN may become a marsh without regular outflow: thus the river lies on a completely formed PENEPLAIN.

sensible temperature cold or heat as felt by the human body, depending not only on actual temperature but also on RELATIVE HUMIDITY and wind. LATENT HEAT.

sensor any apparatus used to detect variations in electromagnetic radiation, acoustic energy or force fields associated with gravity and magnetism at a distance from their source, especially such a device used to gather information about distant objects or phenomena in the ATMOSPHERE-1, BIOSPHERE, HYDROSPHERE, LITHOSPHERE (as in REMOTE SENSING). ELECTROMAGNETIC SPECTRUM, REFLECTION.

separatism the demand of a particular group of people that a particular piece of territory should become separate in territory and political sovereignty from the state within which it lies. PLURAL SOCIETY.

sérac (French) a pinnacle of ice formed in the part of a glacier where crevasses intersect, usually at the point where the glacier breaks on reaching a steep slope.

sere **1.** a developmental series of plant communities resulting from the process of SUCCESSION-2 **2.** any stage in a plant SUCCESSION-2.

sericulture the rearing of silkworms and the production of raw SILK.

service industry TERTIARY INDUSTRY.

services **1.** the products of the paid activities of an employee or professional person **2.** in Britain, utilities (gas, water, electricity) as supplied to a consumer **3.** the product of human activity intended to satisfy a human need or needs but not constituting an item of goods.

sesquioxide an oxide with three oxygen and two metallic elements, in soil mainly iron (Fe_2O_3) and alumina (Al_2O_3).

seston the bioseston (living organisms) and abioseston (inanimate matter, also termed tripton) swimming or floating in a body water.

set aside CAP.

settlement **1.** any form of human habitation, even a single dwelling, although the term is usually applied to a group of dwellings **2.** the act of peopling a formerly uninhabited or under-populated land **3.** a decision or choice made to put an end to a controversy.

settlement hierarchy the ranking of urban places with their associated trade areas, graded according to their functional importance. CENTRAL PLACE HIERARCHY, CENTRAL PLACE THEORY.

Seven Seas the Arctic, Antarctic, North and South Atlantic, North and South Pacific and Indian oceans (OCEAN). In classical literature the term was applied to the seven supposed salt water lagoons on the east coast of Italy, including the lagoon of Venice (cut off from the Adriatic by the LIDO).

Seventh Approximation SOIL CLASSIFICATION.

shade temperature the temperature shown on a THERMOMETER sheltered from the sun's rays, from radiation from surrounding objects (including the ground), from strong wind, and from precipitation, best achieved by putting it in a STEVENSON SCREEN. In climatic statistics the temperature given is shade temperature, unless sun temperature is specified.

shadow effect the effect of a large, well-served urban centre on the transport services of a nearby small centre, the smaller place being relatively ill-provided with direct services. ACCESSIBILITY, INTERVENING LOCATION EFFECT.

shādūf (Arabic) a simple hand-operated irrigation device used for raising water from a river or shallow well. It consists of a long pole with a bucket suspended from a rod, chain or rope at one end and a weight (a stone or pieces of iron etc.) at the other. The pole is mounted and pivots on a vertical support which is fixed firmly in the ground. The pivot point is nearer to the weighted than the bucket end of the pole; the weight acts as a counterbalance to the bucket. The bucket is dipped in the water and the counterpoising weight takes over most of the effort of raising and swinging it round so that the bucket may be emptied into a trough or into a channel by which the water is carried to cultivated land. ARCHIMEDES' SCREW, SAQIA.

shake wave S-wave, the secondary wave (also termed shear wave or transverse wave) produced by an EARTHQUAKE, a body wave in the earth which passes through solids but not liquids, resembling a light wave in that it is a transverse wave, i.e. it displaces particles at right angles to the direction of its own movement. LONGITUDINAL WAVE, MOHOROVICIC DISCONTINUITY, PUSH WAVE, TRANSVERSE WAVE, WAVE.

shale a fine-grained ARGILLACEOUS SEDIMENTARY ROCK, formed from particles of CLAY minerals compressed by overlying rocks, very finely laminated (LAMINATION), in the direction of the BEDROCK, the thin layers easily splitting apart and disintegrating (compare SLATE). CAMBRIAN, SHALE OIL.

shale oil an oil distilled from bituminous SHALE.

shallows an area of little depth of water in a sea, lake, river. CHANNEL-3.

shanty-town, squatter settlement a settlement, lacking services, which consists of a collection of small, crude shacks made of discarded materials and serving as habitations for poor people on the outskirts of towns, especially in South America and parts of Africa, variously termed (South America) FAVELA, rancho, barriades or villas miserias; (Central America) barrio; (Asia) BUSTI or kampong; (Africa) bidonville or shanty-town.

sharecropping an agricultural tenancy system in which the tenant renders rent to the landlord in the form of produce rather than cash. The systems vary, but usually the landlord in addition to providing and being responsible for the land, buildings, drainage and farm roads also provides the sharecropper (the tenant) with machinery, stock, seeds and fertilizer. In return the landlord receives an agreed proportion of the farm produce. LAND TENURE.

sharp sand SAND-1 with angular, as opposed to rounded, GRAINS-4.

shatter belt, shatter-belt a zone of movement in the earth's crust where rocks have been broken into angular fragments, i.e. into FAULT-BRECCIA. It occurs where FAULTS are ragged and extensive, so that a line of weakness develops in the crust, along which WEATHERING and EROSION are facilitated.

shattering a form of physical WEATHERING in which strong mechanical stresses produce fresh fractures in the rocks.

shear 1. in physics, STRESS applied to a body but along one face only of the body (termed shearing stress), or the STRAIN-2 produced by shearing stress, producing a change of shape but not of volume **2.** in geology, a change in the direction of a STRATUM due to lateral pressure. SHEARING.

shearing in geology, the bending, twisting or drawing out, sometimes accompanied by crushing or shattering, of a rock near a FAULT or THRUST-PLANE, due to STRESS with resultant slipping (hence shear-fault, shear-plane, shear-cleavage). The volume of the rock does not alter, but its form does as the two neighbouring parts slide past each other, in some cases causing crushing and shattering along the line of SHEAR-1.

shear wave SHAKE WAVE.

sheep a gregarious animal, domesticated and crossbred for a very long time to produce animals suitable for specific purposes, i.e. for supplying MEAT OR WOOL of varying quality, and milk (especially in the Mediterranean region) for CHEESE. Sheep flourish on land poorer than that required for cattle provided it is not too wet underfoot, those bred for meat needing better fodder than those kept for wool. The fine wool breeds (MERINO) thrive in dry, warm climates, those with medium quality wool in cooler midlatitudes, where they provide meat as well as wool.

sheet erosion very slow EROSION of soil from an extensive, flat, gently sloping area, the result of RUNOFF, most likely to occur in areas where the soil layer is thin. SOIL EROSION.

sheetflood, sheetflow an unhampered, broad expanse of water derived from PRECIPITATION flowing down a slope, occurring where CHANNELS are absent or when the RUNOFF is so great and fast that the existing channels, RILLS, etc. cannot carry it, and thus it overflows. OVERLAND FLOW.

sheeting the splitting away of shells of rock from the upper surface of a massive rock (particularly an IGNEOUS ROCK) resulting from the expansion of the rock by the release of pressure (DILATATION); not to be confused with EXFOLIATION.

sheet lightning a discharge of LIGHTNING within a cloud or between clouds, the brilliance of the flash being diffused by the clouds so that it appears as a sheet of light.

sheet metal metal flattened out to form a thin sheet.

shell sand beach sand consisting mainly of comminuted shell fragments, and therefore highly CALCAREOUS, e.g. MACHAIR.

shelterbelt a windbreak, usually a stand of trees planted to act as a screen against the wind, especially in areas subject to wind EROSION.

shield in geology, a very large rigid mass of PRECAMBRIAN rock, forming a major continental block, relatively stable over a long period of geological time, disturbed only by some slight WARPING, e.g. the Laurentian Shield. GLINT-LINE.

shield volcano a volcano shaped like a shield, i.e. a broad dome, the diameter of the base being large, the angle of slope small, basic LAVA forming the cone, e.g. Mauna Loa in Hawaii. BASALTIC LAVA, HAWAIIAN VOLCANIC ERUPTION, SINK.

shifting cultivation loosely applied to any of the many systems of cultivation where land is cropped and after a few years, with the initial fertility exhausted, abandoned in favour of a new patch. A distinction can be made between the true shifting cultivation of nomadic peoples who do not practise a LAND ROTATION but move on when the soil fertility is exhausted; a regular system of land rotation or BUSH FALLOWING practised by people who usually have a fixed central village; and shifting cultivation associated with certain cash crops whereby land is abandoned when yields begin to drop below an economic level. There are some 150 or so vernacular terms applied to shifting cultivation. FARMING, LAND TENURE, SWIDDEN FARMING.

Shimbel index a measure of the accessibility of a NODE-2 in a NETWORK-2. The shortest-path LINKS between each node and all other nodes in the network are recorded on a MATRIX-5; the lower the index value the better the accessibility.

shingle an accumulation of coarse stones, rounded by water. The term is usually restricted to cover only such an accumulation on a BEACH.

ship canal an artificial waterway large enough for the passage of ocean-going vessels, e.g. the Manchester Ship Canal. SEAWAY.

shipping tonnage a measure of capacity of ships, calculated as follows: **Gross tonnage**, the capacity of the permanently enclosed space between the frame of the vessel and the deck together with any closed-in space above the deck, 2.83 cu m (100 cu ft) being reckoned as 1 ton. **Net or registered tonnage**, gross tonnage less the space occupied by engines, gear, crew's and officers' quarters, i.e. the space available for cargo and passengers, calculated on the same basis as gross tonnage. Dues are usually paid on net or registered tonnage. **Cargo tonnage**, the weight of the cargo carried, calculated by volume; in UK 1.19 cu m (42 cu ft), in USA 1.1 cu m (40 cu ft), being equal to 1 ton. **Deadweight tonnage**, dwt, the total load carried at maximum loadline, including the total weight of cargo, fuel and passengers etc. measured in tonnes. **Displacement tonnage**, the weight of water displaced by the vessel when fully laden, i.e. the weight of the vessel and its contents when calculated on the basis that 0.99 cu m (35 cu ft) of water equals 1 ton. As a rough conversion for a mixed fleet, consisting of tankers and cargo vessels, gross registered tons plus 50 per cent equals deadweight tons; for giant tankers the dwt may be 120 per cent higher than gross tonnage.

shoal 1. a shallow part of a river, sea, lake **2.** an accumulation of sand, mud, pebbles creating such shallow water and in many cases dangerous to navigation **3.** a group of fish.

shoddy 1. a yarn made from the shredded and reconstructed fibre of fabric or fabrics which have already been used **2.** a fabric made from such reconstituted yarn. RECYCLING.

shore 1. loosely applied to the land immediately bordering the sea or other large expanse of water **2.** the meeting of sea and land considered as a boundary of the sea, thus the land as seen from the sea **3.** the area between the lowest water of a SPRING TIDE and the highest point reached by unusually strong waves in a STORM **4.** in law, the ground between the ordinary low and high water marks (LOW WATER, HIGH WATER). BEACH, COAST, FORESHORE, SHORELINE.

shoreline the line where the SHORE meets the water, an imprecise term sometimes regarded as synonymous with COASTLINE (equally imprecise), sometimes applied to the line reached by an ordinary low tide. There is a tendency to regard coastline as the landward limit fixed in position for considerable periods of time, shoreline as a moving phenomenon.

shott (north Africa: Arabic) **1.** a fluctuating shallow brackish or saltwater lake in north Africa, especially in Tunisia and Algeria, dry for much of the year, water-filled in winter **2.** the depression holding such a lake. PLAYA, SALINA.

shoulder 1. a rounded spur on a mountainside **2.** a BENCH on the side of a valley, most likely to occur on the side of a valley deepened by a glacier at the point where the gentle slope of the upper part (unaffected by glacial erosion) changes abruptly to the steep slope of the inner, glaciated valley side. ALP, U-SHAPED VALLEY.

shower a fall of RAIN, HAIL, SLEET or SNOW of brief duration.

shrub a PERENNIAL plant with many persistent woody stems branching from or near the base. HERB-1, TREE.

SI, Système Internationale d'Unités a simplified metric system based on seven basic units, agreed in 1960 by an international committee and now adopted by most countries using the metric system. The seven basic units, from which all other SI units are derived, are the metre (m), kilogram (kg), second (s), ampere (A), the kelvin (K), mole (mol) and candela (cd). Multiples and submultiples preferably separated by the factor of 1000 are used with these basic units, i.e. 10^{12} (prefix tera-, T), 10^{9} (giga-, G), 10^{6} (mega-, M); 10^{3} (kilo-, k); 10^{-3} (milli-, m); 10^{-6} (micro-, µ); 10^{-9} (nano-, n); 10^{-12} (pico-, p); 10^{-15} (femto-, f); 10^{-18} (atto-, a). CENTI-, HECTO-, JOULE, KILO-, MILLI-, NEWTON, PASCAL.

sial *si*lica and *al*umina, granitic rocks (GRANITE-1) of the surface of the earth's continental crust (PLATE TECTONICS), composed mainly of SILICA and ALUMINA, light in colour and density (between 2.65 and 2.70). There is a tendency for the term to be replaced by the less specific term upper crust. GEOTHERMAL GRADIENT, ISOSTASY, SIMA.

Siberian high a persistent anticyclone situated over north central Asia in winter.

sidewalk farmer in USA, a person who lives in an urban area and cultivates land distant in a rural area, the farm equipment being housed on the farm land. The crop raised is usually one which needs little attention in growth, e.g. a CEREAL. SUITCASE FARMER.

sierra (Spanish; Portuguese serra) a high range of mountains with jagged peaks resembling the teeth of a saw. The term was originally applied to such mountains in Spain and Spanish-speaking South America, but it is now extended and applied in Spanish to almost any high mountain range; and in English generally to 'the mountains' or a mountain region.

sieve map, sieve method a series of maps drawn on transparent material (OVERLAY), each showing the distribution of a selected factor. By superimposing the transparencies the factors wanted or not wanted for a particular purpose can be 'sieved out'.

signature the unique pattern of wavebands (ELECTROMAGNETIC SPECTRUM) peculiar to and emitted by an object on the earth's surface. REMOTE SENSING.

significance test a statistic calculated to indicate the likelihood that a characteristic in a SAMPLE-3 reflects accurately the characteristic of the parent POPULATION-4 of that sample, and that it has not occurred by chance in the sampling, e.g. CHI-SQUARED TEST.

significant *adj.* in statistics, unlikely to have occurred by chance.

silage green fodder (e.g. grass, clover, alfalfa, maize plant) packed into a silo, usually with molasses, fermented by ANAEROBIC bacteria to preserve it, and cut into blocks for animal feed when needed.

silica silicon dioxide **1.** the mineral of that composition, e.g. QUARTZ **2.** the silicate mineral content of a rock, commonly expressed chemically as the percentage of silica by weight. Silicate minerals are the largest group of compounds in the earth's crust. SILICATES.

silicate magma MAGMA from which silicate minerals are formed. SILICA-2, SILICATES.

silicates silicate minerals, a group of minerals based around the highly stable SiO_4, constituting (with the SILICA-1 group) some 95 per cent of the earth's crust, and including CLAY minerals, FELDSPAR, QUARTZ etc. as members. SILICA, SILICON.

siliceous *adj.* **1.** of, pertaining to, containing, resembling SILICA-1 **2.** growing in or needing a soil containing silica.

silicon a nonmetallic element occurring, as a brown powder or dark grey crystals, abundantly in nature, always in compounds. It is the second main element in the earth's crust (the first being OXYGEN), comprising by volume some 28 per cent (SIAL). Combined with oxygen it forms SILICA and, with various other oxides, a large group of rocks termed SILICATES. It is used in glass-making, the manufacture of very hard alloys and, in SILICONE compounds, in lacquers, lubricants, water-repellent finishes etc.

silicone any of the large class of synthetic SILICON-containing compounds in which the atoms of SILICON are held together by bonds to OXYGEN atoms which act as bridges.

silk **1.** a fine, strong, protein, thread-like structure secreted by some insects **2.** such a substance secreted by the caterpillar (silkworm) of the moth *Bombyx mori*, so called because it feeds on the leaves of the white mulberry, *Morus alba*. SERICULTURE, SYNTHETIC FIBRE.

sill **1.** an INTRUSION of IGNEOUS ROCK of tabular form, as when a very fluid MAGMA is forced between the bedding planes of sedimentary or volcanic formations, i.e. it is CONCORDANT with the STRATA **2.** a submarine ridge between ocean basins, or between a sea and an ocean or, termed submerged sill, near the entrance to a FJORD. Fig 21.

silt fine particles, larger than those of clay, finer than those of fine sand, diameter 0.002 to 0.02 mm, suspended in, carried or deposited by, water. GRADED SEDIMENTS.

Silurian *adj.* of or pertaining to the third period (time) or system (rock) of the PALAEOZOIC era, preceded by the Ordovician, succeeded by the Devonian. GEOLOGICAL TIMESCALE.

silver a white, stable, malleable, ductile metallic ELEMENT-6, a precious metal, some occurring in silver ores, and a little NATIVE-1 in nature, but more as an impurity in lead ores, particularly galena, considered to be argentiferous (silver-bearing) if it has more than 0.1 per cent of silver. A good conductor of heat and electricity, silver is resistant to oxidation. It is used in electrical apparatus, coins, photography, electroplating, backing mirrors, jewellery, silverware etc.

silviculture a branch of the science of forestry concerned with the breeding, development and cultivation of forest trees.

sima *si*lica and *ma*gnesium, BASALTIC rocks composed of SILICA and MAGNESIUM, forming part of the earth's crust, relatively heavier in density than the SIAL of the CONTINENTAL CRUST which overlies it in places. In areas without sial, sima forms most of the ocean floor. There is thus a tendency for the term sima to be replaced by OCEANIC CRUST. ISOSTASY, PLATE TECTONICS.

simoom, simoon (Arabic) a scorching-hot, heavily dust-laden, swirling wind occurring in the hottest months in the northern Sahara, usually associated with the northward passage of a LOW pressure system. It may carry so much dust and sand that visibility is reduced to zero; and it greatly affects the shape of SAND DUNES in its path.

sink, sinkhole **1.** in general, a hollow in which surface water collects and escapes through a shaft, i.e. a SWALLOW HOLE **2.** specifically, a feature characteristic of limestone country (KARST), a closed depression which is dry or through which water seeps downwards, resembling in shape a basin, funnel or cylinder **3.** a large depression in a SHIELD VOLCANO or lava dome formed when the surface has cooled and solidified but subsides as the underlying molten lava flows away.

sinter a chemical deposit of SILICA formed around a GEYSER or hot spring, the material having been previously held in solution in the water (HYDROTHERMAL).

sirocco, scirrocco a very hot south or southeasterly wind, sometimes dust-laden, dry as it blows over north Africa, sometimes humid by the time it meets the south Italian shore, blowing from the Sahara over the Mediterranean to Malta, Sicily, Italy, ahead of a depression moving eastwards over the Mediterranean. It is common in spring, when it may damage crops, especially blossoming vines and olives. KHAMSIN.

site a fixed position where an object, structure or tissue is placed or where something occurs, e.g. the position on the ground of a place, town, building etc. in the past, present or future. SITUATION.

Site of Special Scientific Interest SSSI.

situation **1.** the PLACE, POSITION or LOCATION of something, e.g. a house, a town, in relation to its surroundings or to another thing **2.** a state of affairs. SITE.

skärgård (Swedish) enclosure by a line of skerries. SKERRY, SKERRY-GUARD.

skerry a small islet, sometimes one of a series lying parallel to the main trend of the coast, usually rocky, sometimes composed of morainic material, over which the sea may break at high tide or in stormy weather.

skerry-guard the area of calm water between a line of skerries (SKERRY) and the mainland. The term should not be applied to the line of skerries itself.

skewness in statistics, asymmetry. FREQUENCY CURVE.

sky the atmosphere enveloping the earth, with or without clouds. Its blue appearance if cloudless in the daytime is due to the scattering of sunlight by the molecules of air. At high altitudes the sky appears to be deep blue because there the short waves of the blue-violet end of the spectrum of solar light are easily scattered by the fine molecules of air present.

Skylab a US SATELLITE-1 that no longer exists. It was launched in 1973 with a crew of three, whose members carried out various experiments and had control over the REMOTE SENSING equipment aboard. LANDSAT, SPACE SHUTTLE, SPACE STATION.

slack **1.** a shallow hollow among coastal sand dunes or mud banks **2.** the state of the TIDE when tidal currents are almost still, commonly about high or low water when there is neither EBB nor FLOW **3.** small pieces of coal, refuse coal.

slash **1.** locally in south and southeast USA, a low, wet, swampy, boggy area, overgrown with bushes, cane etc., favourable for the growth of any one of the slash pines, e.g. *Pinus caribaea* yielding gum and turpentine **2.** the debris of felled trees **3.** part of a forest strewn with such debris. SLASH-AND-BURN.

slash-and-burn a method of clearing land, as in SHIFTING CULTIVATION, by felling trees and burning the SLASH-2. SWIDDEN FARMING.

slate a fine-grained, laminated (LAMINATION), dark grey, METAMORPHIC ROCK derived from SHALES or MUDSTONES subjected to pressure by earth movements. It has the property of being fissile into thin slabs (slates) by the development of minerals such as MICA, the thin flakes of which lie at right angles to the pressure in DYNAMIC METAMORPHISM. Thus the splitting is along the lines of CLEAVAGE, independent of the original BEDDING PLANES, differing from the splitting of SHALE, which takes place along the bedding planes. CLAY-SLATE.

sleet PRECIPITATION-1 consisting of snow and rain mixed, or of partially thawed snow. GLAZED FROST.

slide **1.** a mass of rock or earth falling as a whole, rapidly and sometimes catastrophically down a BEDDING PLANE or JOINT through the force of gravity **2.** the mark on the hillside, caused by such a slide. MASS MOVEMENT, ROCK SLIDE, SLUMPING.

slip **1.** in a FAULT, the actual relative movement along the FAULT PLANE, either in the direction of the STRIKE, termed strike-slip; or in the direction of the DIP of the fault plane, termed dip-slip **2.** a LANDSLIDE in which a mass of rock or surface debris moves as a whole down a slip-plane. MASS MOVEMENT, ROTATIONAL SLIP.

slip-face the leeward side of a SAND DUNE, steeper than the windward side from which sand is blown. DUNE, PLINTH.

slip-off slope of a MEANDER, the gentle slope of the spur on the convex (inside) curve opposite the steep bank or RIVER CLIFF on the concave curve (outer side). POINT BAR. Fig 29.

slope **1.** the upward or downward inclination of a natural or artificial surface, a deviation from the perpendicular or horizontal **2.** the degree or nature of such an incline. The study of the development of slopes on the earth's surface is a complex one, theories abound and many specialized terms are in use apart from those that follow. CONSTANT SLOPE, FREE FACE, GRAVITY SLOPE, HALDENHANG, PARALLEL RETREAT OF SLOPE, SLIP-OFF SLOPE, SLOPE ELEMENTS, SLOPE RETREAT, TRANSPORTATION SLOPE, WANING SLOPE, WAXING SLOPE **3.** on a straight line on a graph, the amount by which the dependent variable (on the vertical axis) increases/decreases for each unit of the independent variable (on the horizontal axis), the slope of a regression line for two variables being the regression coefficient. REGRESSION ANALYSIS.

slope elements of hillside slope, the component parts of the hillside slope profile defined as the WAXING SLOPE (relatively concave and at the top), succeeded downwards by the nearly vertical FREE FACE, the CONSTANT SLOPE (rectilinear in profile), and the WANING SLOPE (relatively concave) at the base. STANDARD HILLSLOPE.

slope length the actual length of the surface of a SLOPE-1, from its highest to its lowest point, not the length projected on to a plane, termed the HORIZONTAL EQUIVALENT, which measures less. GRADIENT.

slope retreat the progressive wearing back of a slope profile by erosion.

slum a rundown settlement or part of a settlement, usually in or near an urban area and characterized by dilapidated buildings or shacks (FAVELA), the poverty of its inhabitants, squalor, the presence of refuse, and overpopulation.

slumping the downward, usually distinctly rotational, slipping under gravity of a mass of rock, torn away from its base, over a curved slip fault (SLIP-1), leaving a scar on the slope surface. It commonly occurs where more massive rocks overlie a weaker layer, e.g. limestone over clay, along an escarpment. It is also believed to occur extensively down the steep slope of the CONTINENTAL SLOPE from the CONTINENTAL PLATFORM, termed submarine slumping. MASS MOVEMENT.

slurry 1. very wet, mobile MUD-1,2 **2.** a mixture of water and insoluble matter, e.g. chalk, clay, coal, lime, sometimes transported by pipeline.

small business in Britain, a firm employing between one and 100 persons.

small circle any hypothetical circle on the earth's surface the plane of which does not pass through the earth's centre, in contrast to the plane of a GREAT CIRCLE. Thus all parallels of LATITUDE to north and south of the equator are small circles, decreasing in size polewards; but the EQUATOR itself is a great circle.

small holding, smallholding, small-holding written as two words, with or without a hyphen, applied generally to any small farm or holding. But as one word given a special legal meaning in Britain under the Agriculture Act 1947, i.e. a holding of less than 20.25 ha (50 acres), having a rental value below a certain level.

smelting the extraction of metal from its ore, usually by a heat process which reduces the oxide of the metal with carbon in a furnace, separating out the metal in a molten state; or by CALCINATION of sulphide ores.

smog (term derived from smoke and fog, 1905) **1.** originally applied to thick, yellow RADIATION FOG, injurious to health, occurring over a built-up area where sooty particles from smoky fuels (SMOKE) formed the nuclei for condensation in the atmosphere, and SULPHUR DIOXIDE added to the POLLUTION. A very dense smog in London in 1952 stimulated a campaign for smoke abatement and was so successful that the use of smoky fuels within specified areas was banned under the clean Air Acts of 1956 and 1968 **2.** the term has since been applied to other foggy air pollution, not necessarily caused by smoke but by the pollution of the air by NITROGEN oxides and HYDROCARBONS from the exhausts of motor vehicles, combined with the chemical change brought about by the action of sunlight, as in the common Los Angeles PHOTOCHEMICAL smog.

smoke fine particles suspended in the ATMOSPHERE-1 and carried by air currents, usually consisting of carbon particles formed from incomplete combustion. SMOG.

snout of a GLACIER, the lower extremity of a VALLEY-GLACIER, sometimes partially hidden by morainic material, but commonly featuring a cave from which MELT-WATER flows. MORAINE.

snow PRECIPITATION-1 in the form of delicate, feather-light, hexagonal, variously patterned, individual ice crystals aggregated to form snowflakes. The ice crystals are formed when water vapour in the atmosphere condenses quickly at a temperature below freezing point, does not liquefy but passes immediately to the solid state, the sparkling whiteness releasing LATENT HEAT, causing a rise in air temperature. Sometimes snow melts in descending, to reach the ground surface as rain; it arrives as snow only if the lower atmosphere is cold enough to prevent melting. Air is trapped between the crystals in snowflakes causing internal reflection of light at the crystal surfaces and giving snow its sparkling whiteness. This trapped air, combined with the air between the flakes, makes snow a good insulator, preventing heat loss by radiation from the surfaces on which it collects. Snow can be dry and powdery, and therefore great in volume (as in cold temperatures, e.g. in the Antarctic); so to make for uniformity in meteorological recording, it is collected (usually in a special cylindrical gauge) and melted, the amount being expressed as the rainfall equivalent, and usually added to the precipitation total. AVALANCHE, RED SNOW.

snow avalanche a swift fall of a mass of snow down a slope (AVALANCHE), distinguished as dry, consisting of newly-fallen snow in winter; wet, caused by spring thaw; wind slab, where the surface layer of the snow is compacted and hard.

snowfield a stretch of permanent snow with a relatively level, smooth surface, occurring in shallow depressions in mountainous areas or on high plateaus.

snow-limit the limit north and south from the equator indicating a zone within which no snow falls and stays unmelted, varying with physical conditions (elevation, proximity to the ocean etc.); not in general use as a technical term.

snow-line the variable lowest level on mountains above which snow never completely disappears, considered to be a permanent level (varying with latitude, altitude, temperature, humidity, precipitation, aspect, steepness of slope) if summer warmth does not melt and remove the winter accumulation. The snow-line in winter is commonly lower than this so-called permanent snow-line.

social *adj.* **1.** relating to human SOCIETY-1 **2.** as applied to human beings, of any BEHAVIOUR or attitude that is influenced or created by experience of the behaviour of other people; or any behaviour or attitude directed towards other people.

social area analysis a technique used to link social structure and urban residential patterns. Widely ranging data, e.g. concerning rank in SOCIAL CLASS, occupation, fertility, size of families, racial and ethnic grouping, are analysed and classified in order to make distinctions between small areas within a city. Sometimes termed SOCIAL ECOLOGY.

social capital in Marxism, state expenditure which, by providing resources that firms themselves would otherwise have to provide, contributes to the profitability of the private sector of an economy.

social class a problematic, disputed concept, widely used in the social sciences, applied to a group of people of similar rank or status in a community (CLASS-3), the basis for the grouping being variable, e.g. determined by education, power, income, wealth, prestige, occupation, or relationship to the MEANS OF PRODUCTION. A distinction may be made between social class (distinguished in relation to the means of production) and SOCIAL STATUS (distinguished on the basis of consumption of goods, of particular life-styles). The Registrar General's classification, UK, distinguishes five categories, from higher managerial or professional through skilled manual to unskilled manual workers. But the term social class is often used loosely in the UK to distinguish upper, middle (also stratified) and WORKING CLASSES without any precise definition of the criteria used. In Marxist theory, social class is related to the ownership or non-ownership of the means of production (MODE OF PRODUCTION), a class comprising individuals with a common behavioural pattern, sharing a common relationship to property and power, the classes being the capitalist, the BOURGEOISIE and the PROLETARIAT. SOCIAL STATUS.

social Darwinism the application of Darwin's theory of the survival of the fittest, and evolution, to human societies.

social distance **1.** the voluntary or enforced separation of distinct social groups for most of their activities **2.** the distance as perceived by individuals or small groups between themselves and other individuals or social groups.

social exclusion the separation of individuals or groups of people from the SOCIETY-3 within which they live.

social geography a branch of geography with a variety of approaches but defined broadly as the analysis of social phenomena in space, and also equated with HUMAN GEOGRAPHY, CULTURAL GEOGRAPHY (in USA), and many of the aspects of URBAN GEOGRAPHY, HUMANISTIC GEOGRAPHY and WELFARE GEOGRAPHY. It deals generally with the interrelationship of people with their environment. But SOCIAL-1,2 implies an individual living and functioning with others in a group or a SOCIETY-1,3; thus social geography emphasizes the importance of studies of population, urban and rural settlements, social activities and problems (including such considerations as SOCIAL JUSTICE, SOCIAL NETWORK, SOCIAL STRUCTURE, SOCIAL WELL-BEING).

social housing PUBLIC HOUSING.

social investment in Marxism, state expenditure which acts mainly as CONSTANT CAPITAL during capitalist production, thereby lowering the costs of private sector investment in the MEANS OF PRODUCTION, i.e. raising the productivity of labour.

socialism a political, social and economic concept which takes various forms. Each form, traditionally, is opposed to uncontrolled CAPITALISM, seeks equality of opportunity for each person, advocates that there should be collective ownership of the means of production and control of distribution, and maintains that in return for contributing to the community, the individual should be entitled to receive the care and protection of that community. To generalize, it may be said that Marxian socialism is concerned largely with economic issues, and stresses the importance of communal ownership and control of the means of production, distribution, exchange. Christian socialism, more concerned with social aspects, sees socialism as a way of life. Democratic socialism stresses the political aspect, and compromises in the economic field between state and private enterprise.

social justice in WELFARE GEOGRAPHY, the fair distribution of benefits and burdens among the members of a SOCIETY-1.

socially necessary labour LABOUR THEORY OF VALUE.

social mobility the movement of people between SOCIAL CLASSES. People moving from unskilled and manual occupations to those that are skilled, non-manual and professional are often considered to be going up (upwardly mobile), those moving in the reverse direction, from what may be considered superior occupations to the inferior, going down (downwardly mobile) in social class position; such mobility is said to be vertical. The term horizontal mobility is applied to movement which involves a change of status (SOCIAL STATUS) and role (particularly in occupation, SOCIAL ROLE) without a change in social class position. MOBILITY, MODERNIZATION, PRE-INDUSTRIAL CITY.

social network the relatives, friends, neighbours with whom an individual person or a family is linked (the persons being represented by nodes, the relationships by connecting links, NETWORK-3).

social overhead capital the public investment in roads, housing, community services necessary for production to take place.

social pathology an approach to social problems which concentrates on the characteristics of problem individuals and communities. It suggests, for example, that the problem of poverty can best be understood by studying any physical and social inadequacies which may be present in the poor themselves.

social physics a mechanistic approach to the study of human geography which uses analogy with physical laws in analysing human behaviour. Introduced in the mid-nineteenth century, it is represented today by, e.g., the GRAVITY MODEL, REILLY'S LAW OF RETAIL GRAVITATION.

social polarization the consequence of such divergence of segregated groups in a society that large discrete groups (representing widely different socio-economic status, cultures etc.) form at the extremities of the social spectrum and there is little or no social contact between them. SEGREGATION-1,2.

social relations the evolving interplay of cultural, economic and political activities that shape and identify individual places but which spread beyond their boundaries.

social relations of production, relations of production in Marxism, the relationships between social groups which are generated by, and form the basis of, a particular MODE OF PRODUCTION.

social role the pattern of behaviour expected by others from an individual in a particular social position, of a particular SOCIAL STATUS, e.g. a mother, doctor, employer, school-teacher.

social segregation the residential grouping, the spatial separation of people within an area, on the basis of social distinctions. SEGREGATION.

social space the space perceived to be homogeneous by those living in it who, in using it, give it its special character, the space itself reflecting their activities, preferences, aspirations, and thereby becoming separate, identifiable by the social group inhabiting it.

social status the social standing of a

person, based on life-style, consumption of goods, the esteem in which that person is held by others. SOCIAL ROLE.

social structure the form, shape, pattern, framework of the interrelationships of people in a SOCIETY-2,3, in a social system, which can be analysed by identifying the roles and sets of roles played by the individual in that society, as well as the rules, constraints, conventions, resources and facilitations which underpin them, any of which may be considered as a social structure in its own right, with structures of its own. That, broadly, is one of the traditional applications of the term social structure in anthropology and analytic (formal) sociology. In human geography it is more commonly applied (as social structures) to the social rules and resources etc. which underlie a social system. CULTURE-1, POST-INDUSTRIAL SOCIETY, SOCIAL ROLE, SOCIAL STATUS, STRUCTURE-1,2.

social well-being, human well-being a state in which the needs and wants of a POPULATION-1 are, in general, satisfied. The identification of these needs and wants is subjective and in many cases historically determined, varying from one CULTURE to another and changing with the passage of time. In western industrialized societies in the twentieth century, it may be said that it is generally accepted that in order to be in a state of ideal social well-being a population should be in good health, have sufficient income for basic needs and thus be free from want, be well fed and clothed, housed with sufficient space in a benign, POLLUTION-free environment; should have command over goods and services, receive all the education desired, be protected by the administration of justice, have social and economic mobility, as well as time and facilities for recreation and leisure; and be able to participate in social affairs in a stable (preferably democratic) administration. QUALITY OF LIFE, STANDARD OF LIVING, WELFARE.

society **1.** the state of living in organized groups **2.** people living and working together and considered as a whole **3.** a large group of people associated together geographically, culturally or otherwise, with collective interests, shared laws, customs etc. and with a particular organization. INDUSTRIAL SOCIETY, POST-INDUSTRIAL SOCIETY, PRE-INDUSTRIAL SOCIETY **4.** an association of people with some special interest, some central discipline.

socioeconomic *adj.* of or relating to social and economic conditions.

sodium a soft, white metallic ELEMENT-6, oxidizing rapidly in air, reacting with water to liberate hydrogen and producing a solution of sodium hydroxide. It is widely distributed in many compounds, the most common being common SALT-2, an essential MICRONUTRIENT. Sodium salts are important in industrial processes. FELDSPAR.

software the set of systems, in the form of programs, which controls the operation of a computer, simplifying and linking the work of computer and user. HARDWARE.

softwood any tree with TIMBER which is soft and relatively light, with an open texture. Most commercial softwoods are CONIFEROUS trees (the timber being known commercially as pine, fir, deal), taken from the great northern forests of Canada, Scandinavia and the CIS; others are grown in plantations elsewhere in Europe and in New Zealand. Softwood is used especially for pulp, cellulose and wood products, and in construction work. HARDWOOD.

soil loosely, the earth or ground, but specifically the loose material of the earth's surface in which terrestrial plants grow, usually formed from weathered rock or REGOLITH changed by chemical, physical and biological processes. Thus the soil may be considered as an entity, quite apart from the rocks below it. It consists partly of mineral particles and partly, to a varying extent, of organic matter (HUMUS). The mineral particles can be graded according to size (GRADED SEDIMENTS); and according to the proportion of the grade present the terms clay soil, sandy soil etc. are applied. A soil is said to be MATURE if it has a fully developed profile (SOIL PROFILE); immature if it lacks a well-developed profile; truncated if it has lost all or part of the upper horizons (SOIL HORIZON). Human beings, by their cultivating activities, have affected the development of many soils and led to the destruction of others (SOIL EROSION). SOIL ASSOCIATION, SOIL CLASSIFICATION and other entries qualified by soil.

soil acidity ACID SOIL, PH.

soil association a term used by some soil scientists but not by others, usually applied to soils, not necessarily with the same profiles (SOIL PROFILE), lying close to each other, but also in UK, a group of SOIL SERIES developed on parent material derived from similar rocks or combinations of rocks.

soil classification a systematic grouping of SOILS. Most soils which develop in the SOLUM fall into one or other of two great groups, the lime-rich PEDOCALS, containing an accumulation of CALCIUM CARBONATE and the lime-poor PEDALFERS containing accumulations of ALUMINIUM and IRON compounds. (The terms pedocal and pedalfer are now little used by soil scientists.) From another viewpoint soils fall into three world groups: ZONAL, soils with profiles which show a dominant influence of climate and vegetation in their development; AZONAL, or skeletal, soils lacking such a profile; and INTRAZONAL, well developed soils with profiles reflecting the influence of some local factor of relief, parent material, or age, rather than those of climate and vegetation (SOIL PROFILE). Most soil scientists recognize the existence of world soil groups at one end of the scale and SOIL SERIES or soil types as the units for description and mapping, but the intermediate soil families and SOIL ASSOCIATIONS are differently interpreted.

In 1960 the Soil Conservation Service of the US Department of Agriculture in *Soil Classification: A Comprehensive System, Seventh Approximation* (commonly termed the Seventh Approximation), later entitled the *Comprehensive Soil Classification System* (CSCS), drew up another classification in which ten major orders were based on the present state of development of soils, divided into sub-orders, great groups, sub-groups, families and soil series. The ten major orders are ALFISOLS, ARIDISOLS, ENTISOLS, HISTOSOLS, INCEPTISOLS, MOLLISOLS, OXISOLS, SPODOSOLS, ULTISOLS, VERTISOLS.

A new soil classification in England and Wales, 1973, groups soils on a consideration of their land use capability. Taking landform, geology, climate and natural vegetation into account, seven major groups emerge: peat soils, and six groups of mineral soils. BROWN SOILS, GLEY SOILS (GLEYSOLS), LITHOMORPHIC SOILS, MAN-MADE SOILS, PELOSOLS and PODZOLIC SOILS. The sub-groups number 108. SOIL HORIZON, SOIL PROFILE.

soil compaction COMPACTION-2.

soil creep the slow downward movement of SOIL under the force of gravity. MASS MOVEMENT, SOLIFLUCTION.

soil erosion the removal of soil by EROSION, the main types being GULLY EROSION, RILL EROSION, SHEET EROSION and wind erosion (DEFLATION), assisted in many cases by human activities and grazing animals, especially in the removal of vegetation acting as a soil protection. ACCELERATED EROSION.

soil horizon a distinctive SOIL layer with features produced by soil-forming processes within the surface layer of the earth's crust. If undisturbed, e.g. by ploughing or similar activities, or by EROSION, soils tend to develop a succession of layers, horizons, commonly identified from the surface downwards as the A HORIZON (often subdivided in soil studies), containing HUMUS, from which material is washed downwards by percolating water (LEACHING) so that it is termed an eluvial horizon (ELUVIATION). Under the A horizon there may or may not be the B HORIZON, a horizon of deposition, of secondary enrichment, an illuvial horizon (ILLUVIATION) into which material (e.g. clay minerals, iron-aluminium oxides from the A horizon) is washed. Underneath is the C HORIZON, with the PARENT MATERIAL for the existing soil, the little altered, though weathered, BEDROCK. The above are the standard horizons in general use, but some soil scientists also distinguish a D horizon, with unweathered rock, underlying the C horizon; an F horizon or layer, a layer of forest soil consisting of partly decomposed plant residues; a G horizon, the layer where GLEY occurs; and H horizon or layer, an organic layer of forest soils with dark-coloured, structureless humus. Inferior numbers are sometimes added to the principal capital letter to indicate small differences within the horizon, e.g. A_2, a leached A layer. Some soil scientists now add inferior lower case letters to the capital letter distinguishing the horizon in order to indicate some special feature, e.g. A_p, an A horizon which is ploughed. Another classification of soil horizons is based on organic (O HORIZON) and mineral (MINERAL HORIZONS) content. A HORIZON, B HORIZON, C HORIZON, DIAGNOSTIC HORIZON, SOIL, SOIL CLASSIFICATION, SOIL PROFILE.

soil moisture the moisture in the pore spaces of a SOIL, important for plant growth.

soil pore POROSITY.

soil profile a vertical section of soil showing the sequence of horizons (SOIL HORIZON) downwards from the surface to the PARENT MATERIAL. It may be cut and studied in the field, or, as a sample section, taken away for study.

soil series a group of soils the members of which, formed from similar PARENT MATERIAL, have horizons (SOIL HORIZON) similar in distinguishing characteristics and arrangement in the SOIL PROFILE, except for the texture of the surface layer and its state of erosion. The series is the group most commonly used as the basic unit in mapping soils (comparable with the SPECIES in biology), and is usually the lowest in the hierarchy of a system of SOIL CLASSIFICATION.

soil structure the character of a soil shown by the ability of its particles to come together and to hold together to form AGGREGATES-3 or CRUMBS and by the way they do so.

soil texture TEXTURE-2.

soil water water held in the soil and available to the roots of plants.

solano (Spanish) LEVANTE.

solar *adj.* of or relating to the SUN.

solar battery, solar cell an apparatus that uses SOLAR RADIATION or its heating effect to produce an electrical current.

solar constant the rate per unit area at which RADIATION from the sun reaches the outer margin of the earth's atmosphere, averaging approximately 2 gramme-calories per sq cm per minute (139.6 mW per sq cm).

solar day the time interval between two successive transits of the sun over the same meridian, varying slightly at different times of the year because the orbit of the EARTH round the sun is elliptical and inclined to the equator. A mean solar day of 24 hours is the calculation commonly used. MEAN SOLAR TIME, SOLAR MONTH, SOLAR YEAR.

solar month one-twelfth of a SOLAR YEAR.

solar radiation ELECTROMAGNETIC WAVES emitted by the sun. The wavelengths range widely outside the earth's atmosphere, but absorption in the STRATOSPHERE ensures that the electromagnetic waves reaching the earth's surface are limited to certain bands. These include the wavelengths of visible light. ELECTROMAGNETIC SPECTRUM, INSOLATION.

solar system the SUN and the celestial bodies orbiting round it under the force of gravity, i.e. the nine PLANETS (in sequence measured in distance from the sun, Mercury, Venus, Earth, Mars, Jupiter, Saturn, Uranus, Neptune, Pluto), the natural SATELLITES which revolve round the planets (e.g. the moon around the earth), the ASTEROIDS, COMETS, METEORS, METEORITES. The orbits of the nine planets are elliptical and approximately in the same plane, the solar system as a whole moving through space at about 18.5 km (11.5 mi) per second.

solar wind a flow of atomic particles from the sun.

solar year the astronomical, equinoctial, natural or tropical year, the average time taken by the earth to complete one orbit round the sun with reference to the vernal EQUINOX as shown by the First Point of Aries now 365.2422 SOLAR DAYS, i.e. 365 days 5 hours 48 minutes 45.51 seconds, decreasing by some 5 seconds in 1000 years.

solfatara (Italian, derived from the name of a small volcano in Phlegraean Fields near Naples) a volcanic vent through which vapours and gases, especially sulphurous gases, gently issue, usually marking a late stage in volcanic activity. FUMAROLE.

solid matter with a definite volume and shape, the structure being determined by the arrangement in space of its molecules, atoms or ions which, unable to move freely, vibrate about a fixed position. FLUID, GAS, LIQUID.

solifluction, solifluxion the slow movement of rock debris, saturated with water and not confined to definite channels, down a slope under the force of gravity. It occurs particularly when thawing releases such surface deposits while the underlying layers are still frozen. Formerly considered to be synonymous with SOIL CREEP, but now usually applied only to saturated deposits. DRY VALLEY, FREEZE-THAW, MASS MOVEMENT.

solonchak (Russian) a saline soil, without structure, occurring in arid and semi-arid regions. SOIL STRUCTURE.

solonetz (Russian) a formerly saline soil from which the salts have been leached. In Russian, solonets, pl. solontsy, is applied to a soil with surface rock salt.

solstice a term conveying the idea of the SUN-2 standing still, i.e. the point in the ECLIPTIC when the sun is farthest from the EQUATOR, either north or south (i.e. approximately 23°30'N, the Tropic of Cancer and 23°30'S, the Tropic of Capricorn). Thus in the northern hemisphere the summer solstice is 21–22 June, the longest day, when at noon the sun is shining vertically over latitude 23°30'N, the Tropic of Cancer; the winter solstice is 21–22 December (the shortest day), when it is shining vertically at noon over latitude 23°30'S, the Tropic of Capricorn. Fig 18.

solum the term applied by soil scientists to the part of the earth's crust influenced by climate and vegetation, i.e. the soil layers above and excluding the PARENT MATERIAL. SOIL, SOIL HORIZON.

solute the solid or gaseous substance which dissolves in the SOLVENT to form a SOLUTION.

solution **1.** a homogeneous mixture of two or more substances, in which a SOLID, LIQUID OR GAS forms a single phase with another solid, liquid or gas (usually a liquid) (SOLUTE, SOLVENT) which has constant physical and chemical properties throughout at any selected concentration up to its SATURATION point. A standard solution is a solution of known concentration. COLLOID, SUSPENSION **2.** the act by which a substance is put into solution, or the state of being put into solution, e.g. in the chemical WEATHERING of rocks (CORROSION) the salts they contain are commonly dissolved by water to form a solution; rainwater charged with CARBON DIOXIDE dissolves (forms a solution with) CALCIUM CARBONATE, to remove it as CALCIUM BICARBONATE (CARBONATION, HYDROLYSIS); and rivers, in their work of transporting debris, carry a variety of substances in solution.

solvent the part of a SOLUTION which is present in greater bulk, i.e. usually the LIQUID in which the SOLUTE is dissolved. If the solvent is not water, this fact is usually noted, e.g. non-aqueous solvent.

sonar Sound Navigation Ranging, echo sounding, a device for locating an underwater object by sending out high frequency sound waves which are reflected from the object and registered on the apparatus, the time delay and nature of the echo giving information about shoals of fish, underwater obstructions, and ocean depths (RADAR). Some animals, e.g. bats (in air), dolphins and whales (underwater) use high frequency sound waves in a similar way, to locate objects and to communicate with each other.

sonde (French) an apparatus designed to measure and record conditions in the atmosphere at certain altitudes, variously equipped with a lifting device and sensing and recording instruments.

sorghum *Sorghum vulgare*, great millet, termed dura (northern Africa), kaffir corn (southern Africa), guinea corn (west Africa), a small-grained CEREAL with seeds larger than, and not storing so well as, those of most MILLETS, regarded by some botanists as a millet, but not by others. Possibly native to Africa, it is cultivated there and in other semi-arid tropical and subtropical regions. The white-grained variety is preferred for cooking, the red-grained for beer making. Sugar is obtained from a sweet variety grown in the USA. The grain of all varieties, some of which enters world trade, is fed to livestock.

sorting separation and putting into groups or classes according to some special quality or kind (shape, size, weight, age etc.), e.g. the sorting of sediments by the natural processes of flowing water and of wind. DEFLATION-I, GRADED SEDIMENTS.

sound **1.** a stretch of water connecting two larger bodies of water, e.g. a sea or large lake with another sea or the ocean, wider than a STRAIT **2.** the channel between an island and the mainland. CONCORDANT COAST **3.** an inlet of the sea **4.** a LAGOON fringing the southern and southeastern coast of USA.

sounding **1.** a method by which the depth of a sea or lake can be determined, formerly by a weighted line (a sounding line) dropped overboard, now usually by an ECHO-SOUNDER (SONAR) **2.** a measure of the depth of water determined by those means.

source of a river, the point at which a RIVER, identifiable as such, begins its flow. This may be at a SPRING-2, or from a lake, glacier, cave, marsh or swamp, or formed from the coalescence of trickles of water in RUNOFF on a hillside.

south **1.** one of the four cardinal points of the COMPASS, directly opposite the north, lying on the right side of a person facing due EAST **2.** towards or facing the south, the southern part, especially of a country, particularly of the southern states (The South) of the USA. BRANDT REPORT.

south *adj.* of, pertaining to, belonging to, situated towards, coming from, the south, e.g. of winds blowing from the south.

southeaster a strong wind or storm coming from southeast of the observer.

southeast trades the TRADE WINDS of the southern hemisphere.

southerly burster, southerly buster a strong, dry cold wind blowing from the south, most frequently in spring and summer, in Australia and New Zealand, in the wake of a trough of LOW pressure. BRICKFIELDER.

Southern Cone a term applied in 1976 by A. P. Whitaker, American historian, to the states of Argentina, Chile and Uruguay. Since then some writers have included Paraguay, or adjacent parts of Paraguay and southern Brazil.

Southern Cross a five-star constellation forming a Christian cross, with a bright star at each extremity, the longer axis pointing towards the SOUTH POLE. It is used to find TRUE SOUTH and LATITUDE from any point in the SOUTHERN HEMISPHERE (comparable with the POLE STAR in the NORTHERN HEMISPHERE).

southern hemisphere the half of the earth south of the EQUATOR. HEMISPHERE.

south magnetic pole MAGNETIC POLE.

South Pole the geographical South Pole, the southern extremity of the earth's axis. MAGNETIC POLE, POLE, TRUE SOUTH.

southwester, sou'wester a strong wind or storm coming from southwest of the observer.

sovereign state a STATE-2 with the supreme authority, the supreme power, held within the state itself, a state which is therefore independent and fully self-governing. NATION, NATION STATE.

soviet (Russian soviet, council) **1.** an elected governing council in the former USSR, at local, provincial and national level, the latter (the Supreme Soviet) comprising delegates from all the Soviet Republics **2.** any of the associated republics of the former USSR. CIS.

soviet *adj.* of, relating to, or pertaining to, the former Union of Soviet Socialist Republics.

sovkhoz (Russian) a state farm (STATE FARMING) in the former USSR, usually large-scale. Under the CIS, state farms became collectivized. KOLKHOZ.

soya bean, soybean *Glycine max*, an ANNUAL leguminous herb (LEGUMINOSAE), susceptible to frost, probably native to southwestern Asia, a wide number of varieties being grown there and in the USA in areas with warm summers. The plant produces highly nutritious seeds (high in protein, low in carbohydrates) which yield oil (used in cooking and industrially for making paints, plastics, etc.), provide flour and 'milk', or are destined (fresh, fermented or dried) for human consumption, as are the young seedlings (bean sprouts). Soy sauce, a piquant sauce, is made from fermented soya beans soaked in brine. The mature plant and the residue remaining after the extraction of oil from the seed are fed to livestock.

spa a watering place, a resort with SPRINGS-2, the water of which contains minerals reputedly of medicinal value. The term is derived from Spa, a watering place near Liège, Belgium, celebrated for the curative properties of its mineral spring water.

space 1. that which objects occupy as a result of their volume, the amount of space occupied being the volume of the object **2.** a part of space, a volume, area or length that may be occupied by something, or may be empty, e.g. an extent or area of the earth's surface, or the distance between two points, or two objects, or two lines on a page **3.** a period of time, e.g. between two events **4.** that which is beyond the limit of the earth's ATMOSPHERE-1. SPATIAL.

space cost curve a CROSS SECTION-1 through a COST SURFACE, related either to total production costs, or to single items in such costs. COST CURVE.

space shuttle a NASA SATELLITE-1 with a crew aboard, designed as a maintenance unit to make repeated return trips between earth and SPACE-4, to act as a platform from which other orbiting satellites (e.g. SPACE STATION) and space research equipment (e.g. the Hubble Telescope) can be serviced and repaired, to carry relief crew members to a space station, and to act as a base for REMOTE SENSING instruments.

space station an artificial SATELLITE-1, serving as a work station, with a crew aboard, orbiting the earth in SPACE-4 equipped with a laboratory and REMOTE SENSING instruments. The crew are engaged in earth observation and in physiological and astronomical research. The station is serviced by crew transported to the station by SPACE SHUTTLE or by rocket. LANDSAT, SKYLAB.

space-time constraints TIME-SPACE CONSTRAINTS.

spalling EXFOLIATION.

Spanish moss *Tillandsia usneoides*, an EPIPHYTE, a stemless HERBACEOUS plant, forming long, loose, grey-green hanging tufts in warm humid conditions in southern USA (e.g. in the Everglades, EVERGLADE) and West Indies; or *Ramalina reticulata*, a LICHEN forming lace-like nets on trees in the humid coastal regions of western USA.

spatial *adj.* **1.** of, pertaining to, or relating to, SPACE **2.** consisting of, or having the character of, space **3.** extending in or occupying space **4.** subject to or controlled by the conditions of space (in contrast to TEMPORAL) **5.** existing in, occurring in, caused by, or involved by, space **6.** as applied to a faculty or a sense, perceiving space.

spatial analysis an approach to geography in which the LOCATIONAL variations of a phenomenon or a series of phenomena are studied and the factors influencing or controlling the patterns of distribution investigated. SPATIAL-1,3 patterns are broken down into simple elements so that measurements can be made of single patterns. This allows the comparison of two or more patterns, e.g. the pattern of a single phenomenon in different areas, or the pattern of different phenomena or VARIABLES in one area; and it allows tests to be developed to show whether a pattern differs significantly from a random pattern.

spatial margin spatial margin to profitability, the limit where the total cost of production of a given volume of output equals the total revenue possible from the sale of that volume of output. It therefore indicates the area within which an activity should be profitable. On a COST SURFACE map the spatial margin of profitability appears as the contour which encloses the area within which production is economically feasible, but outside of which an operation will probably lose money.

spatial patterns the patterns of distribution in SPACE-3, commonly falling into three categories: regular, random, clustered.

spatter cone, driblet cone a small VOLCANIC CONE, between some 3 and 6 m (10 and 20 ft) in height, formed when LAVA erupts violently from the side of a volcanic VENT, or in some cases through a fissure, and spatters (falls in drops) to the ground, where it rapidly congeals.

Spearman's rank correlation coefficient a measure of the strength of the relationship between variables used when only the relative size, or rank, of each variable is known. CORRELATION-2.

species 1. in biology, the smallest unit of classification commonly used (CLASSIFICATION OF ORGANISMS), the group with members having the greatest mutual resemblance, able to breed with each other but not with organisms of other groups. Local differences may occur through reproductive isolation, recognized in classification as a sub-species. The common names of familiar animals and plants often refer to species. INDICATOR PLANT. Most biologists agree that species are groups of organisms that share a common pool of GENES and are separate evolutionary entities. Sexually reproducing organisms maintain the pool by interbreeding; in non-sexually reproducing forms the genetic nature of species is maintained through clonal continuity **2.** in chemistry, entities, such as atoms, molecules, ions, free radicals, and activated atoms or molecules, which are active in chemical reactions.

specific gravity the ratio between the weight, at any chosen place, of a given volume of a substance and the weight of an equal volume of water at 4°C (39.2°F) at the same place. Specific gravity and density are numerically the same, but the former is a relative quantity, DENSITY-2 is absolute.

specific humidity the ratio of the weight of WATER VAPOUR in a parcel of the ATMOSPHERE-1 to the total weight of air (i.e. including water vapour), measured in grams of water vapour per kilogram of air. The specific humidity of very cold dry air is low, that of very warm humid air is high. ABSOLUTE HUMIDITY, HUMIDITY, RELATIVE HUMIDITY.

sphagnum bog moss, a member of *Sphagnum*, a genus of soft MOSSES with erect stems, growing in swamps or in water in cold temperate zones. Minute holes in some of the cell walls of the leaves promote the absorption of water, even when the leaves are dead. The conditions of the environment prevent decay, so as the plant grows upwards the depth of the moss bed increases, becomes compacted, and forms PEAT. BOG.

sphere 1. in geometry, a solid figure made when a circle rotates about a diameter, any point on the surface of this solid figure being equidistant from its centre (the centre of the sphere) **2.** in general, loosely, an object with a shape approximately like that of a sphere, i.e. not necessarily a perfect sphere. SPHEROID **3.** a realm, an area which is limited in extent but within which something is effective. SPHERE OF INFLUENCE **4.** a field of knowledge **5.** the zone in which a CULTURE-1, having spread from its CORE AREA or CULTURAL HEARTH and over its DOMAIN-2, is still effective and influential. OUTLIER-2.

sphere of influence 1. a territory or part of a territory in which a foreign power has political and economic interests, special rights and privileges. SPHERE-3 **2.** in urban studies, an area which depends on an urban centre for various services, or with which it has special relations (the terms UMLAND, URBAN FIELD or URBAN HINTERLAND are now more commonly applied to such an area). CENTRAL PLACE THEORY.

spheroid a figure, a body, resembling a SPHERE-1 (especially an ELLIPSOID). GEOID.

spheroidal weathering onion weathering, a form of underground CHEMICAL WEATHERING occurring particularly in tropical regions in well-jointed rocks such as BASALTS. Water penetrates the intersecting joints, attacking each separate block from all sides simultaneously, breaking off a succession of shells or skins, so that a succession of new surfaces is presented to the weathering solution, leaving a mass of unweathered rock in the centre which, on EXHUMATION, appears at the surface as a rounded mass. The process (HYDROLYSIS) is similar to that involved in EXFOLIATION.

Spilitic suite a PETROGRAPHIC PROVINCE, marked by volcanic action and slow submergence. ATLANTIC SUITE, PACIFIC SUITE, PILLOW LAVA.

spillway 1. the area below a dam or a natural obstruction over which excess water from a reservoir or lake above is allowed to drain away **2.** the part of a dam over which water flows **3.** a passage for the overflow of water from a reservoir etc.

spinifex one of the genus *Tricupsis*, a coarse grass with sharp, pointed, shiny leaves growing in large tufts, separated by bare ground, over large areas of semi-desert in the heart of Australia.

spit a narrow ridge of sand and shingle, resulting from LONGSHORE DRIFT, attached to the seashore at one end, extending some distance seawards, and terminating in open water at the other. The outer end is often deflected landward to form a hook, or a recurved spit; and development of the hook may produce a compound recurved spit or compound hook. BARRIER ISLAND.

spodosols in SOIL CLASSIFICATION, USA, an order of soils associated with a cool and cool-humid climate and forest or heath vegetation, with a leached, acid, ash-grey A HORIZON, low in plant nutrients, overlying a B HORIZON rich in iron oxide and aluminium and enriched by organic material from the A horizon (ELUVIATION, ILLUVIATION), e.g. a PODZOL.

spoil waste material from mining or quarrying operations, piled up in spoil banks, spoil tips, spoil dumps or tip-heaps.

SPOT Satellite Probatoire pour l'Observation de la Terre, a SATELLITE-1 orbiting the earth, launched by France, carrying REMOTE SENSING equipment similar to that aboard LANDSAT satellites in order to gather data on natural resources, environmental conditions, land use etc. needed for certain earth resource development projects. SKYLAB, SPACE SHUTTLE.

spot height a precise point on a map showing the height of the ground at that place measured from a given datum (e.g. in British Ordnance Survey maps, height above OD). It is not necessarily indicated physically on the ground, thus differing from BENCH MARK.

sprawl URBAN SPRAWL.

spread and backwash effect CIRCULAR AND CUMULATIVE GROWTH.

spring 1. one of the SEASONS of the year, occurring between winter and summer in midlatitudes, a period variously defined in the northern hemisphere as being March, April, May or (astronomically) from the spring EQUINOX, 22 March, to the summer SOLSTICE, 21 June **2.** a continuous or INTERMITTENT natural flow of water issuing strongly or seeping gently from the earth's surface under its own pressure, the site being related to the nature and relationship of rocks (especially PERMEABLE and IMPERMEABLE layers), the level of the WATER TABLE, the surface relief. DIKE-SPRING, FAULT SPRING, MINERAL SPRING, SCARP-FOOT SPRING, SPRING-LINE, THERMAL SPRING.

spring-line a line of SPRINGS-2 occurring roughly at the level where, by reason of deposition of the strata, the WATER TABLE reaches the surface, as at the foot of an ESCARPMENT. Where such springs are copious and constant they provided in the past a reliable water supply, one of the factors likely to influence the choice of a site for a village, hence the term spring-line village.

spring-sapping, spring-head sapping the undermining of a hillside at the point of issue of a SPRING-2, caused by the erosive action of swiftly flowing water, leading to small slips and the formation of an alcove or amphitheatre which cuts backwards into the slope and results in the retreat of the valley head.

spring tide a tide with a range (TIDAL RANGE) greater than that of ordinary tides, i.e. the HIGH TIDE is higher, the LOW TIDE lower, occurring twice monthly, when the moon, sun and earth are almost in the same straight line (SYZYGY), either in CONJUNCTION (at the time of the new moon) or in OPPOSITION (the time of the full moon), so that the gravitational effects are reinforced. MOON, NEAP TIDE.

spur a marked projection of land from a

mountain or a ridge. INTERLOCKING SPUR, NESS, TRUNCATED SPUR.

squall **1.** a sudden, violent gusty wind which lasts a minute or two and then subsides, usually accompanied by rain or hail **2.** a storm characterized by a series of squalls, the gusts of wind being short-lived and blowing at speeds half as great again as the average wind speed. LINE SQUALL.

squatter **1.** one who occupies (especially premises) without legal entitlement **2.** in USA, one who occupies government land in order to gain legal title to it **3.** in Australia, a large-scale sheep farmer, or one occupying a tract of grazing land as a crown tenant.

squatter settlement SHANTY-TOWN.

squattocracy (Australia) a class of wealthy landowners who obtained their land in the period before the Free Selection Acts, Australia. Comparable with the squirearchy in Britain.

SSSI Site of Special Scientific Interest, in planning in Britain, an area of land judged by the Nature Conservancy to be of special interest on account of the fauna or flora it supports, or of its special geological or physiographical features.

stabilized dune a coastal DUNE fixed by vegetation and therefore usually not liable to DEFLATION-1. FORE-DUNE, MOBILE DUNE.

stable *adj.* **1.** staying or able to stay unchanged in form, structure, character etc. in conditions which would normally induce changes **2.** not easily decomposing or changing **3.** able to recover, to return to original condition after slight displacement **4.** enduring, permanent.

stable equilibrium the state of the ATMOSPHERE-1 where the ENVIRONMENTAL LAPSE RATE of an air mass is less than the DRY ADIABATIC LAPSE RATE. If a pocket of air at the earth's surface is moved upwards it cools at the dry adiabatic lapse rate, becomes colder (thus denser) than the air around it, and sinks to its original level, conditions typical in an ANTICYCLONE. Contrast UNSTABLE EQUILIBRIUM.

stable population, stationary population a POPULATION-1 in which the proportion of members at each age is constant and the rate of the annual growth or decline is fixed, due to the fact that there is no change in the chance of dying at a particular age or of a female giving birth at a particular age. If the annual rate of growth is zero, the population is termed stationary.

stac (Gaelic) a mass of hard IGNEOUS ROCK, steep-sided, varying in height, especially such a mass forming an offshore island.

stack an isolated mass of rock near a coastline, detached from the main mass usually by marine erosion (especially by wave action), rising steeply from the surrounding sea. It represents the midway stage in the marine erosion cycle: CAVE-1, ARCH, STAC, STUMP, REEF-1.

stadial moraine a RECESSIONAL MORAINE, stadial in this case meaning of or relating to a stage in development.

stages of economic growth model ROSTOW MODEL.

stalactite a cylindrical or conical deposit of mineral matter, hanging from an elevated point, formed by dripping water, i.e. a mass of CALCITE hanging from the roof of a limestone cave, formed from water containing dissolved CALCIUM BICARBONATE (CARBONATION) which has seeped through joints and crevices to drip very slowly from the cave roof. Some of the carbon dioxide in the water is released so that some of the dissolved calcium bicarbonate reverts to CALCIUM CARBONATE; and this process, combined with evaporation, leads to cumulative downward deposits of calcite with every drip of water. STALAGMITE.

stalagmite a mineral mass resembling a STALACTITE and formed by the same process, but markedly conical in shape and rising from the floor as a result of drops of water falling from the roof to the floor. It may be formed by water dripping from a STALACTITE in a limestone cave, leading to the eventual union of stalagmite and stalactite in a pillar connecting the roof to the floor.

standard atmosphere the average condition of the ATMOSPHERE-1 in respect of its pressure and temperature at a selected altitude, used as a unit of measure in aviation in calibrating instruments, assessing aircraft performance etc. ATMOSPHERE-2, ATMOSPHERIC PRESSURE.

standard deviation in statistics, the square root of the VARIANCE, a widely used measure indicating the spread of values on each side of the MEAN in a FREQUENCY DISTRIBUTION. If there is no variation between values in the distribution the standard deviation will be zero; it increases with the increase in variation. A relatively low standard deviation is associated with a close grouping around the MEAN; a relatively high standard deviation, a wide spread of values about the mean.

standard error in statistics, the STANDARD DEVIATION of a whole set of estimates. It owes its existence to the fact that if an infinite number of samples of the same size are drawn from a POPULATION-4 at random with replacement not only the MEANS but also the standard deviations and the VARIANCES etc. of these SAMPLES-3 always vary. STANDARD ERROR OF THE MEAN.

standard error of the mean in statistics, the STANDARD DEVIATION of an infinite set of means of SAMPLES-3, all the samples being the same size and selected at random with like replacements from the same POPULATION-4.

standard hillslope a hillslope comprising four components (SLOPE ELEMENTS), stated by some authors to be the outcome of all types of slope process, independent of climatic influence. But a strong, massive bedrock is essential if all four components are to be supported; if it is weak only the WAXING and the WANING SLOPES may be formed.

standard of living **1.** the conditions in which people live or would like to live **2.** the conditions of living considered to be desirable as defined by national or international convention or agreement for a specific purpose (e.g. in order to establish a minimum wage, working hours etc.).

standard parallel a parallel of latitude selected for its appropriateness for making the necessary calculations and for drawing a particular map projection; or for acting as the horizontal axis of a grid-system. A meridian selected for the same reason and purpose, but providing the vertical axis, is termed a standard meridian.

standard time the mean time of a meridian centrally located over a country or part (zone) of a country (TIME ZONE), and used for the whole of that area in order to avoid the inconvenience resulting from the use of LOCAL TIME. DAYLIGHT SAVING. Fig 43.

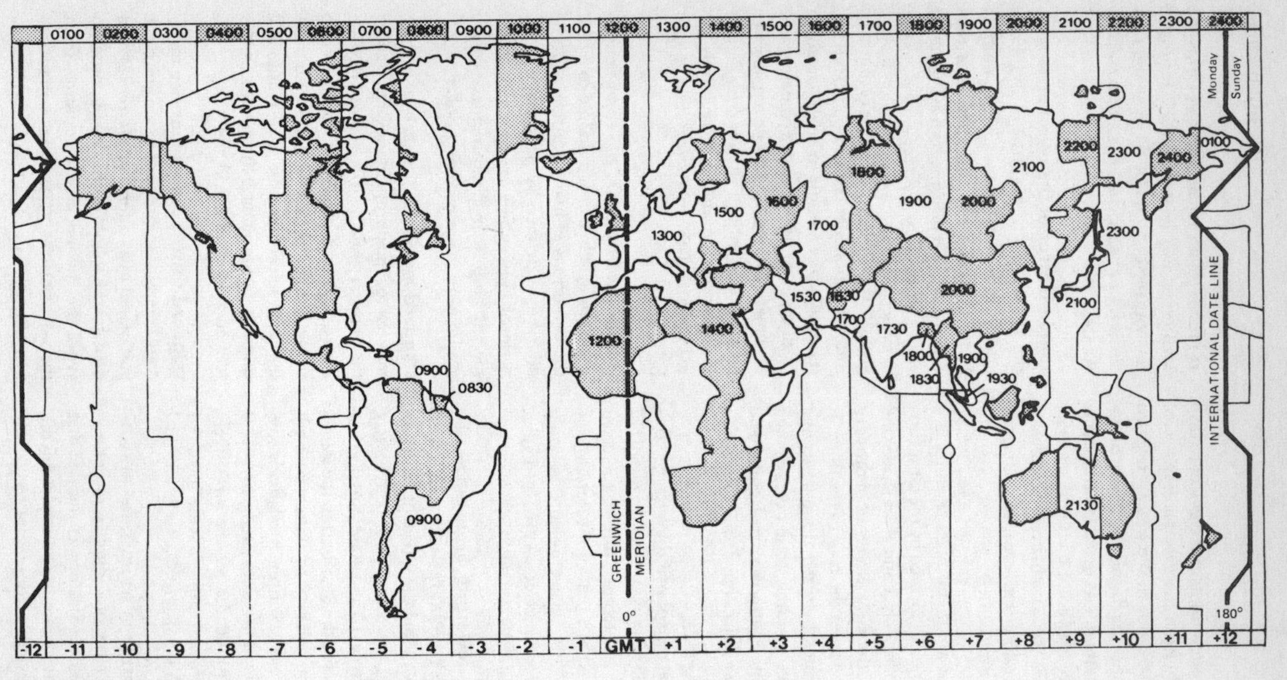

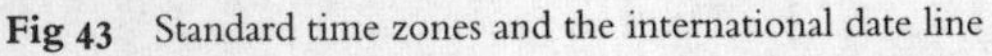

Fig 43 Standard time zones and the international date line

standing wave, stationary wave a wave form produced by the interaction of two wave motions (transverse or longitudinal) of identical amplitude, frequency and velocity, superposed and moving simultaneously through a medium in opposite directions, making a pattern of alternating points of no displacement (at nodes) and most displacement (at antinodes), the intermediate displacement between the two being a smooth curve. LEE WAVE, OSCILLATORY WAVE THEORY OF TIDES.

staple **1.** the main commodity grown, produced, traded in, in a particular area **2.** a commodity in constant demand and regularly kept in stock **3.** a chief ingredient or constituent, e.g. the foodstuff forming the main constituent of the diet of the people of an area.

star dune, star-dune a large, pyramidal, fairly permanent SAND DUNE, with a relatively high peak from which ridges radiate.

state **1.** condition with respect to growth, development, arrangement **2.** a self-governing group of people occupying a defined territory, or the territory thus occupied. NATION STATE, SOVEREIGN STATE **3.** the political organization forming the basis of civil government, the supreme civil power and government (sometimes with initial capital letter).

state capitalism an economic system (or the political doctrine advocating it) in which the STATE-3 owns or controls a major part of the economy, consistently supporting the interests of private capital, or, alternatively, regarding private capital as subordinate to the interests of the state and its own enterprises.

state farming a form of agricultural organization in some countries with a centrally controlled economy, especially in the former USSR (SOVKHOZ), in which the land is owned by the STATE-3 and the farm workers are employed, as wage-earners, by the state. They do not have a share in the produce of the farm (unlike the workers in COLLECTIVE FARMING). LAND TENURE.

state intervention activity by the STATE-3 in which it assumes responsibility for some of the processes usually associated with the market in a capitalist (CAPITALISM) or MARKET ECONOMY, e.g. in a MIXED ECONOMY.

static *adj.* **1.** at rest, not moving in relation to the earth **2.** unchanging **3.** in equilibrium. The opposite of DYNAMIC.

static lapse rate synonymous with ENVIRONMENTAL LAPSE RATE.

static rejuvenation REJUVENATION of a river caused by an increase in its eroding capability, not by a lowering in BASE-LEVEL, e.g. occurring when increased rainfall leads to increased RUNOFF; or the volume of a stream increases owing to RIVER CAPTURE; or there is a decrease in the load carried by a stream, all of which increase the eroding power of the drainage system. DYNAMIC REJUVENATION.

station **1.** a place with its associated buildings on a route, where buses, trains etc. habitually stop to take on and discharge passengers, goods, mail **2.** a place with buildings and equipment for the transmission and reception of radio, television signals, etc. **3.** in Australia, a sheep or cattle run with associated buildings.

statistical test an accepted method of deciding from the given data whether to retain the initial hypothesis (i.e. the NULL HYPOTHESIS) or to set it aside in favour of a specified ALTERNATIVE HYPOTHESIS.

steady state an OPEN SYSTEM in which external and internal relationships produce equilibrium, or internal balance. DYNAMIC EQUILIBRIUM.

steam fog a FOG resulting from the passing of cold air over the surface of warmer freshwater, the moisture from the latter condensing into tiny visible droplets in the air so that its surface seems to steam. In very low temperatures the droplets are converted immediately to ice particles and form ICE FOG. ARCTIC SMOKE.

steam point the point at which pure water at sea-level boils at a standard pressure of ATMOSPHERE-2.

steel any of the many alloys of iron and 0.1 to 1.5 per cent carbon in the form of iron carbide, especially cementite, often alloyed with other metals if a special steel with a particular property is needed. In its solid state steel is hard, with great tensile strength, and it can be cast, rolled, drawn. It is used in construction work, all kinds of machinery, installations, equipment, vehicles, domestic goods etc. It is made by reducing the carbon in cast iron (PIG IRON) or by the diffusion of carbon into CAST IRON. The methods formerly used in steel making (BASIC BESSEMER PROCESS, BESSEMER PROCESS, OPEN HEARTH PROCESS) have now been largely superseded by the use of the electric furnace in large integrated steelworks (ARC FURNACE, BLAST FURNACE, IRON, PIG IRON). Converters lined with dolomite bricks in the basic Bessemer and basic open hearth processes produce basic, or mild, steel; acid steel is produced from non-phosphoric ores processed in converters lined with silica bricks.

step faults a series of parallel faults each with a THROW that projects in the same direction, but to a greater distance than the one above, thus producing step-like changes of level of strata. Rift valleys are often bounded by such a series.

steppe the treeless midlatitude grassland stretching from central Europe to southern Siberia in Asia. It is similar to midlatitude grassland elsewhere (e.g. PAMPAS, PRAIRIE) but if the term is applied to these it is best restricted to their drier parts.

stereoscope a binocular optical instrument designed to give photographs a three-dimensional character, the simplest comprising two lenses mounted horizontally in a frame with legs some 20 cm (8 in) in height, the distance between the eye-pieces being approximately the same as the distance between human eyes. Two photographs of the same area but taken from slightly different angles are viewed through the lenses, with the result that each photograph is viewed by only one eye, giving the impression of a three-dimensional view. Stereoscopes, some very complex, are used especially in the interpretation of aerial photographs.

Stevenson screen a standardized METEOROLOGICAL SCREEN, designed by Thomas Stevenson, an engineer.

stochastic *adj.* pertaining to chance or conjecture; in mathematics, random. SYSTEMATIC-4.

stock **1.** an accumulation of goods held for future use or maintained as a constant source of supply **2.** a RESOURCE **3.** a group of plants or animals having the same line of descent **4.** LIVESTOCK.

stone **1.** rock, hard mineral matter (other than metal) **2.** a piece of rock of imprecise size, but larger than a particle of coarse sand, smaller than a boulder. GRADED SEDIMENTS.

Stone Age the period generally defined as extending from the beginning of the PLEISTOCENE (GEOLOGICAL TIMESCALE) to the beginning of the BRONZE AGE, when stone, bone or wooden (not metal) tools and weapons were used, divided into the culture periods EOLITHIC, PALAEOLITHIC, MESOLITHIC, NEOLITHIC.

store cattle cattle bought and kept for fattening and, as fat cattle, destined for the butcher.

storm 1. any violent disturbance of the ATMOSPHERE-1 and the effects associated with it, e.g. SANDSTORM, THUNDERSTORM **2.** a gale-force wind. BEAUFORT SCALE.

storm-beach a deposit of coarse beach material, including COBBLES and BOULDERS, thrown high up on the shore (usually farther inland than the level reached by a high SPRING TIDE) by unusually strong waves in a storm.

storm-surge an unusual, rapid rise in tide level, above normal heights, caused by atmospheric factors, especially by the passage of an intense atmospheric DEPRESSION-3 producing gale-force (BEAUFORT SCALE) onshore winds.

stoss-and-lee relief a relief feature characteristic of a glaciated region, where small hills are markedly asymmetric, their gentle, smooth slopes (STOSS END) having been abraded by the oncoming ice, their steeper, rougher slopes on the LEE side having been subjected to PLUCKING-1.

stoss end, stossend the side of a prominent crag, hill or knob of rock facing the oncoming movement of an ice sheet or glacier, scratched (STRIAE) by the ice. CRAG AND TAIL, STOSS-AND-LEE RELIEF.

Strahler ordering STREAM ORDER.

strain in physics, the deformation of a body as a result of STRESS, termed homogeneous if the deformation is equal in all directions, heterogeneous if otherwise. Strain is measured by the ratio of the dimensional change produced to the original dimension (in linear measure, area or volume); e.g. the ratio of the change in area to the original area.

strait, straits a narrow passage of water connecting two larger bodies of water. CURRENT-4,5.

strata STRATUM.

strath GLEN.

stratification (*verb* to stratify, to arrange or form in strata, or to become arranged in strata) formation into layers **1.** in geology, the accumulation of SEDIMENTARY ROCKS and of some IGNEOUS ROCKS in layers or STRATA; the condition or manner of being stratified; the arrangement in strata (LAMINATION) **2.** in meteorology, the formation of stable horizontal layers in the ATMOSPHERE-1, occurring when the LAPSE RATE is less than the ADIABATIC LAPSE RATE **3.** in statistics, the division of a POPULATION-4 into layers or strata appropriate to the question at issue in order to be able to draw a representative RANDOM SAMPLE from each layer, thereby increasing the precision of the sample without increasing its size **4.** in oceanography and hydrology, THERMAL STRATIFICATION.

stratified arranged or formed in STRATA, in layers; the antonym is unstratified.

stratified society a SOCIETY-3 in which there are obvious social levels, identifiable by marked differences in attitude and inequality of access to scarce resources. Once defined as scarce by the society, these resources acquire a market value, and a MARKET ECONOMY develops which creates wealth in the society as a whole (but some would add depends on the maintenance of scarcity), and large-scale URBANIZATION becomes possible.

stratiform *adj.* having a stratified arrangement. STRATIFICATION **2.** having the form of STRATUS, applied to all sheet CLOUDS, including ALTOSTRATUS, CIRROSTRATUS, NIMBOSTRATUS, STRATOCUMULUS, STRATUS.

stratigraphy a branch of GEOLOGY concerned with the study of the occurrence, lithology, composition, sequence, fossils and correlation of rock STRATA, and especially with the chronological order of succession of rock formations, by which historical changes in the geography of the earth can be traced.

stratocumulus a continuous extensive sheet of grey, heavy CLOUD, consisting of circular contiguous mounds, usually fairly low (500 m: 1600 ft), but sometimes reaching an altitude of 2400 m (8000 ft). It usually occurs in the northern hemisphere in winter, and does not bring rain.

stratopause the layer in the earth's ATMOSPHERE-1 lying at a varying altitude of some 56 km (35 mi), separating the STRATOSPHERE (below) from the MESOSPHERE (above) and within which the rising temperature reaches some 80°C (176°F). Fig 4.

stratosphere the layer in the atmosphere between the TROPOPAUSE and the STRATOPAUSE, extending upwards from an altitude of some 11 km (about 7 mi), depending on latitude and season and condition of the weather in the TROPOSPHERE. It is a generally tranquil zone in which, at the lower level (sometimes termed the isothermal region), the temperature is relatively constant, moisture content is low, clouds do not form, and large convection currents are absent, ideal flying conditions for aircraft. At greater altitude, nearing the top of the stratosphere, where OZONE absorbs shortwave SOLAR RADIATION, temperatures resemble those on earth; but approaching the STRATOPAUSE the temperature begins to rise. Fig 4.

stratum pl. strata, in geology, a generally distinct, roughly horizontal individual layer of homogeneous material, its surfaces parallel to layers of different material lying above and below. The term is usually restricted to a bed or layer of SEDIMENTARY ROCK or to a layer of pyroclastic material (PYROCLAST). GEOLOGICAL INVERSION, LAW OF SUPERPOSITION, STRATIFICATION, STRATIGRAPHY.

stratus a continuous extensive sheet of grey uniform CLOUD, often persistent, usually low-lying, but occurring at any altitude up to 2400 m (8000 ft), sometimes bringing fine drizzle, but never heavier rainfall. It is termed fractostratus (FRACTUS) if it is fragmented to form irregular patches. STRATIFORM.

straw the stem of certain CEREAL crops remaining after cutting and threshing.

stream **1.** a body of flowing water, permanent or intermittent, from the smallest to the largest, on land, underground, or in the ocean or sea, e.g. Gulf Stream. BED LOAD, DISSOLVED LOAD, RILL, RIVER (and entries qualified by river), SUSPENDED LOAD **2.** the flow or current of a FLUID, or the direction of that flow.

streamline the direction of movement of all parcels of air in an area, measured individually and simultaneously, providing an immediate general picture of air motion; compare TRAJECTORY.

stream order a hierarchical classification of streams based on the magnitude of their channels and position in a DRAINAGE AREA, the outermost tributaries being designated fingertip tributaries or first order streams, two first order streams uniting to form a second order stream, two second order streams joining to form a third order, and so on until the main river or trunk stream, opening to the mouth, is reached. At least two streams of any given order are needed to form a stream of the next higher order. A more recent classification is simpler, using only first order links as an index of magnitude. Fig 44.

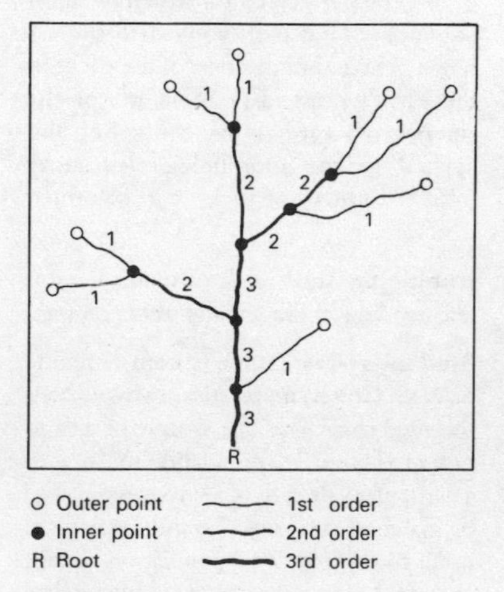

Fig 44 Stream order, according to Strahler (modified)

stream stage the height of the surface of a stream at any particular point in time. BANKFULL STAGE, FLOOD STAGE, OVERBANK STAGE.

strength theory of MIGRATION-1,2, a theory which suggests that people move when they are in a position of economic strength, not weakness, i.e. conditions are satisfactory in the place where they live, but conditions in another (e.g. higher wages, higher social status, more pleasing environment) seem more desirable. PUSH-PULL THEORY, WAGE DIFFERENTIAL THEORY.

stress 1. in physics, the force acting on an object, expressed as force per unit area, measured in newtons per metre squared (i.e. the force needed to give a mass of 1 kg an acceleration of 1 m per second per second, symbol N), calculated by dividing the total force by the area to which it is applied. When stress is applied to a body, e.g. a rock, it produces STRAIN, and the body can be distorted or deformed, according to its ELASTICITY. If two forces press towards each other COMPRESSION or THRUST results; if they pull apart there is TENSION; and if they act in parallel the result is shearing stress (SHEAR) **2.** the state produced by that force **3.** the force exerted by environmental factors on the nervous system of an animal or human being, or the state produced by that force.

striae, striation scratches and narrow grooves on the surface of ice-worn rocks. They are cut as a glacier moves along, overriding boulders and bedrock in its path, by the rocks and the small fragments of rock that are frozen into its underside. Striae therefore indicate the direction of movement of the ice. Similar scratches and grooves on rocks may be the result of other movements, e.g. soil creep, so the distinction 'glacial striae' is sometimes made.

strike the direction of a horizontal line on an inclined rock stratum at right angles to the direction of the TRUE DIP of the rocks. Strike is applied as an *adj.* to features roughly parallel to the strike, e.g.

strike fault, strike joint, strike valley. Figs 16, 42.

strike fault a FAULT of which the STRIKE is parallel to the STRIKE of the strata affected.

strike-slip fault, strike-slip faulting tear fault, a SLIP-1 or FAULT characterized by movement transverse to the STRIKE of the folded strata. Fig 22.

strike valley a valley of which the direction is parallel to the regional STRIKE of the strata in an area, e.g. the valley of a subsequent stream in TRELLIS DRAINAGE.

strip cultivation 1. the system of dividing a large field into strips, each worked by a separate owner or occupant (strip field), as in the FIELD SYSTEM in medieval England and other parts of western Europe **2.** the growing of different crops in contiguous strips along the contours of a hillside, to counteract soil erosion.

strip map a map showing only a narrow band of country in which the user is interested, e.g. on each side of a route.

strip mining (American) OPENCAST MINING or QUARRYING.

Strombolian eruption a volcanic eruption characteristic of Stromboli, a VOLCANO in the Lipari Islands off the coast of Italy. Gases can readily escape from the molten lava in the crater, so pressure does not build up. Instead the volcano ejects incandescent dust (VOLCANIC DUST), SCORIA and bombs (VOLCANIC BOMB) with a little water vapour fairly frequently, the lava being somewhat less BASIC-1 than that typical of the HAWAIIAN VOLCANIC ERUPTION (BASIC LAVA). This is one of the four types of volcanic activity commonly distinguished, the others being HAWAIIAN, VULCANIAN (or VESUVIAN), and PELEAN.

structuralism a movement, an approach to knowledge concerned not so much with the apparent 'surface' STRUCTURES-1,2 of a subject but with the deep structures (e.g. the social and ideological values, the rules, conventions and restraints etc.) which underlie and generate the phenomena being observed.

structure 1. in general, something made of parts fitted or joined together **2.** the way in which the constituent parts (of something) are fitted or joined together or arranged in a way that gives that thing its peculiar character. SOCIAL STRUCTURE **3.** in geology, geological structure, most commonly (and best) applied to the arrangement and disposition of the rocks in the earth's crust, as a result of (or the absence of) earth movements; but also applied to the morphological features (MORPHOLOGY) of rocks, e.g. columnar structure.

stubble the short stalk of CEREAL crops left standing in the ground after reaping.

Student's *t*-test (Student, nom de plume of W. S. Gosset) in statistics, an hypothesis test used to analyse one sample of data in order to compare a population mean with a particular value; or to analyse two paired or matched samples of data (PAIRING) in order to compare two population means; or to analyse two unrelated samples of data in order to compare two population means. The samples of data must be small, and the sample data on INTERVAL SCALE; the population distribution(s) is/are assumed to be normally distributed (NORMAL DISTRIBUTION). The two-sample *t*-test should be used only if it can be assumed that the two population STANDARD DEVIATIONS are approximately equal. For the matched pairs *t*-test the two samples must not only be paired: it should be possible to assume that the population distribution of the differences between the matched pairs is normal. The STANDARD ERROR OF THE MEAN is used to find out if two samples are truly representative of the original POPULATION-4, if the means of two samples differ so much that the samples are unlikely to be drawn from the same population, or if the difference between the sample means is a chance or random occurrence; *t* is given as the ratio of the difference between the sample means to the STANDARD ERROR of the difference between the sample means. The value of the calculation can then be checked in a table of *t* values to assess the probability of its being a chance occurrence.

stump a worn-down STACK, the penultimate stage in the cycle of marine erosion of part of a coastline.

subalpine *adj.* of or relating to the lower Alpine slopes, or to the higher mountain slopes just below the TREE LINE.

subarctic, Sub-Arctic *adj.* **1.** the region in latitudes near or just south of the ARCTIC CIRCLE, and the phenomena associated with it **2.** a group of cold climatic types in KOPPEN'S CLIMATIC CLASSIFICATION, with low precipitation and evaporation, the mean temperature of the warmest month sometimes exceeding 10°C (50°F), of the coldest falling below −3°C (26.6°F).

Sub-Atlantic the climatic phase in which we live, dating from about 500 BC (preceded by the SUB-BOREAL), when summer temperatures became lower than those of the Sub-Boreal, but conditions generally became more mild and humid. In Britain the lime forests declined, but the alder, oak, elm, birch, hornbeam and beech flora (with the latter usually dominant in southern areas) spread. PRE-BOREAL.

Sub-Boreal the climatic period dating from about 3000 BC to 500 BC, when conditions became cooler, reverting to the conditions prevailing before 5500 BC, but being generally drier, the dominant tree of the Atlantic stage, the oak, giving way slowly to ash, birch and pine. PRE-BOREAL.

sub-consequent stream a SECONDARY CONSEQUENT STREAM.

subcontinent 1. a large land mass forming part of a continent, and having a certain geographical entity, e.g. the Indian subcontinent **2.** a very large land mass smaller than one usually termed a continent, e.g. Greenland.

subduction zone ANDESITE, DEEP, OCEANIC TRENCH, PLATE TECTONICS.

subglacial *adj.* of, relating to, formed in or by the underside of a GLACIER. ENGLACIAL, SUPERGLACIAL.

subglacial channel a MELT-WATER CHANNEL formed by melt-water flowing beneath an ice sheet or glacier, and now usually dry or carrying only a small stream.

subglacial moraine debris (MORAINE) frozen into and carried by the ice of a GLACIER at or near its base.

subhumid *adj.* applied to a climate in relation to its PRECIPITATION which (combined with other climatic and physical factors) is neither too little, resulting in arid conditions with the growth of XEROPHYTES, nor adequate in amount and evenness of seasonal fall for tree growth. Such a climate leads to the growth of tall grass. SAVANNA.

subjective *adj.* seen from the viewpoint

of the thinking subject, and therefore conditioned by personal characteristics and feelings. HUMANISTIC GEOGRAPHY, OBJECTIVE.

sublimation a process whereby a SOLID changes directly to a VAPOUR without passing through a LIQUID state; or, similarly, a vapour is changed directly to a solid (e.g. the formation of ice-crystals directly from water vapour when condensation occurs at a temperature lower than that of FREEZING-POINT). LATENT HEAT.

sublittoral zone **1.** the marine zone extending from low tide level to the edge of the CONTINENTAL SHELF, part of the NERITIC ZONE underneath the LITTORAL ZONE **2.** in a lake, the part in which the water is too deep for the growth of rooted plants.

submarine canyon a steep-sided valley in the CONTINENTAL SHELF which in some cases may extend right across the shelf into the CONTINENTAL SLOPE. The origin of these canyons is obscure and has provoked much discussion; some are obviously continuations of valleys of the adjoining land.

submarine ridge OCEANIC RIDGE.

submerged coast, submerged shoreline a coastline formed when a rise of sea level in relation to the land leads to the flooding of the former land surface. DALMATIAN COAST, FJARD, FJORD, RIA, SUBMERGED FOREST.

submerged forest the remains of a former forest, now completely covered by the sea except occasionally at very low tide, resulting from the submergence of the COAST caused either by a rise of sea level relative to the land (EUSTATISM) or by a lowering of the land surface (ISOSTASY). SUBMERGED COAST.

subsequent stream a stream indicating that its development was subsequent to a CONSEQUENT STREAM. A consequent stream flows down the DIP, but a subsequent stream excavates its valley along an outcrop of weak rocks such as clays or shales, or other lines of weakness, e.g. a FAULT. If these lines of weakness occur in the direction of the STRIKE, i.e. more or less at right angles to the valleys of the consequent streams, the subsequent streams will meet the consequents at right angles, resulting in TRELLIS DRAINAGE. ANNULAR DRAINAGE. Fig 42.

subsidence **1.** in general, a sinking in level, settling downwards to a lower position, or returning to a normal state **2.** in the earth's crust it may be readjustment on a large scale, as in a RIFT VALLEY, or on a smaller, more superficial scale, e.g. the collapse of a cave roof in KARST or the collapse of a land surface resulting from mining activity; or, in MASS MOVEMENT, a landslide in which there is downward displacement of relatively dry superficial earth material without a free surface (compare ROCK FALL) and without horizontal displacement **3.** in meteorology, the slow descent of a large air mass, as in an ANTICYCLONE (SUBSIDENCE INVERSION).

subsidence inversion a condition of high level inversion in the ATMOSPHERE-1 caused by the slow descent of a large air mass which, as it approaches the ground, decelerates and spreads horizontally, resulting in ADIABATIC compression, the warming of the descending air, and a stable LAPSE RATE. SUBSIDENCE-3.

subsistence agriculture, subsistence farming farming in which the products are grown or raised primarily (but not necessarily solely) for the support of the farmer and the farmer's dependants, not primarily for sale or trading. The opposite is COMMERCIAL AGRICULTURE which is primarily concerned with the growing of crops or raising of livestock for sale. AGRIBUSINESS, FARMING.

subsistence crop a crop (commonly a FARINACEOUS food crop) grown as the basic item of diet to be eaten by the farmers and their dependants, not for sale or trading. CASH CROP, STAPLE-3.

subsistence economy an economic system in which there is little if any buying and selling, although there may be bartering. MARKET ECONOMY.

subsoil an imprecise term for the SOIL layer consisting of weathered parent material lying immediately below the soil proper (SOLUM, TOPSOIL) and above the BEDROCK, corresponding approximately to the C HORIZON. The term is seldom used by pedologists. PEDOLOGY, SOIL HORIZON, SOIL PROFILE.

sub-surface wash the processes involved in the carrying by water of SOLUTES and particles within the REGOLITH, the flow of water being termed THROUGHFLOW. If the particles are carried down a slope, horizontally to the surface of the regolith, the process is termed lateral ELUVIATION; if the throughflow occurs along definite lines of seepage, the process is termed tunnelling.

subtropical *adj.* applied **1.** imprecisely to climatic conditions that are tropical (TROPICAL CLIMATE) for part of the year, or to those that are 'nearly' tropical throughout the year, to the vegetation growing in, and to the lands with, those conditions, i.e. polewards beyond the TROPICS, merging into the warm TEMPERATE zone **2.** more precisely, to latitudinal zones between the TROPIC OF CANCER and about 40°N and between the TROPIC OF CAPRICORN and about 40°S **3.** to belts of atmospheric high pressure occurring in those zones. SUBTROPICAL HIGH PRESSURE BELTS **4.** to climatic regions where the temperature in any month does not fall below about 6°C (43°F), e.g. MEDITERRANEAN CLIMATE. The term extra-tropical is sometimes preferred when the reference is to something occurring just outside the tropics, e.g. extra-tropical high pressure belt.

subtropical high pressure belts the belts of persistent HIGH atmospheric pressure (ANTICYCLONE) with an east to west trend, centred generally about latitudes 30°N and 30°S. Fig 5.

subtropical jet stream a JET STREAM forming at the TROPOPAUSE immediately over the HADLEY CELL. Fig 5.

suburb, suburbs the outer, socially homogeneous, mainly residential or dormitory part of a continuously built-up urban area, town, or city, distinguished from the inner area by a lower housing density, and characterized by a high level of commuting (COMMUTER) to central locations in the inner area. CONCENTRIC ZONE GROWTH THEORY, CENTRAL BUSINESS DISTRICT, INNER CITY, URBAN-RURAL CONTINUUM.

suburban *adj.* **1.** of or pertaining to the SUBURBS-1,2 **2.** having qualities considered to be characteristic of the suburbs or of the people who inhabit them.

suburbia a synonym for SUBURBS, sometimes used contemptuously.

succession in general, the coming of one thing after another in order or time, a series of things in order, a sequence **1.** in geology, the order, in time, of beds of rock; the vertical sequence of rock in a certain locality **2.** in ecology, the progressive natural development of vegetation from the initial pioneer community (pioneer stage) to the CLIMAX, one community being gradually replaced by another under the influence of physical factors (the living organisms responding to topographical features, and EDAPHIC and climatic factors) and biotic factors (organisms reacting to one another). SERE.

succession and invasion INVASION AND SUCCESSION.

succession phenomenon a problem resulting from the control of pests and diseases, in that by removing one pest or disease another is given the opportunity to operate.

succulent a plant adapted to meet water loss. *adj.* as applied to a xerophytic (XEROPHYTE) plant with enlarged tissue for water storage in leaves and/or stems, which allows it to withstand drought.

sudd (Sudan: Arabic) **1.** floating, compact masses of vegetation in the upper Bahr el-Jebel (White Nile) which obstructs navigation **2.** the marshes in Sudan resulting from this obstruction.

sugar one of a class of crystalline carbohydrates soluble in water, the term generally applied to sucrose, manufactured by all green plants and commonly stored in roots, bulbs, flowers, fruit. The plants from which sugar is refined for commercial use are mainly SUGARCANE (yielding cane sugar) and SUGAR BEET (yielding beet sugar); but some is also regularly obtained from the wild date palm, the sugar palm, the palmyra palm, and the sugar maple. In addition to its use as a foodstuff, sweetener and preservative for other foods sugar (especially that obtained from sugarcane) provides alcohol which can be used as a fuel or a fuel additive; and current experiments indicate that it may be used as a substitute for petrochemicals in various industrial processes, e.g. the making of detergents and plastics.

sugar beet a BIENNIAL plant with a white, conical, swollen root yielding sucrose (SUGAR), native to midlatitude lands in continental Europe, grown there and in similar conditions elsewhere for a sugar supply. The leafy tops and the pulp remaining after the sugar has been extracted are used for cattle feed; molasses, another by-product of the processing of the roots, is also used as stock feed and for making industrial alcohol, the filter cake remaining after the juice has been purified by filtration being used as manure.

sugarcane *Saccharum officinarum*, a species of tall, coarse PERENNIAL grass, native to tropical areas of the OLD WORLD, still cultivated there but even more in similar conditions elsewhere, especially in the West Indies, the USA, Central and South America, and in Australia, for its high yield of sucrose (SUGAR) obtained from the stems of the cane. The plant needs a fertile soil, heavy dressings of manure and fertilizer, a good rainfall, or irrigation.

suitcase farmer (American) an agricultural landholder whose holdings are scattered, and who moves from one holding to another at crucial times, e.g. seed-time, harvest-time, to make the best use of farm machinery. The crop is usually a CEREAL which needs little attention in the growing period. SIDEWALK FARMER.

sukhovey (Russian) a hot, dry wind, blowing during the summer mainly from the southeast in the southeastern part of European CIS and Kazakhstan. The air temperature may rise to 35°C to 40°C (95°F to 105°F) and the RELATIVE HUMID-

ITY may drop to 15 per cent or less, causing excessive evaporation and seriously injuring crops and other vegetation.

sulphur (American sulfur) a nonmetallic element occurring in crystalline or amorphous form, abundant and widespread in nature in a free state or combined as sulphates and sulphides, an essential MACRONUTRIENT for plants. It is deposited near volcanic vents and HOT SPRINGS, occurs in sedimentary rocks and is associated with SALT DOMES.

sulphur (sulfur) dioxide a stable oxide of SULPHUR, gaseous except at very unusual atmospheric pressure, but easily liquefied by pressure, produced by the burning of a wide range of fuels, but especially of coal and heavy fuel oil. It is released, with sulphates (sulfates), in volcanic eruptions. ACID RAIN.

sulphuric (sulfuric) acid a colourless, dense, rather oily liquid (oil of vitriol), dissolving and ionizing with ease in water, used extensively in petroleum refining, in the chemical industry and in making detergents, explosives, dyes, rayon fabrics. It is formed in the atmosphere by the combination of sulphur (sulfur) trioxide with water, resulting in a stable mist of small acid droplets which fall to the ground and, sulphur being an important plant MACRONUTRIENT, contribute to soil fertility. Excessive deposition of sulphur from the atmosphere is, however, undesirable in an area with an acid soil, where it may lead to a high level of acidity in lakes and rivers, and a slowing down of growth, or even the death, of some plants; but it is not so disastrous in an area of ALKALINE SOIL where it may raise soil fertility. The term ACID RAIN is applied to precipitation charged with excessive sulphuric acid (and other acids) derived in part from SULPHUR (sulfur) DIOXIDE emitted in the burning of HYDROCARBON fuels and in other processes in towns and industrial plant.

summer 1. in general, the warmest season of the year, as opposed to winter, the coldest **2.** in the northern hemisphere, the months June, July, August; in the southern December, January, February **3.** in astronomy, the period between the summer solstice (about 21 June) and the autumn equinox (about 22 September) in the northern hemisphere; and between the solstice about 22 December and the equinox about 21 March in the southern. EQUINOX, SEASON, SOLSTICE.

summer solstice SEASON, SOLSTICE, SUMMER.

sun 1. in general, any heavenly body that is the centre of a solar system **2.** in particular, the central body of 'our' SOLAR SYSTEM, lying at the main focus of the orbits of the earth and the other PLANETS, a dwarf yellow star in the spiral arm near the outer edge of the Milky Way, the spiral GALAXY in which the solar system lies. A nearly spherical gaseous body, diameter about 1 392 000 km (865 000 mi), it is thought to be composed of approximately 90 per cent hydrogen, 10 per cent helium mixed with small amounts of all other known elements, the temperature at the core being some 15 mn°C (59 mn°F), at the surface some 5700°C (10 292°F). It rotates once in 24.5 days at its equator, and its mean distance from the earth is 150 mn km (92.9 mn mi) (APHELION, PERIHELION). Only a very small amount of the total SOLAR RADIATION reaches the earth, but it is responsible there, directly or indirectly, for most energy-requiring processes, including the growth of animal and plant life. INSOLATION. Fig 18.

sunrise, sunset the times at which the SUN apparently rises in the morning and sets in the evening below the horizon, varying with latitude and with the sun's declination, due to the earth's rotation on its axis. Sunrise is defined in meteorology as the time at which the upper edge of the sun appears above the apparent horizon on a clear day; sunset, the time at which the upper edge of the sun appears to sink below the apparent horizon on a clear day.

sunspot a dark area on the visible surface of the SUN, consisting of a grey region surrounding a darker centre, causing an increase in the SOLAR RADIATION received on earth, and particularly affecting the earth's magnetic field and IONOSPHERE. Their number usually reaches a maximum every eleven years.

supercooling cooling to a temperature below the normal transition point for a change of state without the occurrence of that change, e.g. when water, if undisturbed, stays liquid at a temperature below 0°C (32°F). This can happen naturally if water droplets in clouds are not disturbed; but they become ice immediately on coming into contact with another body, e.g. an aircraft (which can be most dangerous), or tall trees; and supercooling can also create very large hailstones. SUPERHEATING.

superficial *adj.* of or relating to the surface, not penetrating below the surface.

superficial deposit unconsolidated Pleistocene or Holocene material lying on the surface of the earth's crust, not formed in situ from underlying BEDROCK, but carried and set down in position by the agencies of wind (e.g. LOESS), water (e.g. ALLUVIUM), ice (e.g. GLACIAL DRIFT), or by gravity (e.g. COLLUVIUM). PEAT, formed in situ, is also included; but a SOIL formed in situ through the weathering of the underlying rock is not. CLAY-WITH-FLINTS, of uncertain origin, is usually referred to as a superficial deposit. SURFACE DEPOSIT.

superglacial, supraglacial *adj.* of or relating to the surface or to the environment at the surface, of a glacier. ENGLACIAL, SUBGLACIAL.

superglacial stream a rivulet of meltwater flowing in summer in a deep runnel on the surface of a glacier and descending into a crevasse. ENGLACIAL RIVER, SUBGLACIAL CHANNEL.

superheating heating to a temperature above the normal transition point for a change of state without the occurrence of that change, e.g. when water is heated above boiling point without boiling occurring. SUPERCOOLING.

superimposed (superposed) drainage epigenetic drainage, a drainage pattern appearing to be independent of the structure of the underlying rocks because it was established on a former rock surface, since removed by denudation.

supermarket a large, self-service store stocked with food and (usually) some small household goods displayed on open shelves. HYPERMARKET.

superphosphate any of the various phosphate fertilizers containing soluble phosphorus pentoxide, obtained by treating PHOSPHATE rock with SULPHURIC ACID. PHOSPHORUS.

superposition LAW OF SUPERPOSITION.

supersaturation the state of a SOLUTION having a higher concentration of solute than at saturation (SATURATED), produced by heating and slow, steady, undisturbed cooling or by very fast cooling of a saturated solution. Supersaturation occurs in the ATMOSPHERE-I when a cooling body of air with a RELATIVE HUMIDITY exceeding 100 per cent has enough water vapour to produce condensation, but condensation does not take place unless solid particles (e.g. dust) or negative IONS are available. QUICKSAND.

supply and demand the market forces that govern prices, operating freely in a MARKET ECONOMY, apparent through the price mechanism as responses in quantities offered for sale (the supply) or the quantities that consumers are prepared to buy (demand) when the market price changes.

supraglacial *adj.* sometimes applied to the environment at the surface of a glacier. SUPERGLACIAL.

surazo (Brazil: Portuguese) friagem, a cold wind blowing in winter in the CAMPO of Brazil in the middle Amazon region, produced by an ANTICYCLONE. Temperatures may fall below 10°C (50°F), causing great discomfort.

surf the foaming water produced by a powerful wave as it breaks on rocks on the seashore. LITTORAL CURRENT.

surface deposit unconsolidated material overlying BEDROCK, weathered from the bedrock itself (RESIDUAL DEPOSIT), or weathered elsewhere and carried to the present position by wind (AEOLIAN DEPOSIT), by water (ALLUVIUM), by ice (GLACIAL DRIFT) or by gravity (COLLUVIUM). SUPERFICIAL DEPOSIT.

surface tension the surface force acting on the surface of a liquid with the effect of reducing the surface area to the minimum. This surface force is the result of forces within the liquid acting on the molecules of the surface in the absence of forces acting above the surface, and thereby drawing the surface together. In a small quantity of liquid a shape with the smallest area possible will be found, i.e. a sphere. Surface tension is usually reduced by a rise in temperature. CAPILLARY MOISTURE, POROSITY.

surface wash processes whereby soil and other unconsolidated material are carried by flowing water across the surface of the land. OVERLAND FLOW is the major process, but on sloping ground the action of large raindrops or a heavy downpour in dislodging soil particles is also involved. SOIL EROSION. WANING SLOPE, WAXING SLOPE.

survey **1.** a general inspection or viewing as a whole **2.** a careful consideration, inspection and examination as a whole and in detail, e.g. of a place, building, population, problem, state of affairs, condition etc. **3.** the presentation of the finding of such a survey in written, diagrammatic, cartographic, photographic form **4.** the process of gathering data relating to a chosen area, e.g. by REMOTE SENSING or by the measuring and recording of lines and angles of an area of land in order to make an accurate map of it **5.** an area that has been so measured and recorded etc. **6.** a group of people, or a department, engaged in the surveying indicated in **1,2,3,4** **7.** in statistics, a method for estimating. SAMPLE SURVEY.

suspended load the fine organic and inorganic materials (e.g. silt and clay) consisting of particles with diameters commonly less than 0.2 mm (sometimes termed wash load) carried by a STREAM-I in SUSPENSION (without the aid of SALTATION), as distinct from the BED LOAD and the DISSOLVED LOAD. Measurements of suspended material are usually expressed in milligrams per litre or kilograms per cubic metre, and are used to calculate transport rates.

suspension a two-phase system in which a finely divided insoluble solid is uniformly dispersed in a liquid or gas. In moving water, e.g. in a STREAM-I, small particles are kept buoyant (in suspension) by turbulent upward eddies which prevent the particles sinking under gravity; thus the finest particles may be carried long distances by a stream or river before they sink to the bottom or are carried out to sea. COLLOID, SALTATION, SOLUTION, SUSPENDED LOAD, TRACTION.

sustainable development DEVELOPMENT-I in which the rate of EXPLOITATION-I of RESOURCES-I,2 does not exceed that of the renewal of those resources, or degrade the STOCK-2. MAXIMUM SUSTAINABLE YIELD, RESOURCE MANAGEMENT.

sustained yield forestry any practice or technique (e.g. PATCH CUTTING) used in FORESTRY which aims to conserve the forest and ensure that it remains a flow resource rather than a stock resource (NATURAL RESOURCES). CONSERVATION, RESOURCE CONSERVATION, RESOURCE MANAGEMENT.

swale I. a low, long narrow depression approximately aligned with the coastline and lying between two ridges of shingle on a beach **2.** (American) a marshy or moist depression in level or rolling land, or a long narrow depression lying between the bars of a POINT BAR deposit on the floodplain of a river. Fig 29.

swallow, swallow-hole, swallet I. a deep vertical hole or opening in the earth, especially one produced by the solution of rocks in limestone country (KARST), down which a surface stream or rainfall disappears, in some cases as a waterfall **2.** a hole in a stream bed through which some of the stream water flows and disappears; or a point of no fixed location where a stream may dry up as its water percolates downwards.

swamp I. in general, wet spongy land saturated with water for much of the time, and its associated vegetation **2.** more precisely, the soil-vegetation type in which the normal summer water level is above that of the soil surface, and the characteristic vegetation is woody **3.** in ecological freshwater studies, the last phase of aquatic vegetation before it gives way to land vegetation (the sequence being, from the centre of a body of freshwater to the margins: aquatic, swamp, marsh). BOG, FEN, MANGROVE SWAMP, MARSH.

swash the body of water which rushes up a beach after an ocean WAVE has broken. BACKWASH. Fig 28.

S-wave SHAKE WAVE.

swell I. of a large body of water, especially in the ocean, the regular, undulating motion of the surface, the succession of waves which do not break **2.** on the deep sea floor, a RISE.

swidden farming a term applied by some authors to a type of SHIFTING CULTIVATION in which the existing vegetation is removed by cutting and burning (SLASH-AND-BURN), the cleared area being sown and cropped until the soil nutrients are exhausted and yields begin to fall. The area is then abandoned, the cultivators moving to a new patch which they similarly clear and sow (compare BUSH FALLOWING); but after some time they may return to the abandoned site where the natural vegetation has regenerated and soil fertility has built up naturally, to repeat the process.

symbiosis 1. in biology, the very close association of dissimilar organisms to their mutual benefit. COMMENSALISM, LICHEN, MUTUALISM, MYCORRHIZA, PARASITISM **2.** in human ecology, the very close, mutually advantageous relations among dissimilar members of a human group or among dissimilar groups within a larger group, or among dissimilar institutions and activities. SYMBIOTIC RELATIONSHIP.

symbiotic relationship the relationship of the dissimilar organisms in SYMBIOSIS, applied by analogy to other phenomena, e.g. the mutually beneficial relationship between a city or large town and its surrounding area, or between different groups within a human community when the groups are unlike and the relations complementary.

symbol a sign, shape or object accepted as representing or typifying a thing, person, idea, quality or value. SEMIOTICS. In cartography an explanation of the conventional signs used to represent specific objects commonly appears on the map itself; but if very many symbols are used, reference is usually made to a CHARACTERISTIC SHEET.

symbolic model a MODEL-2 in which selected aspects of reality are all expressed by mathematical expressions. ANALOGUE MODEL, ICONIC MODEL, MATHEMATICAL MODEL.

symmetrical fold a FOLD in which the limbs dip away symmetrically (SYMMETRY) from the AXIS OF FOLD.

symmetry the quality of having a form so regular that one or more axes exist which divide the structure in exactly corresponding and equal parts.

synclinal valley a valley which follows a SYNCLINE in the underlying rocks, i.e. a valley formed by a downfold.

syncline a downfold in the STRATA of the earth's crust, the rocks dipping inwards to a central axis (AXIS OF FOLD), caused by COMPRESSION. ANTICLINE, PITCH-3, SYNCLINORIUM. Fig 1.

synclinorium a broad downfold (SYNCLINE) of the rocks over a considerable tract of country on which numerous minor upfolds (ANTICLINE) and downfolds are superimposed. Fig 1.

synoptic *adj.* providing a general summary or a general view.

synoptic chart a chart giving a summary, a general view of the meteorological conditions (isobars, temperatures, winds etc.) over a large area at a given time, an essential tool in weather forecasting. WEATHER CHART.

synoptic climatology the comprehensive study of the condition of the ATMOSPHERE-1 over a very large area (e.g. one of the hemispheres) at a particular time, of prime importance in weather forecasting. METEOROLOGY, SYNOPTIC METEOROLOGY.

synoptic image a general view of a part of the earth's surface, usually provided by a SENSOR mounted on a high-altitude SATELLITE-1. REMOTE SENSING.

synoptic map a map displaying all the factors to be taken into account in solving a problem. The factors are shown either on one sheet or on a series of OVERLAYS placed over the base map, each overlay showing a separate factor or a selected combination of factors. SIEVE MAP.

synoptic meteorology the branch of METEOROLOGY concerned with the collection of meteorological data (over a smaller area than that involved in SYNOPTIC CLIMATOLOGY), the making of SYNOPTIC CHARTS from the information gathered and the interpretation of the pressure patterns etc. revealed by the charts in order to anticipate changes likely to affect the weather in a particular area. MICROMETEOROLOGY.

synthetic fibre an ARTIFICIAL fibre which can usually be spun, woven, felted etc., made to rival the natural fibres (fine, thread-like structures occurring in plant or animal tissue) of cotton, wool, flax, silk.

synthetic rubber ARTIFICIAL RUBBER made from HYDROCARBONS derived from PETROLEUM.

system 1. a set of related elements organized for a particular purpose, the whole being identifiable by the interconnexion of the elements **2.** a set of things or substances, associated, interdependent, governed by physical laws, and making a whole (as in the SOLAR SYSTEM) **3.** a set of things, structures, processes, activities (e.g. human activities) associated and interconnected, forming and functioning as a complex whole through a regular set of relations (as in the nervous system of an animal) and in many cases forming part of a larger system, the larger system itself forming part of an even larger system **4.** a set of principles linked to form a coherent doctrine **5.** a method of organization, administration, procedure (as in a legal system) **6.** a formal method of classification, nomenclature, notation, governed by well-defined rules **7.** in geology, a division of the succession of stratified rocks deposited during a geological period (GEOLOGICAL TIMESCALE).

systematic *adj.* **1.** working in an orderly, methodical way, in accordance with a SYSTEM-1,2,3,4,5,6 **2.** constituting a SYSTEM-2 **3.** in biology, relating to classification **3.** in statistics, the opposite of random or STOCHASTIC.

systematic error in SAMPLING-2, an error which is in some way biased, occurring predominantly in one direction throughout the sampling measurements, e.g. in a survey of people charged with exceeding the speed limit in driving, most of the accused will underestimate rather than overestimate their speed. A systematic error, unlike a RANDOM ERROR, does not balance out and will BIAS the results of a survey.

systems analysis, systems approach an approach which uses the concept of the SYSTEM-1,2,3 (i.e. broadly, a complex whole with interrelated elements) as an analytical tool. The system is identified by defining its boundaries, its purpose, and (if it is a subsystem), its position, its role, in a larger system. The structure and function of the system are investigated, and the level of abstraction at which it is to be treated defined (MODEL-2). As in GENERAL SYSTEMS THEORY, attempts are made to discover ISOMORPHISMS between systems so that ANALOGUE MODELS can be constructed.

syzygy in astronomy, the point at which two heavenly bodies are in CONJUNCTION or OPPOSITION, especially the MOON and the SUN; applied particularly to the point when the sun, moon and earth are in the same straight line, thus coinciding with the new moon and full moon as seen from the earth.

T

tableland a flat or gently undulating area of high relief, a term generally superseded by PLATEAU, but sometimes applied specifically to a plateau with abrupt cliff-like edges rising sharply from surrounding lowlands, e.g. a MESA.

tabular iceberg a very large, floating ice mass with a flat top, broken off from the Ross ICE SHELF and floating in the Southern Ocean. ICEBERG.

taiga (Russian) cold woodland, the predominantly CONIFEROUS forest stretching in a broad zone in the northern hemisphere, adjacent to the TUNDRA. There does not seem to be any authority for the spelling taïga: etymologically tayga is more correct.

tail dune a SAND DUNE of varying length, formed in the LEE of an obstacle and tapering away from it.

talik (Russian) **1.** a layer of unfrozen ground between the seasonally frozen ground (ACTIVE LAYER) and the PERMAFROST **2.** an unfrozen layer within the permafrost **3.** unfrozen ground between permafrost.

talus 1. commonly applied to a SCREE, a sloping heap of rock debris at the foot of a cliff or mountain slope **2.** applied specifically to the landform produced by such rock debris, the term scree being restricted to the rock debris itself. TALUS CREEP.

talus creep the slow movement of TALUS-1 or SCREE down a slope under the influence of gravity. MASS MOVEMENT.

talweg (German; older spelling thalweg), the LONGITUDINAL-2 profile of a river. LONG PROFILE OF A RIVER, RIVER PROFILE.

tank in Indian subcontinent and Sri Lanka, a reservoir for irrigation, a small lake or pool made by damming the valley of a stream to retain the monsoon rain for later use.

tanker any large rail or road vehicle or ship especially designed with a large vessel (a tank) for the carrying of liquids in bulk.

tapioca starch the very pure starch extracted from the root of the MANIOC (cassava) plant, potentially useful in the making of biodegradable plastic materials. BIODEGRADATION, NATURAL RESOURCES.

tar a dark, thick, viscous substance obtained by the destructive DISTILLATION of coal, wood, or other organic material (COAL TAR). It is used in road surfacing (MACADAM), as a preservative for wood and iron, and in making dyes, antiseptics etc.

tariff 1. a scale of duties imposed by a government on goods exported from or imported into the territory under its jurisdiction; or the classificatory instrument embodying such duties **2.** the duty imposed **3.** a scale of rates or charges, e.g. for hotel accommodation.

tarmac MACADAM.

tarn a small lake among mountains, usually of glacial origin, fed by rainwater from the surrounding steep slopes rather than by a distinct feeder stream; sometimes applied to a nearly circular lake occupying a corrie or CIRQUE.

tar-sand OIL-SAND.

taxonomy the science of classification, usually restricted to the classification of living and extinct plants and animals and covering the principles and methods employed in classifying organisms in hierarchical groups. CLADISTICS, CLASSIFICATION OF ORGANISMS.

tea a product used in making the beverage of the same name, manufactured from the young leaves of a small EVERGREEN tree, *Camellia sinensis*, usually pruned to form a shrub, native to hill lands of southeast Asia, now grown there and elsewhere in Asia, in east Africa and in areas bordering the Black Sea. It flourishes in high rainfall if the soil is well drained, yields well on highly acid soils, withstands frost.

teak a large DECIDUOUS tree, native to the relatively wet areas of tropical regions of southeast Asia, providing very hard, close-grained timber, used especially in furniture-making.

tear fault a STRIKE-SLIP FAULT, sometimes termed a transcurrent fault.

technological determinism in Marxism, the theory that the predominant type of TECHNOLOGY-3 of a society determines its other features, particularly its social organization.

technology **1.** the scientific study concerned with the practical and industrial arts **2.** the practical arts or practical science **3.** the systematic application of scientific knowledge to industrial processes or to the problems arising from the interaction of people with their environment. ALTERNATE, APPROPRIATE, INTERMEDIATE, LOW, and HIGH TECHNOLOGY.

technosphere the part of the physical environment created or modified by human action. PSYCHOSPHERE.

tectonic *adj.* of, relating to, or arising from, the processes which build up or form features of the earth's crust. PLATE TECTONICS.

tectonics a branch of geology concerned with the study of the processes involved in the formation of structural features of the earth's crust. PLATE TECTONICS. Figs 36, 37.

teff a tropical MILLET, the most widely grown food grain in Ethiopia (rarely cultivated elsewhere), yielding a fine white flour.

teleology consequentialism, the study of ends, goals or purposes, the doctrine of final causes. The belief that an explanation of anything (process, object, act, event etc.) can be achieved only if the ends to which it is directed are considered; that explanation restricted to terms of CAUSALITY is insufficient.

temperate *adj.* moderate, without extremes, of equable temperature, especially a climate without extremes of temperature.

temperate glacier WARM GLACIER.

temperate zone term applied in classical times to the latitudinal temperature zone lying between the tropics (23°30'N or S) and the polar circles (66°30'N or S), the midlatitudes. But temperatures of great extremes occur in these midlatitude belts (especially in the interiors of the continents), so in making specific reference to them in terms of a climatic zone the term temperate is inappropriate: it is better to use the *adj.* MIDLATITUDE. Temperate zone does however appear in KOPPEN'S CLIMATIC CLASSIFICATION.

temperature 1. the property of an object which indicates the direction in which heat energy will flow if the object is put in thermal contact with another, heat energy flowing from places of higher to places of lower temperature 2. as a climatic element, the degree of sensible heat or cold in the ATMOSPHERE-1 (SENSIBLE TEMPERATURE), measured on various scales. ABSOLUTE as applied to range, temperature, zero; and DIURNAL RANGE, INVERSION OF TEMPERATURE, TEMPERATURE ANOMALY, THERMOMETER.

temperature anomaly the difference between the mean temperature of a place and the mean temperature along its parallel of latitude (in both cases adjusted to sea-level). If the mean temperature of the place is higher than that of the mean along its parallel of latitude, the anomaly is qualified as positive; if lower, as negative. The British Isles is much warmer than the average of places on the same latitude, and thus has a positive temperature anomaly, especially in winter. ANOMALY, ISANOMAL.

temporal *adj.* of, related to, pertaining to, time (as distinct from SPATIAL).

tension a pulling force, tending to stretch, to cause the extension of a body or to restore the shape of an extended elastic object. In the rocks of the earth's crust tension extends strata, resulting in JOINTS and NORMAL FAULTS. Compare COMPRESSION, STRESS, THRUST.

tenure 1. the act, manner or right of holding office or property, especially landed PROPERTY-2 2. the period of holding this. LAND TENURE.

tephra a collective term for all the solid material ejected into the air from a volcanic vent during an eruption, i.e. PYROCLAST.

teratogenic pollution the kind of POLLUTION which gives rise to the birth of malformed or otherwise defective organisms.

terminal moraine, end moraine a crescent-shaped MORAINE forming a ridge beyond the SNOUT of a GLACIER or at the end of an ice sheet, marking the furthest extent of the ice. If it is large it indicates a long pause in the retreat of the ice.

terrace 1. in agriculture, one of a series of horizontal steps cut into a hillside to provide cultivable land in an area of steep relief, and to reduce soil erosion. TERRACE CULTIVATION 2. in geology, a level or nearly level and horizontal or nearly horizontal strip of land, usually narrow and bordering the sea, a lake, or a river (RIVER TERRACE), lying between a rising slope on one side and a downward, often abrupt, slope on the other. ALLUVIAL TERRACE, CONTINENTAL TERRACE, KAME TERRACE, MEANDER TERRACE; and REJUVENATION, TERRACE GRAVEL, TERRACETTE.

terrace cultivation a system of cultivation which reduces SOIL EROSION, practised in areas of steep relief and scarcity of level land. A series of artificial horizontal steps is cut into the hillside, the soil being retained by stone walls or earth banks, behind which crops are sown. If irrigation is necessary (e.g. in rice cultivation) the water is allowed to move by gravity from the upper to the lower terraces.

terrace gravel a gravel deposit remaining on a RIVER TERRACE after the erosion of the finer alluvium with which it was originally combined.

terracette one of a series of narrow horizontal steps from a few centimetres to 60 cm (up to 2 ft) in height, making a ribbed pattern on a steep, usually grassland, slope in areas accessible to animals. Their origin is disputed. They may owe their existence to SOIL CREEP and, once formed, be used by sheep and other animals; or they may have been formed initially by animals treading the easiest route up the hill. LYNCHET.

terracing the work of making terraces for TERRACE CULTIVATION.

terrain an area of land in respect of its physical characteristics or condition, especially if considered for its fitness or use for a special purpose, e.g. for laying a railway track, or for a military operation.

terra rossa (Italian) a red-coloured thin clay LOAM soil, rich in iron, developed in limestone areas with a warm temperate, seasonally dry climate, occurring especially in KARST in lands bordering the eastern coast of the Adriatic sea, and elsewhere in the Mediterranean region. The fact that it supports GARIGUE suggests that deforestation may have contributed to its existence.

terra roxa (Brazil: Portuguese) a deep, dark red-purple, porous soil, rich in humus, formed on the Parana plateau in Brazil, especially suitable for coffee cultivation.

terrestrial *adj.* **1.** of or pertaining to land, or to the EARTH-1 (as opposed to CELESTIAL) **2.** consisting of land (as opposed to water) **3.** growing on land **4.** living on land. AQUATIC.

terrestrial deposits deposits laid down on land, as opposed to MARINE-1 deposits.

terrestrial magnetism the MAGNETIC FIELD of the earth as a whole, weak and varying in intensity and direction (MAGNETIC DECLINATION, PALAEOMAGNETISM), which causes the needle of a magnetic COMPASS, swinging freely in a horizontal plane, to come to rest, indicating the north and south MAGNETIC POLES. The origin of terrestrial magnetism has not yet been satisfactorily accounted for. MAGNETOSPHERE.

terrestrial radiation long-wave RADIATION given out by the earth.

terrigenous *adj.* derived from the land.

terrigenous deposits inorganic deposits (sand, gravel, pebbles etc.) derived from the denudation of the land and laid down in the LITTORAL ZONE-1 of the sea floor, as distinct from PELAGIC DEPOSITS.

territorial waters the coastal waters with the sea bed below and all that lies or lives therein (and the air space above) over which a coastal state has sovereignty. Various international conventions have suggested the zones shown in Fig 45 (but discussions continue). A line drawn to link major promontories is deemed to be the landward baseline; for the treatment of bays, see BAY-2.

A coastal state has (a) sovereignty over the territorial sea (b) powers of policing the contiguous zone (to prevent infringement of its customs, fiscal, immigration and sanitary regulations etc.) (c) sovereign rights over the sea bed of the so-called continental shelf (to explore and exploit its natural resources). For this purpose the seaward limit of the continental shelf is defined as 370.4 km (200 nautical miles) from the baseline or a limit coinciding with the outer edge of the CONTINENTAL

MARGIN, providing this is not beyond 648.2 km (350 international nautical mi) from the baseline. The exclusive economic zone extends a coastal state's rights of sovereignty over the natural resources of the waters as well as those over the natural resources of the sea bed. Where the distance between coastal states is less than 740.8 km (400 international nautical mi) a median line is usually drawn between the closest points of the baselines of the states concerned.

The traditional freedom of the high seas retains for all nations the right of navigation, overflight and laying of submarine cables in the areas of the so-called continental shelf and the exclusive economic zone. Legal problems relating to rights of ownership of the deep sea bed and of rights of exploitation of its resources have yet to be resolved. Fig 45.

territory 1. the area of land and adjacent seas, and the air space over both, ruled by a sovereign authority **2.** an area dependent on a sovereign state, but having some autonomy, e.g. an area supervised by a sovereign state on behalf of the United Nations **3.** historically, in Australia, Canada, USA, an area not admitted to full rights as a state or province, having a separate legislature under an administrative authority appointed by the central government **4.** any large tract of land, a region, a district, with undefined boundaries **5.** in zoology, the area of the habitat occupied by an individual animal or group of animals which will be defended by them, attacks being made especially against a trespasser or trespassers belonging to the same SPECIES-1 as the occupant(s).

tertiary third in order or rank.

Tertiary *adj.* a much discussed term applied to a division of geological time. It is now generally accepted that it applies to the third period of geological time with its associated system of rocks (following the CRETACEOUS and preceding the Quaternary) consisting of the Palaeocene, Eocene, Oligocene, Miocene, Pliocene epochs (of time) and series (of rocks) when mammals became dominant and mountain

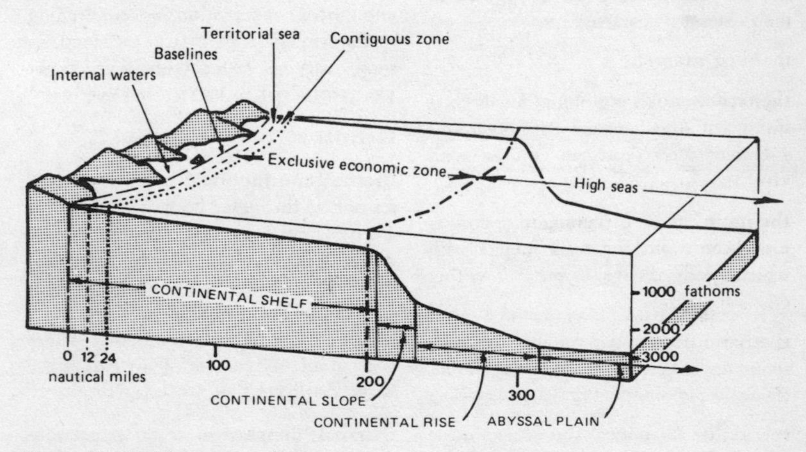

Fig 45 Territorial waters: maritime zones

chains such as the Alps and Himalayas were formed, i.e. the earlier of the two periods of the Cainozoic era (GEOLOGICAL TIME-SCALE).

tertiary industry, tertiary activity, tertiary sector one of the main categories of INDUSTRY, the activity concerned with service to the PRIMARY and SECONDARY INDUSTRIES, to the community and to the individual, e.g. financial, commercial and educational institutions, distributive trades, professions, transport and communications, construction, repairs, maintenance, defence, personal services. QUATERNARY INDUSTRY.

Tethys the name applied to the ocean and the GEOSYNCLINE which it occupied, separating LAURASIA and GONDWANALAND. CONTINENTAL DRIFT, PANGAEA, PLATE TECTONICS.

texture 1. in geology the size, shape, arrangement and distribution of particles constituting a rock or a surface deposit, as opposed to the chemical character of such particles **2.** the arrangement and relationship of particles of a soil as opposed to their chemical character.

thalweg TALWEG.

thanatocoenosis a group of fossils consisting of the remains of organisms assembled after death, in contrast with BIOCOENOSIS-3.

thematic *adj.* of, pertaining to or constituting the topic, the main subject with which a study, discussion, piece of writing etc. is concerned.

thematic map a map on any scale representing a specific spatial distribution, theme, topic or aspect under discussion.

theodolite an optical surveying instrument used to measure vertical and horizontal angles by means of a small telescope, a spirit level, and graduated arcs, mounted on a tripod. TRIANGULATION.

theory 1. in general, an organized body of ideas, an integrated system of HYPOTHESES, put forward as the truth of something, supported by a number of facts relating to it, but sometimes resulting from speculation **2.** scientific, a structure resting on a series of steps of observations and assumptions, each supported by the preceding step, put forward to explain a particular class of phenomena **3.** a process of investigation based on logical or mathematical reasoning rather than on experiment.

thermal a current or updraught of air rising vertically in the ATMOSPHERE-1 (of great advantage to birds and glider-pilots), the result of differential heating by the sun's rays on a sunny day of small parts of the earth's surface. It is the parts that warm up more rapidly which give rise to conductional heating and ABSOLUTE INSTABILITY and thus to rapidly ascending vertical currents. A thermal may rise high enough for condensation to occur, leading to the formation of CUMULUS cloud, in some cases associated with heavy CONVECTION RAIN and a THUNDERSTORM.

thermal *adj.* pertaining to heat.

thermal conduction the process of transfer of heat through a body where there is a temperature gradient, the heat energy diffusing through the body by the action of particles of high KINETIC ENERGY on particles of lower kinetic energy (from a high to a low temperature point). There is no visible movement of any part of the body. CONVECTION, RADIATION.

thermal depression a LOW pressure system, usually intense, varying in size, caused by localized heating of the earth's surface, leading to convectional rising of air, resulting in heavy rainfall and thunderstorms if the warmed air rises high enough for condensation to occur (THERMAL). Small scale thermal depressions lead to DUST DEVILS in hot deserts and the SIMOOM in the Sahara; on a larger scale they are associated with MONSOON conditions.

thermal erosion a type of EROSION occurring in areas of PERMAFROST where, if the organic layer protecting the land surface is removed, the GROUND ICE-2 melts and the land surface breaks up.

thermal (infra-red) sensing REMOTE SENSING in which a thermal scanner, penetrating darkness and cloud, is used rather than a camera to sense the natural RADIATION emitted by features on the earth's surface and, by revealing relative temperature differences, to detect pollution in water, the source of forest fires etc. INFRA-RED RADIATION.

thermal metamorphism contact metamorphism, the alteration of pre-existing rock to form a new, well-defined type of rock, caused by a rise in temperature, usually brought about by an INTRUSION into the pre-existing rock of very hot, molten IGNEOUS ROCK. METAMORPHIC AUREOLE, METAMORPHISM, REGIONAL METAMORPHISM.

thermal pollution the heating of part of the environment by the discharge of substances with TEMPERATURES higher than that of the AMBIENT. The effect is particularly detrimental in freshwater because heating tends to lower its content of free dissolved oxygen, needed by most of the organisms living in it. POLLUTION.

thermal spring hot spring, a continuous flow of hot water from the ground, usually (but not always) associated with present or former volcanic activity. It contrasts with a GEYSER, with its violent, intermittent emission. GEOTHERMAL ENERGY.

thermal stratification the succession of well-defined layers of water of different temperatures lying at various depths in the ocean or a deep lake, the top layer being the warmest, except under ice, where there may be an inversion or, in very cold areas, just one unstratified layer of very cold water. EPILIMNION, HYPOLIMNION, THERMOCLINE.

thermocline the layer of water in an ocean or deep lake, lying between the non-circulating HYPOLIMNION and the warmer EPILIMNION, through which the temperature falls swiftly with increasing depth, commonly exceeding 1°C per metre (about 11°F per foot) of descent. THERMAL STRATIFICATION.

thermokarst, cryokarst a KARST-like landform with irregular depressions formed in periglacial or former periglacial superficial deposits as a result of the melting of GROUND ICE-2 and the subsequent settling or caving of the ground. The term is not widely used by British geomorphologists, but is favoured by Russian authors.

thermometer an instrument used to measure TEMPERATURE on any temperature scale, commonly consisting of a graduated glass tube with a bulb at one end containing mercury or alcohol which is heat-sensitive, expanding with increase in heat, and therefore rising in the tube; contracting with decrease, and falling in the tube (DRY-BULB THERMOMETER, MAXIMUM-MINIMUM THERMOMETER, WET-BULB THERMOMETER). Less common types incorporate heat-sensitive metals, which expand or contract with temperature change, at a known rate; or which, with temperature change, have varying resistance to the passage of electricity. There is also a gas thermometer which measures pressure variations in a gas maintained at constant volume.

thermonuclear *adj*. relating to NUCLEAR FUSION.

thermosphere the layer of the upper atmosphere (above the MESOPAUSE) in which temperature increases with increasing height. Fig 4.

Thiessen polygon DIRICHLET POLYGON.

Third World (from French tiers monde) a term of variously defined origin. It was applied (as Tiers Monde) in the 'cold war' period of the 1950s to countries committed to neither of the two major power blocs, i.e. neither to the generally free market economy Western 'capitalist' bloc (the FIRST WORLD), nor to the centrally controlled economy Eastern 'communist' bloc (the SECOND WORLD). The term Third World is now generally applied to countries considered to be not yet fully DEVELOPED (UNDERDEVELOPED) primarily in economic but also in social terms. Conventionally the Third World includes most countries in Asia, Africa, Latin America; but opinions differ as to whether China, Vietnam, Cuba, South Africa, Israel and the oil-rich nation states of the Middle East should be included; and the United Nations has singled out from the Third World some 25 countries designated as least developed. This group is sometimes termed the Fourth World. BRANDT REPORT.

thorn forest, thorn woodland a general term applied to tropical or subtropical forest or woodland of small, XEROPHILOUS thorny trees, the thorns giving partial protection from browsing animals.

Thornthwaite's climatic classification a classification of climates developed by C. W. Thornthwaite, American climatologist, between 1931 and 1948, based on the effectiveness of climate in the development of plant communities. Identifying rainfall and temperature as the dominant, variable influences, he drew up formulae to assess them. His index for measuring PRECIPITATION in terms of its usefulness to plant growth is:

$$P/E = 11.5 \left(\frac{p}{T - 10}\right) \frac{10}{9}$$

P/E being precipitation efficiency, p the monthly mean precipitation in inches, T the monthly mean temperature. Five major regions were distinguished on that basis. He subdivided these on the basis of thermal efficiency (the heat received by a particular area in relation to the production of POTENTIAL EVAPOTRANSPIRATION). His index for thermal efficiency being:

$$TE = \frac{T - 32}{4}$$

where TE = thermal efficiency, T = mean monthly temperature in °F. Finally he drew up a moisture index, to show whether an area has a positive or negative water balance, expressed as:

$$MI = \frac{100(P - PE)}{PE}$$

where MI = moisture index, P = precipitation, PE = potential evapotranspiration. Using those indices he distinguished the climatic zones: perhumid (A), MI exceeding 100; humid (B, with four subdivisions), MI 20 to 100; moist subhumid (C_2), MI 0 to 20; dry subhumid (C_1), MI

−20 to 0; semi-arid (D), *MI* −40 to −20; arid (E) with *MI* below −40. KOPPEN'S CLIMATIC CLASSIFICATION.

three-field system a system of cultivation common in medieval England whereby the arable land was divided in three large fields (open fields), worked in common by the village, one being fallow or resting each year while the other two were cropped with wheat or rye and barley or oats. The system disappeared with enclosure, the discovery that clovers (LEGUMINOSAE) enrich the soil and that FALLOW is largely unnecessary.

threshold 1. the point at which a stimulus of increasing strength is first perceived or provokes a specific response, the point of transition from one state to another **2.** in CENTRAL PLACE THEORY, the threshold of success, the lowest demand necessary to ensure that any good, service or function will be offered at a CENTRAL PLACE. Demand may be measured in terms of population (THRESHOLD POPULATION) or income per caput.

threshold population the minimum number of people needed in an area to support a function, service or provision of goods. CENTRAL GOOD, CENTRAL PLACE THEORY, THRESHOLD-2.

throughflow the flow of water down a slope through the REGOLITH, as distinct from OVERLAND FLOW. It occurs when the quantity of water falling on the ground surface, or the rate at which it falls, is too great for it all to percolate sufficiently swiftly downwards through the upper SOIL HORIZONS. Lateral ELUVIATION results from the carrying of soil particles by the throughflow. PIPE-4, RUNOFF, SUBSURFACE WASH. Fig 41.

throw of a fault the vertical displacement of strata or rocks in a FAULT, varying from a few millimetres to hundreds of metres in extent. The rocks on one side of the fault-line are termed upthrow, on the other side downthrow, indicating the displacement of each in relation to the other. HEAVE OF FAULT. Fig 23.

thrust 1. a force tending to compress, similar to TENSION but acting in the opposite direction. STRESS **2.** in geology, a compressional force affecting strata in an almost horizontal plane, leading, e.g., to a REVERSE FAULT of very low angle. OVERTHRUST, THRUST FAULT, THRUST PLANE.

thrust fault a low-angled REVERSE FAULT, with the beds of the upper limb pushed far forward over the beds of the lower. FAULT, NAPPE, THRUST PLANE.

thrust plane the surface, usually inclined at a low angle and not strictly a plane, over which the upper strata of a REVERSE FAULT are pushed. FAULT.

thunder the sound produced by the explosive expansion of suddenly heated gases in the ATMOSPHERE-1 resulting from the expending of electrical energy in LIGHTNING. THUNDERSTORM.

thunderstorm a STORM of heavy rain and/or hail and wind, with LIGHTNING and THUNDER, occurring when intense heating of the ground surface leads to strong upward air currents, great ATMOSPHERIC INSTABILITY and the formation of CUMULONIMBUS clouds in which electrical charges become separated, the positive charge collecting in the upper part, the negative charge in the lower, with a small region of positive charge at the base. Eventually the separation of the main negative and positive charges gives rise to a difference of some 100,000 volts, breaking down the insulation of the air, and resulting in a flow of electricity seen as the LIGHTNING flash. Thunderstorms are frequent in tropical and equatorial regions, where air masses are warm and moist; and they are also associated with the passage of a COLD FRONT (termed a frontal thunderstorm).

tidal *adj.* of, or pertaining to, due to, affected by, the TIDE.

tidal barrage a barrier built on the seaward side of the reservoir in a TIDAL POWER STATION.

tidal basin a BASIN-10 filled with water at high tide that can be held and released at low tide, the force of the outflowing water scouring the neighbouring HARBOUR.

tidal current a powerful horizontal movement of seawater in areas affected by the TIDE. It is sometimes regarded as synonymous with TIDAL STREAM; but more specifically it is applied to conditions in a STRAIT where differing tidal regimes result in water levels which differ at each end of the strait, the tidal current effectively equalizing the water level. CURRENT.

tidal dock a DOCK in which the level of water rises and falls with the tide.

tidal flat an area of sand or mud uncovered at low tide.

tidal glacier, tide-water glacier a VALLEY GLACIER which reaches the sea, where part of it may become detached, forming ICEBERGS or ICE FLOES.

tidal power station a coastal power installation which uses the natural POWER of the ebb and flow of tides as they rush out of and into an enclosed reservoir (TIDAL BARRAGE) to drive HYDRAULIC turbines which generate ELECTRICITY.

tidal race RACE-2.

tidal range the difference in the height of the water at high and low tide at a place, varying from day to day. The fortnightly NEAP TIDES have a small range, the SPRING TIDES have a greater range.

tidal stream the normal movement of seawater in a coastal inlet, the inward flow with the FLOOD TIDE, the outward with the EBB TIDE, usually resulting in SCOUR. TIDAL CURRENT.

tidal wave 1. an unusually large wave at high water resulting from tidal movement **2.** popularly, but inaccurately, applied to a giant destructive wave caused by an earthquake, the correct name for which is TSUNAMI.

tide the regular periodic alternating rise and fall of the level of the water in the oceans (often accentuated in adjoining seas, bays, gulfs). The rising of the water is termed the FLOOD TIDE, the falling is the EBB TIDE. The tide flows and ebbs twice in a LUNAR DAY of 24 hours 51 minutes and is caused by the gravitational pull of the SUN and the MOON, the latter being the more powerful. When the sun and moon act together (CONJUNCTION) a higher tide results (SPRING TIDE); when they do not reinforce each other (OPPOSITION) there is a smaller tide (NEAP TIDE). See entries qualified by tidal and tide, as well as APOGEAN TIDE, DOUBLE TIDE, HIGH WATER, LOW WATER, PERIGEAN TIDE.

tied cottage a dwelling house in which a worker (especially a farm worker) can live while employed by the owner.

tierra caliente (Spanish) one of the four

altitudinal zones of low latitudes in the northern Andes, Central America and Mexico (the others being TIERRA FRIA, TIERRA TEMPLADA, TIERRA HELADA). Tierra caliente is the lowest zone, the zone of tropical products, the hot tropical coastland from sea-level to about 1000 m (3000 ft), where the climate is humid, the temperature varying little between 24°C and 27°C (74°F and 80°F) throughout the year, and with the difference between the coldest and warmest months not more than three or four degrees. The natural vegetation is luxuriant and tropical, with dense forests in wetter parts. The crops are bananas, sugar, cocoa; with maize, tobacco and coffee on the mountain slopes.

tierra fria (Spanish) one of the four altitudinal zones of low latitudes in the northern Andes, Central America and Mexico (TIERRA CALIENTE), the tierra fria, the zone of grains, lying higher than the TIERRA TEMPLADA, lower than the TIERRA HELADA, i.e. from about 1800 m to 3000 m (6000 to 10 000 ft). The average annual temperature lies between 12.5°C and 18°C (55°F and 65°F), there is little difference in temperature from one month to another, and the natural vegetation is coniferous forest giving way to scrub and grassland with increasing height and decreasing temperature. The crops are wheat and vegetables; the fruits are those common at sea-level in higher latitudes in the northern hemisphere; and there is much pasture.

tierra helada the highest of the altitudinal zones (TIERRA CALIENTE) in the northern Andes, Central America and Mexico, the permanently snow-covered region of mountain summits, lying higher than the TIERRA FRIA.

tierra templada (Spanish) in tropical Central America, Mexico and the northern Andes, the altitudinal zone known as the zone of coffee (the zone of perpetual spring in the northern Andes), lying between the TIERRA CALIENTE and the TIERRA FRIA, from about 1000 m to 1800 m (3000 to 6000 ft), average annual temperature varying between 18.3°C and 24°C (65°F and 75°F), the range between the coldest and the warmest month rather less than that in the tierra caliente. The natural vegetation is savanna with open forest; the crops are maize, coffee, tobacco; but rainfall is too low for pasture.

till unstratified, unconsolidated DRIFT-1 consisting of a heterogeneous mixture of angular and/or subangular clay, sand, gravel and boulders carried by ice and deposited, with little or no subsequent sorting or transportation by water. The term BOULDER CLAY, long in use, has now been replaced by till; and recently refinements in the nomenclature of till have been proposed, taking into account the processes which form the sediments and the position of their deposition.

tillage 1. the process of cultivating land so as to make it fit for raising crops **2.** land ploughed or hoed in the current year **3.** arable land, excluding ROTATION GRASS and CLOVER.

tilt block a block of rock standing between prominent fault lines and slanted at an angle in such a way that its slopes contrast with those that border it. BASIN AND RANGE, BLOCK FAULTING, TILTING.

tilting the condition of slanting or being slanted from the horizontal or vertical.

timber 1. a wood suitable for or processed for use in construction **2.** a tree yielding such wood. HARDWOOD, LUMBER, SOFTWOOD.

timber-line the altitudinal or latitudinal line or zone beyond which trees sufficiently large to be of use for TIMBER cease to grow. TREE-LINE.

time APPARENT TIME, DAYLIGHT SAVING, GEOLOGICAL TIME, GMT, INTERNATIONAL DATE LINE, LOCAL TIME, MEAN SOLAR TIME, STANDARD TIME; and entries qualified by time.

time geography, time-space geography an approach to geography proposed in the 1960s by T. Hägerstrand, Lund School of Geography, who saw time and space as resources, inextricably linked, together imposing the boundary of the field within which human activities and behaviour are possible (TIME-SPACE CONSTRAINTS). Within this framework he identified (a) capability constraints, those imposed on an individual or group by physical limitations (e.g. need to eat/sleep), or the availability of the facilities required; (b) coupling constraints, the place, time and duration of time available for people to meet, cooperate, join organizations; (c) authority constraints, limitations imposed by law or tradition that determine the place and time when personal contacts are possible, e.g. legal working and shop hours, transport timetables, etc.

time-space compression a force identified by D. Harvey, 1989, as the accelerator in TIME-SPACE CONVERGENCE and TIME-SPACE DISTANCIATION. He attributed it to the mechanism of CAPITALISM in its unceasing search for new markets and a quick turnover of capital for reinvestment.

time-space constraints, space-time constraints the boxing-in of human activity produced by the simultaneous operation of the limitations of available time (which may include hours of daylight, or biological constraints such as the CIRCADIAN RHYTHM) and physical ACCESSIBILITY. For example, it is possible to travel only a certain distance from home to another place, to spend time there in some activity (working, shopping, playing, visiting friends, etc.) and to return home in daylight, or in waking hours. TIME GEOGRAPHY.

time-space convergence the concept that as improvements in transport progress the significance of the travel time between places diminishes. DISTANCE DECAY PHENOMENON, FRICTION OF DISTANCE, TIME-SPACE DISTANCIATION.

time-space distanciation term used by Anthony Giddens to define the social interaction of peoples across space and time by which formerly separate, self-contained social systems, remote from one another, come into contact and become interdependent and integrated, the interaction being facilitated by swift transport and communication systems (TIME-SPACE CONVERGENCE). Giddens distinguishes these 'distanciated' contacts from the routine face-to-face contacts of everyday life, normally confined to a fairly limited locality.

time zone the division represented by 15° longitude (less in small countries) within which the mean time of the central (or near central) meridian is selected to represent the whole division. STANDARD TIME. Fig 43.

tin a silvery-white, soft, MALLEABLE, DUCTILE, stable metallic ELEMENT-6, resistant at ordinary temperatures to the chemical action of air and water. It occurs mainly as the oxide cassiterite, either in PLACER deposits weathered from veins in granite, or directly from LODES. It is used in the making of tinplate, and as an alloy in solder, BRONZE etc.

tithe **1.** from Old English, a tenth or tenth part, an application now obsolete **2.** historically, in England, a tenth part of annual agricultural produce etc. given as an offering or paid as a tax (ultimately in monetary value), especially the tax levied by the church for its support.

tithe barn a large barn where the agricultural produce representing TITHES-2 was delivered and stored.

tombolo (Italian) a bar of sand or shingle linking an island to the mainland, or one island to another.

ton, tonne a measure of weight. In avoirdupois 1 long ton (lgt) is equal to 2240 lb (20 cwt of 112 lb) or 1.016 metric ton (tonne). A metric ton (tonne) is equal to 1000 kilograms or 2204.62 lb or 0.984 long ton, or 1.1 short ton. A short ton (American ton) equals 1000 lb (i.e. 20 cwt of 100 lb) or 0.907 tonne, or 0.892 long ton.

tonnage **1.** the carrying capacity of a vessel or the total carrying capacity of a fleet measured in tons **2.** the duty based on the cargo capacity of a vessel **3.** the charge per ton of cargo carried on canals or in some ports. SHIPPING TONNAGE.

topographic *adj.* of or relating to TOPOGRAPHY.

topographic map a map, usually on a fairly large scale (e.g. 1:50 000), representing surface features, e.g. landforms and other natural phenomena as well as features produced by human activities. The term should not be applied to a map showing only relief features. TOPOGRAPHY.

topography a term which has given rise to so much confusion that geographers today tend to avoid it. Those who do use it apply it in general to the description or representation on a map of all the surface features of an area, natural and ARTIFICIAL (TOPOGRAPHIC MAP). A few restrict it to the relief features, but this use is not generally accepted.

topological *adj.* applied in geographical studies as relating to, pertaining to, concerned with enclosure, order, connectivity, CONTIGUITY and relative position rather than with actual distance and orientation, topological relationships being frequently expressed in terms of NETWORKS. TOPOLOGICAL MAP.

topological diagram or map a diagrammatic map which shows the actual relationship of certain features (e.g. positions of towns) but on which true scale is deformed to accommodate some other consideration(s), e.g. the best way to show a communications system. The diagrammatic maps of underground railway systems are good examples (e.g. of the London Underground, or of the Paris Métro): they show connectivity, i.e. the way in which lines connect the stations, but are not concerned with the correct orientation of the stations, and they are not drawn to scale. GRAPH-2.

topology a branch of mathematics concerned with preserving certain relationships, e.g. closeness. The definition which restricts topology to the study of the properties of a geometrical figure that are unaffected when the figure is continuously transformed or deformed is too narrow in its concept. TOPOLOGICAL, TOPOLOGICAL DIAGRAM.

toponymy the study of place-names.

topophilia the love of place, the coupling

of sentiment with place. MENTAL MAP.

topotype in ecology, a population with characteristics associated with a particular region, which are distinct from the characteristics of the population in another region.

topsoil an imprecise term applied by agriculturalists rather than by soil scientists to the cultivated layer of the mature SOIL, whatever SOIL HORIZONS were originally involved, or to the surface soil as distinct from the SUBSOIL.

tor a prominent, isolated mass of jointed, weathered rock, usually granite, especially one rising from the moorland of Dartmoor, southwest England.

tornado **1.** African tornado, a violent storm over the lands of west Africa, consisting of a SQUALL, usually with torrential rain, sometimes of short duration, but extending over a long front (up to 320 km: 200 mi) associated with a THUNDERSTORM. It occurs most frequently in daytime between the wet and dry seasons when humid monsoon air from the southwest meets the dry northeasterly HARMATTAN from the Sahara **2.** a violent, anticlockwise, very destructive, short-lived revolving storm (sometimes termed a twister in USA), usually accompanied by rain and thunder, associated with an intensely LOW pressure system, with wind velocities estimated to exceed 320 kph (200 mph), in some examples travelling a nearly straight track at between 16 and 80 kph (10 and 50 mph), and with a dark, funnel-shaped cloud (FUNNEL CLOUD), small in diameter, appearing to grow downwards from dark CUMULONIMBUS cloud. Such a tornado is particularly common in the Mississippi basin in the afternoons in spring and early summer where warm humid air from the Gulf of Mexico meets cool, dry air from the north, and when the heating of the land surface is at its greatest. It may travel only a short distance (some 30 km: 18 mi) and last under two hours, but in that time it mows down anything in its path. TROPICAL REVOLVING STORM.

tour a journey made for pleasure or for reasons of business, inspection, education etc. which ends at the place of origin and takes in several places or points of interest on the way. TOURISM.

tourism **1.** the practice of making TOURS for pleasure **2.** synonym for tourist industry, the whole business of providing hotel and other accommodation, facilities and amenities for those travelling, or visiting, or staying in a place for a relatively limited period of time primarily for pleasure. ADVENTURE TOURISM, ECOTOURISM.

tower karst, turmkarst KARST formed in tropical conditions (TROPICAL KARST), with isolated limestone hills, generally flat-topped with steep, forest-covered sides, interspersed with stretches of alluvium or other detrital sand. KEGELKARST.

town in general, a place larger than a VILLAGE consisting of a compact agglomeration of dwellings, shops, offices, public buildings etc., usually with paved roads, street lighting, public services, an organized local government, and a community pursuing a distinctive, URBAN way of life. Specialized functions are commonly defined by qualification (e.g. market town, mining town, railway town) as are locational features (e.g. gap town, seaside town), or special characteristics (e.g. GHOST TOWN, SHANTY TOWN).

township **1.** in Scotland, a CROFTING township, a district in the crofting counties, comprising individually held croft land and the common grazing, the croft land being separated from the common pasture by the township dyke (a stone or turf wall), the number of crofts varying widely (from six to fifty or more) **2.** (American) a term applied in two ways in Public Land Survey, USA, the first being to the congressional townships of 6 mi square, whether it is settled or not, the second to the northern component drawn up for locating and identifying the townships **3.** in Australia, a tract of land laid out with streets and subdivided into lots for future urban development, or a temporary settlement on such a site **4.** in the former Republic of South Africa, separate areas of generally low standard housing reserved for Africans, Asians or 'coloured' people.

towpath a path alongside a navigable river or canal, used originally by draught animals or people towing boats.

trace element one of the chemical elements present in relatively small quantities in the earth's crust, and essential in very small amount for the normal health of plants and animals, e.g. boron, copper, manganese, molybdenum, zinc for higher plants, cobalt for cattle and sheep, iodine for human beings. An insufficiency or lack of intake of trace elements may lead to deficiency diseases. An excess is in some cases harmful. MACRONUTRIENT, MICRONUTRIENT.

trace fossil the trace of plant or animal activities (e.g. tracks or burrows of animals, root passages of plants) preserved in the rocks of the earth's crust. FOSSIL.

traction a process in the transportation of debris by a river, in which the debris rolls and slides along parallel with and close to the bed, other debris being carried in SOLUTION-I and SUSPENSION. SALTATION.

traction load BED LOAD. TRACTION.

trade the business of distributing, selling and exchanging commodities.

trade gap in the trade of a country, the amount by which, over a period of time, the value of imports exceeds that of exports.

trade-off theory of land use, a theory which maintains that consumers are prepared to trade-off (balance out) rents and transport costs, i.e. to trade-off the accessibility offered by city centre sites (where the plots of land are small and the cost of land is high, but transport costs involved in living there are low) and the lower land costs but higher transport costs incurred in living in suburban sites. Trade-off theory thus relies on the ideas of indifference theory (a theory which states that, within certain limits, consumers are indifferent to varying combinations of goods, e.g. of high rents and low transport costs, or low rents and high transport costs), and is used particularly in studies of residential land use. It is similar to bid rent theory (BID PRICE CURVE) in that it explains residential location decision solely in terms of minimizing travel costs (accessibility) and housing costs (location rent). If a graph is drawn to show transport costs, land costs and overall cost in relation to distance from the city centre, there will be various points where the mix of land and transport costs produces the same overall cost. These are the points of access/space trade-off which, in those terms, present the consumer with a choice of best location, the consumer in this case being indifferent to the alternative combinations of factors which make up the total costs.

trade wind (from nautical phrase to blow trade, to blow a regular course) a constant wind which blows (more strongly over the ocean than over the continents) from the tropical HIGH pressure belts towards the equatorial LOW in the northern and southern hemispheres, i.e. from the northeast in the northern, from the southeast in the southern, the typical PLANETARY WINDS. ANTI-TRADE WIND, HADLEY CELL, WESTERLIES. Fig 5.

traffic in endangered species CITES.

traffic principle, transportation principle one of the principles used by W. Christaller to account for the varying levels and distribution of CENTRAL PLACES in a CENTRAL PLACE SYSTEM. Assuming ease and efficiency of transport between central places to be the dominant consideration, as many lower order centres as possible will lie on the traffic routes between the higher order centres (shown by a K-4 hierarchy, where a higher order place serves three adjacent lower order places). There will therefore be a greater number of higher order centres than is accounted for by the MARKETING PRINCIPLE. ADMINISTRATIVE PRINCIPLE, K-VALUE. Fig 9(a).

trajectory the path of an individual parcel of air over some duration of time, in contrast to STREAMLINE.

tramontana, tramontane, tramontanto (Spanish; Italian, north wind) a cold dry north or northeast wind descending towards the sea from cold, dry plateaus in the western Mediterranean area.

transcurrent fault TEAR FAULT.

transect a section taken across a tract of country for the purpose of studying the vegetation in relation to soil and relief.

transfer costs in international trade, TRANSPORT COSTS combined with the costs of overcoming other obstacles to commodity movements, e.g. tariff walls.

transform fault in theory of PLATE TECTONICS, a massive TEAR FAULT that marks the divide where two plates slide past each other, the edges moving jerkily and jostling each other but staying close together. The plates do not dive into the mantle, no material is added to or subtracted from the earth's crust, but the friction arising from the movement of the plates usually causes severe EARTHQUAKES and earth tremors, e.g. the San Francisco earthquake, 1906, caused by the movement along the San Andreas transform fault in California, USA.

transhumance a periodic or seasonal movement of pastoral farmers and livestock seeking fresh pasture between two areas of different climatic condition, e.g. in mountainous areas, the movement from valley floor, the winter location, to mountain pasture for the summer, and the return to the valley in autumn; or the movement from drought-stricken lowlands in summer to cooler higher land, as in Spain. Some authors also use the term as equivalent to NOMADISM, the migration of nomadic pastoralists in search of fresh pasture in a regular, seasonal pattern according to the rainfall regime. MAYEN.

transition zone the zone surrounding the CENTRAL BUSINESS DISTRICT, with residential areas invaded by business and light manufacturing, mainly from the CORE-2, but also from elsewhere.

transit trade in international trade, trade in which freight is exported from one country to cross another, or others, before reaching its destination in the importing country.

translocation in soil science, the transfer of substances in SOLUTION-1 or SUSPEN-

SION from one SOIL HORIZON to another. ILLUVIATION.

transnational capitalism CAPITALISM that is non-territorial, not only crossing state borders in seeking new markets, raw materials, cheaper labour, but concerned especially with speculative finance capital in the world economy. STATE CAPITALISM.

transnational corporation MULTINATIONAL CORPORATION.

transnational movements social, cultural and political networks that cross territorial boundaries of existing states without having a dominant base which expresses the interests of any individual state (or groups within such a state).

transpiration the loss of WATER VAPOUR from a plant mainly through the tiny pores in the outermost layer of cells, i.e. from the stomata of the leaves, resulting in a stream of water with dissolved mineral salts surging up through the plant. Transpiration is unlike EVAPORATION in that it takes place through living tissue under the influence of the physiology of the plant. EVAPOTRANSPIRATION.

transport the act of carrying material or a person from one place to another. In British usage the terms transport and TRANSPORTATION are commonly interchangeable, transport being preferred except in geomorphology in the cases cited below. American usage generally favours transportation.

transportation **1.** (American) the carrying or the conveying of material or a person from one place to another **2.** a phase in the process of DENUDATION concerned with the conveying of loose material of the earth's crust by a natural agent (other than MASS MOVEMENT by GRAVITY) to the site of DEPOSITION, the agents being running water, ice (glaciers and ice sheets), wind, the ocean (waves, tides, currents). The material carried, termed the load (LOAD OF A RIVER), may itself act as an eroding agent (ABRASION, CORRASION), suffering ATTRITION as the particles rub against each other and the surface over which they are being carried.

transportation slope a SLOPE-2 on which at each point the amount of material received from points upslope is balanced by the loss of material passing downslope.

transport costs all the costs involved in the moving of goods from one place to another, e.g. inventory, paperwork, handling, packaging, insurance, freight rates, temporary warehousing en route etc. TRANSFER COSTS.

transverse *adj.* crosswise, lying across, crossing from one side to another.

transverse coast an ATLANTIC TYPE COASTLINE, or DISCORDANT COAST.

transverse dune a DUNE with its crest running at right angles to the direction of the prevailing wind. LONGITUDINAL DUNE.

transverse valley a VALLEY which cuts across a ridge, at right angles to the ridge. LONGITUDINAL VALLEY.

transverse wave SHAKE WAVE.

traverse a surveyed line consisting of a series of observations (legs), measured in distance and direction from a known starting-point, the end of one leg being the beginning of the next, thus incorporating cumulative error. The term closed traverse is applied if the legs are joined to link a known starting-point to a known finishing-point. If the position of only the starting-point or the finishing-point is accurately known, and the legs are joined, the traverse is termed an open traverse. TRIANGULATION.

trawl, trawler a large bag-shaped fishing net with a wide mouth which is dragged along the bottom of the sea by a boat; hence trawler, a fishing boat dragging a trawl for catching fish.

treaty port a sea or river port (later also an inland town) opened by treaty to foreign trade, applied especially in the nineteenth century to certain ports in China, Japan, Korea. In 1842 the first in China were opened with extraterritorial rights to foreigners. The system ended in 1943. EXTRATERRITORIALITY.

tree a woody PERENNIAL rising from the ground with a strong, distinct trunk. HERB-1, SHRUB.

tree-line, tree-limit (synonymous with TIMBER-LINE in American usage) the line or zone beyond which trees do not grow, governed mainly by temperature and water supply, but also by soil, aspect and exposure **1.** the upper limit of tree growth on a mountain (sometimes termed the cold timber-line to distinguish it from the dry timber-line of arid regions) **2.** the lower limit of tree growth on mountains in ARID regions where precipitation decreases with descent down the slope **3.** the latitudinal limit at which tree growth ceases.

trellis drainage, trellised drainage DRAINAGE-2 with a rectilinear pattern, occurring particularly in areas of folded sedimentary rocks, e.g. in scarplands, where CONSEQUENT, SUBSEQUENT, OBSEQUENT and SECONDARY CONSEQUENT STREAMS cut channels through the less resistant rocks at right angles to the initial slope, and thus meet the main streams at right angles. ANGULATE DRAINAGE. Fig 17.

trench 1. a deep elongated submarine trough. DEEP, OCEANIC TRENCH **2.** a long narrow valley between two mountain ranges, especially a RIFT VALLEY or a U-SHAPED VALLEY.

trend line the GRAIN, the pattern of the main structural lines, e.g. of folding (FOLD-2) and faulting (FAULT) in a region.

trend surface analysis a statistical technique particularly useful to the geographer in that it extends REGRESSION ANALYSIS to three dimensions. It entails the fitting of a statistical surface to values which are distributed in space, using a mathematical power function, usually with the aid of a computer. The observations of the DEPENDENT VARIABLE represent a series of sample points on a map, and the INDEPENDENT VARIABLES are the COORDINATES (latitudinal and longitudinal) of those points; but in order to describe the surface accurately some transformation of the independent variables may be necessary.

triangular trade the infamous trade of seventeenth–nineteenth century identified by the export of manufactured goods from western Europe to west Africa where they were bartered for African slaves, who were transported to plantations in South America, the West Indies and North America, the ships returning to Europe laden with sugar, rum and indigo from the West Indies, and tobacco or cotton from Virginia.

triangulation in surveying, the series or network of triangles into which a land area of any size may be divided in a TRIGONOMETRICAL SURVEY in order to provide a geodetic framework for a topographical survey. From a pre-determined BASE

LINE, which serves as one side of the primary triangle, triangles are constructed, their angles being measured with a THEODOLITE, the length of their sides being calculated by TRIGONOMETRY as the equipment is moved from one point that is to be determined to another. Triangulation may be of primary, secondary or tertiary order, in accord with the area of the triangles and the standard of accuracy needed. TRAVERSE.

Triassic *adj.* of or relating to the earliest period (of time) or system (of rocks) of the MESOZOIC era, when reptiles were dominant, gymnosperm plants (plants producing seeds not enclosed in an ovary) appeared, and sandstone and pebble beds, shelly limestone, red sandstones and marls, with layers of rock salt and gypsum, were laid down.

tributary a stream or river flowing into a larger one.

trigonometric(al) survey a survey carried out by TRIANGULATION and trigonometrical calculation.

trigonometry the branch of mathematics that deals with the relationship between the sides and angles of triangles (plane figures bounded by three straight lines).

trip a short journey, especially one of a series of journeys over a particular route.

trophic *adj.* of or related to a food supply, or the obtaining of nutrition.

trophic level **1.** one of the levels in a FOOD CHAIN or FOOD WEB **2.** the nutrient level of a body of water, especially in relation to nitrate and phosphate content. EUTROPHIC.

tropic one of the two parallels of latitude of approximately 23°30'N (TROPIC OF CANCER) and 23°30'S (TROPIC OF CAPRICORN).

tropical *adj.* of or pertaining to the TROPICS, relating either to the specific parallels of latitude, 23°30'N or 23°30'S; or to the zone lying between those two parallels; or to that zone with the adjacent areas, since major climatic and other changes take place more nearly at 30°N or S than at 23°30'N or S. But some authors exclude as distinct the DOLDRUMS or belt of calms (ITCZ) or the EQUATORIAL BELT.

tropical air mass an air mass, symbol T, originating within the SUBTROPICAL HIGH PRESSURE BELTS, either the warm, moist maritime tropical (mT), originating in the TRADE WIND belt and subtropical waters of the ocean, or the hot, very dry, unstable continental tropical (cT), originating in low latitude deserts (especially the Sahara and Australian deserts). POLAR FRONT.

tropical climate any of several types of climate occurring in the tropics (in this case the term tropics is understood to exclude the EQUATORIAL BELT with its EQUATORIAL CLIMATE), i.e. one of the belts which for part of the year comes under the influence of TRADE WINDS but for the rest of the year is subject to convectional rain. There is no cold (winter) season, but in general there are three others, i.e. hot rainy, cool dry, hot dry, with average monthly temperatures exceeding 18°C (64.4°F) and with considerable rainfall, mainly convectional, the maximum falling in the hot rainy period. The two main types are marine, which lacks a pronounced dry season (tropical marine, tropical marine monsoon) and continental, with a pronounced dry season (tropical continental, tropical continental monsoon). HUMID TROPICALITY.

tropical cyclone TROPICAL REVOLVING STORM.

tropical easterly jet stream a JET STREAM moving from east to west at very high altitudes over southeast Asia.

tropical forest the natural vegetation covering the wooded parts of the TROPICS which have a dry season. RAIN FOREST, TROPICAL CLIMATE.

tropical grasslands CAMPO, LLANO, SAVANNA.

tropicality HUMID TROPICALITY.

tropical karst KARST formed in tropical conditions, i.e. high temperature and high rainfall, the two main forms being KEGELKARST and TOWER KARST.

tropical revolving storm a small, localized, very deep LOW pressure area occurring most commonly in late summer or early autumn in tropical latitudes over the western margins of the great oceans, usually moving very slowly along fairly well-defined tracks and causing much destruction. An intense cyclonic circulation is set up about the centre, with violent winds sometimes exceeding 160 kph (100 mph), accompanied by dense dark clouds, heavy rain, sometimes thunder and lightning; but near the centre, the eye, where pressure is at its lowest, there is an area of calm with a clear sky. In the areas affected the storms are given special, local names. CYCLONE, HURRICANE, TYPHOON, WILLY-WILLY.

Tropic of Cancer an imaginary line encircling the earth at approximately 23°30' north of the equator, where the sun's midday rays are vertical about 21 June, the most northerly point of the ECLIPTIC, and the northern limit of the TROPICS.

Tropic of Capricorn an imaginary line encircling the earth at approximately 23°30' south of the equator, where the sun's midday rays are vertical about 21 December, the most southerly point of the ECLIPTIC, and the southern limit of the TROPICS.

Tropics, tropics the zone of the earth's surface lying between the TROPIC OF CANCER and the TROPIC OF CAPRICORN (i.e. between 23°30'N and 23°30'S), where the sun's rays strike vertically at noon on at least two days in the year, termed the torrid zone in classical times. HUMID TROPICALITY (humid tropics), TROPIC, TROPICAL, TROPICAL AIR MASS, TROPICAL CLIMATE and other entries qualified by tropical; and TROPICALITY.

tropopause a zone of the ATMOSPHERE-1 consisting of several, overlapping levels, separating the TROPOSPHERE from the STRATOSPHERE. In the tropopause temperatures cease to decrease with increasing height (as in the troposphere): there is a pause for some kilometres in height before they begin to increase with increasing height (as in the stratosphere). The level of the tropopause varies daily, seasonally and latitudinally, but in general it is some 18 km (11 mi) above the earth's surface at the EQUATOR, some 6 km (4 mi) at the poles. Fig 4.

tropophyte a plant adapted to living in moist or dry conditions according to seasonal variation, e.g. DECIDUOUS trees, which shed their leaves when adequate water is unavailable. HYDROPHYTE, HYGROPHYTE, MESOPHYTE, XEROPHYTE.

troposphere a layer of the ATMOSPHERE-1 in which temperature decreases with height at a mean rate of some 6.5°C per km (3.6°F per 1000 ft), the layer nearest to the earth's surface, i.e. below the TROPOPAUSE. The thickness of the troposphere varies from 7 to 8 km (11 to 13 mi) near the poles to some 16 km (26 mi) over the equator. It contains nearly all the dust and liquid particles, some 90 per cent of the water vapour, and 75 per cent of the total gases in the atmosphere, so most of the WEATHER activity affecting human life on earth takes place in its lower layers. Fig 4.

trough **1.** a DEEP or TRENCH in the ocean floor **2.** an elongated U-shaped valley **3.** a SYNCLINE **4.** of a WAVE, the position of displacement or disturbance opposite to the position of a CREST, the depression between any two crests of a regular wave motion **5.** in meteorology, a narrow, elongated region of low barometric pressure between two areas of higher pressure.

trough end a steep wall of rock at the head of a glaciated valley.

trough lake a lake occupying part of the floor of a trench made by an ALPINE GLACIER.

truck farming (American) the cultivation of vegetables and fruit for market. Truck farming differs from British MARKET GARDENING mainly in that it is on a larger scale, there is a concentration on one or more crops, and the distance of the enterprise from the market is greater.

true dip in geology, DIP-2, i.e. the maximum inclination of a stratum, as distinct from APPARENT DIP. STRIKE. Fig 16.

true north, true south the direction determined by the geographical NORTH POLE or SOUTH POLE of the earth, i.e. the direction of the geographical North or South Pole from the observer, i.e. along the MERIDIAN passing through the observer, as distinct from MAGNETIC NORTH and GRID north.

true origin the point on which the GRID-1 system on a map is based, at the intersection of the projection axes (the central meridian and a line drawn at right angles to it). FALSE ORIGIN.

true south TRUE NORTH.

truncated soil a SOIL which has lost all or part of its upper horizons by erosion.

truncated spur a SPUR which projected into the side of a pre-glacial valley until the valley became glaciated, when it was sharply cut and shortened by the glacier as it moved down the valley. U-SHAPED VALLEY.

trust territory, trusteeship a territory (which is not self-governing) under the authority of the United Nations, or an authority deputed by the United Nations Trusteeship Council. It may be a former MANDATED TERRITORY not yet independent, a territory taken away from a state after the Second World War, a territory placed under the authority of the United Nations by the state governing it. The Charter of the Trusteeship Council of the United Nations provides a system to safeguard the interests of the inhabitants of the territories not yet fully self-governing.

trypanosomiasis any disease caused by a trypanosome, a microscopic organism, e.g. sleeping sickness.

tsetse fly a member of *Glossina*, family Glossinidae, a bloodsucking fly of central and south Africa, carrying trypanosomes which cause nagana, a disease fatal to domestic animals, notably cattle, and sleeping sickness and other diseases affecting human beings. The presence of tsetse fly restricts the areas where cattle rearing is possible in Africa.

tsunami (Japanese) a large-scale seismic ocean wave (incorrectly termed a TIDAL WAVE) caused by a submarine earthquake or a volcanic eruption. It travels at great speed in the open ocean (between some 600 and 1000 kph: 370 and 1600 mph), with enough energy in some cases to travel halfway round the world. The wave height is low in the open ocean, but on entering shallow water the energy is concentrated, resulting in a wave of great height (up to 15 m: 50 ft), inundating low-lying areas on the shore.

tuba cloud a CLOUD in the shape of a cone or column emerging from a CLOUD BASE, e.g. a FUNNEL CLOUD.

tuber the swollen food-storing part of an underground stem (e.g. POTATO) or root of a plant.

tube well a WELL dug or drilled to reach a deep-seated supply of water, lined with a pipe (a tube).

tufa (Italian) calc tufa, a soft porous SEDIMENTARY ROCK composed of CALCIUM CARBONATE or SILICA, deposited by evaporation of circulating GROUND WATER, or water in lakes, or water near the point of issue of a SPRING-2 or THERMAL SPRING.

tuff a rock formed from compacted or cemented PYROCLASTIC material (fine volcanic dust, ash etc. thrown out of a VOLCANO in eruption), with particles smaller than 4 mm in diameter.

tumulus **1.** an ancient burial mound. BARROW **2.** a small mound or hummock of solid lava on a lava flow, in some cases up to 9 m (30 ft) in height and 18 m (60 ft) across at the base, produced by the resistance of the lava surface to the spreading of a more fluid lava below, thus resembling a LACCOLITH in origin.

tundra (Russian) a treeless region and its associated vegetation north of the northern latitudinal TREE-LINE, characterized by long, very cold winters and PERMAFROST, and supporting a vegetation of MOSSES, LICHENS, HERBS-1 and dwarf SHRUBS, infested with insects (blackflies, midges, mosquitoes) in the short summer. The mean monthly summer temperature lies below 10°C (50°F), warm enough to thaw the snow and the surface of the permafrost, providing ideal marshy conditions in the ill-drained soils for insect-breeding. Köppen included tundra in his polar zone. KOPPEN'S CLIMATIC CLASSIFICATION, TUNDRA SOIL.

tundra soil a dark-coloured shallow soil with a highly organic surface layer and usually (but not necessarily) with frozen subsoil. Comparatively little moisture is available but where it is the typical TUNDRA plants in decay give rise to a surface layer of organic material which accumulates and forms a peaty layer over an ANAEROBIC layer. In general soil profiles are not well developed, and the soils range from polar desert soil in the arid north to Arctic brown earths in the more humid upland regions with better drainage.

turbidity cloudiness in a FLUID caused by disturbance which results in the holding in SUSPENSION of finely-divided particles.

turbidity current a current occurring in the ocean when sediment is locally churned up (e.g. by an earthquake, or by material sliding down the CONTINENTAL SLOPE), raising the density of the water to higher values than that of the surrounding clear water. The heavier water flows very swiftly, under the influence of gravity, down any available slope, spreading out on a horizontal floor. TURBULENCE due to the flow tends to keep the sediment in SUSPENSION until the flow itself ceases; the sediment is then deposited. Turbidity currents near the ocean floor have an erosional effect and are thought to have cut deep valleys in the CONTINENTAL SLOPE. SUBMARINE CANYON.

turbulence movement of a FLUID in which the flow is not in smooth, parallel layers, but in eddies (EDDY), so that mixing occurs, in contrast to a smooth LAMINAR FLOW. The term is used especially in meteorology in connexion with the flow and mixing of air (DIFFUSION-2). Turbulence in river flow helps to carry material in SUSPENSION; and it is a characteristic feature of ocean DRIFT-2. HYDRAULIC FORCE.

turnpike historically, a spiked barrier fixed across a passage for defence purposes; later a gate placed across a road (hence turnpike road) where a toll was paid by anyone wishing to use the road. In Britain in the eighteenth century the roads were privately owned (especially by 'turnpike trusts') and maintained by these tolls. Later the term turnpike was applied to the road itself; and later still revived, especially in the USA, and applied to motorways subject to a charge or toll.

tussock grass BUNCH GRASS.

twilight **1.** the faint light of the sun reflected from the upper ATMOSPHERE-1 on to the earth before the sun itself rises above the horizon in the morning (dawn or sunrise) and after it has sunk below the horizon (sunset) in the evening, its duration depending on the date and latitude, i.e. brief in tropical regions, longer in higher latitudes. SUNRISE, SUNSET **2.** astronomical twilight begins in the morning when the centre of the sun is 18° below the horizon, lasting until dawn; in the evening from sunset until the centre of the sun is 18° below the horizon. About the time of the SOLSTICE in high latitudes the sun's centre never sinks to 18° below the horizon, so twilight is continuous from sunset to sunrise for some nights, the number of such nights increasing towards the poles. **3.** civil twilight is classified as commencing or ending when the sun's centre is 6° below the horizon, when the light is judged to be sufficient for outdoor work **4.** nautical twilight commences or ends when the sun's centre is 12° below the horizon, when the light should be sufficient to allow vague shapes to be seen.

twilight area the part of a town where old buildings in need of repair and inadequate facilities lead to poor living conditions.

twister a TORNADO-2 in USA, a WATERSPOUT.

two-field system a simple system of cultivation practised in parts of medieval England, in which half the land was cultivated, half left fallow each season. THREE-FIELD SYSTEM.

typhoon a violent TROPICAL REVOLVING STORM in the China Sea and adjacent regions, particularly around the Philippines, commonly occurring in the period from July to October.

U

ubac (French dialect; Italian opaco; German Schattenseite) the shady side of a valley facing away from the sun, in contrast to ADRET.

ubiquitous materials the raw materials used in MANUFACTURING INDUSTRY which are available anywhere, not localized, and therefore do not influence the selection of location of the industry concerned. RESOURCE ORIENTATION.

Uinta structure (USA: from the Uinta mountains, Utah) a broad, flattened ANTICLINAL FLEXURE from which STRATA descend sharply on each flank before resuming their horizontal state. In the Uinta mountains there is a classic example of subsidence and uplift on a giant scale, a flattened anticlinal flexure being there raised up in Cretaceous times, later extensively denuded to expose Precambrian rocks, uplifted again at the end of the Eocene, with large faults on its north and south flanks, and again uplifted in the late Pliocene-Pleistocene epochs. GEOLOGICAL TIMESCALE.

ultisols in SOIL CLASSIFICATION, USA, an order of soils which are deeply weathered and relatively infertile, lacking base minerals, having red and yellow CLAY constituents, associated with humid temperate to tropical climates. Characteristically the A HORIZON is marked by residual iron oxides and the B HORIZON has accumulations of clay.

ultrabasic rock an IGNEOUS ROCK, generally PLUTONIC and containing very little QUARTZ or FELDSPAR, with a SILICA content less than 45 per cent and a BASIC-1 oxide content more than 55 per cent, largely FERROMAGNESIAN MINERALS, metallic oxides, sulphides. Ultrabasic rocks usually occur in association with other basic rocks in layered igneous INTRUSIONS.

ultraviolet radiation very short ELECTROMAGNETIC WAVES emanating from the SUN, with wavelengths between those of X-RAYS and those of the violet end of visible light. Most ultraviolet radiation is absorbed by OZONE molecules in the upper atmosphere, but some reaches the earth's surface in sunlight, especially in high mountainous areas. It plays an important part as a PHOTOCHEMICAL agent in some life processes, e.g. it acts on the skin of some animals, including people, as a factor in helping to produce vitamin D; but an excess of ultraviolet radiation can be lethal to organic life. ELECTROMAGNETIC SPECTRUM.

umlak the traditional Inuit open boat made from skins stretched tightly over a wooden frame, paddled especially by women and children.

umland (German) an area which is culturally, economically and politically related to a particular town or city. URBAN FIELD is now the preferred term.

uncertainty 1. the possibility that several outcomes will occur as a consequence of a decision or action, the form but not the probability of each being known. Thus uncertainty is incalculable, while RISK is calculable **2.** in statistics, the degree to which a sample statistic (e.g. the MEAN or MEDIAN) does not agree with (is in error from) the 'true' value which would have been found if the whole POPULATION-4 had been measured. The degree of uncertainty is sometimes termed the limits of error.

unconformable *adj.* applied to an overlying rock STRATUM which does not conform in dip and strike to the underlying strata. CONFORMABLE, UNCONFORMITY.

unconformity in geology, a break or gap in the continuity of a stratigraphical sequence, between two beds that are in contact, where the overlying younger rocks have been laid down on a surface resulting from a very long period of denudation, the older, lower set of beds having been laid down then uplifted, tilted, warped or folded and denuded to a greater or less degree before the deposition of the upper, younger series. The plane or the division between two such sets is the unconformity, and it implies a break (of any duration) in a geological record.

underclass an imprecise term, variously used, but in general applied to disadvantaged people who suffer MULTIPLE DEPRIVATION, who do not share in the benefits of the society in which they live, who may fall outside a prevailing welfare system, and who live in POVERTY-3. POVERTY CYCLE.

undercliff a large mass of unstable rock debris lying below a cliff, consisting of material which has slipped as a result of weathering, occurring particularly if chalk overlies clay.

undercut slope the steeper slope on the outside curve of a meandering stream, the opposite to SLIP-OFF SLOPE. UNDERCUTTING.

undercutting the carving away, eroding, of material from the undersurface as by **1.** the current of a meandering stream, cutting into its bank on the outside of a bend. MEANDER **2.** a sand-laden wind in deserts, eroding the base of exposed rocks **3.** wave action against cliffs along sea coasts.

underdeveloped, underdeveloped lands, underdevelopment terms that came into prominence in 1949 as a result of President Truman's inaugural address in the USA, applied to those countries to which the adjective 'backward' had previously been applied but to which exception was naturally taken by their people. Underdeveloped (of a country or region) came to be applied to one not achieving the level that could be reached, given its natural and economic resources, if the necessary capital, skills, machinery etc. were available, i.e. underdevelopment was assessed in economic and technical terms. From an economist's viewpoint, underdeveloped can be applied to a country which could use more capital, labour or more available resources (or all of these) to support its present population on a higher living standard or, if its per caput income is already fairly high, could support a larger population on a living standard which would not be lower. In time the narrow economic view of underdevelopment came to be regarded as unsatisfactory: it did not take into account social conditions in the country or region concerned. It is thus now usual to include a consideration of social (as well as economic and technical) elements in references to development or underdevelopment. Over time other terms have been introduced relating to underdevelopment, e.g. in relation to poverty or to the degree of realized potential, less developed country (LDC), moderately developed country (MDC) or highly developed country (HDC). Other terms refer to the stage reached in INDUSTRIALIZATION, e.g. advanced industrial country (AIC) and newly industrializing country (NIC). BRANDT REPORT, DEVELOPED, DEVELOPING, DEVELOPMENT, INTERNATIONAL DIVISION OF LABOUR, THIRD WORLD.

underfit river, underfit stream MISFIT RIVER.

undergrowth the low plants (HERBS-1, SHRUBS and saplings) under the trees in a forest or in woodland.

underpopulation too few people in relation to resources, a term applied to the population in an area where the available RESOURCES-1 are not used so fully as they might be because very few people live there. OVERPOPULATION.

undertow a strong current flowing near the bottom of the sea close to the shore, pulling away from or aligned with the coastline. It is caused by the flowing seawards of water thrown up on the beach by a wave.

undiscovered reserves RESERVES which are thought, on the basis of current scientific knowledge, to exist; or are known to exist but of which the extent and quantity are uncertain. RESOURCES.

unemployment 1. lack of employment, the state of being unable to secure paid employment **2.** the inability to find a paid job at the current wage rate, commonly classified as demand deficient (no vacant jobs of any type), structural (the skills of the unemployed do not fit the jobs available; the unemployed in this category can usually get jobs only by retraining or by accepting work needing less skill than they possess, DE-SKILLING), or frictional (vacant jobs available and the unemployed have the appropriate skills but for some reason, e.g. immobility or lack of information, the unemployed do not fill them).

UNEP United Nations Environment Programme, an agency of the United Nations formed in 1972 to coordinate intergovernmental measures for the monitoring and protection of the environment.

Unesco United Nations Educational, Scientific and Cultural Organization, an agency of the United Nations established in 1946 to promote international collaboration in education, science and culture in the cause of peace and security.

uneven development the condition at any time in the DEVELOPMENT-1 of an economic and/or social system, or in compared regions or territories, when some parts progress and prosper while others are static or in relative decline. Some writers maintain that increased global competition and the mobility of capital (and the collapse of state socialism) encourage uneven development and spatial differentiation. AREAL DIFFERENTIATION, GLOBALIZATION, INTERNATIONAL DIVISION OF LABOUR, UNDERDEVELOPMENT.

uniclinal *adj.* applied by some authors to strata dipping steadily and uniformly in one direction, hence uniclinal structure.

uniclinal shifting the process of asymmetrical development of a valley due to a stream which, following the line of the geological STRIKE, tends to cut sideways in the direction of the DIP, i.e. the valley is gradually eroded and moved laterally down the dip. Some authors apply the term monoclinal shifting to this process, but others consider this confusing (UNICLINAL), preferring uniclinal because it implies a uniform dip in one direction.

uniformitarianism the principle that the processes and natural laws existing in the past, which steadily and slowly brought about changes in and on the surface of the earth, can still be seen at work today. The opposite view is expressed in CATASTROPHE THEORY.

unimproved *adj.* not used well, not made better, not enhanced in value, e.g. as applied to land which has not been cultivated.

uninverted relief a landscape where the surface relief reflects the underlying geological structure, where hill ridges coincide with ANTICLINES and valleys with SYNCLINES, the opposite of INVERTED RELIEF.

unitary system of government, a state system in which authoritative control is vested in a central or national government which delegates power to local government authorities. FEDERAL SYSTEM.

United Nations an association of countries which by signing the Charter pledge themselves to maintain international peace and so to help the political, social and economic progress of the world. The organization is not authorized to intervene in the domestic affairs of any state. It originated during the Second World War, statesmen from the UK, USA and USSR (including Winston Churchill, Franklin Roosevelt and Joseph Stalin) agreeing that such a conflict must never happen again. They met at Washington DC in August–September 1944, talks with China came later, and the United Nations came formally into existence on 24 October 1945, headquarters New York, the working languages being English, French and (in the General Assembly) Arabic, Chinese, Spanish and Russian. The official languages are English, French, Russian, Spanish and Chinese. The principal organs are the General Assembly, the Security Council, the Economic and Social Council, the Trusteeship Council, the International Court of Justice, the Secretariat. A very full account of the United Nations, its operations and agencies appears in *The Statesman's Year-Book,* The Macmillan Press Ltd.

United Nations Conference on Trade and Development UNCTAD.

unstable air mass an AIR MASS with a high WATER VAPOUR content, liable to spontaneous convectional activity (CONVECTION) which results in heavy showers and THUNDERSTORMS.

unstable equilibrium the state of the ATMOSPHERE-1 where the ENVIRONMENTAL LAPSE RATE of an AIR MASS exceeds the DRY ADIABATIC LAPSE RATE. Such an air mass, being warmer and lighter than the air around it, will rise and go on rising. If it is very moist it will cool very slowly, at the SATURATED LAPSE RATE, and be even more unstable, leading to the formation oflarge CUMULUS clouds which may be associated with heavy rainfall, HAIL, THUNDERSTORMS. It will stop rising only when its temperature equals that of the surrounding air. STABLE EQUILIBRIUM.

upland, uplands a general, unspecific term applied to higher ground, in contrast to lowland or lowlands. It usually implies an area of relatively subdued relief inland, away from the coast.

upland, uplands *adj.* situated or living on, growing in, UPLAND, UPLANDS.

upper atmosphere EXOSPHERE.

upwelling in a deep body of water, the upward movement of colder water from lower layers to the warmer zone above, caused by a CURRENT-1, of economic importance in the ocean because the colder water, rich in PLANKTON and other nutrients, attracts fish. Many of the world's most important fisheries are associated with such upwelling, e.g. the cold Benguela current off southwest Africa.

uranium a radioactive, hard, white metallic element in the CHROMIUM group, the heaviest of the elements occurring in nature, found in PITCHBLENDE and some other minerals. Natural uranium consists of isotopes U 238, which converts into plutonium, U 235 and some U 234. Plutonium and U 235 are used as a source of NUCLEAR ENERGY.

urban *adj.* **1.** relating to, belonging to, characteristic of, constituting, forming part of, a town or city, the opposite of RURAL, i.e. applied to any settlement in which most of the working inhabitants (some authorities specify over 60 per cent) are engaged in non-agricultural occupations (retail and wholesale trades, handicrafts, manufacturing industries and commerce, with associated service occupations, etc.) **2.** relating to, belonging to, people living in such a place. URBAN SETTLEMENT.

urban area specifically (Office of Population Censuses and Surveys, EEC), 'a continuous area of urban land extending for 20 ha (50 acres) or more. Urban land includes permanent structures, roads and transport-related land use, mineral working and quarries and any area completely surrounded by built-up sites' (in practice nearly every settlement with 2000 or more inhabitants).

urban climate the LOCAL CLIMATE of a built-up area, where the buildings affect temperature (by their heated interiors) and the pattern and speed of winds (by their layout), and they together with the paved surfaces create an impermeable layer which increases and speeds up RUNOFF; and where the emissions from the burning of HYDROCARBON fuels in motor vehicles combined with those and other emissions from industrial plant may pollute the atmosphere, affecting cloud formation and precipitation. HEAT ISLAND, POLLUTION, SMOG, VENTURI EFFECT.

urban economic base a two-fold classification of urban economic function, based on space relationships, providing a means of identifying economic ties between a large URBAN SETTLEMENT and other areas, of classifying and drawing up a comparative analysis of settlements, and of classifying individual economic activities within the urban area itself. The two categories are: basic activities (functions), which serve areas outside the TOWN or metropolitan area (METROPOLIS-2), and thus link the activities of the town/metropolitan area with other parts of the earth's surface; and non-basic activities (functions), which serve the inhabitants of the town or metropolitan area and form links within the settlement itself. The extent to which a given function is 'basic' is measured by calculating the ratio of people engaged in that activity to the total population of the town/metropolitan area, and comparing this with a similar ratio for the country as a whole. BASIC ACTIVITY, NON-BASIC ACTIVITY.

urban field, urban region the sphere of influence of a town, the territory around a town with which it is functionally linked. UMLAND.

urban fringe the area of social change around a town or metropolitan area (METROPOLIS-2), where urban development impinges on agricultural land, population density increases and land values rise. RURBAN.

urban geography the branch of geography concerned with the site, evolution, morphology (URBAN MORPHOLOGY) and classification of villages, towns and cities, their location in relation to a region or the country, the general processes (economic, political, social) at work within them, and the pattern of their relationship to other urban areas.

urban hearth the place of origin of URBAN development, of urban culture, e.g. the land between the rivers Tigris and Euphrates, Mesopotamia; the Nile valley; the Indus valley in the Indian subcontinent; the valley of the Hwang Ho, northern China. HYDRAULIC HYPOTHESIS.

urban hinterland the hexagonal trade areas in CENTRAL PLACE THEORY. ISOTROPIC SURFACE, SPHERE OF INFLUENCE-2.

urbanism **1.** town character, the typical condition of a town, or way of life characteristic of a town **2.** sometimes used as an alternative term for URBANIZATION. ANTI-URBANISM.

urbanization **1.** the continuous process of transformation from being of RURAL to being of URBAN character, and the continuous change within the urban area itself as it grows by natural increase and by MIGRATION-1,2 from other (usually rural) areas. The result is that an increasing proportion of the population of an extensive area is concentrated in defined urban places, with resulting changes in land use, landscape, way of life, economic activities etc. In the process most urban places grow but the population tends to concentrate most quickly and in the greatest numbers in the largest places **2.** the state reached in the process. ANTI-URBANISM, COUNTER-URBANIZATION, ECONOMIES OF URBANIZATION, URBANISM.

urban land the land on which an URBAN SETTLEMENT is built.

urban renewal a process in which the obsolete fabric of an urban area is restored, renovated and improved in order to meet contemporary needs or standards. In most cases an attempt is made to retain its original external character, but if this is impossible, a certain amount of redevelopment may be included. Redevelopment involves the total destruction of all or part of the obsolete fabric before the work of new building, creation of open spaces, possibly new roads etc. begins.

urban-rural continuum rural-urban continuum, the merging of town and country, a term used in recognition of the fact that in general there is rarely, either physically or socially, a sharp division, a clearly marked boundary, between the two, with one part of the population wholly urban, the other wholly rural. RURAL, RURBAN, RURBAN FRINGE.

urban settlement a term loosely applied to a relatively densely built-up area with its associated open spaces where the majority of the economically occupied inhabitants are engaged in activities mainly concerned with SECONDARY, TERTIARY, QUATERNARY INDUSTRIES, i.e. the definition is based on the function, not on the number, of the inhabitants. URBAN, URBAN LAND.

urban sprawl an irregular, unplanned, untidy spread of buildings around a town, sometimes linking-up with similar development around a neighbouring town, and usually consisting of residential areas, small shopping centres, small industrial enterprises.

urban village VILLAGE-2.

urstromtal (German) a wide, shallow valley excavated by a melt-water stream flowing in front of a continental ice sheet in the North European Plain, corresponding to static periods in the northwards retreat of the edge of the Scandinavian ice sheet.

U-shaped valley a glaciated valley, a valley which in cross section has the shape of a U, the floor being generally flat, the sides usually steep, due to the work of a VALLEY GLACIER moving down the V-SHAPED VALLEY of a pre-glacial river. The glacier gouges out the floor and erodes the valley sides up to the level of the surface of the ice. If the ice does not fill the valley a prominent SHOULDER-2 commonly occurs where the steepened sides meet the more gentle slopes of the pre-glacial valley. The glacier straightens the valley, shortens projecting spurs (TRUNCATED SPUR) and creates HANGING VALLEYS. The head of the valley may end in a steep wall (TROUGH END); and PATERNOSTER LAKES may be formed. Post-glacial alluvial deposits may have enhanced the flatness of the floor.

utilitarianism in ethics, a doctrine (expounded broadly by David Hume, 1711–76, but fully developed by Jeremy Bentham, 1748–1832) that the greatest good is the greatest happiness of the greatest number; and that therefore the moral and political rightness of an action is determined by its UTILITY, i.e. its contribution to the greatest happiness.

utility **1.** in economic theory, VALUE-1, the capacity (especially the capacity of goods or services) to satisfy human wants. The worth to the consumer is determined by the extent to which these satisfy the wants; and it is reflected in the price which the consumer is prepared to pay **2.** PUBLIC UTILITY.

V

vadose water water wandering in the ground above the permanent WATER TABLE, varying in amount and position as it moves through PERMEABLE rock. GRAVITY WATER, PHREATIC WATER.

vale imprecise term applied to a broad, flat, extensive valley (e.g. the Vale of Aylesbury), or simply a gently undulating lowland (e.g. the Vale of Glamorgan). It is also used poetically, and is best avoided as a geographical term.

valley an elongated depression, usually with an outlet, sloping down to an area of inland drainage, a lake, or to the sea, sometimes (but not always) occupied by a river. See valley qualified by BEADED, DRY, HANGING, LONGITUDINAL, TRANSVERSE, U-SHAPED, V-SHAPED.

valley glacier alpine glacier, mountain glacier, outlet glacier, a GLACIER which occupies an existing valley, i.e. a preglacial valley, termed an alpine glacier if it is formed by the merging of several CIRQUE GLACIERS; an outlet glacier if it originates from the margin of an ice cap or ice sheet. TIDAL GLACIER.

valley wind, valley breeze a general term applied to cold air draining down a valley especially by night (KATABATIC WIND) or a wind blowing up a valley (ANABATIC WIND) by day, the result of the differential heating of the mountains above and the low land below.

value 1. the measure of how much something is wanted for its special quality (e.g. beauty, rarity, UTILITY) expressed in terms of the money, effort etc. someone is prepared to expend in order to acquire, hold in possession, preserve it **2.** a quality, principle etc. that excites such a desire (e.g. moral values) **3.** in economics, the monetary equivalent of a product or a factor of production which satisfies three criteria, i.e. it is capable of being owned, it has UTILITY (satisfies needs or desires), and is in limited supply **4.** in mathematics, the amount represented by a symbol or expression, or **5.** the category of a VARIABLE-2.

value judgement a judgement that attributes worth or goodness, evil, beauty or some other VALUE-1,2, to something; or which asserts that some action ought or ought not to occur.

Van Allen radiation belt either of the two layers of intense ionizing radiation, with high energy particles, which envelop the earth in its outer atmosphere, the inner occurring at some 3000 km (1865 mi) above the earth's surface, the outer at some 13–19 000 km (8080 to 11 800 mi). The movement of the particles is influenced by the MAGNETIC FIELD of the earth rather than by GRAVITATION. MAGNETOSPHERE.

vapour a substance in the gaseous state (GAS-2) which separates into two phases when compressed. CONDENSATION, CRITICAL TEMPERATURE-1.

vapour pressure the pressure exerted by the vapour of a substance, e.g. the pressure exerted by water vapour in the atmosphere. If the air is SATURATED the term saturated vapour pressure is applied. ATMOSPHERIC PRESSURE.

vapour trail CONDENSATION TRAIL.

vardarac a cold wind, similar to the MISTRAL, which blows down the valleys of Macedonia, including that of the river Vardar, to the Aegean sea in winter.

variable 1. something VARIABLE-1 *adj.* **2.** in mathematics, a quantity which may take two or more values, may take any one of a specified set of VALUES-4 (also applied to denote non-measurable characteristics, e.g. sex is a variable in that any human individual may take one of two 'values', i.e. male or female), or a symbol for such a quantity. DEPENDENT VARIABLE, INDEPENDENT VARIABLE.

variable *adj.* **1.** having the quality of being able to change, able to be changed **2.** in biology, not true to type. CLASSIFICATION OF ORGANISMS **3.** in mathematics, characteristic of a quantity which may have different values. VARIABLE-2.

variable capital in Marxism, the living labour expended in the production process, as distinct from CONSTANT CAPITAL which represents 'dead' labour. It is termed variable because, being produced by living labour, the only creator of new value, its value increases during the production process. LABOUR THEORY OF VALUE, SOCIAL CAPITAL.

variable cost analysis an approach to the explanation of industrial (or other) location based on spatial variation in production costs. COMPARATIVE COST ANALYSIS, COST BENEFIT ANALYSIS, VARIABLE REVENUE ANALYSIS.

variable costs 1. costs that vary with volume of output, unlike FIXED COSTS **2.** in spatial economic analysis, costs that are subject to spatial variation. COST SURFACE, VARIABLE COST ANALYSIS.

variable revenue analysis an approach to the explanation of industrial location based on spatial variation in revenue. VARIABLE COST ANALYSIS.

variance in statistics, the square of the STANDARD DEVIATION, a statistic which measures the VARIABILITY-2 in a set of observations. COVARIANCE.

variate in statistics, an individual observation, one member of a set of values for one variable. A variate is a quantity that may take any of the range of values of a specified set with a specified relative FREQUENCY-3 or PROBABILITY-2, and is therefore sometimes termed a random variable. It is particularly associated with frequency (probability) function and expresses how often those values appear in the situation being observed.

variety in CLASSIFICATION OF ORGANISMS, a taxonomic group below a subspecies, i.e. a group of organisms with certain qualities in common which distinguish the group from others in the subspecies. The qualities may be, but are not necessarily, inherited.

Variscan orogeny in geology, a phase of the ARMORICAN orogeny, of late Carboniferous, early Permian times (GEOLOGICAL TIMESCALE). The term is used by some authors as a synonym for HERCYNIAN, but by others is applied only to the eastern arc of the Hercynian orogeny.

varve (Swedish) a distinct two-layered sediment deposited annually in a lake or other body of still water, especially in lakes near the margins of retreating ice sheets, where, of the laminated sediment, the lower, thicker layer, light in colour, consists of coarser material (deposited by meltwater from the rapidly thawing ice in summer), the upper, thinner layer, darker in colour, of very fine-grained material (settling during the slow melting of ice in winter). Each varve thus represents a year, so by counting the varves the time involved in the formation of the sediment can be estimated; and by correlations over a fairly extensive area, a glacial chronology be established. The sediments are termed varve clays, varved clays, varved sediments.

vauclusian spring (from Fontaine de Vaucluse, southern France) a gushing spring, a large spring, the resurgence of an actively eroding underground stream, commonly occurring in limestone country and varying greatly in output. RESURGENCE.

vector in biology, an agent carrying a disease-producing agent (e.g. a BACTERIUM) from one organism to another.

veering of wind a change of direction of wind, in a clockwise direction, i.e. from north through east to south in the northern hemisphere, equivalent to a BACKING WIND in the southern hemisphere.

vegetable oils the various oils obtained from plants, used mainly in food products, cosmetics, soap-making and other industrial processes, the residue left after the extraction in many cases being fed to livestock (OIL CAKE). Vegetable oils can be classified as DRYING OILS, oils which in drying form a thin elastic film (e.g. linseed, soya bean oil); semi-drying, those which form a soft film only after long exposure (e.g. cottonseed oil); non-drying, those which remain liquid at ordinary temperatures (e.g. OLIVE oil); and vegetable fats or tallow which are more or less solid at ordinary temperatures (e.g. COCONUT oil). Allied are the waxes, harder than oil, occurring on leaf surface (e.g. carnauba wax); and lather-forming products of leaves and stems used as soap-substitutes (e.g. cultivated soapwort, *Saponaria officinalis*).

vegetation a general term for the total plant cover in an area or on the surface of the earth as a whole. CLIMATIC FORMATIONS, EDAPHIC FORMATIONS, NATURAL VEGETATION.

vein LODE, a general term for a crack or fissure in the earth's crust in which highly heated waters from below have deposited CRYSTALLINE minerals (especially vein quartz) from solution and, under special circumstances, metallic minerals of economic importance.

veld (Afrikaans) any unenclosed country in South Africa with vegetation suitable for pasture. Many different types are distinguished: high, middle, low, or mountain veld according to elevation; sour or sweet veld according to lime content, sour veld being deficient; bush grass, karoo veld according to vegetation; sand veld, hardeveld, according to soil condition.

velocity the speed of movement in a certain direction, absolute or relative swiftness, high speed of action or operation.

vent an orifice in the earth's surface, especially that of a volcano, through which molten material erupts during volcanic activity. In a volcano it may become choked as the lava solidifies to form a PLUG or VOLCANIC NECK.

vent d'Autan a strong, hot dry wind blowing from southern France towards the centres of low pressure which come from the ocean into the Bay of Biscay.

ventifact glyptolith or rillstone, a PEBBLE with several flat facets which meet at fairly sharp angles, worn and polished by wind-blown sand, usually in desert conditions. If only three facets are present the term dreikanter is preferable; and some authors use the term einkanter if there is only one facet, zweikanter if there are two.

Venturi effect the effect produced by the narrowing of the channel or opening along or through which a gas or a liquid is passing, the flow of either being speeded up, e.g. as indicated by the gusts of wind in narrow streets and passages between tall buildings in towns. URBAN CLIMATE.

vertebrate *adj.* having a segmented spinal column. INVERTEBRATE.

vertical exaggeration the increase necessary in the vertical scale in comparison with the horizonal scale in order fairly to represent the height of an ELEVATION-1 in a MODEL-1 or a SECTION-2.

vertical interval the difference in vertical height between two points. HORIZONTAL EQUIVALENT.

vertical temperature gradient LAPSE RATE.

vertical zone altitudinal zone, in South and Central America, TIERRA CALIENTE, TIERRA FRIA, TIERRA HELADA, TIERRA TEMPLADA.

vertisols in SOIL-CLASSIFICATION, USA, an order of clay-rich soils which swell and crack in seasonally alternating wet and dry conditions, thereby mixing or inverting their horizons.

vesicle in geology, a small, generally rounded or oval cavity in a mineral or rock, particularly in glassy volcanic rock, formed by the trapping of steam or gas bubbles in the molten material as it solidified.

vesicular *adj.* covered with vesicles, or resembling a VESICLE in form or structure, applied especially in geology to the TEXTURE-1 of rock.

Vesuvian eruption VULCANIAN ERUPTION.

village **1.** a collection of dwelling houses and other buildings, especially in rural surroundings; a nucleated settlement as contrasted with dispersed habitations **2.** a close-knit, small community forming an 'island' in an URBAN environment and sometimes termed an urban village, often situated in the INNER CITY or the TRANSITION ZONE. People of similar ethnic or cultural characteristics may cluster in some urban villages.

vineyard a plot of land devoted to the cultivation of GRAPE vines.

viscosity **1.** the property or quality of a FLUID which makes it resistant to flow **2.** the degree to which a LIQUID is resistant to flow.

viscous *adj.* having VISCOSITY, being sticky, slow-flowing, e.g. the lava forming a volcanic CONE.

visibility **1.** the state or fact of being visible **2.** the range of vision of an observer, which depends on the time of day, the quality of light, the clarity of the atmosphere (presence or absence of dust, fog, mist), the height above sea-level at which the observer stands. HORIZON.

visible light the part of the ELECTROMAGNETIC SPECTRUM which may be perceived by the human eye, the wavelengths ranging between the limits of INFRA-RED and ULTRAVIOLET radiation.

viticulture the cultivation of the GRAPE vine for the production of grapes and WINE.

voe (Scottish: Shetland and Orkney dialect) **1.** a narrow gully cut in a cliff, in many cases ending in a cave or tunnel with a BLOWHOLE **2.** a bay, creek, inlet, specifically in the Shetland and Orkney islands, but the term is applied elsewhere.

volcanic *adj.* of, pertaining to, or like a VOLCANO covering all types of extrusive IGNEOUS activity, as distinct from intrusive PLUTONIC activity.

volcanic ash the unconsolidated PYROCLASTIC material consisting of finely comminuted fragments of rock and lava which have been ejected explosively from a volcano. The term ASH, which dates from the time when a volcano was thought to be a 'burning mountain', is a misnomer.

volcanic cone CONE-1.

volcanic dust the finest particles thrown out of a volcano in eruption. It may be shot high in the air by the explosive force and carried great distances by wind, e.g. dust from the 1883 eruption of Krakatoa in Indonesia is said to have been carried round the earth three times before settling. OOZE.

volcanic eruption ERUPTION. The various types are: HAWAIIAN, PELEAN, STROMBOLIAN, VULCANIAN.

volcanic neck or plug strictly the orifice of a volcano through which lava reaches the surface, and in which the lava eventually solidifies as a plug, the plug itself also being termed a neck. This may later stand in isolation if the material of the surrounding cone is denuded. DENUDATION, PIPE-1, VENT.

volcanic rock an IGNEOUS ROCK formed by volcanic action at the earth's surface, consisting of solidified material which has issued in the molten state from the depths of the earth, i.e. EXTRUSIVE ROCK, in contrast to HYPABYSSAL and PLUTONIC rocks. That is the most common application, but some authors include rock formed in association with intrusive activity (INTRUSION), and therefore include some hypabyssal rocks; and others also include plutonic rocks. MAGMA.

volcano a rift or vent in the earth's crust through which molten material is erupted and solidifies on the surface as LAVA or through which the molten rocks, charged with gases and vapours, are ejected with explosive force and fall back as VOLCANIC ASH and VOLCANIC DUST etc. (PYROCLAST). A volcano may be a central type (the eruption taking place through a more or less cylindrical pipe) or a fissure type

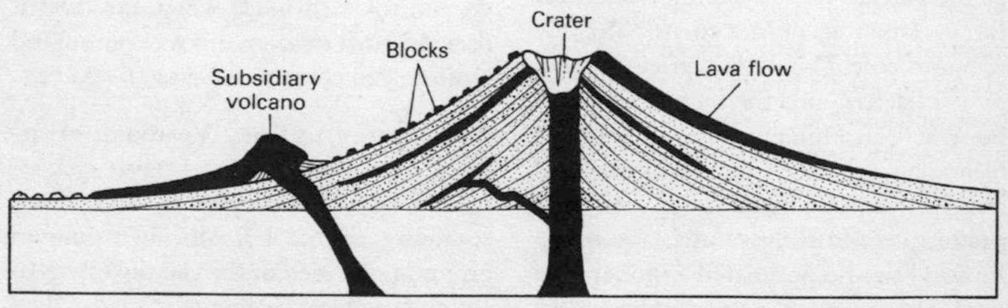

Fig 46 Diagrammatic section through a volcano

(the lava issuing through a line of weakness in the earth's crust, FISSURE ERUPTION). It is described as active whilst in eruption or liable to eruption; dormant during a long period of inactivity; extinct after all eruptions are presumed to have ceased. Four types of eruption are identified: HAWAIIAN, PELEAN, STROMBOLIAN, VULCANIAN. Figs 21, 46.

Von Thünen model a model devised by J. H. Von Thünen, published 1826 in his *Der Isolierte Staat*, to explain the principles which govern the prices of agricultural products and the way in which these variable prices control the pattern of agricultural land use. Using the ISOLATED STATE as his basic assumption and applying his theory of ECONOMIC RENT, he further assumed that all farmers would produce the crop giving them individually the maximum net profit (LAND RENT). To accomplish this each farmer would produce a crop or adopt an agricultural system for which the location of the farm land in relation to the market was most advantageous. Von Thünen also assumed that the value of a unit of produce to the farmer would be equal to its market price less the cost of transport to the market. The cost of transport of agricultural produce from farm to market was the only variable in the model. Land rent for any product declined with increasing distance from the market point, but the rate of decline varied for each product according to its particular transport cost. The market price for each product determined the highest land rent possible. Von Thünen also considered the technology available for production and transport, and the kinds and quantities of produce needed by the central, large town. From all this he postulated a model with concentric rings or zones of agricultural land use centred on the large town, the central market (ISOLATED STATE). In this simple, original model (reflecting the needs, conditions, equipment and technology of 1826), the zone nearest to the town was devoted to market gardening and milk production, the zone beyond this to forestry (for fuel, building timber, wood products), the next one to intensive crop rotation (without FALLOW), the next to crop farming, fallow and pasture, the next to a three-field system, the next to LIVESTOCK farming, which gave way to waste. This original model thus concentrated on the distance from MARKET-I as the governing, independent variable. Later Von Thünen considered the effects of varying soil fertility on production costs, and modified his assumptions of transport costs being uniform in all directions.

voralp (German) the lower pastures of an alpine valley, i.e. those above the valley floor but lower than the ALP-I proper. MAYEN.

V-shaped valley a valley eroded by a RIVER, V-shaped in cross-section in contrast to the U-shape common to a glacially modified valley. In W. M. Davis's CYCLE OF EROSION the V-shape is cited as evidence of youth in the stage of river erosion; but nearly all river valleys, if not subjected to glaciation, are V-shaped in the upper course of the river. Among other factors the valley shape will be influenced by the type of rocks through which the river is flowing, their resistance to weathering and erosion, and climate. U-SHAPED VALLEY.

Vulcanian eruption, Vesuvian eruption a VOLCANIC ERUPTION characterized by Vesuvius, the active volcano southeast of Naples, the first known eruption of which destroyed the towns of Pompeii and Herculaneum in 79 AD. The eruptions are less frequent, the magma more viscous, the ejected lava less basic than the phenomena associated with a STROMBOLIAN ERUPTION; but the ejected material is not so sticky and acid as that of a PELEAN ERUPTION. ACID LAVA, BASIC LAVA, PLINIAN ERUPTION.

W

wadi (Arabic) a stream course or valley in hot desert or semi-arid areas, especially in north Africa, usually dry but occasionally carrying a stream following heavy rain. ARROYO, NALA.

wage differential theory of MIGRATION-1,2 a theory which asserts that differences in wages in different places are the main cause of migration. PUSH-PULL THEORY, STRENGTH THEORY.

wake dune a SAND DUNE formed in the LEE of a larger dune, trailing away DOWNWIND.

Wallace's line a 'line' drawn by the English naturalist and geographer, Alfred Russell Wallace, 1823–1913, modified by Huxley, separating the distinct flora and fauna of southeast Asia and Australasia. In the Oriental region all the characteristic mammals are placental (as are the great majority of all living mammals); in the Australian region the characteristic mammals are marsupial (e.g. kangaroo) and monotreme (e.g. duck-billed platypus). Fig 47.

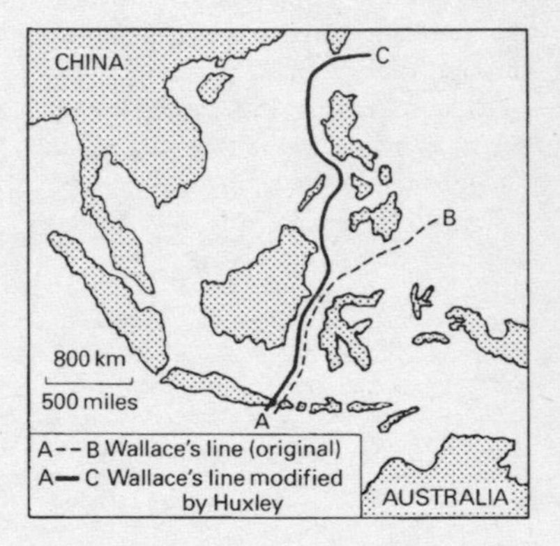

Fig 47 Wallace's line, modified by Huxley

wall-sided glacier a VALLEY GLACIER projecting from the valley it is occupying on to an adjoining plain so that its steep sides are visible, unrestricted by the valley sides.

waning slope, waxing slope in the study of hillslopes (SLOPE-2) four elements are commonly identified: the waxing slope near the top which becomes steeper as the hillside is worn back; the FREE FACE; the CONSTANT SLOPE; and the waning slope (sometimes termed the wash slope, or wash controlled slope) at the foot which becomes less steep as it develops by accumulating debris brought down by SURFACE WASH from above, merging into the widening valley floor. STANDARD HILLSLOPE.

warm front the boundary zone between the advancing mass of warm air forming part of a DEPRESSION-3 and the colder air which it is overriding and overtaking. In advance of the passing of a warm front over a fixed point, pressure falls, the cloud base lowers, the wind backs (BACKING OF WIND). A characteristic series of cloud forms accompanies its approach and passing (CIRRUS, CIRROSTRATUS, ALTOSTRATUS, STRATUS, NIMBOSTRATUS), intermittent DRIZZLE and then steady, heavy precipitation accompanying the nimbostratus. The wind veers (VEERING

OF WIND) as the front passes over, and the precipitation dies away. COLD FRONT, FRONT, OCCLUSION, WARM SECTOR.

warm glacier temperate glacier, a moving ice mass, the surface of which melts through THERMAL CONDUCTION, the resultant surface water percolating through the ice mass, releasing LATENT HEAT in refreezing, thus raising the temperature of the ice mass itself (the summer temperature of the whole mass is about 0°C: 32°F but in winter the surface is colder). COLD GLACIER.

warm occlusion an OCCLUSION where the cold, overtaking air is not so cold as the air mass ahead of it.

warm sector a region, a bulge, of warm air in a DEPRESSION-3, where the air temperature and RELATIVE HUMIDITY rise. It is preceded by the WARM FRONT and followed by a COLD FRONT. OCCLUSION.

warping 1. the process (natural or ARTIFICIAL-1) whereby the low-lying land of a tidal estuary is flooded, leading to deposition of silt, mud or clay **2.** a gentle, slow deformation of the earth's crust over a wide area, resulting in the raising or lowering of the surface.

Warsaw Pact a treaty of defence and mutual assistance signed in 1955 by Albania, Bulgaria, Czechoslovakia, East Germany, Hungary, Poland, Romania and the USSR in response to the formation of NATO. Albania withdrew in 1968 when the USSR invaded Czechoslovakia.

wash 1. the surging movement of the sea or other large body of water; the surge of water up a beach following the breaking of a wave, in contrast to the BACKWASH **2.** an area of sand and mud washed by the tide **3.** fine material moved down a slope. SUB-SURFACE WASH, SURFACE WASH, WASH SLOPE.

wash load SUSPENDED LOAD.

wash slope, wash-controlled slope GRAVITY SLOPE, HALDENHANG, WANING SLOPE, WAXING SLOPE.

waste, waste land 1. commonly, any wild, uncultivated, uninhabited land **2.** formerly, the little-used common land, usually on light soil, which failed to yield a return to the medieval and later cultivator. Now rarely applied to such land because in so many cases it is valued as an open space **3.** now applied to land previously used but abandoned, for which further use has yet to be found.

water buffalo the common domesticated buffalo, widely used as a draught and dairy animal in the warm parts of Asia.

water cycle HYDROLOGICAL CYCLE.

waterfall a sudden, steep or perpendicular descent of water in the bed of a river, occurring where the flow of the river is broken by a nearly horizontal bed of hard rock overlying easily eroded soft rock; or by the sharp edge of a plateau; or by the abrupt end of a HANGING VALLEY, high on the slopes of a U-SHAPED VALLEY; or by a FAULT-LINE SCARP; or by the edge of a coastal cliff. RAPIDS.

water-gap a cutting or gap, a low-lying valley made by a river as it flows across a ridge. Water-gaps tend to be associated with ANTECEDENT DRAINAGE and may be cut by either CONSEQUENT or OBSEQUENT STREAMS. Unlike a WIND-GAP, a water-gap is occupied by a stream which continues to form part of the drainage system.

water hole, waterhole 1. a hollow, natural or ARTIFICIAL-1, where water gathers, especially in savanna and hot desert lands, in some cases fed by a spring **2.** a depression in the bed of an intermittent stream **3.** a hole on the surface of ice.

watering place a term formerly applied to **1.** a place to which animals were taken for drinking **2.** a place at which ships called to lay in a supply of water **3.** a spa or locality with a mineral spring to which people went to drink the waters or to bathe.

waterlogged *adj.* applied in soil science to the state of a soil when it is SATURATED-2 with water.

water meadow a low-lying meadow by the side of a stream, artificially irrigated by flooding in the early part of the year to encourage an early growth of grass, a practice especially common in the chalk valleys of southern England where the lime in the water was beneficial. Most of these English water meadows have now fallen into disuse on account of the high cost (especially the labour cost) of maintaining the elaborate series of miniature canals and drains.

water power, waterpower the ENERGY of moving water converted into mechanical energy, formerly used directly to drive water mills etc. but now nearly always used to generate ELECTRICITY. HYDRAULIC, HYDROELECTRIC POWER, POWER.

watershed 1. (British) a water parting, the elevated line which may or may not be sharply defined, separating two contiguous drainage areas from which the headstreams flow in different directions, into different river systems or basins **2.** (American) a water parting, as in British usage; but also the whole catchment area or DRAINAGE BASIN of a single river system. ANOMALOUS WATERSHED, NORMAL WATERSHED.

waterspout the product of an intense, localized small scale cyclonic storm (CYCLONE) occurring over the ocean or a lake, usually in tropical and subtropical regions. From the underside of a CUMULONIMBUS cloud a spinning FUNNEL CLOUD (carrying water droplets formed by condensation) descends to meet spray thrown up from the water surface by whirling winds, the combination forming a rotating column of mist, water and spray which is sometimes vertical and straight, sometimes bent (when the top part moves faster than the base), moving swiftly over the surface of the water. TORNADO.

water table, water-table the surface below which PERMEABLE rocks are SATURATED-2 with water. In areas with PERVIOUS soil and pervious subsoil-rocks it tends to follow generally, but not in detail, the form of the land surface. Where the water table lies below the land surface its height corresponds to the level of water in wells (GROUND WATER, PHREATIC WATER), and similarly fluctuates seasonally. Where the water table reaches the land surface a SPRING-2 results; fluctuations in the water table account for the intermittent flow of BOURNES. A permanent marsh or lake results when the theoretical water table is above the land surface. In some circumstances there is no regular water table, e.g. where underlying rocks are irregularly fissured, as in the ancient METAMORPHIC plateau of Africa. In other cases there is a PERCHED WATER TABLE. ARTESIAN WELL, CAPILLARY FRINGE, PIEZOMETRIC LEVEL, VADOSE WATER, WELL.

water vapour in the earth's ATMOSPHERE-1, water in the VAPOUR state and below the CRITICAL TEMPERATURE for water. HUMIDITY.

waterway 1. a navigable stretch of inland water, i.e. of a lake, river, canal, which is or can be used for transport **2.** the route followed by inland water traffic.

wave in a body of water, particularly in the ocean, the rise and fall in the forward movement in the surface area of the water, due to the oscillation of water particles, usually caused by friction of wind on the water surface. The motion of the water particles is perpendicular to the direction of the movement of the water, each particle moving up to the CREST and falling back almost to its original position in the TROUGH-4. The size of wave depends on the cause, e.g. speed and direction of wind (FETCH); the height constitutes the distance between the trough and the crest; length, the distance between two successive crests; steepness, the ratio of height to length (BREAKER); velocity, the speed of movement of an individual crest. BACKWASH, CONSTRUCTIVE WAVE, DOMINANT WAVE, HYDRAULIC FORCE, LONGSHORE DRIFT, SWASH, TIDAL WAVE, TSUNAMI, and the entries, qualified by wave, which follow. Fig 28. ELECTROMAGNETIC WAVE.

wave-base the greatest depth at which sea floor sediment can be just slightly moved by oscillating water.

wave-built terrace a TERRACE-2 formed by marine deposition seawards from a WAVE-CUT BENCH.

wave-cut bench, wave-cut beach bench a marine erosion plane formed at the base of a sea cliff, sloping down towards the sea, in some cases merging imperceptibly into a WAVE-BUILT TERRACE. ABRASION PLATFORM, BENCH.

waxing slope WANING SLOPE.

wealth consuming sector of an economy, the services sector.

wealth creating sector of an economy, the manufacturing sector.

weather a general term for the conditions prevailing in the ATMOSPHERE-1, especially in the layer near the ground (TROPOSPHERE), over a short period of time (in contrast to CLIMATE) or at a specific time, at any one place, and as affecting human beings. Temperature, sunshine, pressure and wind, humidity, amount of cloud, precipitation (rain, sleet, hail, snow), the presence of fog or mist are all taken into account. ANALOGUE, LONG-RANGE WEATHER FORECAST, METEOROLOGY, SYNOPTIC CHART, WEATHER CHART.

weather chart, weather map a chart or map showing weather details for a selected area at a specific time. Data collected at observation posts and transmitted to meteorological stations provides the basis for a series of charts relating to pressure conditions, etc. affecting the selected area at set, regular intervals; and from this series the final maps, giving a summary of isobars, temperature, winds etc. at a selected time, are drawn. Some authors prefer to use the term SYNOPTIC CHART, as indicating the summary character of the final chart.

weathering 1. the action of the WEATHER on objects exposed to it **2.** in geology, the mechanical or physical, chemical and biological processes (CHEMICAL WEATHERING, MECHANICAL WEATHERING, ORGANIC WEATHERING) by which rocks are decomposed or disintegrated by exposure at or near the earth's surface to water, the atmosphere, organic matter (DENUDATION), a mantle of rock debris being produced in situ. Transport (except by gravity) is not involved (EROSION). The main mechanical or physical agents are SHATTERING, frost action and temperature change, assisted by the biological processes, the organic agents being plant roots, mosses, lichens, the burrowing of animals. The chemical processes include CARBONATION, HYDRATION, HYDROLYSIS, OXIDATION, SOLUTION. CORROSION.

weathering front the boundary between weathered and unweathered rock.

weather vane wind vane, an apparatus for indicating wind direction. A broad, thin strip, usually of metal, is fixed to a pivoted, freely rotating support, so that it may swing round easily in an air current.

Weberian analysis a theory of the optimum location of firms (manufacturing enterprises) formulated by Alfred Weber, 1909, German economist, who maintained that transport costs were the major factor determining location; that optimum location was primarily the point where the costs of the transport of raw materials to the factory and of supplying goods to the necessary market were at their lowest; but that if variations in other costs (e.g. of labour) were high enough, location determined solely by transport costs might not be the optimum one. LEAST COST LOCATION, LOCATION, LOCATION THEORY, MINIMAX LOCATION, OPTIMUM LOCATION.

wedge of high pressure a region of HIGH atmospheric pressure, indicated by a V-shaped pattern of isobars, narrower than a RIDGE OF HIGH PRESSURE, occurring between two DEPRESSIONS-3, bringing a brief period of fine weather in a generally rainy period.

weir an obstruction built across a river to impound or raise the level of the water for fishing purposes, for creating a head for a water mill, for the control of the current and maintenance of the water depth to aid navigation, for irrigation, or to divert the flow. The term is limited to small, low constructions over which the water may flow, the larger being termed DAMS and BARRAGES.

welfare 1. the state or condition of being well, thriving, happy, prospering **2.** work organization to bring about this state in needy members of a community.

welfare geography an approach in HUMAN GEOGRAPHY concerned with social inequality, which considers the areal differentiation and spatial organization of human activity from the perspective of the WELFARE-1 of the people involved. It touches on everything, positive or negative, contributing to the quality of human life, covering everything differentiating one state of society from another, the 'good' and the 'bad' things consumed in society, what these are, to whom and where they are distributed; and how the observed differences arise (i.e. who gets what, where and how). QUALITY OF LIFE, SOCIAL GEOGRAPHY, WELFARE.

welfare state a state with a political system based on the principle that the protection, social security and WELFARE of the individual is the responsibility of the community as represented in the state. The ideal welfare state therefore provides the facilities and services necessary to bring this about (e.g. by providing medical care, education, public housing, pay for the unemployed and the aged etc.), financed by taxation and compulsory contributions from the population.

well originally a natural SPRING-2 or pool fed from a spring. The term is now restricted to a deep hole, usually cylindrical, or a shaft, dug in the ground to obtain water, oil or gas. A well sunk for water is usually lined with brick or masonry, but may be unlined (e.g. if sunk through hard rock) and normally fills with water up to the level of the WATER TABLE (PHREATIC WATER), the surface of water fluctuating seasonally with the height of the water table. A well sunk into an ARTESIAN BASIN taps water held under considerable pressure. ARTESIAN WELL, OIL WELL, TUBE WELL.

Wentworth scale a scale devised by C. K. Wentworth, 1922, to measure the size of particles in sediments, a geometric scale of factor 2. This scale ranges from clay particles of 0.004 mm diameter, through silt, sand, granule, pebble, cobble to boulder, exceeding 256 mm (10 in) diameter. GRADED SEDIMENTS.

west 1. one of the four cardinal points of the COMPASS, directly opposite the EAST on the side of someone facing due NORTH, i.e. the direction of the setting sun at the EQUINOX **2.** towards or facing the west **3.** the western part, especially of a country, particularly the states west of the Mississippi, USA **4.** the West, the countries of western Europe and North America as distinct from those of eastern Europe and Asia.

west *adj.* of, pertaining to, belonging to, situated towards, coming from, the west, e.g. of winds blowing from the west.

Westerlies winds which blow frequently from the subtropical high pressure area to the temperate low pressure area, between 35°N and 65°N and 35°S and 65°S, blowing predominantly from the southwest in the northern hemisphere, predominantly from the northwest in the southern hemisphere. In winter in the northern hemisphere their presence makes the North Atlantic ocean one of the stormiest regions in the world; and in winter too they move southwards, carrying winter rain to the Mediterranean region. In the northern hemisphere their force and direction vary, and they are associated with the succession of DEPRESSIONS-3 and ANTICYCLONES characteristic of the weather of the area in which they blow. But in the southern hemisphere they blow strongly and with greater regularity throughout the year over the great expanse of ocean, giving the region the name ROARING FORTIES. Westerlies gain strength with height, evolving into JET STREAMS. The old term applied to the Westerlies, the ANTI-TRADE WINDS, is misleading and no longer used. Fig 5.

wet adiabatic lapse rate SATURATED ADIABATIC LAPSE RATE.

wet-bulb thermometer a THERMOMETER with a bulb covered by wet muslin, thereby being cooled by evaporation. The temperature recorded is accordingly lower than that shown by a DRY-BULB THERMOMETER; and the two different readings, combined with reference to a set of statistical tables, enable DEW-POINT, RELATIVE HUMIDITY and the VAPOUR PRESSURE of the air to be calculated.

wet-day in UK, officially, a day of 24 hours beginning at 0900 hours during which at least 1 mm (0.04 in) of rain falls. PRECIPITATION-DAY.

wet dock a large BASIN-9 in which the water level is maintained at the level of HIGH TIDE so that the vessel in it stays afloat. DOCK.

wetland, wetlands a general term applied to an ECOSYSTEM intermediate between the TERRESTRIAL-2 and the AQUATIC, a natural or artificial landscape in which fresh or salt water plays a key role, i.e. where the soil is waterlogged, the WATER TABLE is at or near the surface, or the land is covered occasionally, periodically or permanently, by shallow fresh or salt water (e.g. BOG, FEN, MARSH, SWAMP, flooded pasture land, intertidal mud flats).

wet-point settlement a settlement the site of which was related to the availability of a water supply, especially to a constant SPRING-2, in contrast to a DRY-POINT SETTLEMENT in lands liable to flood.

wet spell in UK, officially, an unbroken succession of 15 or more consecutive WET-DAYS, a definition not accepted internationally. DRY SPELL.

wharf a landing stage to which barges and ships may be moored while loading and unloading.

wheat any of the grasses of the genus *Triticum*, probably native to southwestern Asia and eastern Mediterranean regions, an important grain crop, now grown especially in the former grasslands of the midlatitudes, providing the STAPLE-3 food in temperate climates, the grain being ground into FLOUR used in the making of bread, biscuits, cakes etc. *Triticum aestivum* is the most widely grown for this purpose. Hard wheat, durum wheat, *Triticum durum*, with small hard grains, grown in dry regions in the Mediterranean region, the CIS, Asia, North and South America, is the best for making pasta.

Most wheats need about 100 days to grow and ripen between the last killing frost of spring and the first killing frost of autumn, and this sets the northern limit in Canada and the CIS where, however, wheat is being bred to ripen in 90 days. Wheat needs a good firm moisture-holding soil, such as a heavy loam. The black earths or CHERNOZEMS are particularly favourable. Climatically the best conditions are a cool moist spring, which causes the grain to produce a number of stalks capable of bearing a head of grain, followed by warm sunny weather when the heads have formed, and some rain or moisture just before harvest to swell the grain. A total rainfall of between 375 and 875 mm (15 and 35 in) is about right. In countries with a very cold winter (e.g. Canada) wheat is sown in spring; in countries with a mild winter (e.g. Britain) it can be sown in the autumn or fall, the seeds remaining in the ground during the winter to sprout as soon as the temperature rises in spring. Hence the distinction between 'spring' wheat and 'winter' or 'fall' wheat. In tropical regions (e.g. in Egypt or in parts of the Indian subcontinent) wheat is grown as a winter crop to be reaped before the heat of summer.

whirlpool a circular eddy or current in a river or the sea produced by the configuration of the channel, by the effect of winds on tides, by the meeting of currents, or by similar phenomena.

whirlwind a rapidly rotating column of air revolving around a local centre of low pressure, caused by local surface heating and exceptionally strong CONVECTION. It is limited in extent, formed round a vertical or slightly inclined axis, the inward and upward spiral movement of the lower air spreading to an outward and upward spiral, the whole moving progressively over land or water. CYCLONE, DUST-DEVIL, TORNADO, WATERSPOUT.

white box approach BLACK BOX APPROACH.

white-collar worker a person who works in an office, and is employed in non-manual work (usually excluding those engaged in the professions). BLUE-COLLAR WORKER.

white-out a condition in a BLIZZARD when the snow cover is extensive and the falling snow so great that visibility is reduced to the minimum and finding direction almost impossible.

WHO World Health Organization, an international body established in April 1948, headquarters Geneva, with an executive board consisting of technically qualified health experts, to foster the highest possible level of health in the world. Its work includes dealing with matters of international health, helping governments to strengthen health services (especially so that primary health care reaches the maximum number of people, and that diseases endemic in under-developed areas are combated); promoting maternal and child care; stimulating work and research in mental health, medical research, the prevention of accidents and the eradication of disease; and encouraging the improvement of standards in teaching and training in the health professions etc.

wildcat a test well for petroleum or natural gas bored as a speculation without any detailed geological evidence of the existence of either of them.

wilderness **1.** an uncultivated, uninhabited region **2.** in nature conservation 'wilderness areas' are those left in a wild state as natural habitats, in contrast to those nature reserves which may need careful management to maintain small communities of plants and animals.

willy-willy, willi-willi (Australian) a type of TROPICAL REVOLVING STORM originating off the coast of western Australia, in some cases moving on to the land.

wilt a condition of plants in which the cells lose their turgidity, so that the leaves, young stems and tops of older stems become limp. It is usually caused by an excess of water loss through TRANSPIRATION in relation to water absorption (WILTING POINT); or it may be due to functional disorder or the action of fungus parasites.

wilting point in soil science, the point below which the amount of water stored in the soil cannot be absorbed by plants quickly enough to meet their needs, causing WILT in any plant not adapted to drought. Wilting point is used as a measure of storage capacity (FIELD CAPACITY) of a soil.

wind air in motion, usually restricted to natural horizontal movement, varying in strength from light to hurricane (BEAUFORT SCALE). The term CURRENT-1 is usually applied to the vertical movement of air (THERMAL). There are very many names for local winds (e.g. BORA, CHINOOK, FOHN, MISTRAL, NORTHER). ANABATIC, DEFLATION, DOMINANT WIND, KATABATIC, PLANETARY WINDS, PREVAILING WIND, TRADE WIND.

windbreak, wind-break something (especially a line of trees or a thick hedge or a hurdle) designed to break the force of the wind and provide shelter for animals or, more often, for growing plants. A windbreak is particularly important where a cold wind, e.g. the MISTRAL, would damage unprotected crops. DUST BOWL.

wind chill the cooling power of wind and temperature on shaded dry human skin. It was originally measured as the product of wind speed in metres per second and air temperature in degrees Centigrade below zero; but a later formula measures the cooling power of wind and temperature in complete shade and regardless of evaporation.

wind erosion DEFLATION, EROSION.

wind-gap, air-gap a dry gap, a notch or gap in the crest of a hill range, or a pass, originally cut by a stream, from which the water has disappeared, e.g. a dry COL in an escarpment through which a CONSEQUENT STREAM may have flowed before RIVER CAPTURE. In many cases a wind-gap lies at a higher level than that of a neighbouring WATER-GAP.

windmill a mill operated by rotating sails which are turned by the wind. WIND POWER.

window atmospheric window, one of the bands in the ELECTROMAGNETIC SPECTRUM within which TERRESTRIAL RADIATION escapes into outer space because, cloud cover being thin or non-existent, it is not absorbed by WATER VAPOUR and CARBON DIOXIDE present in the clouds in the atmosphere.

wind power mechanical or electrical POWER generated by the rotor of a WINDMILL.

wind pump a pump activated by the wind's POWER in rotating a propeller wheel composed of vanes (blade-like, thin, flat strips, often curved).

wind rose a diagram with radiating arms constructed to show the frequency (and usually the speed as well) of winds blowing from the eight chief points (but sometimes from twelve points) of the COMPASS. The length of each arm shows the frequency recorded over a specific period of time, and gradations on the arms show the frequency of wind speeds.

windward the direction from which the wind is blowing, facing into the wind, as opposed to LEEWARD.

wine an alcoholic drink made from the fermented juice of the grape (but the term is also applied to other juices fermented and containing alcohol). Wine is produced in nearly all grapevine-growing countries (GRAPE), where most of it is consumed, although much enters international trade. Even in a small area variations in soil and microclimate, as well as in the weather, type of grape and manufacturing processes, produce great differences in yield and the quality of the wine.

winnowing the act of blowing chaff (the outer husk) free of GRAIN-5.

winter **1.** the colder part of the year, in contrast to summer, the hotter **2.** loosely, the cold season; in tropical regions the term winter is usually dropped, the term cool season being preferred **3.** one of the seasons in mid- and high-latitudes, popularly December, January, February in the northern hemisphere (the other seasons being SPRING, SUMMER, AUTUMN), or June, July, August in the southern **4.** astronomically, from the winter SOLSTICE to the spring EQUINOX, i.e. from about 22 December (also paradoxically termed midwinter day) to 20 March in the northern hemisphere, 22 June to 21 September in the southern hemisphere.

winter solstice 21–22 December in the northern hemisphere, 21–22 June in the southern. SEASON, SOLSTICE, WINTER.

wood **1.** the hard, fibrous vascular tissue of mature plants, forming stems, roots and the trunks of trees, providing mechanical support, and through which water containing dissolved mineral salts passes **2.** with indefinite article, i.e. a wood, or pl. woods, imprecise terms applied to a piece of ground (small in relation to that supporting a FOREST-1) covered with relatively widely-spaced trees growing naturally (as distinct from a PLANTATION-1), with or without undergrowth. HANGER, WOODLAND.

woodland, woodlands land covered with trees, sometimes defined as an open stand of widely-spaced trees without a continuous canopy of overhead foliage (sometimes specifically as a canopy coverage between 25 and 60 per cent). FOREST, FORESTRY, WOOD-2.

wood pulp the fibre of wood processed by mechanical means and chemicals to form a mixture of water and cellulose fibres, used as raw material in making paper or artificial fibres.

wool the fibrous growth on the skin of some animal species, especially of the SHEEP. The fibres are covered with overlapping scales which hook into each other when the fibre is spun into yarn, entrapping air and making any fabrics or articles made from it warm in wear. WOOLLEN.

woollen *adj*. applied to a fabric made from WOOL.

World Bank an agency of the United Nations (inaugurated in 1945 as the International Bank for Reconstruction and Development) with the aim of assisting postwar reconstruction in Europe. It extended its role by making large sums available in ambitious development projects in Third World countries, in many cases involving them in very great debt. The Bank has now scaled down some of its credit advances, and is supporting smaller, more viable projects. MICROFINANCE.

world city, global city a CITY-1 characterized by the range of its economic, financial, cultural and political power and influence on a global scale. It may, or may not, be a large or a capital city, but its role in the world is dynamic and dominating.

World Health Organization WHO.

world-island a term applied by H. J. Mackinder to the world's largest landmass, the combined continents of Europe, Asia and Africa. Being surrounded by water this vast landmass is, by conventional definition, an island. In the same way the two Americas are an island, as are the continents of Australia and Antarctica. HEARTLAND.

wrought iron cast iron, PIG IRON, STEEL.

WTO GATT.

xeno- (Greek) a stranger, a foreigner, a prefix used in that sense in many scientific terms.

xenoparasite a PARASITE which lives on an organism which is not its usual host; or which can live only by invading an injured organism.

xenophobia a fear or dislike of individuals or groups thought of as strangers or foreigners.

xeric *adj.* **1.** having a low or inadequate supply of moisture to sustain plant life **2.** adjusted to arid conditions, applied particularly to a plant or animal having such a quality. HYDRIC, MESIC.

xero- (Greek) dry, a prefix used in that sense in many scientific terms.

xerophyte a plant adapted to a dry habitat (in desert conditions, or in an alkaline, acid, salt or dry soil) and able to withstand prolonged drought (XEROPHYTIC CONIFEROUS FOREST). HYDROPHYTE, HYGROPHYTE, MESOPHYTE, TROPOPHYTE.

xerophytic, xerophilous *adj.* of, pertaining to, characteristic of, a XEROPHYTE.

xerophytic coniferous forest forest occurring at high elevations in semi-arid zones, e.g. in southwestern USA, the species including juniper. HYDROPHYTE, HYGROPHYTE, MESOPHYTE, TROPOPHYTE.

xerothermic *adj.* related to both dryness and heat.

X-rays extremely short wavelength (high frequency), high energy radiation in the ELECTROMAGNETIC SPECTRUM.

Y

yardang a narrow, steep-sided crest in desert, particularly in central Asia, separated from others lying parallel to it by grooves or corridors cut in the desert floor by wind carrying sand. A yardang may reach 6 m (20 ft) high and from 9 to 36 m (30 to 120 ft) in width. CORRASION.

yardland in Britain, in agricultural history, virgate, an imprecise unit of land measurement, representing a tenement varying in size measured in customary acres (ACRE), including arable with adjoining meadow and pasture. In some cases it might be divided into two ox-gangs (an ox-gang being half a virgate) or four ferlings (quarter-virgate), again of inexact area. In areas under Danish influence the equivalent to ox-gang was bovate (the HIDE being termed carucate, the HUNDRED being termed wapentake). In south-eastern England the equivalent of hide was sulung (subdivided into four yokes).

yazoo a DEFERRED JUNCTION of a tributary, the name derived from the Yazoo river, the type example of a tributary which flows for some distance parallel to the main river (the lower Mississippi, in the case of the Yazoo) before merging with it.

year the period of time taken by the earth to complete one orbit round the SUN. SOLAR YEAR (365 days 5 hours 48 minutes 45.51 seconds).

yellow ground KIMBERLITE.

yield **1.** in agriculture, output, product, amount of produce, result, e.g. output or production expressed in relation to units of land or livestock or to units of capital or labour applied **2.** from investment, the rate of return from the investment of CAPITAL-2, e.g. with a 5 per cent yield the capital invested should be recouped in 20 years (5 × 20 = 100); with a 20 per cent yield in 5 years (20 × 5 = 100).

young *adj.* in the early stages of development, not far advanced. MATURE, OLD AGE, SENILE, YOUTH, YOUTHFUL.

young fold mountains FOLD MOUNTAINS of the Alpine orogeny, in contrast to Armorican, Caledonian and other earlier orogenies. OROGENESIS.

young mountains mountains so recently formed that their surface configuration of jagged peaks etc. has not yet been smoothed by the agents of erosion.

youth the first stage of development, applied specifically (as are YOUNG and YOUTHFUL) in the CYCLE OF EROSION to the first stage in the development of landforms, i.e. when the original upland surfaces are undissected, not yet attacked by CORRASION or EROSION, when RIVERS flow swiftly, slopes are steep and gradients irregular. MATURE, OLD AGE, SENILITY.

youthful *adj.* of or pertaining to youth, e.g. applied to a landform which has suffered little erosion. MATURE, OLD AGE, SENILE, YOUNG, YOUTH.

Z

zenith the point where the line joining the earth's centre to the observer cuts the CELESTIAL SPHERE, the opposite of NADIR.

zinc a hard, blue-white, corrosion-resistant metallic element, a TRACE ELEMENT, an essential MICRONUTRIENT, often occurring in association with lead and silver. It is used especially in coating sheet iron to prevent rust (galvanized iron), in alloys (with copper to make brass), in electric cells; in its oxide form, zinc oxide, as a white pigment (zinc white); and as a filler in ointments.

zonal flow atmospheric circulation in which the dominant airflow follows the lines of latitude, e.g. TRADE WINDS, WESTERLIES, in contrast to MERIDIONAL FLOW.

zonal soil a soil with a profile showing a dominant influence of climate and vegetation in its development, as contrasted with an AZONAL SOIL. SOIL CLASSIFICATION.

zone 1. frequently applied more or less loosely to a region, belt, tract or area of the earth (i.e. of the atmosphere, lithosphere, hydrosphere, or of any place or space), with or without defined limits, with some characteristic or characteristics or activity particular to it (e.g. of climate, rocks, soil, plant and animal life, condition), indicated by a qualifying word or phrase which differentiates it from other regions, belts, tracts or areas etc. **2.** in classical times, one of the latitudinal climatic belts into which the earth's surface was divided, i.e. frigid, temperate, torrid zones **3.** in geology, a group of strata of limited but variable thickness, characterized by a definite assemblage of FOSSILS which distinguishes it from all other deposits, the zone being named after one of the characteristic species **4.** a layer or part of the earth's crust (e.g. zone of weathering or, deeper in the crust, zone of fracture; and, deeper still, zone of flow) **5.** in land use planning, an area designated (zoned) for a specific purpose.

zone of assimilation, zone of discard related concepts indicating the movement of city centre growth and decline. City centres tend to grow outwards systematically; thus the zone of new growth becomes the zone of assimilation, and the zone relatively in decline is termed the zone of discard.

zone of indifference in CENTRAL PLACE THEORY, the area between the hinterlands of competing centres within which no one centre exerts a dominant influence. THRESHOLD-3.

zoning in land use planning, the designation of specific sites for specific uses, e.g. for residential use, for industrial use etc. ZONE-5.

zoogeography the scientific study of the natural distribution of animals.

zoonosis a disease of animals naturally transmitted between VERTEBRATE animals and human beings.

zoophyte an animal resembling a plant in appearance or growth, e.g. CORAL.

zooplankton minute animals, many microscopic and including the larvae of molluscs and other INVERTEBRATES, which float or very feebly swim in bodies of fresh or salt water. PHYTOPLANKTON, PLANKTON.

***z*-test** in statistics, an hypothesis test used either to analyse one sample of data in order to compare a population mean with a particular value, or to analyse two unrelated samples of data in order to compare two population means. The sample data must be interval (INTERVAL SCALE) and the sample sizes must be large, the assumption being that the sampling distribution of the mean is approximately normal (NORMAL DISTRIBUTION); and, for the two sample test, the data must be totally unrelated.

zymogenous *adj.* applied to soil organisms whose metabolic and reproductive rates increase if organic material is added to the soil.

RANDOM NUMBER TABLE

Random digits produced by throws of a ten-sided dice marked 0, 1 . . . 9.

39 56 14	02 45 65	16 86 78	90 46 39	58 62 66	96 12 56
32 53 16	30 76 36	80 52 65	02 10 07	81 40 80	33 18 70
98 43 67	05 82 06	19 24 86	24 30 44	06 15 54	29 00 60
53 08 00	94 46 80	60 94 01	83 94 45	42 43 55	52 27 23
28 21 05	43 60 40	73 70 75	33 10 74	91 83 95	25 43 89
89 79 63	50 98 53	56 42 12	76 48 56	34 46 82	02 58 68
61 48 17	25 59 95	19 14 31	68 94 23	83 40 83	53 36 90
41 98 20	72 70 69	39 46 17	37 70 37	81 75 23	82 31 79
51 08 35	35 16 20	92 94 25	05 04 01	65 33 82	87 28 54
73 97 76	94 92 07	24 89 41	98 35 91	96 52 82	62 63 42
43 74 49	01 59 38	60 29 94	61 02 11	61 86 36	95 57 95
94 94 39	87 49 44	54 02 52	56 28 49	34 49 25	35 65 55
52 10 65	11 34 68	68 65 58	90 17 33	98 36 82	93 87 17
54 42 73	62 51 54	80 63 36	65 12 44	52 16 12	64 41 70
73 27 51	94 71 14	37 55 00	05 32 36	59 89 86	79 08 65
77 69 59	62 33 99	26 67 95	72 77 16	02 28 96	75 17 45
08 19 98	26 68 06	02 05 57	21 73 55	35 07 79	91 04 44
50 83 92	60 44 28	52 83 25	39 83 60	92 71 10	34 33 73
16 89 30	82 48 70	63 82 71	48 72 82	77 37 56	22 90 95
21 41 74	65 08 73	82 94 72	22 67 92	34 74 33	69 86 14
99 08 47	77 43 94	17 07 76	57 93 68	61 15 97	78 76 99
20 02 69	70 87 44	57 23 35	99 94 16	63 40 99	72 64 82
93 95 15	81 21 75	71 39 23	31 06 43	87 44 21	81 55 34
10 91 65	40 88 43	50 57 83	50 82 34	12 78 80	00 34 07
91 72 35	36 80 19	49 49 37	17 40 98	02 53 59	18 91 30
23 82 82	20 56 34	76 49 27	40 78 29	99 07 22	01 40 97
21 02 08	25 07 15	36 45 19	21 30 48	30 76 99	24 46 39
82 45 49	85 02 33	58 84 03	74 63 52	15 47 04	09 50 45
44 33 94	98 75 51	62 00 17	59 00 42	09 39 66	86 57 76
96 00 26	82 60 22	02 60 69	99 09 67	01 12 01	88 58 15
20 67 56	12 77 16	78 04 36	38 95 35	71 26 49	34 20 46
64 60 21	12 41 60	04 63 93	45 25 52	75 50 35	51 13 61
64 76 41	17 07 54	01 29 86	41 93 16	55 54 40	32 80 30
93 46 82	67 64 48	91 74 85	94 40 51	30 93 08	42 35 24
82 64 44	58 45 94	30 39 86	19 64 84	35 30 19	04 77 69
61 46 40	89 21 47	20 85 91	90 56 67	40 31 46	30 97 14
92 80 33	89 23 96	24 33 16	80 45 20	35 36 00	76 31 13
45 65 20	02 56 40	21 35 17	71 33 07	36 71 90	71 56 81
40 99 02	66 37 59	24 79 35	21 61 29	96 50 01	27 51 87
50 31 47	84 44 30	70 33 12	63 54 86	63 08 62	63 07 30